Fundamentals of Digital Communication

This textbook presents the fundamental concepts underlying the design of modern digital communication systems, which include the wireline, wireless, and storage systems that pervade our everyday lives. Using a highly accessible, lecture style exposition, this rigorous textbook first establishes a firm grounding in classical concepts of modulation and demodulation, and then builds on these to introduce advanced concepts in synchronization, noncoherent communication, channel equalization, information theory, channel coding, and wireless communication. This up-to-date textbook covers turbo and LDPC codes in sufficient detail and clarity to enable hands-on implementation and performance evaluation, as well as "just enough" information theory to enable computation of performance benchmarks to compare them against. Other unique features include the use of complex baseband representation as a unifying framework for transceiver design and implementation; wireless link design for a number of modulation formats, including space–time communication; geometric insights into noncoherent communication; and equalization. The presentation is self-contained, and the topics are selected so as to bring the reader to the cutting edge of digital communications research and development.

Numerous examples are used to illustrate the key principles, with a view to allowing the reader to perform detailed computations and simulations based on the ideas presented in the text.

With homework problems and numerous examples for each chapter, this textbook is suitable for advanced undergraduate and graduate students of electrical and computer engineering, and can be used as the basis for a one or two semester course in digital communication. It will also be a valuable resource for practitioners in the communications industry.

Additional resources for this title, including instructor-only solutions, are available online at www.cambridge.org/9780521874144.

Upamanyu Madhow is Professor of Electrical and Computer Engineering at the University of California, Santa Barbara. He received his Ph.D. in Electrical Engineering from the University of Illinois, Urbana-Champaign, in 1990, where he later served on the faculty. A Fellow of the IEEE, he worked for several years at Telcordia before moving to academia.

Fundamentals of Digital Communication

Upamanyu Madhow
University of California, Santa Barbara

CAMBRIDGE
UNIVERSITY PRESS

University Printing House, Cambridge CB2 8BS, United Kingdom

Published in the United States of America by Cambridge University Press, New York

Cambridge University Press is part of the University of Cambridge.

It furthers the University's mission by disseminating knowledge in the pursuit of education, learning and research at the highest international levels of excellence.

www.cambridge.org
Information on this title: www.cambridge.org/9780521874144

© Cambridge University Press 2008

This publication is in copyright. Subject to statutory exception and to the provisions of relevant collective licensing agreements, no reproduction of any part may take place without the written permission of Cambridge University Press.

First published 2008

A catalogue record for this publication is available from the British Library

ISBN 978-0-521-87414-4 Hardback

Cambridge University Press has no responsibility for the persistence or accuracy of URLs for external or third-party internet websites referred to in this publication, and does not guarantee that any content on such websites is, or will remain, accurate or appropriate.

To my family

To my family.

Contents

Preface	*page* xiii
Acknowledgements	xvi
1 Introduction	**1**
1.1 Components of a digital communication system	2
1.2 Text outline	5
1.3 Further reading	6
2 Modulation	**7**
2.1 Preliminaries	8
2.2 Complex baseband representation	18
2.3 Spectral description of random processes	31
2.3.1 Complex envelope for passband random processes	40
2.4 Modulation degrees of freedom	41
2.5 Linear modulation	43
2.5.1 Examples of linear modulation	44
2.5.2 Spectral occupancy of linearly modulated signals	46
2.5.3 The Nyquist criterion: relating bandwidth to symbol rate	49
2.5.4 Linear modulation as a building block	54
2.6 Orthogonal and biorthogonal modulation	55
2.7 Differential modulation	57
2.8 Further reading	60
2.9 Problems	60
2.9.1 Signals and systems	60
2.9.2 Complex baseband representation	62
2.9.3 Random processes	64
2.9.4 Modulation	66
3 Demodulation	**74**
3.1 Gaussian basics	75
3.2 Hypothesis testing basics	88

3.3	**Signal space concepts**	94
3.4	**Optimal reception in AWGN**	102
3.4.1	Geometry of the ML decision rule	106
3.4.2	Soft decisions	107
3.5	**Performance analysis of ML reception**	109
3.5.1	Performance with binary signaling	110
3.5.2	Performance with M-ary signaling	114
3.6	**Bit-level demodulation**	127
3.6.1	Bit-level soft decisions	131
3.7	**Elements of link budget analysis**	133
3.8	**Further reading**	136
3.9	**Problems**	136
3.9.1	Gaussian basics	136
3.9.2	Hypothesis testing basics	138
3.9.3	Receiver design and performance analysis for the AWGN channel	140
3.9.4	Link budget analysis	149
3.9.5	Some mathematical derivations	150
4	**Synchronization and noncoherent communication**	**153**
4.1	**Receiver design requirements**	155
4.2	**Parameter estimation basics**	159
4.2.1	Likelihood function of a signal in AWGN	162
4.3	**Parameter estimation for synchronization**	165
4.4	**Noncoherent communication**	170
4.4.1	Composite hypothesis testing	171
4.4.2	Optimal noncoherent demodulation	172
4.4.3	Differential modulation and demodulation	173
4.5	**Performance of noncoherent communication**	175
4.5.1	Proper complex Gaussianity	176
4.5.2	Performance of binary noncoherent communication	181
4.5.3	Performance of M-ary noncoherent orthogonal signaling	185
4.5.4	Performance of DPSK	187
4.5.5	Block noncoherent demodulation	188
4.6	**Further reading**	189
4.7	**Problems**	190
5	**Channel equalization**	**199**
5.1	**The channel model**	200
5.2	**Receiver front end**	201
5.3	**Eye diagrams**	203
5.4	**Maximum likelihood sequence estimation**	204
5.4.1	Alternative MLSE formulation	212
5.5	**Geometric model for suboptimal equalizer design**	213
5.6	**Linear equalization**	216

5.6.1	Adaptive implementations	223
5.6.2	Performance analysis	226
5.7	**Decision feedback equalization**	**228**
5.7.1	Performance analysis	230
5.8	**Performance analysis of MLSE**	**231**
5.8.1	Union bound	232
5.8.2	Transfer function bound	237
5.9	**Numerical comparison of equalization techniques**	**240**
5.10	**Further reading**	**242**
5.11	**Problems**	**243**
5.11.1	MLSE	243

6 Information-theoretic limits and their computation — 252

6.1	**Capacity of AWGN channel: modeling and geometry**	**253**
6.1.1	From continuous to discrete time	256
6.1.2	Capacity of the discrete-time AWGN channel	257
6.1.3	From discrete to continuous time	259
6.1.4	Summarizing the discrete-time AWGN model	261
6.2	**Shannon theory basics**	**263**
6.2.1	Entropy, mutual information, and divergence	265
6.2.2	The channel coding theorem	270
6.3	**Some capacity computations**	**272**
6.3.1	Capacity for standard constellations	272
6.3.2	Parallel Gaussian channels and waterfilling	277
6.4	**Optimizing the input distribution**	**280**
6.4.1	Convex optimization	281
6.4.2	Characterizing optimal input distributions	282
6.4.3	Computing optimal input distributions	284
6.5	**Further reading**	**287**
6.6	**Problems**	**287**

7 Channel coding — 293

7.1	**Binary convolutional codes**	**294**
7.1.1	Nonrecursive nonsystematic encoding	295
7.1.2	Recursive systematic encoding	297
7.1.3	Maximum likelihood decoding	298
7.1.4	Performance analysis of ML decoding	303
7.1.5	Performance analysis for quantized observations	309
7.2	**Turbo codes and iterative decoding**	**311**
7.2.1	The BCJR algorithm: soft-in, soft-out decoding	311
7.2.2	Logarithmic BCJR algorithm	320
7.2.3	Turbo constructions from convolutional codes	325
7.2.4	The BER performance of turbo codes	328

7.2.5	Extrinsic information transfer charts	329
7.2.6	Turbo weight enumeration	336
7.3	**Low density parity check codes**	**342**
7.3.1	Some terminology from coding theory	343
7.3.2	Regular LDPC codes	345
7.3.3	Irregular LDPC codes	347
7.3.4	Message passing and density evolution	349
7.3.5	Belief propagation	352
7.3.6	Gaussian approximation	354
7.4	**Bandwidth-efficient coded modulation**	**357**
7.4.1	Bit interleaved coded modulation	358
7.4.2	Trellis coded modulation	360
7.5	**Algebraic codes**	**364**
7.6	**Further reading**	**367**
7.7	**Problems**	**369**

8 Wireless communication — 379

8.1	**Channel modeling**	380
8.2	**Fading and diversity**	387
8.2.1	The problem with Rayleigh fading	387
8.2.2	Diversity through coding and interleaving	390
8.2.3	Receive diversity	393
8.3	**Orthogonal frequency division multiplexing**	397
8.4	**Direct sequence spread spectrum**	406
8.4.1	The rake receiver	409
8.4.2	Choice of spreading sequences	413
8.4.3	Performance of conventional reception in CDMA systems	415
8.4.4	Multiuser detection for DS-CDMA systems	417
8.5	**Frequency hop spread spectrum**	426
8.6	**Continuous phase modulation**	428
8.6.1	Gaussian MSK	432
8.6.2	Receiver design and Laurent's expansion	433
8.7	**Space–time communication**	439
8.7.1	Space–time channel modeling	440
8.7.2	Information-theoretic limits	443
8.7.3	Spatial multiplexing	447
8.7.4	Space–time coding	448
8.7.5	Transmit beamforming	451
8.8	**Further reading**	451
8.9	**Problems**	453

Appendix A Probability, random variables, and random processes — 474

A.1	**Basic probability**	474
A.2	**Random variables**	475

Contents

A.3	**Random processes**	**478**
A.3.1	Wide sense stationary random processes through LTI systems	478
A.3.2	Discrete-time random processes	479
A.4	**Further reading**	**481**

Appendix B The Chernoff bound — **482**

Appendix C Jensen's inequality — **485**

References — 488
Index — 495

Preface

The field of digital communication has evolved rapidly in the past few decades, with commercial applications proliferating in wireline communication networks (e.g., digital subscriber loop, cable, fiber optics), wireless communication (e.g., cell phones and wireless local area networks), and storage media (e.g., compact discs, hard drives). The typical undergraduate and graduate student is drawn to the field because of these applications, but is often intimidated by the mathematical background necessary to understand communication theory. A good lecturer in digital communication alleviates this fear by means of examples, and covers only the concepts that directly impact the applications being studied. The purpose of this text is to provide such a lecture style exposition to provide an accessible, yet rigorous, introduction to the subject of digital communication. This book is also suitable for self-study by practitioners who wish to brush up on fundamental concepts.

The book can be used as a basis for one course, or a two course sequence, in digital communication. The following topics are covered: complex baseband representation of signals and noise (and its relation to modern transceiver implementation); modulation (emphasizing linear modulation); demodulation (starting from detection theory basics); communication over dispersive channels, including equalization and multicarrier modulation; computation of performance benchmarks using information theory; basics of modern coding strategies (including convolutional codes and turbo-like codes); and introduction to wireless communication. The choice of material reflects my personal bias, but the concepts covered represent a large subset of the tricks of the trade. A student who masters the material here, therefore, should be well equipped for research or cutting edge development in communication systems, and should have the fundamental grounding and sophistication needed to explore topics in further detail using the resources that any researcher or designer uses, such as research papers and standards documents.

Organization

Chapter 1 provides a quick perspective on digital communication. Chapters 2 and 3 introduce modulation and demodulation, respectively, and contain material that I view as basic to an understanding of modern digital communication systems. In addition, a review of "just enough" background in signals and systems is woven into Chapter 2, with a special focus on the complex baseband representation of passband signals and systems. The emphasis is placed on complex baseband because it is key to algorithm design and implementation in modern digital transceivers. In a graduate course, many students will have had a first exposure to digital communication, hence the instructor may choose to discuss only a few key concepts in class, and ask students to read the chapter as a review. Chapter 3 focuses on the application of detection and estimation theory to the derivation of optimal receivers for the additive white Gaussian noise (AWGN) channel, and the evaluation of performance as a function of E_b/N_0 for various modulation strategies. It also includes a glimpse of soft decisions and link budget analysis.

Once students are firmly grounded in the material of Chapters 2 and 3, the remaining chapters more or less stand on their own. Chapter 4 contains a framework for estimation of parameters such as delay and phase, starting from the derivation of the likelihood ratio of a signal in AWGN. Optimal non-coherent receivers are derived based on this framework. Chapter 5 describes the key ideas used in channel equalization, including maximum likelihood sequence estimation (MLSE) using the Viterbi algorithm, linear equalization, and decision feedback equalization. Chapter 6 contains a brief treatment of information theory, focused on the *computation* of performance benchmarks. This is increasingly important for the communication system designer, now that turbo-like codes provide a framework for approaching information-theoretic limits for virtually any channel model. Chapter 7 introduces channel coding, focusing on the shortest route to conveying a working understanding of basic turbo-like constructions and iterative decoding. It includes convolutional codes, serial and parallel concatenated turbo codes, and low density parity check (LDPC) codes. Finally, Chapter 8 contains an introduction to wireless communication, and includes discussion of channel models, fading, diversity, common modulation formats used in wireless systems, such as orthogonal frequency division multiplexing, spread spectrum, and continuous phase modulation, as well as multiple antenna, or space–time, communication. Wireless communication is a richly diverse field to which entire books are devoted, hence my goal in this chapter is limited to conveying a subset of the concepts underlying link design for existing and emerging wireless systems. I hope that this exposition stimulates the reader to explore further.

How to use this book

My view of the dependencies among the material covered in the different chapters is illustrated in Figure 1, as a rough guideline for course design or self-study based on this text. Of course, an instructor using this text

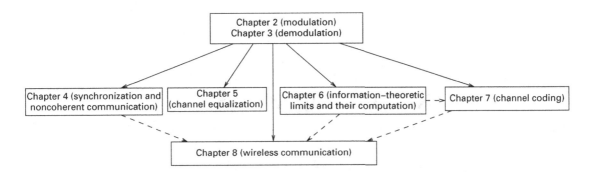

Figure 1 Dependencies among various chapters. Dashed lines denote weak dependencies.

may be able to short-circuit some of these dependencies, especially the weak ones indicated by dashed lines. For example, much of the material in Chapter 7 (coding) and Chapter 8 (wireless communication) is accessible without detailed coverage of Chapter 6 (information theory).

In terms of my personal experience with teaching the material at the University of California, Santa Barbara (UCSB), in the introductory graduate course on digital communication, I cover the material in Chapters 2, 3, 4, and 5 in one quarter, typically spending little time on the material in Chapter 2 in class, since most students have seen some version of this material. Sometimes, depending on the pace of the class, I am also able to provide a glimpse of Chapters 6 and 7. In a follow-up graduate course, I cover the material in Chapters 6, 7, and 8. The pace is usually quite rapid in a quarter system, and the same material could easily take up two semesters when taught in more depth, and at a more measured pace.

An alternative course structure that is quite appealing, especially in terms of systematic coverage of fundamentals, is to cover Chapters 2, 3, 6, and part of 7 in an introductory graduate course, and to cover the remaining topics in a follow-up course.

Acknowledgements

This book is an outgrowth of graduate and senior level digital communication courses that I have taught at the University of California, Santa Barbara (UCSB) and the University of Illinois at Urbana-Champaign (UIUC). I would, therefore, like to thank students over the past decade who have been guinea pigs for my various attempts at course design at both of these institutions. This book is influenced heavily by my research in communication systems, and I would like to thank the funding agencies who have supported this work. These include the National Science Foundation, the Office of Naval Research, the Army Research Office, Motorola, Inc., and the University of California Industry-University Cooperative Research Program.

A number of graduate students have contributed to this book by generating numerical results and plots, providing constructive feedback on draft chapters, and helping write solutions to problems. Specifically, I would like to thank the following members and alumni of my research group: Bharath Ananthasubramaniam, Noah Jacobsen, Raghu Mudumbai, Sandeep Ponnuru, Jaspreet Singh, Sumit Singh, Eric Torkildson, and Sriram Venkateswaran. I would also like to thank Ibrahim El-Khalil, Jim Kleban, Michael Sander, and Sheng-Luen Wei for pointing out typos. I would also like to acknowledge (in order of graduation) some former students, whose doctoral research influenced portions of this textbook: Dilip Warrier, Eugene Visotsky, Rong-Rong Chen, Gwen Barriac, and Noah Jacobsen.

I would also like to take this opportunity to acknowledge the supportive and stimulating environment at the University of Illinois at Urbana-Champaign (UIUC), which I experienced both as a graduate student and as a tenure-track faculty. Faculty at UIUC who greatly enhanced my graduate student experience include my thesis advisor, Professor Mike Pursley (now at Clemson University), Professor Bruce Hajek, Professor Vince Poor (now at Princeton University), and Professor Dilip Sarwate. Moreover, as a faculty at UIUC, I benefited from technical interactions with a number of other faculty in the communications area, including Professor Dick Blahut, Professor Ralf Koetter, Professor Muriel Medard, and Professor Andy Singer. Among my

Acknowledgements

UCSB colleagues, I would like to thank Professor Ken Rose for his helpful feedback on Chapter 6, and I would like to acknowledge my collaboration with Professor Mark Rodwell in the electronics area, which has educated me on a number of implementation considerations in communication systems. Past research collaborators who have influenced this book indirectly include Professor Mike Honig and Professor Sergio Verdu.

I would like to thank Dr. Phil Meyler at Cambridge University Press for pushing me to commit to writing this textbook. I also thank Professor Venu Veeravalli at UIUC and Professor Prakash Narayan at the University of Maryland, College Park, for their support and helpful feedback regarding the book proposal that I originally sent to Cambridge University Press.

Finally, I would like to thank my family for always making life unpredictable and enjoyable at home, regardless of the number of professional commitments I pile on myself.

CHAPTER 1

Introduction

We define communication as information transfer between different points in space or time, where the term *information* is loosely employed to cover standard formats that we are all familiar with, such as voice, audio, video, data files, web pages, etc. Examples of communication between two points in space include a telephone conversation, accessing an Internet website from our home or office computer, or tuning in to a TV or radio station. Examples of communication between two points in time include accessing a storage device, such as a record, CD, DVD, or hard drive. In the preceding examples, the information transferred is directly available for human consumption. However, there are many other communication systems, which we do not directly experience, but which form a crucial part of the infrastructure that we rely upon in our daily lives. Examples include high-speed packet transfer between routers on the Internet, inter- and intra-chip communication in integrated circuits, the connections between computers and computer peripherals (such as keyboards and printers), and control signals in communication networks.

In *digital* communication, the information being transferred is represented in digital form, most commonly as binary digits, or *bits*. This is in contrast to *analog* information, which takes on a continuum of values. Most communication systems used for transferring information today are either digital, or are being converted from analog to digital. Examples of some recent conversions that directly impact consumers include cellular telephony (from analog FM to several competing digital standards), music storage (from vinyl records to CDs), and video storage (from VHS or beta tapes to DVDs). However, we typically consume information in analog form; for example, reading a book or a computer screen, listening to a conversation or to music. Why, then, is the world going digital? We consider this issue after first discussing the components of a typical digital communication system.

1.1 Components of a digital communication system

Consider the block diagram of a digital communication link depicted in Figure 1.1. Let us now briefly discuss the roles of the blocks shown in the figure.

Source encoder Information theory tells us that any information can be efficiently represented in digital form up to arbitrary precision, with the number of bits required for the representation depending on the required fidelity. The task of the source encoder is to accomplish this in a practical setting, reducing the redundancy in the original information in a manner that takes into account the end user's requirements. For example, voice can be intelligibly encoded into a 4 kbit/s bitstream for severely bandwidth constrained settings, or sent at 64 kbit/s for conventional wireline telephony. Similarly, audio encoding rates have a wide range – MP3 players for consumer applications may employ typical bit rates of 128 kbit/s, while high-end digital audio studio equipment may require around ten times higher bit rates. While the preceding examples refer to lossy source coding (in which a controlled amount of information is discarded), lossless compression of data files can also lead to substantial reductions in the amount of data to be transmitted.

Channel encoder and modulator While the source encoder eliminates unwanted redundancy in the information to be sent, the channel encoder introduces redundancy in a controlled fashion in order to combat errors that may arise from channel imperfections and noise. The output of the channel encoder is a codeword from a channel code, which is designed specifically for the anticipated channel characteristics and the requirements dictated by higher network layers. For example, for applications that are delay insensitive, the channel code may be optimized for error detection, followed by a request for retransmission. On the other hand, for real-time applications for which retransmissions are not possible, the channel code may be optimized for error correction. Often, a combination of error correction and detection may be employed. The modulator translates the discrete symbols output by the channel code into an analog waveform that can be transmitted over the

Figure 1.1 Block diagram of a digital communication link.

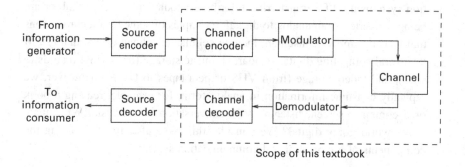

physical channel. The physical channel for an 802.11b based wireless local area network link is, for example, a band of 20 MHz width at a frequency of approximately 2.4 GHz. For this example, the modulator translates a bitstream of rate 1, 2, 5.5, or 11 Mbit/s (the rate varies, depending on the channel conditions) into a waveform that fits within the specified 20 MHz frequency band.

Channel The physical characteristics of communication channels can vary widely, and good channel models are critical to the design of efficient communication systems. While receiver thermal noise is an impairment common to most communication systems, the channel distorts the transmitted waveform in a manner that may differ significantly in different settings. For wireline communication, the channel is well modeled as a linear time-invariant system, and the transfer function in the band used by the modulator can often be assumed to be known at the transmitter, based on feedback obtained from the receiver at the link set-up phase. For example, in high-speed digital subscriber line (DSL) systems over twisted pairs, such channel feedback is exploited to send more information at frequencies at which the channel gain is larger. On the other hand, for wireless mobile communication, the channel may vary because of relative mobility between the transmitter and receiver, which affects both transmitter design (accurate channel feedback is typically not available) and receiver design (the channel must either be estimated, or methods that do not require accurate channel estimates must be used). Further, since wireless is a broadcast medium, multiple-access interference due to simultaneous transmissions must be avoided either by appropriate resource sharing mechanisms, or by designing signaling waveforms and receivers to provide robust performance in the presence of interference.

Demodulator and channel decoder The demodulator processes the analog received waveform, which is a distorted and noisy version of the transmitted waveform. One of its key tasks is synchronization: the demodulator must account for the fact that the channel can produce phase, frequency, and time shifts, and that the clocks and oscillators at the transmitter and receiver are not synchronized a priori. Another task may be channel equalization, or compensation of the intersymbol interference induced by a dispersive channel. The ultimate goal of the demodulator is to produce tentative decisions on the transmitted symbols to be fed to the channel decoder. These decisions may be "hard" (e.g., the demodulator guesses that a particular bit is 0 or 1), or "soft" (e.g., the demodulator estimates the likelihood of a particular bit being 0 or 1). The channel decoder then exploits the redundancy in the channel to code to improve upon the estimates from the demodulator, with its final goal being to produce an estimate of the sequence of information symbols that were the input to the channel encoder. While the demodulator and decoder operate independently in traditional receiver designs, recent advances in coding and communication theory show that iterative information exchange between the demodulator and the decoder can dramatically improve performance.

Source decoder The source decoder converts the estimated information bits produced by the channel decoder into a format that can be used by the end user. This may or may not be the same as the original format that was the input to the source encoder. For example, the original source encoder could have translated speech into text, and then encoded it into bits, and the source decoder may then display the text to the end user, rather than trying to reproduce the original speech.

We are now ready to consider why the world is going digital. The two key advantages of the digital communication approach to the design of transmission and storage media are as follows:

Source-independent design Once information is transformed into bits by the source encoder, it can be stored or transmitted without interpretation: as long as the bits are recovered, the information they represent can be reconstructed with the same degree of precision as originally encoded. This means that the storage or communication medium can be independent of the source characteristics, so that a variety of information sources can share the same communication medium. This leads to significant economies of scale in the design of individual communication links as well as communication networks comprising many links, such as the Internet. Indeed, when information has to traverse multiple communication links in a network, the source encoding and decoding in Figure 1.1 would typically be done at the end points alone, with the network transporting the information bits put out by the source encoder without interpretation.

Channel-optimized design For each communication link, the channel encoder or decoder and modulator or demodulator can be optimized for the specific channel characteristics. Since the bits being transported are regenerated at each link, there is no "noise accumulation."

The preceding framework is based on a separation of source coding and channel coding. Not only does this *separation principle* yield practical advantages as mentioned above, but we are also reassured by the source–channel separation theorem of information theory that it is theoretically optimal for point-to-point links (under mild conditions). While the separation approach is critical to obtaining the economies of scale driving the growth of digital communication systems, we note in passing that joint source and channel coding can yield superior performance, both in theory and practice, in certain settings (e.g., multiple-access and broadcast channels, or applications with delay or complexity constraints).

The scope of this textbook is indicated in Figure 1.1: we consider modulation and demodulation, channel encoding and decoding, and channel modeling.

Source encoding and decoding are not covered. Thus, we implicitly restrict attention to communication systems based on the separation principle.

1.2 Text outline

The objective of this text is to convey an understanding of the principles underlying the design of a modern digital communication link. An introduction to modulation techniques (i.e., how to convert bits into a form that can be sent over a channel) is provided in Chapter 2. We emphasize the important role played by the complex baseband representation for passband signals in both transmitter and receiver design, describe some common modulation formats, and discuss how to determine how much bandwidth is required to support a given modulation format. An introduction to demodulation (i.e., how to estimate the transmitted bits from a noisy received signal) for the classical additive white Gaussian noise (AWGN) channel is provided in Chapter 3. Our starting point is the theory of hypothesis testing. We emphasize the geometric view of demodulation first popularized by the classic text of Wozencraft and Jacobs, introduce the concept of soft decisions, and provide a brief exposure to link budget analysis (which is used by system designers for determining parameters such as antenna gains and transmit powers). Mastery of Chapters 2 and 3 is a prerequisite for the remainder of this book. The remaining chapters essentially stand on their own. Chapter 4 contains a framework for estimation of parameters such as delay and phase, starting from the derivation of the likelihood ratio of a signal in AWGN. Optimal noncoherent receivers are derived based on this framework. Chapter 5 describes the key ideas used in channel equalization, including maximum likelihood sequence estimation (MLSE) using the Viterbi algorithm, linear equalization, and decision feedback equalization. Chapter 6 contains a brief treatment of information theory, focused on the *computation* of performance benchmarks. This is increasingly important for the communication system designer, now that turbo-like codes provide a framework for approaching information-theoretic limits for virtually any channel model. Chapter 7 introduces error-correction coding. It includes convolutional codes, serial and parallel concatenated turbo codes, and low density parity check (LDPC) codes. It also provides a very brief discussion of how algebraic codes (which are covered in depth in coding theory texts) fit within modern communication link design, with an emphasis on Reed–Solomon codes. Finally, Chapter 8 contains an introduction to wireless communication, including channel modeling, the effect of fading, and a discussion of some modulation formats commonly used over the wireless channel that are not covered in the introductory treatment in Chapter 2. The latter include orthogonal frequency division multiplexing (OFDM), spread spectrum communication, continuous phase modulation, and space–time (or multiple antenna) communication.

1.3 Further reading

Useful resources for getting a quick exposure to many topics on communication systems are *The Communications Handbook* [1] and *The Mobile Communications Handbook* [2], both edited by Gibson. Standards for communication systems are typically available online from organizations such as the Institute for Electrical and Electronics Engineers (IEEE). Recently published graduate-level textbooks on digital communication include Proakis [3], Benedetto and Biglieri [4], and Barry, Lee, and Messerschmitt [5]. Undergraduate texts on communications include Haykin [6], Proakis and Salehi [7], Pursley [8], and Ziemer and Tranter [9]. Classical texts of enduring value include Wozencraft and Jacobs [10], which was perhaps the first textbook to introduce signal space design techniques, Viterbi [11], which provides detailed performance analysis of demodulation and synchronization techniques, Viterbi and Omura [12], which provides a rigorous treatment of modulation and coding, and Blahut [13], which provides an excellent perspective on the concepts underlying digital communication systems.

We do not cover source coding in this text. An information-theoretic treatment of source coding is provided in Cover and Thomas [14], while a more detailed description of compression algorithms is found in Sayood [15].

Finally, while this text deals with the design of individual communication links, the true value of these links comes from connecting them together to form communication networks, such as the Internet, the wireline phone network, and the wireless cellular communication network. Two useful texts on communication networks are Bertsekas and Gallager [16] and Walrand and Varaiya [17]. On a less technical note, Friedman [18] provides an interesting discussion on the immense impact of advances in communication networking on the global economy.

CHAPTER 2

Modulation

Modulation refers to the representation of digital information in terms of analog waveforms that can be transmitted over physical channels. A simple example is depicted in Figure 2.1, where a sequence of bits is translated into a waveform. The original information may be in the form of bits taking the values 0 and 1. These bits are translated into symbols using a bit-to-symbol map, which in this case could be as simple as mapping the bit 0 to the symbol $+1$, and the bit 1 to the symbol -1. These symbols are then mapped to an analog waveform by multiplying with translates of a transmit waveform (a rectangular pulse in the example shown): this is an example of *linear modulation*, to be discussed in detail in Section 2.5. For the bit-to-symbol map just described, the bitstream encoded into the analog waveform shown in Figure 2.1 is 01100010100.

While a rectangular timelimited transmit waveform is shown in the example of Figure 2.1, in practice, the analog waveforms employed for modulation are often constrained in the frequency domain. Such constraints arise either from the physical characteristics of the communication medium, or from external factors such as government regulation of spectrum usage. Thus, we typically classify channels, and the signals transmitted over them, in terms of the frequency bands they occupy. In this chapter, we discuss some important modulation techniques, after first reviewing some basic concepts regarding frequency domain characterization of signals and systems. The material in this chapter is often covered in detail in introductory digital communication texts,

Figure 2.1 A simple example of binary modulation.

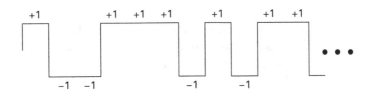

but we emphasize some specific points in somewhat more detail than usual. One of these is the complex baseband representation of passband signals, which is a crucial tool both for understanding and implementing modern communication systems. Thus, the reader who is familiar with this material is still encouraged to skim through this chapter.

Map of this chapter In Section 2.1, we review basic notions such as the frequency domain representation of signals, inner products between signals, and the concept of baseband and passband signals. While currents and voltages in a circuit are always real-valued, both baseband and passband signals can be treated under a unified framework by allowing baseband signals to take on complex values. This complex baseband representation of passband signals is developed in Section 2.2, where we point out that manipulation of complex baseband signals is an essential component of modern transceivers. While the preceding development is for deterministic, finite energy signals, modeling of signals and noise in digital communication relies heavily on finite power, random processes. We therefore discuss frequency domain description of random processes in Section 2.3. This completes the background needed to discuss the main theme of this chapter: modulation. Section 2.4 briefly discusses the degrees of freedom available for modulation, and introduces the concept of bandwidth efficiency. Section 2.5 covers linear modulation using two-dimensional constellations, which, in principle, can utilize all available degrees of freedom in a bandlimited channel. The Nyquist criterion for avoidance of intersymbol interference (ISI) is discussed, in order to establish guidelines relating bandwidth to bit rate. Section 2.6 discusses orthogonal and biorthogonal modulation, which are nonlinear modulation formats optimized for power efficiency. Finally, Section 2.7 discusses differential modulation as a means of combating phase uncertainty. This concludes our introduction to modulation. Several other modulation formats are discussed in Chapter 8, where we describe some modulation techniques commonly employed in wireless communication.

2.1 Preliminaries

This section contains a description of just enough material on signals and systems for our purpose in this text, including the definitions of inner product, norm and energy for signals, convolution, Fourier transform, and baseband and passband signals.

Complex numbers A complex number z can be written as $z = x + jy$, where x and y are real numbers, and $j = \sqrt{-1}$. We say that $x = \text{Re}(z)$ is the real part of z and $y = \text{Im}(z)$ is the imaginary part of z. As depicted in Figure 2.2, it is often advantageous to interpret the complex number z as

2.1 Preliminaries

Figure 2.2 A complex number z represented in the two-dimensional real plane.

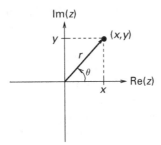

a two-dimensional real vector, which can be represented in rectangular form as $(x, y) = (\text{Re}(z), \text{Im}(z))$, or in polar form as

$$r = |z| = \sqrt{x^2 + y^2},$$
$$\theta = \arg(z) = \tan^{-1} \frac{y}{x}.$$

Euler's identity We routinely employ this to decompose a complex exponential into real-valued sinusoids as follows:

$$e^{j\theta} = \cos\theta + j\sin\theta. \quad (2.1)$$

A key building block of communication theory is the relative geometry of the signals used, which is governed by the inner products between signals. Inner products for continuous-time signals can be defined in a manner exactly analogous to the corresponding definitions in finite-dimensional vector space.

Inner product The inner product for two $m \times 1$ complex vectors $\mathbf{s} = (s[1], \ldots, s[m])^T$ and $\mathbf{r} = (r[1], \ldots, r[m])^T$ is given by

$$\langle \mathbf{s}, \mathbf{r} \rangle = \sum_{i=1}^{m} s[i] r^*[i] = \mathbf{r}^H \mathbf{s}. \quad (2.2)$$

Similarly, we define the inner product of two (possibly complex-valued) signals $s(t)$ and $r(t)$ as follows:

$$\langle s, r \rangle = \int_{-\infty}^{\infty} s(t) r^*(t) \, dt. \quad (2.3)$$

The inner product obeys the following linearity properties:

$$\langle a_1 s_1 + a_2 s_2, r \rangle = a_1 \langle s_1, r \rangle + a_2 \langle s_2, r \rangle,$$
$$\langle s, a_1 r_1 + a_2 r_2 \rangle = a_1^* \langle s, r_1 \rangle + a_2^* \langle s, r_2 \rangle,$$

where a_1, a_2 are complex-valued constants, and s, s_1, s_2, r, r_1, r_2 are signals (or vectors). The complex conjugation when we pull out constants from the second argument of the inner product is something that we need to remain aware of when computing inner products for complex signals.

Energy and norm The *energy* E_s of a signal s is defined as its inner product with itself:

$$E_s = ||s||^2 = \langle s, s \rangle = \int_{-\infty}^{\infty} |s(t)|^2 dt, \qquad (2.4)$$

where $||s||$ denotes the *norm* of s. If the energy of s is zero, then s must be zero "almost everywhere" (e.g., $s(t)$ cannot be nonzero over any interval, no matter how small its length). For continuous-time signals, we take this to be equivalent to being zero everywhere. With this understanding, $||s|| = 0$ implies that s is zero, which is a property that is true for norms in finite-dimensional vector spaces.

Cauchy–Schwartz inequality The inner product obeys the *Cauchy–Schwartz* inequality, stated as follows:

$$|\langle s, r \rangle| \leq ||s|| \, ||r||, \qquad (2.5)$$

with equality if and only if, for some complex constant a, $s(t) = ar(t)$ or $r(t) = as(t)$ almost everywhere. That is, equality occurs if and only if one signal is a scalar multiple of the other. The proof of this inequality is given in Problem 2.4.

Convolution The convolution of two signals s and r gives the signal

$$q(t) = (s * r)(t) = \int_{-\infty}^{\infty} s(u) r(t-u) du.$$

Here, the convolution is evaluated at time t, while u is a "dummy" variable that is integrated out. However, it is sometimes convenient to abuse notation and use $q(t) = s(t) * r(t)$ to denote the convolution between s and r. For example, this enables us to state compactly the following linear time invariance (LTI) property:

$$(a_1 s_1(t-t_1) + a_2 s_2(t-t_2)) * r(t) = a_1(s_1 * r)(t-t_1) + a_2(s_2 * r)(t-t_2),$$

for any complex gains a_1 and a_2, and any time offsets t_1 and t_2.

Delta function The delta function $\delta(t)$ is defined via the following "sifting" property: for any finite energy signal $s(t)$, we have

$$\int_{-\infty}^{\infty} \delta(t - t_0) s(t) dt = s(t_0). \qquad (2.6)$$

In particular, this implies that convolution of a signal with a shifted version of the delta function gives a shifted version of the signal:

$$\delta(t - t_0) * s(t) = s(t - t_0). \qquad (2.7)$$

Equation (2.6) can be shown to imply that $\delta(0) = \infty$ and $\delta(t) = 0$ for $t \neq 0$. Thus, thinking of the delta function as a signal is a convenient abstraction, since it is not physically realizable.

2.1 Preliminaries

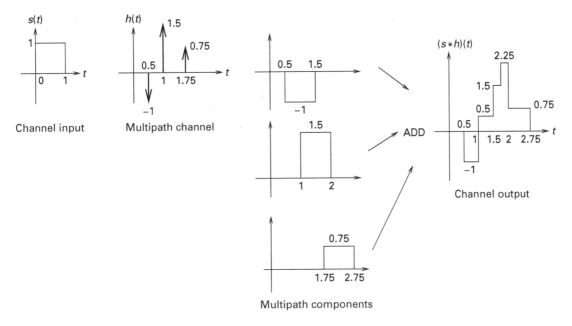

Figure 2.3 A signal going through a multipath channel.

Convolution plays a fundamental role in both modeling and transceiver implementation in communication systems, as illustrated by the following examples.

Example 2.1.1 (Modeling a multipath channel) The *channel* between the transmitter and the receiver is often modeled as an LTI system, with the received signal y given by

$$y(t) = (s * h)(t) + n(t),$$

where s is the transmitted waveform, h is the channel impulse response, and $n(t)$ is receiver thermal noise and interference. Suppose that the channel impulse response is given by

$$h(t) = \sum_{i=1}^{M} a_i \delta(t - t_i).$$

Ignoring the noise, a signal $s(t)$ passing through such a channel produces an output

$$y(t) = (s * h)(t) = \sum_{i=1}^{M} a_i s(t - t_i).$$

This could correspond, for example, to a wireless *multipath* channel in which the transmitted signal is reflected by a number of scatterers, each of which gives rise to a copy of the signal with a different delay and scaling. Typically, the results of propagation studies would be used to obtain statistical models for the number of multipath components M, the delays $\{t_i\}$, and the gains $\{a_i\}$.

Example 2.1.2 (Matched filter) For a complex-valued signal $s(t)$, the matched filter is defined as a filter with impulse response $s_{MF}(t) = s^*(-t)$; see Figure 2.4 for an example. Note that $S_{MF}(f) = S^*(f)$. If the input to the matched filter is $x(t)$, then the output is given by

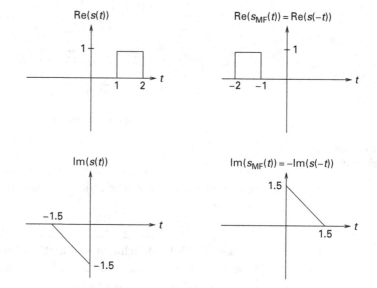

Figure 2.4 Matched filter for a complex-valued signal.

$$y(t) = (x * s_{MF})(t) = \int_{-\infty}^{\infty} x(u) s_{MF}(t-u) du = \int_{-\infty}^{\infty} x(u) s^*(u-t) du. \quad (2.8)$$

The matched filter, therefore, computes the inner product between the input x and all possible time translates of the waveform s, which can be interpreted as "template matching." In particular, the inner product $\langle x, s \rangle$ equals the output of the matched filter at time 0. Some properties of the matched filter are explored in Problem 2.5. For example, if $x(t) = s(t - t_0)$ (i.e., the input is a time translate of s), then, as shown in Problem 2.5, the magnitude of the matched filter output is maximum at $t = t_0$. We can, then, intuitively see how the matched filter would be useful, for example, in delay estimation using "peak picking." In later chapters, a more systematic development is used to reveal the key role played by the matched filter in digital communication receivers.

2.1 Preliminaries

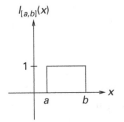

Figure 2.5 The indicator function of an interval has a boxcar shape.

Indicator function We use I_A to denote the indicator function of a set A, defined as

$$I_A(x) = \begin{cases} 1, & x \in A, \\ 0, & \text{otherwise.} \end{cases}$$

For example, the indicator function of an interval has a boxcar shape, as shown in Figure 2.5.

Sinc function The sinc function is defined as

$$\text{sinc}(x) = \frac{\sin(\pi x)}{\pi x},$$

where the value at $x = 0$, defined as the limit as $x \to 0$, is set as $\text{sinc}(0) = 1$. The sinc function is shown in Figure 2.19. Since $|\sin(\pi x)| \leq 1$, we have that $|\text{sinc}(x)| \leq (1/\pi x)$. That is, the sinc function exhibits a sinusoidal variation, with an envelope that decays as $1/x$. We plot the sinc function later in this chapter, in Figure 2.19, when we discuss linear modulation.

Fourier transform Let $s(t)$ denote a signal, and $S(f) = \mathcal{F}(s(t))$ denote its Fourier transform, defined as

$$S(f) = \int_{-\infty}^{\infty} s(t) e^{-j2\pi ft} \, dt. \tag{2.9}$$

The inverse Fourier transform is given by

$$s(t) = \int_{-\infty}^{\infty} S(f) e^{j2\pi ft} \, df. \tag{2.10}$$

Both $s(t)$ and $S(f)$ are allowed to take on complex values. We denote the relationship that $s(t)$ and $S(f)$ are a *Fourier transform pair* by $s(t) \leftrightarrow S(f)$.

Time–frequency duality in Fourier transform From an examination of the expressions (2.9) and (2.10), we obtain the following duality relation: if $s(t)$ has Fourier transform $S(f)$, then the signal $r(t) = S(t)$ has Fourier transform $R(f) = s(-f)$.

Important Fourier transform pairs

(i) The boxcar and the sinc functions form a pair:

$$s(t) = I_{[-\frac{T}{2}, \frac{T}{2}]}(t) \leftrightarrow S(f) = T\text{sinc}(fT). \tag{2.11}$$

(ii) The delta function and the constant function form a pair:

$$s(t) = \delta(t) \leftrightarrow S(f) \equiv 1. \tag{2.12}$$

We list only two pairs here, because most of the examples that we use in our theoretical studies can be derived in terms of these, using time–frequency duality and the properties of the Fourier transform below. On the other hand, closed form analytical expressions are not available for many waveforms encountered in practice, and the Fourier or inverse Fourier transform is computed numerically using the discrete Fourier transform (DFT) in the sampled domain.

Basic properties of the Fourier transform Some properties of the Fourier transform that we use extensively are as follows (it is instructive to derive these starting from the definition (2.9)):

(i) Complex conjugation in the time domain corresponds to conjugation and reflection around the origin in the frequency domain, and vice versa;

$$\begin{aligned} s^*(t) &\leftrightarrow S^*(-f), \\ s^*(-t) &\leftrightarrow S^*(f). \end{aligned} \qquad (2.13)$$

(ii) A signal $s(t)$ is real-valued (i.e., $s(t) = s^*(t)$) if and only if its Fourier transform is conjugate symmetric (i.e., $S(f) = S^*(-f)$). Note that conjugate symmetry of $S(f)$ implies that $\text{Re}(S(f)) = \text{Re}(S(-f))$ (real part is symmetric) and $\text{Im}(S(f)) = -\text{Im}(S(-f))$ (imaginary part is antisymmetric).

(iii) Convolution in the time domain corresponds to multiplication in the frequency domain, and vice versa;

$$\begin{aligned} s(t) = (s_1 * s_2)(t) &\leftrightarrow S(f) = S_1(f)S_2(f), \\ s(t) = s_1(t)s_2(t) &\leftrightarrow S(f) = (S_1 * S_2)(f). \end{aligned} \qquad (2.14)$$

(iv) Translation in the time domain corresponds to multiplication by a complex exponential in the frequency domain, and vice versa;

$$\begin{aligned} s(t-t_0) &\leftrightarrow S(f)e^{-j2\pi f t_0}, \\ s(t)e^{j2\pi f_0 t} &\leftrightarrow S(f-f_0). \end{aligned} \qquad (2.15)$$

(v) Time scaling leads to reciprocal frequency scaling;

$$s(at) \leftrightarrow \frac{1}{|a|} S\left(\frac{f}{a}\right). \qquad (2.16)$$

(vi) *Parseval's identity* The inner product of two signals can be computed in either the time or frequency domain, as follows:

$$\langle s_1, s_2 \rangle = \int_{-\infty}^{\infty} s_1(t) s_2^*(t) dt = \int_{-\infty}^{\infty} S_1(f) S_2^*(f) df = \langle S_1, S_2 \rangle. \qquad (2.17)$$

Setting $s_1 = s_2 = s$, we obtain the following expression for the energy E_s of a signal $s(t)$:

$$E_s = ||s||^2 = \int_{-\infty}^{\infty} |s(t)|^2 dt = \int_{-\infty}^{\infty} |S(f)|^2 df. \qquad (2.18)$$

2.1 Preliminaries

Energy spectral density The energy spectral density $E_s(f)$ of a signal $s(t)$ can be defined operationally as follows. Pass the signal $s(t)$ through an ideal narrowband filter with transfer function;

$$H_{f_0}(f) = \begin{cases} 1, & f_0 - \frac{\Delta f}{2} < f < f_0 + \frac{\Delta f}{2}, \\ 0, & \text{else.} \end{cases}$$

The energy spectral density $E_s(f_0)$ is defined to be the energy at the output of the filter, divided by the width Δf (in the limit as $\Delta f \to 0$). That is, the energy at the output of the filter is approximately $E_s(f_0)\Delta f$. But the Fourier transform of the filter output is

$$Y(f) = S(f)H(f) = \begin{cases} S(f), & f_0 - \frac{\Delta f}{2} < f < f_0 + \frac{\Delta f}{2}, \\ 0, & \text{else.} \end{cases}$$

By Parseval's identity, the energy at the output of the filter is

$$\int_{-\infty}^{\infty} |Y(f)|^2 \, df = \int_{f_0 - \frac{\Delta f}{2}}^{f_0 + \frac{\Delta f}{2}} |S(f)|^2 \, df \approx |S(f_0)|^2 \, \Delta f,$$

assuming that $S(f)$ varies smoothly and Δf is small enough. We can now infer that the energy spectral density is simply the magnitude squared of the Fourier transform:

$$E_s(f) = |S(f)|^2. \tag{2.19}$$

The integral of the energy spectral density equals the signal energy, which is simply a restatement of Parseval's identity.

Autocorrelation function The inverse Fourier transform of the energy spectral density $E_s(f)$ is termed the autocorrelation function $R_s(\tau)$, since it measures how closely the signal s matches delayed versions of itself. Since $|S(f)|^2 = S(f)S^*(f) = S(f)S_{MF}(f)$, where $s_{MF}(t) = s^*(-t)$ is the matched filter for s introduced earlier. We therefore have that

$$E_s(f) = |S(f)|^2 \leftrightarrow R_s(\tau) = (s * s_{MF})(\tau) = \int_{-\infty}^{\infty} s(u)s^*(u - \tau) \, du. \tag{2.20}$$

Thus, $R_s(\tau)$ is the outcome of passing the signal s through its matched filter, and sampling the output at time τ, or equivalently, correlating the signal s with a complex conjugated version of itself, delayed by τ.

While the preceding definitions are for finite energy deterministic signals, we revisit these concepts in the context of finite power random processes later in this chapter.

Baseband and passband signals A signal $s(t)$ is said to be *baseband* if

$$S(f) \approx 0, \quad |f| > W \tag{2.21}$$

for some $W > 0$. That is, the signal energy is concentrated in a band around DC. Similarly, a channel modeled as a linear time-invariant system is said to be baseband if its transfer function $H(f)$ satisfies (2.21).

16 Modulation

A signal $s(t)$ is said to be *passband* if

$$S(f) \approx 0, \quad |f \pm f_c| > W \qquad (2.22)$$

where $f_c > W > 0$. A channel modeled as a linear time-invariant system is said to be passband if its transfer function $H(f)$ satisfies (2.22).

Examples of baseband and passband signals are shown in Figures 2.6 and 2.7, respectively. We consider real-valued signals, since any signal that has a physical realization in terms of a current or voltage must be real-valued. As shown, the Fourier transforms can be complex-valued, but they must satisfy the conjugate symmetry condition $S(f) = S^*(-f)$. The bandwidth B is defined to be the size of the frequency interval occupied by $S(f)$, where we consider only the spectral occupancy for the positive frequencies

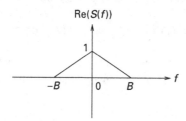
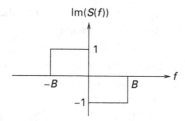

Figure 2.6 Example of the spectrum $S(f)$ for a real-valued baseband signal. The bandwidth of the signal is B.

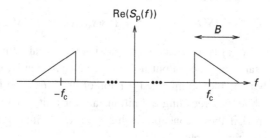
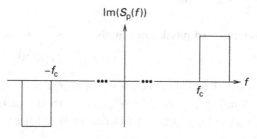

Figure 2.7 Example of the spectrum $S(f)$ for a real-valued passband signal. The bandwidth of the signal is B. The figure shows an arbitrarily chosen frequency f_c within the band in which $S(f)$ is nonzero. Typically, f_c is much larger than the signal bandwidth B.

for a real-valued signal $s(t)$. This makes sense from a physical viewpoint: after all, when the FCC allocates a frequency band to an application, say, around 2.4 GHz for unlicensed usage, it specifies the positive frequencies that can be occupied. However, in order to be clear about the definition being used, we occasionally employ the more specific term *one-sided* bandwidth, and also define the *two-sided* bandwidth based on the spectral occupancy for both positive and negative frequencies. For real-valued signals, the two-sided bandwidth is simply twice the one-sided bandwidth, because of the conjugate symmetry condition $S(f) = S^*(-f)$. However, when we consider the complex baseband representation of real-valued passband signals in the next section, the complex-valued signals which we consider do not, in general, satisfy the conjugate symmetry condition, and there is no longer a deterministic relationship between the two-sided and one-sided bandwidths. As we show in the next section, a real-valued passband signal has an equivalent representation as a complex-valued baseband signal, and the (one-sided) bandwidth of the passband signal equals the two-sided bandwidth of its complex baseband representation.

In Figures 2.6 and 2.7, the spectrum is shown to be exactly nonzero outside a well defined interval, and the bandwidth B is the size of this interval. In practice, there may not be such a well defined interval, and the bandwidth depends on the specific definition employed. For example, the bandwidth might be defined as the size of an appropriately chosen interval in which a specified fraction (say 99%) of the signal energy lies.

Example 2.1.3 (Fractional energy containment bandwidth) Consider a rectangular time domain pulse $s(t) = I_{[0,T]}$. Using (2.11) and (2.15), the Fourier transform of this signal is given by $S(f) = T\text{sinc}(fT)e^{-j\pi fT}$, so that

$$|S(f)|^2 = T^2 \text{sinc}^2(fT).$$

Clearly, there is no finite frequency interval that contains all of the signal energy. Indeed, it follows from a general *uncertainty* principle that strictly timelimited signals cannot be strictly bandlimited, and vice versa. However, most of the energy of the signal is concentrated around the origin, so that $s(t)$ is a baseband signal. We can now define the (one-sided) fractional energy containment bandwidth B as follows:

$$\int_{-B}^{B} |S(f)|^2 \, df = a \int_{-\infty}^{\infty} |S(f)|^2 \, df, \qquad (2.23)$$

where $0 < a \leq 1$ is the fraction of energy contained in the band $[-B, B]$. The value of B must be computed numerically, but there are certain simplifications that are worth pointing out. First, note that T can be set to any convenient value, say $T = 1$ (equivalently, one unit of time is redefined to be T). By virtue of the scaling property (2.16), time scaling leads to reciprocal frequency scaling. Thus, if the bandwidth for $T = 1$ is B_1, then the bandwidth for arbitrary T must be $B_T = B_1/T$. This holds regardless of the specific notion of bandwidth used, since the scaling property can be viewed simply as redefining the unit of frequency in a consistent manner with the change in our unit for time. The second observation is that the right-hand side of (2.23) can be evaluated in closed form using Parseval's identity (2.18). Putting these observations together, it is left as an exercise for the reader to show that (2.23) can be rewritten as

$$\int_{-B_1}^{B_1} \text{sinc}^2 f \, df = a, \tag{2.24}$$

which can be further simplified to

$$\int_0^{B_1} \text{sinc}^2 f \, df = \frac{a}{2}, \tag{2.25}$$

using the symmetry of the integrand around the origin. We can now evaluate B_1 numerically for a given value of a. We obtain $B_1 = 10.2$ for $a = 0.99$, and $B_1 = 0.85$ for $a = 0.9$. Thus, while the 90% energy containment bandwidth is moderate, the 99% energy containment bandwidth is large, because of the slow decay of the sinc function. For an arbitrary value of T, the 99% energy containment bandwidth is $B = 10.2/T$.

A technical note: (2.24) could also be inferred from (2.23) by applying a change of variables, replacing fT in (2.23) by f. This change of variables is equivalent to the scaling argument that we invoked.

2.2 Complex baseband representation

We often employ passband channels, which means that we must be able to transmit and receive passband signals. We now show that all the information carried in a real-valued passband signal is contained in a corresponding complex-valued baseband signal. This baseband signal is called the *complex baseband representation*, or *complex envelope*, of the passband signal. This equivalence between passband and complex baseband has profound practical significance. Since the complex envelope can be represented accurately in discrete time using a much smaller sampling rate than the corresponding passband signal $s_p(t)$, modern communication transceivers can implement complicated signal processing algorithms digitally on complex baseband signals, keeping the analog processing of passband signals to a minimum. Thus, the transmitter encodes information into the complex baseband waveform using encoding, modulation and filtering performed using digital signal processing (DSP). The complex baseband waveform is then *upconverted* to the corresponding passband signal to be sent on the channel. Similarly, the passband received waveform is *downconverted* to complex baseband by the receiver, followed

2.2 Complex baseband representation

by DSP operations for synchronization, demodulation, and decoding. This leads to a modular framework for transceiver design, in which sophisticated algorithms can be developed in complex baseband, independent of the physical frequency band that is ultimately employed for communication.

We now describe in detail the relation between passband and complex baseband, and the relevant transceiver operations. Given the importance of being comfortable with complex baseband, the pace of the development here is somewhat leisurely. For a reader who knows this material, quickly browsing this section to become familiar with the notation should suffice.

Time domain representation of a passband signal Any passband signal $s_p(t)$ can be written as

$$s_p(t) = \sqrt{2} s_c(t) \cos 2\pi f_c t - \sqrt{2} s_s(t) \sin 2\pi f_c t, \tag{2.26}$$

where $s_c(t)$ ("c" for "cosine") and $s_s(t)$ ("s" for "sine") are real-valued signals, and f_c is a frequency reference typically chosen in or around the band occupied by $S_p(f)$. The factor of $\sqrt{2}$ is included only for convenience in normalization (more on this later), and is often omitted in the literature.

In-phase and quadrature components The waveforms $s_c(t)$ and $s_s(t)$ are also referred to as the in-phase (or I) component and the quadrature (or Q) component of the passband signal $s_p(t)$, respectively.

Example 2.2.1 (Passband signal) The signal

$$s_p(t) = \sqrt{2} I_{[0,1]}(t) \cos 300\pi t - \sqrt{2}(1-|t|) I_{[-1,1]}(t) \sin 300\pi t$$

is a passband signal with I component $s_c(t) = I_{[0,1]}(t)$ and Q component $s_s(t) = (1-|t|) I_{[-1,1]}(t)$. Like Example 2.1.3, this example also illustrates that we do not require strict bandwidth limitations in our definitions of passband and baseband: the I and Q components are timelimited, and hence cannot be bandlimited. However, they are termed baseband signals because most of their energy lies in the baseband. Similarly, $s_p(t)$ is termed a passband signal, since most of its frequency content lies in a small band around 150 Hz.

Complex envelope The complex envelope, or complex baseband representation, of $s_p(t)$ is now defined as

$$s(t) = s_c(t) + j s_s(t). \tag{2.27}$$

In the preceding example, the complex envelope is given by $s(t) = I_{[0,1]}(t) + j(1-|t|) I_{[-1,1]}(t)$.

Time domain relationship between passband and complex baseband We can rewrite (2.26) as
$$s_p(t) = \text{Re}(\sqrt{2}s(t)e^{j2\pi f_c t}). \quad (2.28)$$
To check this, plug in (2.27) and Euler's identity (2.1) on the right-hand side to obtain the expression (2.26).

Envelope and phase of a passband signal The complex envelope $s(t)$ can also be represented in polar form, defining the *envelope* $e(t)$ and *phase* $\theta(t)$ as
$$e(t) = |s(t)| = \sqrt{s_c^2(t) + s_s^2(t)}, \quad \theta(t) = \tan^{-1}\frac{s_s(t)}{s_c(t)}. \quad (2.29)$$
Plugging $s(t) = e(t)e^{j\theta(t)}$ into (2.28), we obtain yet another formula for the passband signal s:
$$s_p(t) = e(t)\cos(2\pi f_c t + \theta(t)). \quad (2.30)$$
The equations (2.26), (2.28) and (2.30) are three different ways of expressing the same relationship between passband and complex baseband in the time domain.

Example 2.2.2 (Modeling frequency or phase offsets in complex baseband) Consider the passband signal s_p (2.26), with complex baseband representation $s = s_c + js_s$. Now, consider a phase-shifted version of the passband signal
$$\tilde{s}_p(t) = \sqrt{2}s_c(t)\cos(2\pi f_c t + \theta(t)) - \sqrt{2}s_s(t)\sin(2\pi f_c t + \theta(t)),$$
where $\theta(t)$ may vary slowly with time. For example, a carrier frequency offset a and a phase offset b corresponds to $\theta(t) = 2\pi at + b$. We wish to find the complex envelope of $\tilde{s}_p$ with respect to f_c. To do this, we write $\tilde{s}_p$ in the standard form (2.28) as follows:
$$\tilde{s}_p(t) = \text{Re}(\sqrt{2}s(t)e^{j(2\pi f_c t + \theta(t))}).$$
Comparing with the desired form
$$\tilde{s}_p(t) = \text{Re}(\sqrt{2}\tilde{s}(t)e^{j2\pi f_c t}),$$
we can read off
$$\tilde{s}(t) = s(t)e^{j\theta(t)}. \quad (2.31)$$
Equation (2.31) relates the complex envelopes before and after a phase offset. We can expand out this "polar form" representation to obtain the corresponding relationship between the I and Q components. Suppressing time dependence from the notation, we can rewrite (2.31) as
$$\tilde{s}_c + j\tilde{s}_s = (s_c + js_s)(\cos\theta + j\sin\theta)$$

2.2 Complex baseband representation

> using Euler's formula. Equating real and imaginary parts on both sides, we obtain
>
> $$\begin{aligned} \tilde{s}_c &= s_c \cos\theta - s_s \sin\theta, \\ \tilde{s}_s &= s_c \sin\theta + s_s \cos\theta. \end{aligned} \quad (2.32)$$
>
> This is a typical example of the advantage of working in complex baseband. Relationships between passband signals can be compactly represented in complex baseband, as in (2.31). For signal processing using real-valued arithmetic, these complex baseband relationships can be expanded out to obtain relationships involving real-valued quantities, as in (2.32).

Orthogonality of I and Q channels The passband waveform $x_c(t) = \sqrt{2}s_c(t)\cos 2\pi f_c t$ corresponding to the I component, and the passband waveform $x_s(t) = \sqrt{2}s_s(t)\sin 2\pi f_c t$ corresponding to the Q component, are orthogonal. That is,

$$\langle x_c, x_s \rangle = 0. \quad (2.33)$$

Since what we know about s_c and s_s (i.e., they are baseband) is specified in the frequency domain, we prove this result by computing the inner product in the frequency domain, using Parseval's identity (2.17):

$$\langle x_c, x_s \rangle = \langle X_c, X_s \rangle = \int_{-\infty}^{\infty} X_c(f) X_s^*(f) \, df.$$

We now need expressions for X_c and X_s. Since $\cos\theta = \frac{1}{2}(e^{j\theta} + e^{-j\theta})$ and $\sin\theta = \frac{1}{2j}(e^{j\theta} - e^{-j\theta})$ we have

$$x_c(t) = \frac{1}{\sqrt{2}}(s_c(t)e^{j2\pi f_c t} + s_c(t)e^{-j2\pi f_c t}) \leftrightarrow X_c(f) = \frac{1}{\sqrt{2}}(S_c(f-f_c) + S_c(f+f_c)),$$

$$x_s(t) = \frac{1}{\sqrt{2}j}(s_s(t)e^{j2\pi f_c t} - s_s(t)e^{-j2\pi f_c t}) \leftrightarrow X_s(f) = \frac{1}{\sqrt{2}j}(S_s(f-f_c) - S_s(f+f_c)).$$

The inner product can now be computed as follows:

$$\langle X_c, X_s \rangle = \frac{1}{2j} \int_{-\infty}^{\infty} [S_c(f-f_c) + S_c(f+f_c)] \, [S_s^*(f-f_c) - S_s^*(f+f_c)] df. \quad (2.34)$$

We now look more closely at the integrand above. Since f_c is assumed to be larger than the bandwidth of the baseband signals S_c and S_s, the translation of $S_c(f)$ to the right by f_c has zero overlap with a translation of $S_s^*(f)$ to the left by f_c. That is, $S_c(f-f_c)S_s^*(f+f_c) \equiv 0$. Similarly, $S_c(f+f_c)S_s^*(f-f_c) \equiv 0$. We can therefore rewrite the inner product in (2.34) as

$$\langle X_c, X_s \rangle = \frac{1}{2j} \left[\int_{-\infty}^{\infty} S_c(f-f_c) S_s^*(f-f_c) df - \int_{-\infty}^{\infty} S_c(f+f_c) S_s^*(f+f_c) df \right]$$

$$= \frac{1}{2j} \left[\int_{-\infty}^{\infty} S_c(f) S_s^*(f) df - \int_{-\infty}^{\infty} S_c(f) S_s^*(f) df \right] = 0, \quad (2.35)$$

where we have used a change of variables to show that the integrals involved cancel out.

Exercise 2.2.1 Work through the details of an alternative, shorter, proof of (2.33) as follows. Show that

$$u(t) = x_c(t)x_s(t) = 2s_c(t)s_s(t)\cos 2\pi f_c t \sin 2\pi f_c t$$

is a passband signal (around what frequency?), and thus infer that

$$\int_{-\infty}^{\infty} u(t)dt = U(0) = 0.$$

Passband and complex baseband inner products For real passband signals a_p and b_p with complex envelopes a and b, respectively, the inner product satisfies

$$\langle u_p, v_p \rangle = \langle u_c, v_c \rangle + \langle u_s, v_s \rangle = \text{Re}(\langle u, v \rangle). \tag{2.36}$$

To show the first equality, we substitute the standard form (2.26) for u_p and v_p and use the orthogonality of the I and Q components. For the second equality, we write out the complex inner product $\langle u, v \rangle$,

$$\langle u, v \rangle = \int_{-\infty}^{\infty} (u_c(t) + ju_s(t))(v_c(t) - jv_s(t)) \, dt$$
$$= (\langle u_c, v_c \rangle + \langle u_s, v_s \rangle) + j(-\langle u_c, v_s \rangle + \langle u_s, v_c \rangle), \tag{2.37}$$

and note that the real part gives the desired term.

Energy of complex envelope Specializing (2.36) to the inner product of a signal with itself, we infer that the energy of the complex envelope is equal to that of the corresponding passband signal (this is a convenient consequence of the specific scaling we have chosen in our definition of the complex envelope). That is,

$$||s||^2 = ||s_p||^2. \tag{2.38}$$

To show this, set $u = v = s$ and $u_p = v_p = s_p$ in (2.36), noting that $\text{Re}(\langle s, s \rangle) = \text{Re}(||s||^2) = ||s||^2$.

Frequency domain relationship between passband and complex baseband We first summarize the results relating $S_p(f)$ and $S(f)$. Let $S_p^+(f) = S_p(f)I_{\{f>0\}}$ denote the segment of $S_p(f)$ occupying positive frequencies. Then the complex envelope is specified as

$$S(f) = \sqrt{2}S_p^+(f + f_c). \tag{2.39}$$

Conversely, given the complex envelope $S(f)$ in the frequency domain, the passband signal is specified as

$$S_p(f) = \frac{S(f - f_c) + S^*(-f - f_c)}{\sqrt{2}}. \tag{2.40}$$

2.2 Complex baseband representation

We now derive and discuss these relationships. Define

$$v(t) = \sqrt{2}s(t)e^{j2\pi f_c t} \leftrightarrow V(f) = \sqrt{2}S(f - f_c). \qquad (2.41)$$

By the time domain relationship between s_p and s, we have

$$s_p(t) = \text{Re}(v(t)) = \frac{v(t) + v^*(t)}{2} \leftrightarrow S_p(f) = \frac{V(f) + V^*(-f)}{2}$$

$$= \frac{S(f - f_c) + S^*(-f - f_c)}{\sqrt{2}}. \qquad (2.42)$$

If $S(f)$ has energy concentrated in the baseband, then the energy of $V(f)$ is concentrated around f_c, and the energy of $V^*(-f)$ is concentrated around $-f_c$. Thus, $S_p(f)$ is indeed passband. We also see from (2.42) that the symmetry condition $S_p(f) = S_p^*(-f)$ holds, which implies that $s_p(t)$ is real-valued. This is, of course, not surprising, since our starting point was the time domain expression (2.28) for a real-valued signal $s_p(t)$.

Figure 2.8 shows the relation between the passband signal $S_p(f)$, its scaled version $V(f)$ restricted to positive frequencies, and the complex baseband signal $S(f)$. As this example emphasizes, all of these spectra can, in general, be complex-valued. Equation (2.41) corresponds to starting with an arbitrary

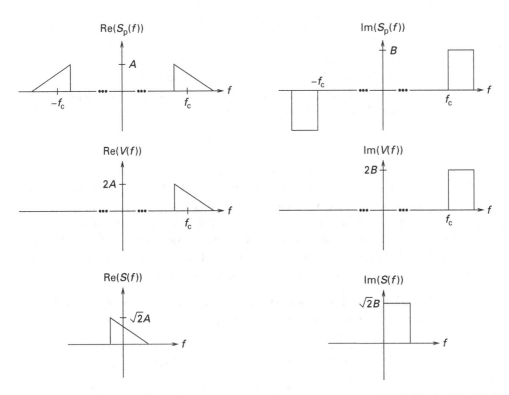

Figure 2.8 Frequency domain relationship between a real-valued passband signal and its complex envelope. The figure shows the spectrum $S_p(f)$ of the passband signal, its scaled restriction to positive frequencies $V(f)$, and the spectrum $S(f)$ of the complex envelope.

baseband signal $S(f)$ as in the bottom of the figure, and constructing $V(f)$ as depicted in the middle of the figure. We then use $V(f)$ to construct a conjugate symmetric passband signal $S_p(f)$, proceeding from the middle of the figure to the top. This example also shows that $S(f)$ does not, in general, obey conjugate symmetry, so that the baseband signal $s(t)$ is complex-valued. However, by construction, $S_p(f)$ is conjugate symmetric, and hence the passband signal $s_p(t)$ is real-valued.

General applicability of complex baseband representation We have so far seen that, given a complex baseband signal (or equivalently, a pair of real baseband signals), we can generate a real-valued passband signal using (2.26) or (2.28). But do these baseband representations apply to any real-valued passband signal? To show that they indeed do apply, we simply reverse the frequency domain operations in (2.41) and (2.42). Specifically, suppose that $s_p(t)$ is an arbitrary real-valued passband waveform. This means that the conjugate symmetry condition $S_p(f) = S_p^*(-f)$ holds, so that knowing the values of S_p for positive frequencies is enough to characterize the values for all frequencies. Let us therefore consider an appropriately scaled version of the segment of S_p for positive frequencies, defined as

$$V(f) = 2S_p^+(f) = \begin{cases} 2S_p(f), & f > 0, \\ 0, & \text{else}. \end{cases} \qquad (2.43)$$

By the definition of V, and using the conjugate symmetry of S_p, we see that (2.42) holds. Note also that, since S_p is passband, the energy of V is concentrated around $+f_c$. Now, let us define the complex envelope of S_p by inverting the relation (2.41), as follows:

$$S(f) = \frac{1}{\sqrt{2}} V(f + f_c). \qquad (2.44)$$

Since $V(f)$ is concentrated around $+f_c$, $S(f)$, which is obtained by translating it to the left by f_c, is baseband. Thus, starting from an arbitrary passband signal $S_p(f)$, we have obtained a baseband signal $S(f)$ that satisfies (2.41) and (2.42), which are equivalent to the time domain relationship (2.28). We refer again to Figure 2.8 to illustrate the relation between $S_p(f)$, $V(f)$ and $S(f)$. However, we now go from top to bottom: starting from an arbitrary conjugate symmetric $S_p(f)$, we construct $V(f)$, and then $S(f)$.

Upconversion and downconversion Equation (2.26) immediately tells us how to *upconvert* from baseband to passband. To *downconvert* from passband to baseband, consider

$$\sqrt{2} s_p(t) \cos(2\pi f_c t) = 2s_c(t) \cos^2 2\pi f_c t - 2s_s(t) \sin 2\pi f_c t \cos 2\pi f_c t$$
$$= s_c(t) + s_c(t) \cos 4\pi f_c t - s_s(t) \sin 4\pi f_c t.$$

The first term on the extreme right-hand side is the I component, a baseband signal. The second and third terms are passband signals at $2f_c$, which we can

2.2 Complex baseband representation

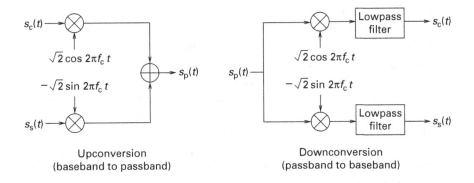

Figure 2.9 Upconversion from baseband to passband and downconversion from passband to baseband.

get rid of by lowpass filtering. Similarly, we can obtain the Q component by lowpass filtering $-\sqrt{2}s_p(t)\sin 2\pi f_c t$. The upconversion and downconversion operations are depicted in Figure 2.9.

Information resides in complex baseband The complex baseband representation corresponds to subtracting out the rapid, but predictable, phase variation due to the fixed reference frequency f_c, and then considering the much slower amplitude and phase variations induced by baseband modulation. Since the phase variation due to f_c is predictable, it cannot convey any information. Thus, all the information in a passband signal is contained in its complex envelope.

Example 2.2.3 (Linear modulation) Suppose that information is encoded into a complex number $b = b_c + jb_s = re^{j\theta}$, where b_c, b_s are real-valued numbers corresponding to its rectangular form, and $r \geq 0$, θ are real-valued and correspond to its polar form. Let $p(t)$ denote a baseband pulse (for simplicity, assume that p is real-valued). Then the *linearly modulated* complex baseband waveform $s(t) = bp(t)$ can be used to convey the information in b over a passband channel by upconverting to an arbitrary carrier frequency f_c. The corresponding passband signal is given by

$$s_p(t) = \text{Re}\left(\sqrt{2}s(t)e^{j2\pi f_c t}\right) = \sqrt{2}\left(b_c p(t)\cos 2\pi f_c t - b_s p(t)\sin 2\pi f_c t\right)$$
$$= \sqrt{2}r\cos(2\pi f_c t + \theta).$$

Thus, linear modulation in complex baseband by a complex symbol b can be viewed as separate amplitude modulation (by b_c, b_s) of the I component and the Q component, or as amplitude and phase modulation (by r, θ) of the overall passband waveform. In practice, we encode information in a stream of complex symbols $\{b[n]\}$ that linearly modulate time shifts of a basic waveform, and send the complex baseband waveform $\sum_n b[n]p(t-nT)$. Linear modulation is discussed in detail in Section 2.5.

Complex baseband equivalent of passband filtering We now state another result that is extremely relevant to transceiver operations; namely, any passband filter can be implemented in complex baseband. This result applies to filtering operations that we desire to perform at the transmitter (e.g., to conform to spectral masks), at the receiver (e.g., to filter out noise), and to a broad class of channels modeled as linear filters. Suppose that a passband signal $s_p(t)$ is passed through a passband filter with impulse response $h_p(t)$. Denote the filter output (which is clearly also passband) by $y_p(t) = (s_p * h_p)(t)$. Let y, s and h denote the complex envelopes for y_p, s_p and h_p, respectively, with respect to a common frequency reference f_c. Since real-valued passband signals are completely characterized by their behavior for positive frequencies, the passband filtering equation $Y_p(f) = S_p(f)H_p(f)$ can be separately (and redundantly) written out for positive and negative frequencies, because the waveforms are conjugate symmetric around the origin, and there is no energy around $f = 0$. Thus, focusing on the positive frequency segments $Y_+(f) = Y_p(f)I_{\{f>0\}}$, $S_+(f) = S_p(f)I_{\{f>0\}}$, $H_+(f) = H_p(f)I_{\{f>0\}}$, we have $Y_+(f) = S_+(f)H_+(f)$, from which we conclude that the complex envelope of y is given by

$$Y(f) = \sqrt{2}Y_+(f+f_c) = \sqrt{2}S_+(f+f_c)H_+(f+f_c) = \frac{1}{\sqrt{2}}S(f)H(f).$$

Figure 2.10 The relationship between passband filtering and its complex baseband analog.

Figure 2.10 depicts the relationship between the passband and complex baseband waveforms in the frequency domain, and supplies a pictorial proof of the preceding relationship. We now restate this important result in the time domain:

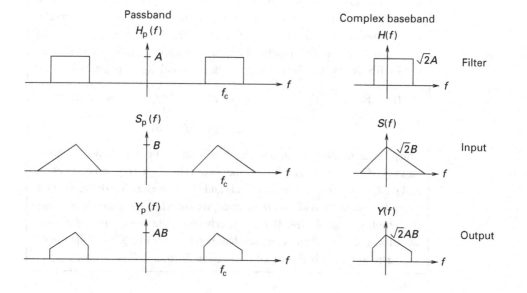

2.2 Complex baseband representation

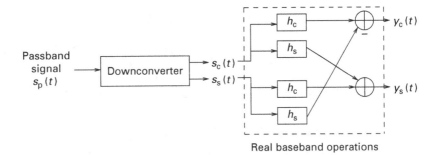

Figure 2.11 Complex baseband realization of passband filter. The constant scale factors of $1/\sqrt{2}$ have been omitted.

$$y(t) = \frac{1}{\sqrt{2}}(s * h)(t). \qquad (2.45)$$

That is, passband filtering can be implemented in complex baseband, using the complex baseband representation of the desired filter impulse response. As shown in Figure 2.11, this requires four real baseband filters: writing out the real and imaginary parts of (2.45), we obtain

$$y_c = \frac{1}{\sqrt{2}}(s_c * h_c - s_s * h_s), \quad y_s = \frac{1}{\sqrt{2}}(s_s * h_c + s_c * h_s). \qquad (2.46)$$

Remark 2.2.1 (Complex baseband in transceiver implementations) Given the equivalence of passband and complex baseband, and the fact that key operations such as linear filtering can be performed in complex baseband, it is understandable why, in typical modern passband transceivers, most of the intelligence is moved to baseband processing. For moderate bandwidths at which analog-to-digital and digital-to-analog conversion can be accomplished inexpensively, baseband operations can be efficiently performed in DSP. These digital algorithms are independent of the passband over which communication eventually occurs, and are amenable to a variety of low-cost implementations, including very large scale integrated circuits (VLSI), field programmable gate arrays (FPGA), and general purpose DSP engines. On the other hand, analog components such as local oscillators, power amplifiers, and low noise amplifiers must be optimized for the bands of interest, and are often bulky. Thus, the trend in modern transceivers is to accomplish as much as possible using baseband DSP algorithms. For example, complicated filters shaping the transmitted waveform to a spectral mask dictated by the FCC can be achieved with baseband DSP algorithms, allowing the use of relatively sloppy analog filters at passband. Another example is the elimination of analog phase locked loops for carrier synchronization in many modern receivers; the receiver instead employs a fixed analog local oscillator for downconversion, followed by a digital phase locked loop implemented in complex baseband.

28 Modulation

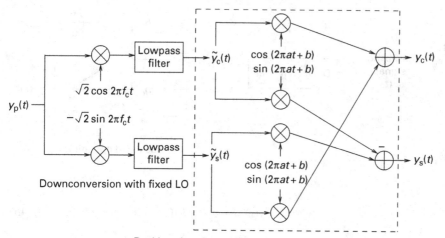

Figure 2.12 Undoing frequency and phase offsets in complex baseband after downconverting using a local oscillator at a fixed carrier frequency f_c. The complex baseband operations are expanded out into real arithmetic as shown.

> **Example 2.2.4 (Handling carrier frequency and phase offsets in complex baseband)** As shown in Figure 2.12, a communication receiver uses a local oscillator with a fixed carrier frequency f_c to demodulate an incoming passband signal
>
> $$y_p(t) = \sqrt{2}[y_c(t)\cos(2\pi(f_c+a)t+b) - y_s(t)\sin(2\pi(f_c+a)t+b)],$$
>
> where a, b, are carrier frequency and phase offsets, respectively. Denote the I and Q components at the output of the downconverter as $\tilde{y}_c$, $\tilde{y}_s$, respectively, and the corresponding complex envelope as $\tilde{y} = \tilde{y}_c + j\tilde{y}_s$. We wish to recover y_c, y_s, the I and Q components relative to a reference that accounts for the offsets a and b. Typically, the receiver would estimate a and b using the downconverter output $\tilde{y}$; an example of an algorithm for such frequency and phase synchronization is discussed in the next chapter. Assuming that such estimates are available, we wish to specify baseband operations using real-valued arithmetic for obtaining y_c, y_s from the downconverter output. Equivalently, we wish to recover the complex envelope $y = y_c + jy_s$ from $\tilde{y}$. We can relate y and $\tilde{y}$ via (2.31) as in Example 2.2.2, and obtain
>
> $$\tilde{y}(t) = y(t)e^{j(2\pi at+b)}.$$
>
> This relation can now be inverted to get y from $\tilde{y}$:
>
> $$y(t) = \tilde{y}(t)e^{-j(2\pi at+b)}.$$

2.2 Complex baseband representation

Plugging in I and Q components explicitly and using Euler's formula, we obtain

$$y_c(t) + jy_s(t) = (\tilde{y}_c(t) + j\tilde{y}_s(t))(\cos(2\pi at + b) - j\sin(2\pi at + b)).$$

Equating real and imaginary parts, we obtain equations involving real-valued quantities alone:

$$\begin{aligned} y_c &= \tilde{y}_c \cos(2\pi at + b) + \tilde{y}_s \sin(2\pi at + b) \\ y_s &= -\tilde{y}_c \sin(2\pi at + b) + \tilde{y}_s \cos(2\pi at + b). \end{aligned} \quad (2.47)$$

These computations are depicted in Figure 2.12.

Example 2.2.5 (Coherent and noncoherent reception) We see in the next two chapters that a fundamental receiver operation is to compare a noisy received signal against noiseless copies of the received signals corresponding to the different possible transmitted signals. This comparison is implemented by a correlation, or inner product. Let $y_p(t) = \sqrt{2}\mathrm{Re}(y(t)e^{j2\pi f_c t})$ denote the noisy received passband signal, and $s_p(t) = \sqrt{2}\mathrm{Re}(s(t)e^{j2\pi f_c t})$ denote a noiseless copy that we wish to compare it with, where $y = y_c + jy_s$ and $s = s_c + js_s$ are the complex envelopes of y_p and s_p, respectively. A coherent receiver (which is a building block for the optimal receivers in Chapter 3) for s implements the inner product $\langle y_p, s_p \rangle$. In terms of complex envelopes, we know from (2.36) that this can be written as

$$\langle y_p, s_p \rangle = \mathrm{Re}(\langle y, s \rangle) = \langle y_c, s_c \rangle + \langle y_s, s_s \rangle. \quad (2.48)$$

Clearly, when $y = As$ (plus noise), where $A > 0$ is an arbitrary amplitude scaling, the coherent receiver gives a large output. However, coherent reception assumes carrier phase synchronization (in order to separate out and compute inner products with the I and Q components of the received passband signal). If, on the other hand, the receiver is not synchronized in phase, then (see Example 2.2.2) the complex envelope of the received signal is given by $y = Ae^{j\theta}s$ (plus noise), where $A > 0$ is the amplitude scale factor, and θ is an unknown carrier phase. Now, the coherent receiver gives the output

$$\begin{aligned} \langle y_p, s_p \rangle &= \mathrm{Re}(\langle Ae^{j\theta}s, s \rangle) \text{ (plus noise)} \\ &= A\cos\theta \|s\|^2 \text{ (plus noise)}. \end{aligned}$$

In this case, the output can be large or small, depending on the value of θ. Indeed, for $\theta = \pi/2$, the signal contribution to the inner product becomes zero. The noncoherent receiver deals with this problem by using the magnitude, rather than the real part, of the complex inner product $\langle y, s \rangle$. The signal contribution to this noncoherent correlation is given by

$$|\langle y, s \rangle| = |\langle Ae^{j\theta}s, s \rangle| \text{ (plus noise)} | \approx A\|s\|^2 \text{ (ignoring noise)},$$

where we have omitted the terms arising from the nonlinear interaction between noise and signal in the noncoherent correlator output. The noiseless output from the preceding computation shows that we do get a large signal contribution regardless of the value of the carrier phase θ. It is convenient to square the magnitude of the complex inner product for computation. Substituting the expression (2.37) for the complex inner product, we obtain that the squared magnitude inner product computed by a noncoherent receiver requires the following real baseband computations:

$$|\langle y, s \rangle|^2 = (\langle u_c, v_c \rangle + \langle u_s, v_s \rangle)^2 + (-\langle u_c, v_s \rangle + \langle u_s, v_c \rangle)^2. \qquad (2.49)$$

We see in Chapter 4 that the preceding computations are a building block for optimal noncoherent demodulation under carrier phase uncertainty. The implementations of the coherent and noncoherent receivers in complex baseband are shown in Figure 2.13.

Remark 2.2.2 (Bandwidth) Given the scarcity of spectrum and the potential for interference between signals operating in neighboring bands, determining the spectral occupancy of signals accurately is an important part of communication system design. As mentioned earlier, the spectral occupancy of a physical (and hence real-valued) signal is the smallest band of positive frequencies that contains most of the signal content. Negative frequencies are not included in the definition, since they contain no information beyond that already contained in the positive frequencies ($S(-f) = S^*(f)$ for real-valued $s(t)$). For complex baseband signals, however, information resides in both positive and negative frequencies, since the complex baseband representation is a translated version of the corresponding passband signal restricted to

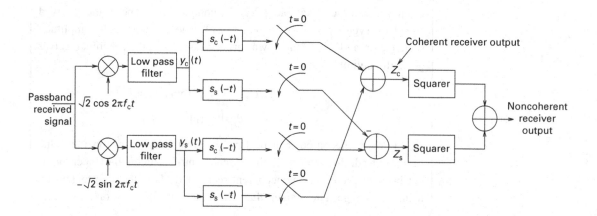

Figure 2.13 Complex baseband implementations of coherent and noncoherent receivers. The real-valued correlations are performed using matched filters sampled at time zero.

positive frequencies. We therefore define the spectral occupancy of a complex baseband signal as the smallest band around the origin, including both positive and negative frequencies, that contains most of the signal content. These definitions of bandwidth are consistent: from Figure 2.8, it is evident that the bandwidth of a passband signal (defined based on positive frequencies alone) is equal to the bandwidth of its complex envelope (defined based on both positive and negative frequencies). Thus, it suffices to work in complex baseband when determining the spectral occupancy and bandwidth of passband signals.

2.3 Spectral description of random processes

So far, we have considered deterministic signals with finite energy. From the point of view of communication system design, however, it is useful to be able to handle random signals, and to allow the signal energy to be infinite. For example, consider the binary signaling example depicted in Figure 2.1. We would like to handle bitstreams of arbitrary length within our design framework, and would like our design to be robust to which particular bitstream was sent. We therefore model the bitstream as random (and demand good system performance averaged over these random realizations), which means that the modulated signal is modeled as a random process. Since the bitstream can be arbitrarily long, the energy of the modulated signal is unbounded. On the other hand, when averaged over a long interval, the power of the modulated signal in Figure 2.1 is finite, and tends to a constant, regardless of the transmitted bitstream. It is evident from this example, therefore, that we must extend our discussion of baseband and passband signals to random processes. Random processes serve as a useful model not only for modulated signals, but also for noise, interference, and for the input–output response of certain classes of communication channels (e.g., wireless mobile channels).

For a finite-power signal (with unbounded energy), a time-windowed realization is a deterministic signal with finite energy, so that we can employ our existing machinery for finite-energy signals. Our basic strategy is to define properties of a finite-power signal in terms of quantities that can be obtained as averages over a time window, in the limit as the time window gets large. These time averaged properties can be defined for any finite-power signal. However, we are interested mainly in scenarios where the signal is a realization of a random process, and we wish to ensure that properties we infer as a time average over one realization apply to most other realizations as well. In this case, a time average is meaningful as a broad descriptor of the random process only under an *ergodicity* assumption that the time average along a realization equals a corresponding statistical average across realizations. Moreover, while the time average provides a definition that has operational significance (in terms of being implementable by measurement or computer simulation), when a suitable notion of ergodicity holds, it is often analytically tractable to compute the statistical average. In the forthcoming discussions, we discuss both time averages and statistical averages for several random processes of interest.

Power spectral density As with our definition of energy spectral density, let us define the power spectral density (PSD) for a finite-power signal $s(t)$ in operational terms. Pass the signal $s(t)$ through an ideal narrowband filter with transfer function

$$H_{f_0}(f) = \begin{cases} 1, & f_0 - \frac{\Delta f}{2} < f < f_0 + \frac{\Delta f}{2}, \\ 0, & \text{else.} \end{cases}$$

The PSD evaluated at f_0, $S_s(f_0)$, can now be defined to be the measured power at the filter output, divided by the filter width Δf (in the limit as $\Delta f \to 0$).

The preceding definition directly leads to a procedure for computing the PSD based on empirical measurements or computer simulations. Given a physical signal or a computer model, we can compute the PSD by time-windowing the signal and computing the Fourier transform, as follows. Define the time-windowed version of s as

$$s_{T_o}(t) = s(t) I_{[-\frac{T_o}{2}, \frac{T_o}{2}]}(t), \tag{2.50}$$

where T_o is the length of the observation interval. (The observation interval need not be symmetric about the origin, in general.) Since T_o is finite, $s_{T_o}(t)$ has finite energy if $s(t)$ has finite power, and we can compute its Fourier transform

$$S_{T_o}(f) = \mathcal{F}(s_{T_o}).$$

The energy spectral density of s_{T_o} is given by $|S_{T_o}(f)|^2$, so that an estimate of the PSD of s is obtained by averaging this over the observation interval. We thus obtain an estimated PSD

$$\hat{S}_s(f) = \frac{|S_{T_o}(f)|^2}{T_o}. \tag{2.51}$$

The computations required to implement (2.51) are often referred to as a periodogram. In practice, the signal s is sampled, and the Fourier transform is computed using a DFT. The length of the observation interval determines the frequency resolution, while the sampling rate is chosen to be large enough to avoid significant aliasing. The multiplication by a rectangular time window in (2.50) corresponds to convolution in the frequency domain with the sinc function, which can lead to significant spectral distortion. It is common, therefore, to employ time windows that taper off at the edges of the observation interval, so as to induce a quicker decay of the frequency domain signal being convolved with. Finally, multiple periodograms can be averaged in order to get a less noisy estimate of the PSD.

2.3 Spectral description of random processes

Autocorrelation function As with finite-energy signals, the inverse Fourier transform of (2.51) has the interpretation of autocorrelation. Specifically, using (2.20), we have that the inverse Fourier transform of (2.51) is given by

$$\hat{S}_s(f) = \frac{|S_{T_o}(f)|^2}{T_o} \leftrightarrow \hat{R}_s(\tau) = \frac{1}{T_o} \int_{-\infty}^{\infty} s_{T_o}(u) s_{T_o}^*(u-\tau) \, du$$

$$= \frac{1}{T_o} \int_{-\frac{T_o}{2}+\max(0,\tau)}^{\frac{T_o}{2}+\min(0,\tau)} s_{T_o}(u) s_{T_o}^*(u-\tau) \, du$$

$$\approx \frac{1}{T_o} \int_{-\frac{T_o}{2}}^{\frac{T_o}{2}} s_{T_o}(u) s_{T_o}^*(u-\tau) \, du, \tag{2.52}$$

where the last approximation neglects edge effects as T_o gets large (for fixed τ). An alternative method for computing PSD, therefore, is first to compute the empirical autocorrelation function (again, this is typically done in discrete time), and then to compute the DFT. While these methods are equivalent in theory, in practice, the properties of the estimates depend on a number of computational choices, discussion of which is beyond the scope of this book. The interested reader may wish to explore the various methods for estimating PSD available in MATLAB or similar programs.

Formal definitions of PSD and autocorrelation function In addition to providing a procedure for computing the PSD, we can also use (2.51) to provide a formal definition of PSD by letting the observation interval get large:

$$\bar{S}_s(f) = \lim_{T_o \to \infty} \frac{|S_{T_o}(f)|^2}{T_o}. \tag{2.53}$$

Similarly, we can take limits in (2.52) to obtain a formal definition of the autocorrelation function as follows:

$$\bar{R}_s(\tau) = \lim_{T_o \to \infty} \frac{1}{T_o} \int_{-\frac{T_o}{2}}^{\frac{T_o}{2}} s_{T_o}(u) s_{T_o}^*(u-\tau) \, du, \tag{2.54}$$

where the overbar notation denotes time averages along a realization. As we see shortly, we can also define the PSD and autocorrelation function as statistical averages across realizations; we drop the overbar notation when we consider these. More generally, we adopt the shorthand $\bar{f}(t)$ to denote the time average of $f(t)$. That is,

$$\bar{f}(t) = \lim_{T_o \to \infty} \frac{1}{T_o} \int_{-\frac{T_o}{2}}^{\frac{T_o}{2}} f(u) \, du.$$

Thus, the definition (2.54) can be rewritten as

$$\bar{R}_s(\tau) = \overline{s(u)s^*(u-\tau)}.$$

Baseband and passband random processes A random process is baseband if its PSD is baseband, and it is passband if its PSD is passband. Since the PSD is defined as a time average along a realization, this also means (by assumption) that the realizations of baseband and passband random processes are modeled as deterministic baseband and passband signals, respectively. In the next section, this assumption enables us to use the development of the complex baseband representation for deterministic passband signals in our discussion of the complex baseband representation of passband random processes.

Crosscorrelation function For finite-power signals s_1 and s_2, we define the crosscorrelation function as the following time average:

$$\bar{R}_{s_1,s_2}(\tau) = \overline{s_1(u)s_2^*(u-\tau)}. \tag{2.55}$$

The cross-spectral density is defined as the Fourier transform of the crosscorrelation function:

$$\bar{S}_{s_1,s_2}(f) = \mathcal{F}(\bar{R}_{s_1,s_2}(f)). \tag{2.56}$$

Example 2.3.1 (Autocorrelation function and power spectral density for a complex waveform) Let $s(t) = s_c(t) + js_s(t)$ be a complex-valued, finite-power, signal, where s_c and s_s are real-valued. Then the autocorrelation function of s can be computed as

$$\bar{R}_s(\tau) = \overline{s(t)s^*(t-\tau)} = \overline{(s_c(t) + js_s(t))(s_c(t-\tau) - js_s(t-\tau))}.$$

Simplifying, we obtain

$$\bar{R}_s(\tau) = [\bar{R}_{s_c}(\tau) + \bar{R}_{s_s}(\tau)] + j[\bar{R}_{s_s,s_c}(\tau) - \bar{R}_{s_c,s_s}(\tau)]. \tag{2.57}$$

Taking Fourier transforms, we obtain the PSD

$$\bar{S}_s(f) = [\bar{S}_{s_c}(f) + \bar{S}_{s_s}(f)] + j[\bar{S}_{s_s,s_c}(f) - \bar{S}_{s_c,s_s}(f)]. \tag{2.58}$$

We use this result in our discussion of the complex baseband representation of passband random processes in the next section.

Example 2.3.2 (Power spectral density of a linearly modulated signal) The modulated waveform shown in Figure 2.1 can be written in the form

$$s(t) = \sum_{n=-\infty}^{\infty} b[n]p(t-nT),$$

where the bits $b[n]$ take the values ± 1, and $p(t)$ is a rectangular pulse. Let us try to compute the PSD for this signal using the definition (2.53).

2.3 Spectral description of random processes

Anticipating the discussion of linear modulation in Section 2.5, let us consider a generalized version of Figure 2.1 in the derivation, allowing the symbols $b[n]$ to be complex-valued and $p(t)$ to be an arbitrary pulse in the derivation. Consider the signal $\hat{s}(t)$ restricted to the observation interval $[0, NT]$, given by

$$s_{T_o}(t) = \sum_{n=0}^{N-1} b[n] p(t - nT).$$

The Fourier transform of the time-windowed waveform is given by

$$S_{T_o}(f) = \sum_{n=0}^{N-1} b[n] P(f) e^{-j2\pi f nT} = P(f) \sum_{n=0}^{N-1} b[n] e^{-j2\pi f nT}.$$

The estimate of the power spectral density is therefore given by

$$\frac{|S_{T_o}(f)|^2}{T_o} = \frac{|P(f)|^2 |\sum_{n=0}^{N-1} b[n] e^{-j2\pi f nT}|^2}{NT}. \quad (2.59)$$

Let us now simplify the preceding expression and take the limit as $T_o \to \infty$ (i.e., $N \to \infty$). Define the term

$$A = \left| \sum_{n=0}^{N-1} b[n] e^{-j2\pi f nT} \right|^2 = \sum_{n=0}^{N-1} b[n] e^{-j2\pi f nT} \left(\sum_{m=0}^{N-1} b[m] e^{-j2\pi f mT} \right)^*$$

$$= \sum_{n=0}^{N-1} \sum_{m=0}^{N-1} b[n] b^*[m] e^{-j2\pi f (n-m)T}.$$

Setting $k = n - m$, we can rewrite the preceding as

$$A = \sum_{n=0}^{N-1} |b[n]|^2 + \sum_{k=1}^{N-1} e^{-j2\pi f kT} \sum_{n=k}^{N-1} b[n] b^*[n-k]$$

$$+ \sum_{k=-(N-1)}^{-1} e^{-j2\pi f kT} \sum_{n=0}^{N-1+k} b[n] b^*[n-k].$$

Now, suppose that the symbols are uncorrelated, in that the time average of $b[n] b^*[n-k]$ is zero for $k \neq 0$. Also, denote the empirical average of $|b[n]|^2$ by σ_b^2. Then the limit becomes

$$\lim_{N \to \infty} \frac{A}{N} = \sigma_b^2.$$

Substituting into (2.59), we can now infer that

$$\bar{S}_s(f) = \lim_{T_o \to \infty} \frac{|S_{T_o}(f)|^2}{T_o} = \lim_{N \to \infty} \frac{|P(f)|^2 A}{NT} = \sigma_b^2 \frac{|P(f)|^2}{T}.$$

Thus, we have shown that the PSD of a linearly modulated signal scales as the magnitude squared of the spectrum of the modulating pulse.

The time averages discussed thus far interpret a random process $s(t)$ as a collection of deterministic signals, or *realizations*, evolving over time t, where the specific realization is chosen randomly. Next, we discuss methods for computing the corresponding statistical averages, which rely on an alternate view of $s(t)$, for fixed t, as a random variable which takes a range of values across realizations of the random process. If $\mathcal{T}$ is the set of allowable values for the index t, which is interpreted as time for our purpose here (e.g., $\mathcal{T} = (-\infty, \infty)$ when the time index can take any real value), then $\{s(t), t \in \mathcal{T}\}$ denotes a collection of random variables over a common probability space. The term *common probability space* means that we can talk about the joint distribution of these random variables.

In particular, the statistical averages of interest to us are the autocorrelation function and PSD for wide sense stationary and wide sense cyclostationary random processes (defined later). Since most of the signal and noise models that we encounter fall into one of these two categories, these techniques form an important part of the communication system designer's toolkit. The practical utility of a statistical average in predicting the behavior of a particular realization of a random process depends, of course, on the ergodicity assumption (discussed in more detail later) that time averages equal statistical averages for the random processes of interest.

Mean, autocorrelation, and autocovariance functions For a random process $s(t)$, the mean function is defined as

$$m_s(t) = E[s(t)] \tag{2.60}$$

and the autocorrelation function as

$$R_s(t_1, t_2) = E[s(t_1) s^*(t_2)]. \tag{2.61}$$

The autocovariance function of s is the autocorrelation function of the zero mean version of s, and is given by

$$C_s(t_1, t_2) = E[(s(t_1) - E[s(t_1)])(s(t_2) - E[s(t_2)])^*] = R_s(t_1, t_2) - m_s(t_1) m_s^*(t_2). \tag{2.62}$$

Crosscorrelation and crosscovariance functions For random processes s_1 and s_2 defined on a common probability space (i.e., we can talk about the joint distribution of samples from these random processes), the crosscorrelation function is defined as

$$R_{s_1,s_2}(t_1, t_2) = E[s_1(t_1) s_2^*(t_2)] \tag{2.63}$$

and the crosscovariance function is defined as

$$C_{s_1,s_2}(t_1, t_2) = E[(s_1(t_1) - E[s_1(t_1)])(s_2(t_2) - E[s_2(t_2)])^*]$$
$$= R_{s_1,s_2}(t_1, t_2) - m_{s_1}(t_1) m_{s_2}^*(t_2). \tag{2.64}$$

2.3 Spectral description of random processes

Stationary random process A random process $s(t)$ is said to be stationary if it is statistically indistinguishable from a delayed version of itself. That is, $s(t)$ and $s(t-d)$ have the same statistics for any delay $d \in (-\infty, \infty)$.

For a stationary random process s, the mean function satisfies

$$m_s(t) = m_s(t-d)$$

for any t, regardless of the value of d. Choosing $d = t$, we infer that

$$m_s(t) = m_s(0).$$

That is, the mean function is a constant. Similarly, the autocorrelation function satisfies

$$R_s(t_1, t_2) = R_s(t_1 - d, t_2 - d)$$

for any t_1, t_2, regardless of the value of d. Setting $\tau = t_2$, we have

$$R_s(t_1, t_2) = R_s(t_1 - t_2, 0).$$

That is, the autocorrelation function depends only on the difference of its arguments.

Stationarity is a stringent requirement that is not always easy to verify. However, the preceding properties of the mean and autocorrelation functions can be used as the defining characteristics for a weaker property termed wide sense stationarity.

Wide sense stationary (WSS) random process A random process s is said to be WSS if

$$m_s(t) \equiv m_s(0) \quad \text{for all } t$$

and

$$R_s(t_1, t_2) = R_s(t_1 - t_2, 0) \quad \text{for all } t_1, t_2.$$

In this case, we change notation and express the autocorrelation function as a function of $\tau = t_1 - t_2$ alone. Thus, for a WSS process, we can define the autocorrelation function as

$$R_s(\tau) = E[s(t)s^*(t-\tau)] \quad \text{for } s \text{ WSS} \tag{2.65}$$

with the understanding that the expectation is independent of t.

Power spectral density for a WSS process We define the PSD of a WSS process s as the Fourier transform of its autocorrelation function, as follows:

$$S_s(f) = \mathcal{F}(R_s(\tau)). \tag{2.66}$$

We sometimes also need the notion of joint wide sense stationarity of two random processes.

Jointly wide sense stationary random processes The random processes X and Y are said to be jointly WSS if (a) X is WSS, (b) Y is WSS, (c) the crosscorrelation function $R_{X,Y}(t_1, t_2) = E[X(t_1)Y^*(t_2)]$ depends on the time difference $\tau = t_1 - t_2$ alone. In this case, we can redefine the crosscorrelation function as $R_{X,Y}(\tau) = E[X(t)Y^*(t-\tau)]$.

Ergodicity A stationary random process s is ergodic if time averages along a realization equal statistical averages across realizations. For WSS processes, we are primarily interested in ergodicity for the mean and autocorrelation functions. For example, for a WSS process s that is ergodic in its autocorrelation function, the definitions (2.54) and (2.65) of autocorrelation functions give the same result, which gives us the choice of computing the autocorrelation function (and hence the PSD) as either a time average or a statistical average. Intuitively, ergodicity requires having "enough randomness" in a given realization so that a time average along a realization is rich enough to capture the statistics across realizations. Specific technical conditions for ergodicity are beyond our present scope, but it is worth mentioning the following intuition in the context of the simple binary modulated waveform depicted in Figure 2.1. If all bits take the same value over a realization, then the waveform is simply a constant taking value $+1$ or -1: clearly, a time average across such a degenerate realization does not yield "typical" results. Thus, we need the bits in a realization to exhibit enough variation to obtain ergodicity. In practice, we often use line codes or scramblers specifically designed to avoid long runs of zeros or ones, in order to induce enough transitions for proper operation of synchronization circuits. It is fair to say, therefore, that there is typically enough randomness in the kinds of waveforms we encounter (e.g., modulated waveforms, noise and interference) that ergodicity assumptions hold.

Example 2.3.3 Armed with these definitions, let us revisit the binary modulated waveform depicted in Figure 2.1, or more generally, a linearly modulated waveform of the form

$$s(t) = \sum_{n=-\infty}^{\infty} b[n]p(t - nT). \quad (2.67)$$

When we delay this waveform by d, we obtain

$$s(t - d) = \sum_{n=-\infty}^{\infty} b[n]p(t - nT - d).$$

Let us consider the special case $d = kT$, where k is an integer. We obtain

$$s(t - kT) = \sum_{n=-\infty}^{\infty} b[n]p(t - (n+k)T) = \sum_{n=-\infty}^{\infty} b[n-k]p(t - nT), \quad (2.68)$$

where we have replaced $n + k$ by n in the last summation. Comparing (2.67) and (2.68), we note that the only difference is that the symbol sequence

2.3 Spectral description of random processes

> $\{b[n]\}$ is replaced by a delayed version $\{b[n-k]\}$. If the symbol sequence is stationary, then it has the same statistics as its delayed version, which implies that $s(t)$ and $s(t-kT)$ are statistically indistinguishable. However, this is a property that only holds for delays that are integer multiples of the symbol time. For example, for the binary signaling waveform in Figure 2.1, it is immediately evident by inspection that $s(t)$ can be distinguished easily from $s(t-T/2)$ (e.g., from the location of the symbol edges). Slightly more sophisticated arguments can be used to show similar results for pulses that are more complicated than the rectangular pulse. We conclude, therefore, that a linearly modulated waveform of the form (2.67), with a stationary symbol sequence $\{b[n]\}$, is a *cyclostationary* random process, where the latter is defined formally below.

Cyclostationary random process The random process $s(t)$ is *cyclostationary* with respect to time interval T if it is statistically indistinguishable from $s(t-kT)$ for any integer k.

As with the concept of stationarity, we can relax the notion of cyclostationarity by considering only the first and second order statistics.

Wide sense cyclostationary random process The random process $s(t)$ is *wide sense cyclostationary* with respect to time interval T if the mean and autocorrelation functions satisfy the following:

$$m_s(t) = m_s(t-T) \quad \text{for all } t,$$

$$R_s(t_1, t_2) = R_s(t_1 - T, t_2 - T) \quad \text{for all } t_1, t_2.$$

We now state the following theorem regarding cyclostationary processes; this is proved in Problem 2.14.

Theorem 2.3.1 (Stationarizing a cyclostationary process) *Let $s(t)$ be a cyclostationary random process with respect to the time interval T. Suppose that D is a random variable that is uniformly distributed over $[0, T]$, and independent of $s(t)$. Then $s(t-D)$ is a stationary random process. Similarly, if $s(t)$ is wide sense cyclostationary, then $s(t-D)$ is a WSS random process.*

The random process $s(t-D)$ is a "stationarized" version of $s(t)$, with the random delay D transforming the periodicity in the statistics of $s(t)$ into time invariance in the statistics of $s(t-D)$. We can now define the PSD of s to be that of its stationarized version, as follows.

Computation of PSD for a (wide sense) cyclostationary process as a statistical average For $s(t)$ (wide sense) cyclostationary with respect to time interval T, we define the PSD as

$$S_s(f) = \mathcal{F}(R_s(\tau)),$$

where R_s is the "stationarized" autocorrelation function,
$$R_s(\tau) = E[s(t-D)s^*(t-D-\tau)],$$
with the random variable D chosen as in Theorem 2.3.1.

In Problem 2.14, we discuss why this definition of PSD for cyclostationary processes is appropriate when we wish to relate statistical averages to time averages. That is, when a cyclostationary process satisfies intuitive notions of ergodicity, then its time averaged PSD equals the statistically averaged PSD of the corresponding stationarized process. We then rederive the PSD for a linearly modulated signal, obtained as a time average in Example 2.3.2 and as a statistical average in Problem 2.22.

2.3.1 Complex envelope for passband random processes

For a passband random process $s_p(t)$ with PSD $S_{s_p}(f)$, we know that the time-windowed realizations are also approximately passband. We can therefore define the complex envelope for these time-windowed realizations, and then remove the windowing in the limit to obtain a complex baseband random process $s(t)$. Since we have *defined* this relationship on the basis of the deterministic time-windowed realizations, the random processes s_p and s obey the same upconversion and downconversion relationships (Figure 2.9) as deterministic signals. It remains to specify the relation between the PSDs of s_p and s, which we again infer from the relationships between the time-windowed realizations. For notational simplicity, denote by $\hat{s}_p(t)$ a realization of $s_p(t)$ windowed by an observation interval of length T_o; that is, $\hat{s}_p(t) = s_p(t)I_{[-\frac{T_o}{2},\frac{T_o}{2}]}$. Let $\hat{S}_p(f)$ denote the Fourier transform of $\hat{s}_p$, $\hat{s}(t)$ the complex envelope of $\hat{s}_p$, and $\hat{S}(f)$ the Fourier transform of $\hat{s}(t)$. We know that the PSD of s_p and s can be approximated as follows:

$$S_{s_p}(f) \approx \frac{|\hat{S}_p(f)|^2}{T_o}, \quad S_s(f) \approx \frac{|\hat{S}(f)|^2}{T_o}. \tag{2.69}$$

Furthermore, we know from the relationship (2.42) between deterministic passband signals and their complex envelopes that the following spectral relationships hold:

$$\hat{S}_p(f) = \frac{1}{\sqrt{2}}\left(\hat{S}(f-f_c) + \hat{S}^*(-f-f_c)\right).$$

Since the $\hat{S}(f)$ is (approximately) baseband, the right translate $\hat{S}(f-f_c)$ and the left translate $\hat{S}^*(-f-f_c)$ do not overlap, so that

$$|\hat{S}_p(f)|^2 = \frac{1}{2}\left(|\hat{S}(f-f_c)|^2 + |\hat{S}^*(-f-f_c)|^2\right).$$

Combining with (2.69), and letting the observation interval T_o get large, we obtain

$$S_{s_p}(f) = \frac{1}{2}\left(S_s(f-f_c) + S_s(-f-f_c)\right). \tag{2.70}$$

In a similar fashion, starting from the passband PSD and working backward, we can infer that
$$S_s(f) = 2S_{s_p}^+(f+f_c),$$
where $S_{s_p}^+(f) = S_{s_p}(f)I_{\{f>0\}}$ is the "right half" of the passband PSD.

As with deterministic signals, the definition of bandwidth for real-valued passband random processes is based on occupancy of positive frequencies alone, while that for complex baseband random processes is based on occupancy of both positive and negative frequencies. For a given passband random process, both definitions lead to the same value of bandwidth.

2.4 Modulation degrees of freedom

While analog waveforms and channels live in a continuous time space with uncountably infinite dimensions, digital communication systems employing such waveforms and channels can be understood in terms of vector spaces with finite, or countably infinite, dimensions. This is because the dimension, or degrees of freedom, available for modulation is limited when we restrict the time and bandwidth of the signaling waveforms to be used. Let us consider signaling over an ideally bandlimited passband channel spanning $f_c - W/2 \leq f \leq f_c + W/2$. By choosing f_c as a reference, this is equivalent to an ideally bandlimited complex baseband channel spanning $[-W/2, W/2]$. That is, modulator design corresponds to design of a set of complex baseband transmitted waveforms that are bandlimited to $[-W/2, W/2]$. We can now invoke Nyquist's sampling theorem, stated below.

Theorem 2.4.1 (Nyquist's sampling theorem) *Any signal $s(t)$ bandlimited to $[-W/2, W/2]$ can be described completely by its samples $\{s(n/W)\}$ at rate W. Furthermore, $s(t)$ can be recovered from its samples using the following interpolation formula:*
$$s(t) = \sum_{n=-\infty}^{\infty} s\left(\frac{n}{W}\right) p\left(t - \frac{n}{W}\right), \quad (2.71)$$
where $p(t) = \text{sinc}(Wt)$.

By the sampling theorem, the modulator need only specify the samples $\{s(n/W)\}$ to specify a signal $s(t)$ bandlimited to $[-W/2, W/2]$. If the signals are allowed to span a large time interval T_o (large enough that they are still approximately bandlimited), the number of complex-valued samples that the modulator must specify is approximately WT_o. That is, the set of possible transmitted signals lies in a finite-dimensional complex subspace of dimension WT_o, or equivalently, in a real subspace of dimension $2WT_o$. To summarize, the dimension of the complex-valued signal space (i.e., the number of degrees of freedom available to the modulator) equals the time–bandwidth product.

The interpolation formula (2.71) can be interpreted as linear modulation (which has been introduced informally via several examples, and is considered in detail in the next section) at rate W using the samples $\{s(n/W)\}$ as the symbols, and the sinc pulse as the modulating pulse $g_{TX}(t)$. Linear modulation with the sinc pulse has the desirable characteristic, therefore, of being able to utilize all of the degrees of freedom available in a bandlimited channel. As we show in the next section, however, the sinc pulse has its problems, and in practice, it is necessary to back off from utilizing all available degrees of freedom, using modulating pulses that have less abrupt transitions in the frequency domain than the brickwall Fourier transform of the sinc pulse.

Bandwidth efficiency Bandwidth efficiency for a modulation scheme is defined to be the number of bits conveyed per degree of freedom. Thus, M-ary signaling in a D-dimensional signal space has bandwidth efficiency

$$\eta_B = \frac{\log_2 M}{D}. \tag{2.72}$$

The number of degrees of freedom in the preceding definition is taken to be the maximum available. Thus, for a bandlimited channel with bandwidth W, we would set $D = WT_o$ to obtain the number of complex degrees of freedom available to a modulator over a large time interval T_o. In practice, the number of effective degrees of freedom is smaller owing to a variety of implementation considerations, as mentioned for the example of linear modulation in the previous paragraph. We do not include such considerations in our definition of bandwidth efficiency, in order to get a number that fundamentally characterizes a modulation scheme, independent of implementation variations.

To summarize, the set of possible transmitted waveforms in a timelimited and bandwidth-limited system lies in a finite-dimensional *signal space*. The broad implication of this observation is that we can restrict attention to discrete-time signals, or vectors, for most aspects of digital communication system design, even though the physical communication mechanism is based on sending continuous-time waveforms over continuous-time channels. In particular, we shall see in Chapter 3 that signal space concepts play an important role in developing a geometric understanding of receiver design. Signal space concepts are also useful for describing modulation techniques, as we briefly describe below (postponing a more detailed development to Chapter 3).

Signal space description of modulation formats Consider a modulation format in which one of M signals, $s_1(t), \ldots, s_M(t)$, is transmitted. The *signal space* spanned by these signals is of dimension $n \leq M$, so we can represent each signal $s_i(t)$ by an n-dimensional vector $\mathbf{s}_i = (\mathbf{s}_i[1], \ldots, \mathbf{s}_i[n])^T$, with respect to some orthonormal basis $\psi_1(t), \ldots, \psi_n(t)$ satisfying $\langle \psi_k, \psi_l \rangle = \delta_{kl}$, $1 \leq k, l \leq n$. That is, we have

$$s_i(t) = \sum_{l=1}^{n} \mathbf{s}_i[l]\psi_l(t), \qquad \mathbf{s}_i[l] = \langle s_i, \psi_l \rangle = \int s_i(t)\psi_l^*(t)dt. \qquad (2.73)$$

By virtue of (2.73), we can describe a modulation format by specifying either the signals $s_i(t)$, $1 \leq i \leq M$, or the vectors $\mathbf{s}_i$, $1 \leq i \leq M$. More importantly, the *geometry* of the signal set is preserved when we go from continuous-time to vectors, in the sense that inner products, and Euclidean distances, are preserved: $\langle s_i, s_j \rangle = \langle \mathbf{s}_i, \mathbf{s}_j \rangle$ for $1 \leq i, j \leq M$. As we shall see in Chapter 3, it is this geometry that determines performance over the AWGN channel, which is the basic model upon which we build when designing most communication systems. Thus, we can design vectors with a given geometry, depending on the performance characteristics we desire, and then map them into continuous-time signals using a suitable orthonormal basis $\{\psi_k\}$. This implies that the same vector space design can be reused over very different physical channels, simply by choosing an appropriate basis matched to the channel's time–bandwidth constraints. An example of signal space construction based on linear modulation is provided in Section 2.5.4.

2.5 Linear modulation

We now know that we can encode information to be transmitted over a passband channel into a complex-valued baseband waveform. For a physical baseband channel, information must be encoded into a real-valued baseband waveform. We focus on more general complex baseband (i.e., passband) systems, with physical real baseband systems automatically included as a special case.

As the discussion in the previous section indicates, *linear modulation* is a technique of fundamental importance for communication over bandlimited channels. We have already had sneak previews of this modulation technique in Figure 2.1 and Examples 2.3.2, 2.2.3, and we now build on these for a more systematic exposition. The complex baseband transmitted waveform for linear modulation can be written as

$$u(t) = \sum_n b[n] g_{TX}(t - nT). \qquad (2.74)$$

Here $\{b[n]\}$ are the transmitted *symbols*, typically taking values in a fixed symbol *alphabet*, or *constellation*. The modulating pulse $g_{TX}(t)$ is a fixed baseband waveform. The *symbol rate*, or *baud rate* is $1/T$, and T is termed the *symbol interval*.

2.5.1 Examples of linear modulation

We now discuss some commonly used linear modulation formats for baseband and passband channels.

Baseband line codes Linear modulation over physical baseband channels is a special case of (2.74), with all quantities constrained to be real-valued, since $u(t)$ is actually the physical waveform transmitted over the channel. For such real baseband channels, methods of mapping bits to (real-valued) analog waveforms are often referred to as *line codes*. Examples of some binary line codes are shown in Figure 2.14 and can be interpreted as linear modulation with either a $\{-1, +1\}$ or a $\{0, 1\}$ alphabet.

If a clock is not sent in parallel with the modulated data, then bit timing must be extracted from the modulated signal. For the non return to zero (NRZ) formats shown in Figure 2.14, a long run of zeros or ones can lead to loss of synchronization, since there are no transitions in voltage to demarcate bit boundaries. This can be alleviated by precoding the transmitted data so that it has a high enough rate of transitions from 0 to 1, and vice versa. Alternatively, transitions can be guaranteed through choice of modulating pulse: the Manchester code shown in Figure 2.14 has transitions that are twice as fast as the bit rate. The spectral characteristics of baseband line codes are discussed further in Problem 2.23.

Linear memoryless modulation is not the only option The line codes in Figure 2.14 can be interpreted as memoryless linear modulation: the waveform corresponding to a bit depends only on the value of the bit, and is a translate of a single basic pulse shape. We note at this point that this is certainly not the only way to construct a line code. Specifically, the Miller code, depicted in Figure 2.15, is an example of a line code employing memory and nonlinear modulation. The code uses two different basic pulse shapes, $\pm s_1(t)$ to send 1, and $\pm s_0(t)$ to send 0. A sign change is enforced when 0 is followed by 0, in order to enforce a transition. For the sequences 01, 10 and 11, a transition is ensured because of the transition within $s_1(t)$. In this case, the sign of the waveform is chosen to delay the transition as much as possible; it is intuitively

Figure 2.14 Some baseband line codes using memoryless linear modulation.

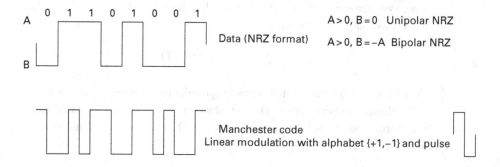

Figure 2.15 The Miller code is a nonlinear modulation format with memory.

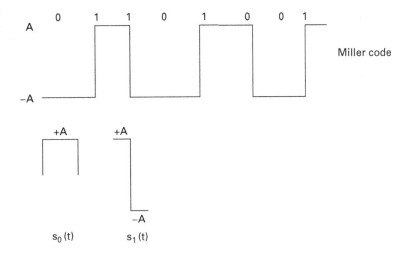

plausible that this makes the modulated waveform smoother, and reduces its spectral occupancy.

Passband linear modulation For passband linear modulation, the symbols $\{b[n]\}$ in (2.74) are allowed to be complex-valued, so that they can be represented in the two-dimensional real plane. Thus, we often use the term *two-dimensional* modulation for this form of modulation. The complex baseband signal $u(t) = u_c(t) + ju_s(t)$ is upconverted to passband as shown in Figure 2.9.

Two popular forms of modulation are phase shift keying (PSK) and quadrature amplitude modulation (QAM). Phase shift keying corresponds to choosing $\arg(b[n])$ from a constellation where the modulus $|b[n]|$ is constant. Quadrature amplitude modulation allows both $|b[n]|$ and $\arg(b[n])$ to vary, and often consists of varying $\text{Re}(b[n])$ and $\text{Im}(b[n])$ independently. Assuming, for simplicity, that $g_{TX}(t)$ is real-valued, we have

$$u_c(t) = \sum_n \text{Re}(b[n]) g_{TX}(t-nT), \quad u_s(t) = \sum_n \text{Im}(b[n]) g_{TX}(t-nT).$$

The term QAM refers to the variations in the amplitudes of I and Q components caused by the modulating symbol sequence $\{b[n]\}$. If the sequence $\{b[n]\}$ is real-valued, then QAM specializes to pulse amplitude modulation (PAM). Figure 2.16 depicts some well-known constellations, where we plot $\text{Re}(b)$ on the x-axis, and $\text{Im}(b)$ on the y-axis, as b ranges over all possible values for the signaling alphabet. Note that rectangular QAM constellations can be interpreted as modulation of the in-phase and quadrature components using PAM (e.g., 16-QAM is equivalent to I and Q modulation using 4-PAM).

Each symbol in a constellation of size M can be uniquely mapped to $\log_2 M$ bits. For a symbol rate of $1/T$ symbols per unit time, the *bit rate*

Figure 2.16 Some constellations for two-dimensional linear modulation.

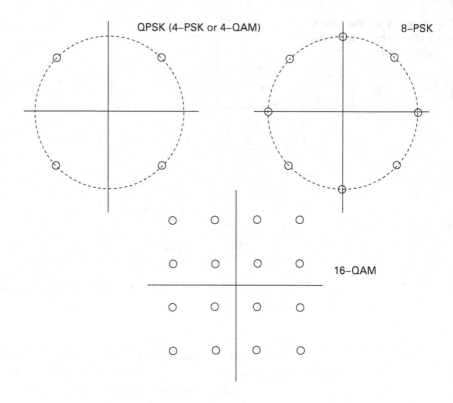

is therefore $(\log_2 M)/T$ bits per unit time. Since the transmitted bits often contain redundancy because of a channel code employed for error correction or detection, the *information rate* is typically smaller than the bit rate.

Design choices Some basic choices that a designer of a linearly modulated system must make are: the transmitted pulse shape g_{TX}, the symbol rate $1/T$, the signaling constellation, the mapping from bits to symbols, and the channel code employed, if any. We now show that the symbol rate and pulse shape are determined largely by the available bandwidth, and by implementation considerations. The background needed to make the remaining choices is built up as we progress through this book. In particular, it will be seen later that the constellation size M and the channel code, if any, should be chosen based on channel quality measures such as the signal-to-noise ratio.

2.5.2 Spectral occupancy of linearly modulated signals

From Example 2.3.3, we know that the linearly modulated signal u in (2.74) is a cyclostationary random process if the modulating symbol sequence $\{b[n]\}$ is a stationary random process. Problem 2.22 discusses computation of the PSD for u as a statistical average across realizations, while Example 2.3.2

2.5 Linear modulation

discusses computation of the PSD as a time average. We now summarize these results in the following theorem.

Theorem 2.5.1 (Power spectral density of a complex baseband linearly modulated signal) *Consider a linearly modulated signal*

$$u(t) = \sum_{n=-\infty}^{\infty} b[n] g_{TX}(t - nT).$$

Assume that the symbol stream $b[n]$ is uncorrelated and has zero mean. That is, $E[b[n]b^*[m]] = E[|b[n]|^2]\delta_{nm}$ and $E[b[n]] = 0$ (the expectation is replaced by a time average when the PSD is defined as a time average). Then the PSD of u is given by

$$S_u(f) = \frac{\mathbb{E}[|b[n]|^2]}{T} |G_{TX}(f)|^2. \qquad (2.75)$$

Figure 2.17 shows the PSD (as a function of normalized frequency fT) for linear modulation using a rectangular timelimited pulse, as well as the cosine-shaped timelimited pulse used for minimum shift keying, which is discussed in Problem 2.24. The smoother shape of the cosine pulse leads to a faster decay of the PSD beyond the main lobe.

Theorem 2.5.1 implies that, for uncorrelated symbols, the shape of the PSD of a linearly modulated signal is determined completely by the spectrum of the modulating pulse $g_{TX}(t)$. A generalization of this theorem for correlated symbol sequences is considered in Problem 2.22, which also discusses the use of such correlations in spectrally shaping the transmitted signal. Another

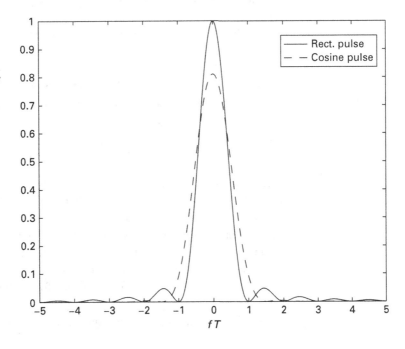

Figure 2.17 PSD for linear modulation using rectangular and cosine timelimited pulses. The normalization is such that the power (i.e., the area under the PSD) is the same in both cases.

generalization of this theorem, discussed in Problem 2.23, is when the symbols $\{b[n]\}$ have nonzero mean, as is the case for some baseband line codes. In this case, the PSD has spectral lines at multiples of the symbol rate $1/T$, which can be exploited for symbol synchronization.

The preceding result, and the generalizations in Problems 2.22 and 2.23, do not apply to nonlinear modulation formats such as the Miller code shown in Figure 2.15. However, the basic concepts of analytical characterization of PSD developed in these problems can be extended to more general modulation formats with a Markovian structure, such as the Miller code. The details are straightforward but tedious, hence we do not discuss them further.

Once the PSD is known, the bandwidth of u can be characterized using any of a number of definitions. One popular concept (analogous to the energy containment bandwidth for a finite-energy signal) is the $1-\epsilon$ power containment bandwidth, where ϵ is a small number: this is the size of the smallest contiguous band that contains a fraction $1-\epsilon$ of the signal power. The fraction of the power contained is often expressed in terms of a percentage: for example, the 99% power containment bandwidth corresponds to $\epsilon = 0.01$. Since the PSD of the modulated signal u is proportional to $|G_{TX}(f)|^2$, the fractional power containment bandwidth is equal to the fractional energy containment bandwidth for $G_{TX}(f)$. Thus, the $1-\epsilon$ power containment bandwidth B satisfies

$$\int_{-\frac{B}{2}}^{\frac{B}{2}} |G_{TX}(f)|^2 df = (1-\epsilon) \int_{-\infty}^{\infty} |G_{TX}(f)|^2 df. \qquad (2.76)$$

We use the two-sided bandwidth B for the complex baseband signal to quantify the signaling bandwidth needed, since this corresponds to the physical (one-sided) bandwidth of the corresponding passband signal. For real-valued signaling over a *physical* baseband channel, the one-sided bandwidth of u would be used to quantify the physical signaling bandwidth.

Normalized bandwidth Time scaling the modulated waveform $u(t)$ preserves its shape, but corresponds to a change of symbol rate. For example, we can double the symbol rate by using a time compressed version $u(2t)$ of the modulated waveform in (2.74):

$$u_2(t) = u(2t) = \sum_n b[n] g_{TX}(2t - nT) = \sum_n b[n] g_{TX}\left(2\left(t - n\frac{T}{2}\right)\right).$$

Time compression leads to frequency dilation by a factor of two, while keeping the signal power the same. It is intuitively clear that the PSD $S_{u_2}(f) = 1/2 S_u(f/2)$, regardless of what definition we use to compute it. Thus, whatever our notion of bandwidth, changing the symbol rate in this fashion leads to a proportional scaling of the required bandwidth. This has the following important consequence. Once we have arrived at a design for a given symbol rate $1/T$, we can reuse it without any change for a different symbol rate a/T, simply by replacing $g_{TX}(t)$ with $g_{TX}(at)$ (i.e., $G_{TX}(f)$ with a scaled version of $G_{TX}(f/a)$). If the bandwidth required was B, then the new bandwidth required is aB. Thus, it makes sense to consider the *normalized bandwidth* BT, which is invariant to the specific symbol rate employed, and depends only on the *shape* of the modulating pulse $g_{TX}(t)$. In doing this, it is also convenient to consider the *normalized time* t/T and normalized frequency fT. Equivalently, we can, without loss of generality, set $T = 1$ to compute the normalized bandwidth, and then simply scale the result by the desired symbol rate.

> **Example 2.5.1 (Fractional power containment bandwidth with time-limited pulse)** We wish to determine the 99% power containment bandwidth when signaling at 100 Mbps using 16-QAM, using a rectangular transmit pulse shape timelimited over the symbol interval. Since there are $\log_2 16 = 4$ bit/symbol, the symbol rate is given by
>
> $$1/T = \frac{100 \text{ Mbps}}{4 \text{ bit/symbol}} = 25 \text{ Msymbol/s}.$$
>
> Let us first compute the normalized bandwidth B_1 for $T = 1$. The transmit pulse is $g_{TX}(t) = I_{[0,1]}(t)$, so that
>
> $$|G_{TX}(f)|^2 = |\text{sinc}(f)|^2.$$
>
> We can now substitute into (2.76) to compute the power containment bandwidth B. We have actually already solved this problem in Example 2.1.3, where we computed $B_1 = 10.2$ for 99% energy containment. We therefore find that the bandwidth required is
>
> $$B = \frac{B_1}{T} = 10.2 \times 25 \text{ MHz} = 260 \text{ MHz}.$$
>
> This is clearly very wasteful of bandwidth. Thus, if we are concerned about strict power containment within the allocated band, we should not be using rectangular timelimited pulses. On the other hand, if we are allowed to be sloppier, and can allow 10% of the power to spill outside the allocated band, then the required bandwidth is less than 25 MHz ($B_1 = 0.85$ for $a = 0.9$, from Example 2.1.3).

2.5.3 The Nyquist criterion: relating bandwidth to symbol rate

Typically, a linearly modulated system is designed so as to avoid intersymbol interference at the receiver, assuming an ideal channel, as illustrated in Figure 2.18, which shows symbols going through a transmit filter, a channel (also modeled as a filter), and a receive filter (noise is ignored for now). Since symbols are being fed into the transmit filter at rate $1/T$, it is natural to expect that we can process the received signal such that, in the absence of channel distortions and noise, samples at rate $1/T$ equal the transmitted symbols. This expectation is fulfilled when the cascade of the transmit filter, the channel

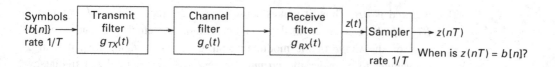

Figure 2.18 Set-up for applying Nyquist criterion.

filter, and the receive filter satisfy the Nyquist criterion for ISI avoidance, which we now state.

From Figure 2.18, the noiseless signal at the output of the receive filter is given by

$$z(t) = \sum_n b[n]x(t-nT), \qquad (2.77)$$

where

$$x(t) = (g_{TX} * g_C * g_{RX})(t)$$

is the overall system response to a single symbol. The Nyquist criterion answers the following question: when is $z(nT) = b[n]$? That is, when is there no ISI in the symbol-spaced samples? The answer is stated in the following theorem.

Theorem 2.5.2 (Nyquist criterion for ISI avoidance) *Intersymbol interference can be avoided in the symbol-spaced samples, i.e.,*

$$z(nT) = b[n] \quad \text{for all } n \qquad (2.78)$$

if

$$x(mT) = \delta_{m0} = \begin{cases} 1, & m = 0, \\ 0, & m \neq 0. \end{cases} \qquad (2.79)$$

Letting $X(f)$ denote the Fourier transform of $x(t)$, the preceding condition can be equivalently written as

$$1/T \sum_{k=-\infty}^{\infty} X\left(f + \frac{k}{T}\right) = 1 \quad \text{for all } f. \qquad (2.80)$$

Proof of Theorem 2.5.2 It is immediately obvious that the time domain condition (2.79) gives the desired ISI avoidance (2.78). It can be shown that this is equivalent to the frequency domain condition (2.80) by demonstrating that the sequence $\{x(-mT)\}$ is the Fourier series for the periodic waveform

$$B(f) = 1/T \sum_{k=-\infty}^{\infty} X\left(f + \frac{k}{T}\right)$$

obtained by summing all the aliased copies $X(f+k/T)$ of the Fourier transform of x. Thus, for the sequence $\{x(mT)\}$ to be a discrete delta, the periodic function $B(f)$ must be a constant. The details are developed in Problem 2.15. □

2.5 Linear modulation

A pulse $x(t)$ or $X(f)$ is said to be Nyquist at rate $1/T$ if it satisfies (2.79) or (2.80), where we permit the right-hand sides to be scaled by arbitrary constants.

Minimum bandwidth Nyquist pulse The minimum bandwidth Nyquist pulse is

$$X(f) = \begin{cases} T, & |f| \leq \frac{1}{2T}, \\ 0, & \text{else}, \end{cases}$$

corresponding to the time domain pulse

$$x(t) = \text{sinc}\left(\frac{t}{T}\right).$$

The need for excess bandwidth The sinc pulse is not used in practice because it decays too slowly: the $1/t$ decay implies that the signal $z(t)$ in (2.77) can exhibit arbtrarily large fluctuations, depending on the choice of the sequence $\{b[n]\}$. It also implies that the ISI caused by sampling errors can be unbounded (see Problem 2.21). Both of these phenomena are related to the divergence of the series $\sum_{n=1}^{\infty} 1/n$, which determines the worst-case contribution from "distant" symbols at a given instant of time. Since the series $\sum_{n=1}^{\infty} 1/n^a$ converges for $a > 1$, these problems can be fixed by employing a pulse $x(t)$ that decays as $1/t^a$ for $a > 1$. A faster time decay implies a slower decay in frequency. Thus, we need *excess bandwidth*, beyond the minimum bandwidth dictated by the Nyquist criterion, to fix the problems associated with the sinc pulse. The (fractional) excess bandwidth for a linear modulation scheme is defined to be the fraction of bandwidth over the minimum required for ISI avoidance at a given symbol rate.

Raised cosine pulse An example of a pulse with a fast enough time decay is the frequency domain raised cosine pulse shown in Figure 2.20, and specified as

$$S(f) = \begin{cases} T, & |f| \leq \frac{1-a}{2T}, \\ \frac{T}{2}\left[1 - \sin((|f| - \frac{1}{2T})\frac{\pi T}{a})\right], & \frac{1-a}{2T} \leq |f| \leq \frac{1+a}{2T}, \\ 0, & |f| > \frac{1+a}{2T}, \end{cases}$$

where a is the fractional excess bandwidth, typically chosen in the range where $0 \leq a < 1$. As shown in Problem 2.16, the time domain pulse $s(t)$ is given by

$$s(t) = \text{sinc}\left(\frac{t}{T}\right) \frac{\cos \pi a \frac{t}{T}}{1 - \left(\frac{2at}{T}\right)^2}.$$

This pulse inherits the Nyquist property of the sinc pulse, while having an additional multiplicative factor that gives an overall $(1/t^3)$ decay with

Figure 2.19 Sinc pulse for minimum bandwidth ISI-free signaling at rate $1/T$. Both time and frequency axes are normalized to be dimensionless.

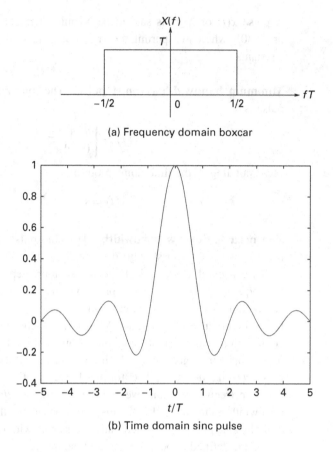

(a) Frequency domain boxcar

(b) Time domain sinc pulse

time. The faster time decay compared with the sinc pulse is evident from a comparison of Figures 2.20(b) and 2.19(b).

The Nyquist criterion applies to the cascade of the transmit, channel, and receive filters. How is Nyquist signaling done in practice, since the channel is typically not within our control? Typically, the transmit and receive filters are designed so that the cascade $G_{TX}(f)G_{RX}(f)$ is Nyquist, and the ISI introduced by the channel, if any, is handled separately. A typical choice is to set G_{TX} and G_{RX} to be square roots (in the frequency domain) of a Nyquist pulse. Such a pulse is called a *square root Nyquist* pulse. For example, the square root raised cosine (SRRC) pulse is often used in practice. Another common choice is to set G_{TX} to be a Nyquist pulse, and G_{RX} to be a wideband filter whose response is flat over the band of interest.

We had argued in Section 2.4, using Nyquist's sampling theorem, that linear modulation using the sinc pulse takes up all of the degrees of freedom in a bandlimited channel. The Nyquist criterion for ISI avoidance may be viewed loosely as a converse to the preceding result, saying that if there are not enough degrees of freedom, then linear modulation incurs ISI. The relation between these two observations is not accidental: both Nyquist's sampling

2.5 Linear modulation

Figure 2.20 Raised cosine pulse for minimum bandwidth ISI-free signaling at rate $1/T$, with excess bandwidth a. Both time and frequency axes are normalized to be dimensionless.

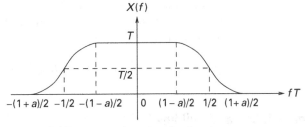

(a) Frequency domain raised cosine

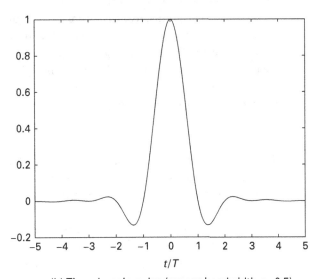

(b) Time domain pulse (excess bandwidth $a = 0.5$)

theorem and the Nyquist criterion are based on the Fourier series relationship between the samples of a waveform and its aliased Fourier transform.

Bandwidth efficiency We define the bandwidth efficiency of linear modulation with an M-ary alphabet as

$$\eta_B = \log_2 M \text{ bit/symbol.}$$

This is consistent with the definition (2.72) in Section 2.4, since one symbol in linear modulation takes up one degree of freedom. Since the Nyquist criterion states that the minimum bandwidth required equals the symbol rate, knowing the bit rate R_b and the bandwidth efficiency η_B of the modulation scheme, we can determine the symbol rate, and hence the minimum required bandwidth $B_{\min}$.

$$B_{\min} = \frac{R_b}{\eta_B}.$$

This bandwidth would then be expanded by the excess bandwidth used in the modulating pulse, which (as discussed already in Section 2.4) is not included in our definition of bandwidth efficiency, because it is a highly variable quantity dictated by a variety of implementation considerations. Once we decide on the fractional excess bandwidth a, the actual bandwidth required is

$$B = (1+a)B_{\min} = (1+a)\frac{R_b}{\eta_B}.$$

2.5.4 Linear modulation as a building block

Linear modulation can be used as a building block for constructing more sophisticated waveforms, using a square root Nyquist pulse as the modulating waveform. To see this, let us first describe the square root Nyquist property in the time domain. Suppose that $\psi(t)$ is square root Nyquist at rate $1/T_c$. This means that $Q(f) = |\Psi(f)|^2 = \Psi(f)\Psi^*(f)$ is Nyquist at rate $1/T_c$. Note that $\Psi^*(f)$ is simply the frequency domain representation of $\psi_{\mathrm{MF}}(t) = \psi^*(-t)$, the matched filter for $\psi(t)$. This means that

$$Q(f) = \Psi(f)\Psi^*(f) \leftrightarrow q(t) = (\psi * \psi_{\mathrm{MF}})(t) = \int \psi(s)\psi^*(s-t)\,ds. \quad (2.81)$$

That is, $q(t)$ is the autocorrelation function of $\psi(t)$, obtained by passing ψ through its matched filter. Thus, ψ is square root Nyquist if its autocorrelation function q is Nyquist. That is, the autocorrelation function satisfies $q(kT_c) = \delta_{k0}$ for integer k.

We have just shown that the translates $\{\psi(t - kT_c)\}$ are orthonormal. We can now use these as a basis for signal space constructions. Representing a signal $s_i(t)$ in terms of these basis functions is equivalent to linear modulation at rate $1/T_c$ as follows:

$$s_i(t) = \sum_{k=0}^{N-1} s_i[k]\psi(t - kT_c), \quad i = 1, \ldots, M,$$

where $\mathbf{s}_i = (s_i[0], \ldots, s_i[N-1])$ is a code vector that is mapped to continuous time by linear modulation using the waveform ψ. We often refer to $\psi(t)$ as the *chip waveform*, and $1/T_c$ as the *chip rate*, where N chips constitute a single symbol. Note that the continuous-time inner product between the signals thus constructed is determined by the discrete-time inner product between the corresponding code vectors:

$$\langle s_i, s_j \rangle = \sum_{k=0}^{N-1}\sum_{l=0}^{N-1} s_i[k]s_j^*[l] \int \psi(t-kT_c)\psi^*(t-lT_c)\,dt = \sum_{k=0}^{N-1} s_i[k]s_j^*[k] = \langle \mathbf{s}_i, \mathbf{s}_j \rangle,$$

where we have used the orthonormality of the translates $\{\psi(t-kT_c)\}$.

Examples of square root Nyquist chip waveforms include a rectangular pulse timelimited to an interval of length T_c, as well as bandlimited pulses such as the square root raised cosine. From Theorem 2.5.1, we see that the PSD of the modulated waveform is proportional to $|\Psi(f)|^2$ (it is typically a good approximation to assume that the chips $\{s_i[k]\}$ are uncorrelated). That is, the bandwidth occupancy is determined by that of the chip waveform ψ.

In the next section, we apply the preceding construction to obtain waveforms for orthogonal modulation. In Chapter 8, I discuss direct sequence spread spectrum systems based on this construction.

2.6 Orthogonal and biorthogonal modulation

The number of possible transmitted signals for orthogonal modulation equals the number of degrees of freedom M available to the modulator, since we can fit only M orthogonal vectors in an M-dimensional signal space. However, as discussed below, whether we need M real degrees of freedom or M complex degrees of freedom depends on the notion of orthogonality required by the receiver implementation.

Frequency shift keying A classical example of orthogonal modulation is frequency shift keying (FSK). The complex baseband signaling waveforms for M-ary FSK over a signaling interval of length T are given by

$$s_i(t) = e^{j2\pi f_i t} I_{[0,T]}, \quad i = 1, \ldots, M,$$

where the frequency shifts $|f_i - f_j|$ are chosen to make the M waveforms orthogonal. The bit rate for such a system is therefore given by $(\log_2 M)/T$, since $\log_2 M$ bits are conveyed over each interval of length T. To determine the bandwidth needed to implement such an FSK scheme, we must determine the minimal frequency spacing such that the $\{s_i\}$ are orthogonal. Let us first discuss what orthogonality means.

We have introduced the concepts of coherent and noncoherent reception in Example 2.2.5, where we correlated the received waveform against copies of the possible noiseless received waveforms corresponding to different transmitted signals. In practical terms, therefore, orthogonality means that, if s_i is sent, and we are correlating the received signal against s_j, $j \neq i$, then the output of the correlator should be zero (ignoring noise). This criterion leads to two different notions of orthogonality, depending on the assumptions we make on the receiver's capabilities.

Orthogonality for coherent and noncoherent systems Consider two complex baseband waveforms $u = u_c + ju_s$ and $v = v_c + jv_s$, and their passband equivalents $u_p(t) = \text{Re}(\sqrt{2}u(t)e^{j2\pi f_c t})$ and $v_p(t) = \text{Re}(\sqrt{2}v(t)e^{j2\pi f_c t})$, respectively. From (2.36), we know that

$$\langle u_p, v_p \rangle = \text{Re}(\langle u, v \rangle) = \langle u_c, v_c \rangle + \langle u_s, v_s \rangle. \tag{2.82}$$

Thus, one concept of orthogonality between complex baseband waveforms is that their passband equivalents (with respect to a common frequency and phase reference) are orthogonal. This requires that $\text{Re}(\langle u, v \rangle) = 0$. In the inner product $\text{Re}(\langle u, v \rangle)$, the I and Q components are correlated separately and then summed up. At a practical level, extracting the I and Q components from a passband waveform requires a *coherent* system, in which an accurate frequency and phase reference is available for downconversion.

Now, suppose that we want the passband equivalents of u and v to remain orthogonal in *noncoherent* systems in which an accurate phase reference may not be available. Mathematically, we want $u_p(t) = \text{Re}(\sqrt{2}u(t)e^{j2\pi f_c t})$ and $\hat{v}_p(t) = \text{Re}(\sqrt{2}v(t)e^{j(2\pi f_c t + \theta)})$ to remain orthogonal, regardless of the value of θ. The complex envelope of $\hat{v}_p$ with respect to f_c is $\hat{v}(t) = v(t)e^{j\theta}$, so that, applying (2.82), we have

$$\langle u_p, \hat{v}_p \rangle = \text{Re}(\langle u, \hat{v} \rangle) = \text{Re}(\langle u, v \rangle e^{-j\theta}). \tag{2.83}$$

It is easy to see that the preceding inner product is zero for all possible θ if and only if $\langle u, v \rangle = 0$; set $\theta = 0$ and $\theta = \pi/2$ in (2.83) to see this.

We therefore have two different notions of orthogonality, depending on which of the inner products (2.82) and (2.83) is employed:

$$\begin{aligned} \text{Re}(\langle s_i, s_j \rangle) &= 0 \quad \textbf{Coherent orthogonality criterion} \\ \langle s_i, s_j \rangle &= 0 \quad \textbf{Noncoherent orthogonality criterion}. \end{aligned} \tag{2.84}$$

It is left as an exercise (Problem 2.25) to show that a tone spacing of $1/2T$ provides orthogonality in coherent FSK, while a tone spacing of $1/T$ is required for noncoherent FSK. The bandwidth for coherent M-ary FSK is therefore approximately $M/2T$, which corresponds to a time–bandwidth product of approximately $M/2$. This corresponds to a complex vector space of dimension $M/2$, or a real vector space of dimension M, in which we can fit M orthogonal signals. On the other hand, M-ary noncoherent signaling requires M complex dimensions, since the complex baseband signals must remain orthogonal even under multiplication by complex-valued scalars. This requirement doubles the bandwidth requirement for noncoherent orthogonal signaling.

Bandwidth efficiency We can conclude from the example of orthogonal FSK that the bandwidth efficiency of orthogonal signaling is $\eta_B = (\log_2 M + 1)/M$ bit/complex dimension for coherent systems, and $\eta_B = (\log_2 M)/M$ bit/complex dimension for noncoherent systems. This is a general observation that holds for any realization of orthogonal signaling. In a signal space of complex dimension D (and hence real dimension $2D$), we can fit $2D$ signals satisfying the coherent orthogonality criterion, but only D signals satisfying the noncoherent orthogonality criterion. As M gets large, the bandwidth efficiency tends to zero. In compensation, as we see in Chapter 3, the power efficiency of orthogonal signaling for large M is the "best possible."

Orthogonal Walsh–Hadamard codes Section 2.5.4 shows how to map vectors to waveforms while preserving inner products, by using linear modulation with a square root Nyquist chip waveform. Applying this construction, the problem of designing orthogonal waveforms $\{s_i\}$ now reduces to designing orthogonal code vectors $\{\mathbf{s}_i\}$. Walsh–Hadamard codes are a standard construction employed for this purpose, and can be constructed recursively as follows: at the nth stage, we generate 2^n orthogonal vectors, using the 2^{n-1} vectors constructed in the $n-1$ stage. Let $\mathbf{H}_n$ denote a matrix whose rows are 2^n orthogonal codes obtained after the nth stage, with $H_0 = (1)$. Then

$$\mathbf{H}_n = \begin{pmatrix} \mathbf{H}_{n-1} & \mathbf{H}_{n-1} \\ \mathbf{H}_{n-1} & -\mathbf{H}_{n-1} \end{pmatrix}.$$

We therefore get

$$H_1 = \begin{pmatrix} 1 & 1 \\ 1 & -1 \end{pmatrix}, \quad H_2 = \begin{pmatrix} 1 & 1 & 1 & 1 \\ 1 & -1 & 1 & -1 \\ 1 & 1 & -1 & -1 \\ 1 & -1 & -1 & 1 \end{pmatrix}, \quad \text{etc.}$$

The signals $\{s_i\}$ obtained above can be used for noncoherent orthogonal signaling, since they satisfy the orthogonality criterion $\langle s_i, s_j \rangle = 0$ for $i \neq j$. However, just as for FSK, we can fit twice as many signals into the same number of degrees of freedom if we used the weaker notion of orthogonality required for coherent signaling, namely $\text{Re}(\langle s_i, s_j \rangle) = 0$ for $i \neq j$. It is easy to check that for M-ary Walsh–Hadamard signals $\{s_i, i = 1, \ldots, M\}$, we can get $2M$ orthogonal signals for coherent signaling: $\{s_i, js_i, i = 1, \ldots, M\}$. This construction corresponds to independently modulating the I and Q components with a Walsh–Hadamard code.

Biorthogonal modulation Given an orthogonal signal set, a biorthogonal signal set of twice the size can be obtained by including a negated copy of each signal. Since signals s and $-s$ cannot be distinguished in a noncoherent system, biorthogonal signaling is applicable to coherent systems. Thus, for an M-ary Walsh–Hadamard signal set $\{s_i\}$ with M signals obeying the noncoherent orthogonality criterion, we can construct a coherent orthogonal signal set $\{s_i, js_i\}$ of size $2M$, and hence a biorthogonal signal set of size $4M$, e.g., $\{s_i, js_i, -s_i, -js_i\}$.

2.7 Differential modulation

Differential modulation uses standard PSK constellations, but encodes the information in the phase transitions between successive symbols rather than in the absolute phase of one symbol. This allows recovery of the information even when there is no absolute phase reference.

Differential modulation is useful for channels in which the amplitude and phase may vary over time (e.g., for a wireless mobile channel), or if there is a residual carrier frequency offset after carrier synchronization. To see why, consider linear modulation of a PSK symbol sequence $\{b[n]\}$. Under ideal Nyquist signaling, the samples at the output of the receive filter obey the model

$$r[n] = h[n]b[n] + \text{noise}$$

where $h[n]$ is the channel gain. If the phase of $h[n]$ can vary arbitrarily fast with n, then there is no hope of conveying any information in the carrier phase. However, if $h[n]$ varies slowly enough that we can approximate it as piecewise constant over at least two symbol intervals, then we can use phase transitions to convey information. Figure 2.21 illustrates this for two successive noiseless received samples for a QPSK alphabet, comparing $b[n]b^*[n-1]$ with $r[n]r^*[n-1]$. We see that, ignoring noise, these two quantities have the same phase. Thus, even when the channel imposes an arbitrary phase shift, as long as the phase shift is roughly constant over two consecutive symbols, the phase *difference* is unaffected by the channel, and hence can be used to convey information. On the other hand, we see from Figure 2.21 that the *amplitude* of $b[n]b^*[n-1]$ differs from that of $r[n]r^*[n-1]$. Thus, some form of explicit amplitude estimation or tracking is required in order to generalize differential modulation to QAM constellations. How best to design differential modulation for QAM alphabets is still a subject of ongoing research, and we do not discuss it further.

Figure 2.22 shows an example of how two information bits can be mapped to phase transitions for differential QPSK. For example, if the information bits at time n are $\mathbf{i}[n] = 00$, then $b[n]$ has the same phase as $b[n-1]$. If

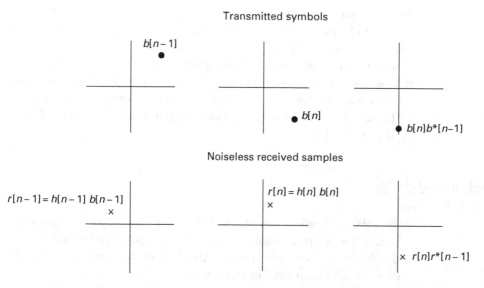

Figure 2.21 Ignoring noise, the phase transitions between successive symbols remain unchanged after an arbitrary phase offset induced by the channel. This motivates differential modulation as a means of dealing with unknown or slowly time-varying channels.

2.7 Differential modulation

Figure 2.22 Mapping information bits to phase transitions in differential QPSK.

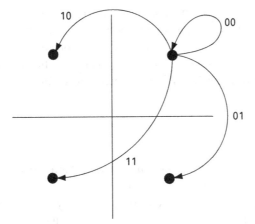

$i[n] = 10$, then $b[n] = e^{j\pi/2}b[n-1]$, and so on, where the symbols $b[n]$ take values in $\{e^{j\pi/4}, e^{j3\pi/4}, e^{j5\pi/4}, e^{j7\pi/4}\}$.

We now discuss the special case of binary differential PSK (DPSK), which has an interesting interpretation as orthogonal modulation. For a BPSK alphabet, suppose that the information bits $\{i[n]\}$ take values in $\{0, 1\}$, the transmitted symbols $\{b[n]\}$ take values in $\{-1, +1\}$, and the encoding rule is as follows:

$$b[n] = b[n-1] \text{ if } i[n]=0,$$
$$b[n] = -b[n-1] \text{ if } i[n]=1.$$

If we think of the signal corresponding to $i[n]$ as $\mathbf{s}[n] = (b[n-1], b[n])$, then $\mathbf{s}[n]$ can take the following values:

$$\text{for } i[n] = 0, \mathbf{s}[n] = \pm\mathbf{s}_0, \text{ where } \mathbf{s}_0 = (+1, +1),$$
$$\text{for } i[n] = 1, \mathbf{s}[n] = \pm\mathbf{s}_1, \text{ where } \mathbf{s}_1 = (+1, -1).$$

The signals $\mathbf{s}_0$ and $\mathbf{s}_1$ are orthogonal. Note that $s[n] = (b[n-1], b[n]) = b[n-1](1, b[n]/b[n-1])$. Since $b[n]/b[n-1]$ depends only on the information bit $i[n]$, the direction of $s[n]$ depends only on $i[n]$, while there is a sign ambiguity due to $b[n-1]$. Not knowing the channel $h[n]$ would impose a further phase ambiguity. Thus, binary DPSK can be interpreted as binary noncoherent orthogonal signaling, with the signal duration spanning two symbol intervals. However, there is an important distinction from standard binary noncoherent orthogonal signaling, which conveys one bit using two complex degrees of freedom. Binary DPSK uses the available degrees of freedom more efficiently by employing overlapping signaling intervals for sending successive information bits: the signal $(b[n], b[n-1])$ used to send $i[n]$ has one degree of freedom in common with the signal $(b[n+1], b[n])$ used to send $i[n+1]$. In particular, we need $n+1$ complex degrees of freedom to send n bits. Thus, for large enough n, binary DPSK needs one complex degree of freedom per information bit, so that its bandwidth efficiency is twice that of standard binary noncoherent orthogonal signaling.

A more detailed investigation of noncoherent communication and differential modulation is postponed to Chapter 4, after we have developed the tools for handling noise and noncoherent processing.

2.8 Further reading

Additional modulation schemes (and corresponding references for further reading) are described in Chapter 8, when we discuss wireless communication. Analytic computations of PSD for a variety of modulation schemes, including line codes with memory, can be found in the text by Proakis [3]. Averaging techniques for simulation-based computation of PSD are discussed in Chapter 8, Problem 8.29.

2.9 Problems

2.9.1 Signals and systems

Problem 2.1 A signal $s(t)$ and its matched filter are shown in Figure 2.4.

(a) Sketch the real and imaginary parts of the output waveform $y(t) = (s * s_{MF})(t)$ when $s(t)$ is passed through its matched filter.
(b) Draw a rough sketch of the magnitude $|y(t)|$. When is the output magnitude the largest?

Problem 2.2 For $s(t) = \text{sinc}(t)\text{sinc}(2t)$:

(a) Find and sketch the Fourier transform $S(f)$.
(b) Find and sketch the Fourier transform $U(f)$ of $u(t) = s(t)\cos(100\pi t)$ (sketch real and imaginary parts separately if $U(f)$ is complex-valued).

Problem 2.3 For $s(t) = (10 - |t|)I_{[-10,10]}(t)$:

(a) Find and sketch the Fourier transform $S(f)$.
(b) Find and sketch the Fourier transform $U(f)$ of $u(t) = s(t)\sin(1000\pi t)$ (sketch real and imaginary parts separately if $U(f)$ is complex-valued).

Problem 2.4 In this problem, we prove the Cauchy–Schwartz inequality (2.5), restated here for convenience,

$$|\langle s, r \rangle| = \left|\int s(t) r^*(t)\, dt\right| \leq \|s\| \|r\|,$$

2.9 Problems

for any complex-valued signals $s(t)$ and $r(t)$, with equality if and only if one signal is a scalar multiple of the other. For simplicity, we assume in the proof that $s(t)$ and $r(t)$ are real-valued.

(a) Suppose we try to approximate $s(t)$ by $ar(t)$, a scalar multiple of r, where a is a real number. That is, we are trying to approximate s by an element in the subspace spanned by r. Then the error in the approximation is the signal $e(t) = s(t) - ar(t)$. Show that the energy of the error signal, as a function of the scalar a, is given by

$$J(a) = ||e||^2 = ||s||^2 + a^2||r||^2 - 2a\langle s, r\rangle.$$

(b) Note that $J(a)$ is a quadratic function of a with a global minimum. Find the minimizing argument a_{min} by differentiation and evaluate $J(a_{min})$. The Cauchy–Schwartz inequality now follows by noting that the minimum error energy is nonnegative. That is, it is a restatement of the fact that $J(a_{min}) \geq 0$.

(c) Infer the condition for equality in the Cauchy–Schwartz inequality.

Note For a rigorous argument, the case when $s(t) = 0$ or $r(t) = 0$ almost everywhere should be considered separately. In this case, it can be verified directly that the Cauchy–Schwartz condition is satisfied with equality.

(d) Interpret the minimizing argument a_{min} as follows: the signal $a_{min}r(t)$ corresponds to the projection of $s(t)$ along a unit "vector" in the direction of $r(t)$. The Cauchy–Schwartz inequality then amounts to saying that the error incurred in the projection has nonnegative energy, with equality if $s(t)$ lies in the subspace spanned by $r(t)$.

Problem 2.5 Let us now show why using a matched filter makes sense for delay estimation, as asserted in Example 2.1.2. Suppose that $x(t) = As(t - t_0)$ is a scaled and delayed version of s. We wish to design a filter h such that, when we pass x through h, we get a peak at time t_0, and we wish to make this peak as large as possible. Without loss of generality, we scale the filter impulse response so as to normalize it as $||h|| = ||s||$.

(a) Using the Cauchy–Schwartz inequality, show that the output y is bounded at all times as follows:

$$|y(t)| \leq |A|\,||s||^2.$$

(b) Using the condition for equality in Cauchy–Schwartz, show that $y(t_0)$ attains the upper bound in (a) if and only if $h(t) = s^*(-t)$ (we are considering complex-valued signals in general, so be careful with the complex conjugates). This means two things: $y(t)$ must have a peak at $t = t_0$, and this peak is an upper bound for the output of any other choice of filter (subject to the normalization we have adopted) at any time.

Note We show in the next chapter that, under a suitable noise model, the matched filter is the optimal form of preprocessing in a broad class of digital communication systems.

2.9.2 Complex baseband representation

Problem 2.6 Consider a real-valued passband signal $x_p(t)$ whose Fourier transform for positive frequencies is given by

$$\text{Re}(X_p(f)) = \begin{cases} \sqrt{2}, & 20 \leq f \leq 22, \\ 0, & 0 \leq f < 20, \\ 0, & 22 < f < \infty, \end{cases}$$

$$\text{Im}(X_p(f)) = \begin{cases} \frac{1}{\sqrt{2}}(1-|f-22|), & 21 \leq f \leq 23, \\ 0, & 0 \leq f < 21, \\ 0, & 23 < f < \infty. \end{cases}$$

(a) Sketch the real and imaginary parts of $X_p(f)$ for both positive and negative frequencies.
(b) Specify the time domain waveform that you get when you pass $\sqrt{2}x_p(t)\cos(40\pi t)$ through a low pass filter.

Problem 2.7 Let $v_p(t)$ denote a real passband signal, with Fourier transform $V_p(f)$ specified as follows for negative frequencies:

$$V_p(f) = \begin{cases} f+101, & -101 \leq f \leq -99, \\ 0, & f < -101 \text{ or } -99 < f \leq 0. \end{cases}$$

(a) Sketch $V_p(f)$ for both positive and negative frequencies.
(b) Without explicitly taking the inverse Fourier transform, can you say whether $v_p(t) = v_p(-t)$ or not?
(c) Choosing $f_0 = 100$, find real baseband waveforms $v_c(t)$ and $v_s(t)$ such that

$$v_p(t) = \sqrt{2}(v_c(t)\cos 2\pi f_0 t - v_s(t)\sin 2\pi f_0 t).$$

(d) Repeat (c) for $f_0 = 101$.

Problem 2.8 Consider the following two passband signals:

$$u_p(t) = \sqrt{2}\,\text{sinc}(2t)\cos 100\pi t$$

and

$$v_p(t) = \sqrt{2}\,\text{sinc}(t)\sin\left(101\pi t + \frac{\pi}{4}\right).$$

2.9 Problems

(a) Find the complex envelopes $u(t)$ and $v(t)$ for u_p and v_p, respectively, with respect to the frequency reference $f_c = 50$.
(b) What is the bandwidth of $u_p(t)$? What is the bandwidth of $v_p(t)$?
(c) Find the inner product $\langle u_p, v_p \rangle$, using the result in (a).
(d) Find the convolution $y_p(t) = (u_p * v_p)(t)$, using the result in (a).

Problem 2.9 Let $u(t)$ denote a real baseband waveform with Fourier transform for $f > 0$ specified by

$$U(f) = \begin{cases} e^{j\pi f} & 0 < f < 1, \\ 0 & f > 1. \end{cases}$$

(a) Sketch $\text{Re}(U(f))$ and $\text{Im}(U(f))$ for both positive and negative frequencies.
(b) Find $u(t)$.

Now, consider the bandpass waveform $v(t)$ generated from $u(t)$ as follows:

$$v(t) = \sqrt{2} u(t) \cos 200\pi t.$$

(c) Sketch $\text{Re}(V(f))$ and $\text{Im}(V(f))$ for both positive and negative frequencies.
(d) Let $y(t) = (v * h_{hp})(t)$ denote the result of filtering $v(t)$ using a high pass filter with transfer function

$$H_{hp}(f) = \begin{cases} 1 & |f| \geq 100 \\ 0 & \text{else}. \end{cases}$$

Find real baseband waveforms y_c, y_s such that

$$y(t) = \sqrt{2}(y_c(t) \cos 200\pi t - y_s(t) \sin 200\pi t).$$

(e) Finally, pass $y(t) \cos 200\pi t$ through an ideal low pass filter with transfer function

$$H_{lp}(f) = \begin{cases} 1 & |f| \leq 1 \\ 0 & \text{else}. \end{cases}$$

How is the result related to $u(t)$?

Remark It is a good idea to draw pictures of what is going on in the frequency domain to get a good handle on this problem.

Problem 2.10 Consider a passband filter whose transfer function for $f > 0$ is specified by

$$H_p(f) = \begin{cases} 1 & f_c - 2 \leq f \leq f_c \\ 1 - f + f_c & f_c \leq f \leq f_c + 1 \quad (f_c \gg 1) \\ 0 & \text{else}. \end{cases} \quad (2.85)$$

Let $y_p(t)$ denote the output of the filter when fed by a passband signal $u_p(t)$. We would like to generate $y_p(t)$ from $u_p(t)$ using baseband processing in the system shown in Figure 2.23.

Figure 2.23 Implementation of a passband filter using downconversion, baseband operations and upconversion (Problem 2.10).

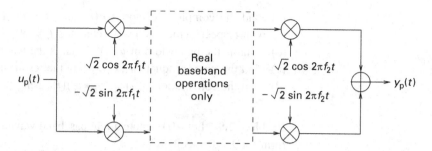

(a) For $f_1 = f_2 = f_c$, sketch the baseband processing required, specifying completely the transfer function of all baseband filters used. Be careful with signs.

(b) Repeat (a) for $f_1 = f_c + 1/2$ and $f_2 = f_c - 1/2$.

Hint The inputs to the black box are the real and imaginary parts of the complex baseband representation for $u(t)$ centered at f_1. Hence, we can use baseband filtering to produce the real and imaginary parts for the complex baseband representation for the output $y(t)$ using f_1 as center frequency. Then use baseband processing to construct the real and imaginary parts of the complex baseband representation for $y(t)$ centered at f_2. These will be the output of the black box.

Problem 2.11 Consider a pure sinusoid $s_p(t) = \cos 2\pi f_c t$, which is the simplest possible example of a passband signal with finite power.

(a) Find the time-averaged PSD $\bar{S}_s(f)$ and autocorrelation function $\bar{R}_s(\tau)$, proceeding from the definitions. Check that the results conform to your intuition.

(b) Find the complex envelope $s(t)$, and its time-averaged PSD and autocorrelation function. Check that the relation (2.70) holds for the passband and baseband PSDs.

2.9.3 Random processes

Problem 2.12 Consider a passband random process $n_p(t) = \text{Re}(\sqrt{2}n(t) e^{j2\pi f_c t})$ with complex envelope $n(t) = n_c(t) + jn_s(t)$.

(a) Given the time-averaged PSD for n_p, can you find the time-averaged PSD for n? Specify any additional information you might need.

(b) Given the time-averaged PSD for n_p, can you find the time-averaged PSDs for n_c and n_s? Specify any additional information you might need.

(c) Now, consider a statistical description of n_p. What are the conditions on n_c and n_s for n_p to be WSS? Under these conditions, what are the relations between the statistically averaged PSDs of n_p, n, n_c and n_s?

Problem 2.13 We discuss passband white noise, an important noise model used extensively in Chapter 3, in this problem. A passband random process

2.9 Problems

$n_p(t) = \text{Re}(\sqrt{2}n(t)e^{j2\pi f_c t})$ with complex envelope $n(t) = n_c(t) + jn_s(t)$ has PSD

$$S_n(f) = \begin{cases} \frac{N_0}{2}, & |f - f_c| \leq \frac{W}{2} \text{ or } |f + f_c| \leq \frac{W}{2}, \\ 0, & \text{else,} \end{cases}$$

where n_c, n_s are independent and identically distributed zero mean random processes.

(a) Find the PSD $S_n(f)$ for the complex envelope n.
(b) Find the PSDs $S_{n_c}(f)$ and $S_{n_s}(f)$ if possible. If this is not possible from the given information, say what further information is needed.

Problem 2.14 In this problem, we prove Theorem 2.3.1 regarding (wide sense) stationarization of a (wide sense) cyclostationary process. Let $s(t)$ be (wide sense) cyclostationary with respect to the time interval T. Define $v(t) = s(t - D)$, where D is uniformly distributed over $[0, T]$ and independent of s.

(a) Suppose that s is cyclostationary. Use the following steps to show that v is stationary; that is, for any delay a, the statistics of $v(t)$ and $v(t - a)$ are indistinguishable.

  (i) Show that $a + D = kT + \tilde{D}$, where k is an integer, and $\tilde{D}$ is a random variable which is independent of s, and uniformly distributed over $[0, T]$.
  (ii) Show that the random process $\tilde{s}$ defined by $\tilde{s}(t) = s(t - kT)$ is statistically indistinguishable from s.
  (iii) Show that the random process $\tilde{v}$ defined by $\tilde{v}(t) = v(t-a) = \tilde{s}(t - \tilde{D})$ is statistically indistinguishable from v.

(b) Now, suppose that s is wide sense cyclostationary. Use the following steps to show that u is WSS.

  (i) Show that $m_v(t) = 1/T \int_0^T m_s(\nu) \, d\nu$ for all t. That is, the mean function of v is constant.
  (ii) Show that

  $$R_v(t_1, t_2) = 1/T \int_0^T R_s(t + t_1 - t_2, t) \, dt.$$

  This implies that the autocorrelation function of v depends only on the time difference $t_1 - t_2$.

(c) Now, let us show that, under an intuitive notion of ergodicity, the autocorrelation function for s, computed as a time average along a realization, equals the autocorrelation function computed as a statistical average for its stationarized version u. This means, for example, that it is the stationarized version of a cyclostationary process which is relevant for computation of PSD as a statistical average.

(i) Show that $s(t)s^*(t-\tau)$ has the same statistics as $s(t+T)s^*(t+T-\tau)$ for any t.

(ii) Show that the time-averaged autocorrelation estimate

$$\hat{R}_s(\tau) = \frac{1}{T_o}\int_{-\frac{T_o}{2}}^{\frac{T_o}{2}} s(t)s^*(t-\tau)dt$$

can be rewritten, for $T_o = KT$ (K a large integer), as

$$\hat{R}_s(\tau) \approx 1/T \int_0^T \frac{1}{K}\sum_{k=-K/2}^{K/2} s(t+kT)s^*(t+kT-\tau)dt.$$

(iii) Invoke the following intuitive notion of ergodicity: the time average of the identically distributed random variables $\{s(t+kT)s^*(t+kT-\tau)\}$ equals its statistical average $E[s(t)s^*(t-\tau)]$. Infer that

$$\hat{R}_s(\tau) \to 1/T \int_0^T R_s(t, t-\tau)dt = R_v(\tau)$$

as K (and T_o) becomes large.

2.9.4 Modulation

Problem 2.15 In this problem, we derive the Nyquist criterion for ISI avoidance. Let $x(t)$ denote a pulse satisfying the time domain Nyquist condition for signaling at rate $1/T$: $x(mT) = \delta_{m0}$ for all integer m. Using the inverse Fourier transform formula, we have

$$x(mT) = \int_{-\infty}^{\infty} X(f)e^{j2\pi fmT}df.$$

(a) Observe that the integral can be written as an infinite sum of integrals over segments of length $1/T$:

$$x(mT) = \sum_{k=-\infty}^{\infty} \int_{\frac{k-\frac{1}{2}}{T}}^{\frac{k+\frac{1}{2}}{T}} X(f)e^{j2\pi fmT}df.$$

(b) In the integral over the kth segment, make the substitution $\nu = f - k/T$. Simplify to obtain

$$x(mT) = T\int_{-\frac{1}{2T}}^{\frac{1}{2T}} B(\nu)e^{-j2\pi\nu mT}d\nu,$$

where $B(f) = 1/T\sum_{k=-\infty}^{\infty} X(f+k/T)$.

(c) Show that $B(f)$ is periodic in f with period $P = 1/T$, so that it can be written as a Fourier series involving complex exponentials:

$$B(f) = \sum_{m=-\infty}^{\infty} a[m]e^{j2\pi\frac{m}{P}f},$$

where the Fourier series coefficients $\{a[m]\}$ are given by

$$a[m] = \frac{1}{P}\int_{-\frac{P}{2}}^{\frac{P}{2}} B(f)\,e^{-j2\pi\frac{m}{P}f}\,df.$$

2.9 Problems

(d) Conclude that $x(mT) = a[-m]$, so that the Nyquist criterion is equivalent to $a[m] = \delta_{m0}$. This implies that $B(f) \equiv 1$, which is the desired frequency domain Nyquist criterion.

Problem 2.16 In this problem, we derive the time domain response of the frequency domain raised cosine pulse. Let $R(f) = I_{[-\frac{1}{2},\frac{1}{2}]}(f)$ denote an ideal boxcar transfer function, and let $C(f) = \pi/2a \cos(\pi/af) I_{[-\frac{a}{2},\frac{a}{2}]}$ denote a cosine transfer function.

(a) Sketch $R(f)$ and $C(f)$, assuming that $0 < a < 1$.
(b) Show that the frequency domain raised cosine pulse can be written as

$$S(f) = (R * C)(f).$$

(c) Find the time domain pulse $s(t) = r(t)c(t)$. Where are the zeros of $s(t)$? Conclude that $s(t/T)$ is Nyquist at rate $1/T$.
(d) Sketch an argument that shows that, if the pulse $s(t/T)$ is used for BPSK signaling at rate $1/T$, then the magnitude of the transmitted waveform is always finite.

Problem 2.17 Consider a pulse $s(t) = \text{sinc}(at)\text{sinc}(bt)$, where $a \geq b$.

(a) Sketch the frequency domain response $S(f)$ of the pulse.
(b) Suppose that the pulse is to be used over an ideal real baseband channel with one-sided bandwidth 400 Hz. Choose a and b so that the pulse is Nyquist for 4-PAM signaling at 1200 bit/s and exactly fills the channel bandwidth.
(c) Now, suppose that the pulse is to be used over a passband channel spanning the frequencies 2.4–2.42 GHz. Assuming that we use 64-QAM signaling at 60 Mbit/s, choose a and b so that the pulse is Nyquist and exactly fills the channel bandwidth.
(d) Sketch an argument showing that the magnitude of the transmitted waveform in the preceding settings is always finite.

Problem 2.18 Consider the pulse

$$p(t) = \begin{cases} 1 - \frac{|t|}{T}, & 0 \leq |t| \leq T, \\ 0, & \text{else.} \end{cases}$$

Let $P(f)$ denote the Fourier transform of $p(t)$.

(a) **True or False** The pulse $p(t)$ is Nyquist at rate $1/T$.
(b) **True or False** The pulse $p(t)$ is square root Nyquist at rate $1/T$ (i.e., $|P(f)|^2$ is Nyquist at rate $1/T$).

Modulation

Problem 2.19 Consider the pulse $p(t)$, whose Fourier transform satisfies:

$$P(f) = \begin{cases} 1, & 0 \leq |f| \leq A, \\ \frac{B-|f|}{B-A}, & A \leq |f| \leq B, \\ 0, & \text{else}, \end{cases}$$

where $A = 250\,\text{kHz}$ and $B = 1.25\,\text{MHz}$.

(a) **True or False** The pulse $p(t)$ can be used for Nyquist signaling at rate 3 Mbps using an 8-PSK constellation.

(b) **True or False** The pulse $p(t)$ can be used for Nyquist signaling at rate 4.5 Mbps using an 8-PSK constellation.

Problem 2.20 True or False Any pulse timelimited to duration T is square root Nyquist (up to scaling) at rate $1/T$.

Problem 2.21 (Effect of timing errors) Consider digital modulation at rate $1/T$ using the sinc pulse $s(t) = \text{sinc}(2Wt)$, with transmitted waveform

$$y(t) = \sum_{n=1}^{100} b_n s(t - (n-1)T),$$

where $1/T$ is the symbol rate and $\{b_n\}$ is the bitstream being sent (assume that each b_n takes one of the values ± 1 with equal probability). The receiver makes bit decisions based on the samples $r_n = y((n-1)T)$, $n = 1, \ldots, 100$.

(a) For what value of T (as a function of W) is $r_n = b_n$, $n = 1, \ldots, 100$?

Remark In this case, we simply use the sign of the nth sample r_n as an estimate of b_n.

(b) For the choice of T as in (a), suppose that the receiver sampling times are off by $0.25T$. That is, the nth sample is given by $r_n = y((n-1)T + 0.25T)$, $n = 1, \ldots, 100$. In this case, we do have ISI of different degrees of severity, depending on the bit pattern. Consider the following bit pattern:

$$b_n = \begin{cases} (-1)^{n-1} & 1 \leq n \leq 49, \\ (-1)^n & 50 \leq n \leq 100. \end{cases}$$

Numerically evaluate the 50th sample r_{50}. Does it have the same sign as the 50th bit b_{50}?

Remark The preceding bit pattern creates the worst possible ISI for the 50th bit. Since the sinc pulse dies off slowly with time, the ISI contributions from the 99 bits other than the 50th sample sum up to a number larger in magnitude, and opposite in sign, relative to the contribution due to b_{50}. A decision on b_{50} based on the sign of r_{50} would therefore be wrong. This sensitivity to timing error is why the sinc pulse is seldom used in practice.

(c) Now, consider the digitally modulated signal in (a) with the pulse $s(t) = \text{sinc}(2Wt)\text{sinc}(Wt)$. For ideal sampling as in (a), what are the two values of T such that $r_n = b_n$?

(d) For the smaller of the two values of T found in (c) (which corresponds to faster signaling, since the symbol rate is $1/T$), repeat the computation in (b). That is, find r_{50} and compare its sign with b_{50} for the bit pattern in (b).

(e) Find and sketch the frequency response of the pulse in (c). What is the excess bandwidth relative to the pulse in (a), assuming Nyquist signaling at the same symbol rate?

(f) Discuss the impact of the excess bandwidth on the severity of the ISI due to timing mismatch.

Problem 2.22 (PSD for linearly modulated signals) Consider the linearly modulated signal

$$s(t) = \sum_{n=-\infty}^{\infty} b[n]p(t-nT).$$

(a) Show that s is cyclostationary with respect to the interval T if $\{b[n]\}$ is stationary.

(b) Show that s is wide sense cyclostationary with respect to the interval T if $\{b[n]\}$ is WSS.

(c) Assume that $\{b[n]\}$ is zero mean, WSS with autocorrelation function $R_b[k] = \mathbb{E}[b[n]b^*[n-k]]$. The z-transform of R_b is denoted by $S_b(z) = \sum_{k=-\infty}^{\infty} R_b[k]z^{-k}$. Let $v(t) = s(t-D)$ denote the stationarized version of s, where D is uniform over $[0, T]$ and independent of s. Show that the PSD of v is given by

$$S_v(f) = S_b(e^{j2\pi fT})\frac{|P(f)|^2}{T}. \tag{2.86}$$

For uncorrelated symbols with equal average energy (i.e., $R_b[k] = \sigma_b^2 \delta_{k0}$), we have $S_b(z) \equiv \sigma_b^2$, and the result reduces to Theorem 2.5.1.

(d) **Spectrum shaping via line coding** We can design the sequence $\{b[n]\}$ using a line code so as to shape the PSD of the modulated signal v. For example, for physical baseband channels, we might want to put a null at DC. For example, suppose that we wish to send i.i.d. symbols $\{a[n]\}$ which take values ± 1 with equal probability. Instead of sending $a[n]$

directly, we can send $b[n] = a[n] - a[n-1]$. The transformation from $a[n]$ to $b[n]$ is called a line code.

(i) What is the range of values taken by $b[n]$?
(ii) Show that there is a spectral null at DC.
(iii) Find a line code of the form $b[n] = a[n] + ka[n-1]$ which puts a spectral null at $f = 1/2T$.

Remark The preceding line code can be viewed as introducing ISI in a controlled fashion, which must be taken into account in receiver design. The techniques for dealing with controlled ISI (introduced by a line code) and uncontrolled ISI (introduced by channel distortion) operate on the same principles. Methods for handling ISI are discussed in Chapter 5.

Problem 2.23 (Linear modulation using alphabets with nonzero mean)
Consider again the linearly modulated signal

$$s(t) = \sum_{n=-\infty}^{\infty} b[n]p(t-nT),$$

where $\{b[n]\}$ is WSS, but with nonzero mean $\bar{b} = \mathbb{E}[b[n]]$.

(a) Show that we can write s as a sum of a deterministic signal $\bar{s}$ and a zero mean random signal $\tilde{s}$ as follows:

$$s(t) = \bar{s}(t) + \tilde{s}(t),$$

where

$$\bar{s}(t) = \bar{b} \sum_{n=-\infty}^{\infty} p(t-nT)$$

and

$$\tilde{s}(t) = \sum_{n=-\infty}^{\infty} \tilde{b}[n]p(t-nT),$$

where $\tilde{b}[n] = b[n] - \bar{b}$ is zero mean, WSS with autocorrelation function $R_{\tilde{b}}[k] = C_b[k]$, where $C_b[k]$ is the autocovariance function of the symbol sequence $\{b[n]\}$.

(b) Show that the PSD of s is the sum of the PSDs of $\bar{s}$ and $\tilde{s}$, by showing that the two signals are uncorrelated.

(c) Note that the PSD of $\tilde{s}$ can be found using the result of Problem 2.22(c). It remains to find the PSD of $\bar{s}$. Note that $\bar{s}$ is periodic with period T. It can therefore be written as a Fourier series

$$\bar{s}(t) = \sum_k a[k] e^{j2\pi kt/T},$$

where

$$a[n] = 1/T \int_0^T \bar{s}(t) e^{-j2\pi nt/T} dt.$$

2.9 Problems

Argue that the PSD of $\tilde{s}$ is given by

$$S_{\tilde{s}}(f) = \sum_k |a[k]|^2 \delta\left(f - \frac{k}{T}\right).$$

(d) Find the PSD for the unipolar NRZ baseband line code in Figure 2.14 (set $A = 1$ and $B = 0$ in the NRZ code in the figure).

Problem 2.24 (OQPSK and MSK) Linear modulation with a bandlimited pulse can perform poorly over nonlinear passband channels. For example, the output of a passband hardlimiter (which is a good model for power amplifiers operating in a saturated regime) has constant envelope, but a PSK signal employing a bandlimited pulse has an envelope that passes through zero during a 180 degree phase transition, as shown in Figure 2.24. One way to alleviate this problem is to not allow 180 degree phase transitions. Offset QPSK (OQPSK) is one example of such a scheme, where the transmitted signal is given by

$$s(t) = \sum_{n=-\infty}^{\infty} b_c[n] g_{TX}(t - nT) + j b_s[n] g_{TX}\left(t - nT - \frac{T}{2}\right), \quad (2.87)$$

where $\{b_c[n]\}$, $b_s[n]$ are ± 1 BPSK symbols modulating the I and Q channels, with the I and Q signals being staggered by half a symbol interval. This leads to phase transitions of at most 90 degrees at integer multiples of the *bit time* $T_b = T/2$. Minimum shift keying (MSK) is a special case of OQPSK with timelimited modulating pulse

$$g_{TX}(t) = \sin\left(\frac{\pi t}{T}\right) I_{[0,T]}(t). \quad (2.88)$$

(a) Sketch the I and Q waveforms for a typical MSK signal, clearly showing the timing relationship between the waveforms.
(b) Show that the MSK waveform has constant envelope (an extremely desirable property for nonlinear channels).
(c) Find an analytical expression for the PSD of an MSK signal, assuming that all bits sent are i.i.d., taking values ± 1 with equal probability. Plot the PSD versus normalized frequency fT.
(d) Find the 99% power containment normalized bandwidth of MSK. Compare with the minimum Nyquist bandwidth, and the 99% power containment bandwidth of OQPSK using a rectangular pulse.

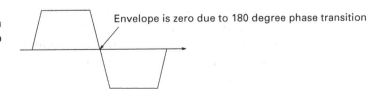

Figure 2.24 The envelope of a PSK signal passes through zero during a 180 degree phase transition, and gets distorted over a nonlinear channel.

(e) Recognize that Figure 2.17 gives the PSD for OQPSK and MSK, and reproduce this figure, normalizing the area under the PSD curve to be the same for both modulation formats.

Problem 2.25 (FSK tone spacing) Consider two real-valued passband pulses of the form

$$s_0(t) = \cos(2\pi f_0 t + \phi_0), \ 0 \le t \le T,$$
$$s_1(t) = \cos(2\pi f_1 t + \phi_1), \ 0 \le t \le T,$$

where $f_1 > f_0 \gg 1/T$. The pulses are said to be *orthogonal* if $\langle s_0, s_1 \rangle = \int_0^T s_0(t) s_1(t) dt = 0$.

(a) If $\phi_0 = \phi_1 = 0$, show that the minimum frequency separation such that the pulses are orthogonal is $f_1 - f_0 = 1/2T$.

(b) If ϕ_0 and ϕ_1 are arbitrary phases, show that the minimum separation for the pulses to be orthogonal regardless of ϕ_0, ϕ_1 is $f_1 - f_0 = 1/T$.

Remark The results of this problem can be used to determine the bandwidth requirements for coherent and noncoherent FSK, respectively.

Problem 2.26 (Walsh–Hadamard codes)

(a) Specify the Walsh–Hadamard codes for 8-ary orthogonal signaling with noncoherent reception.

(b) Plot the baseband waveforms corresponding to sending these codes using a square root raised cosine pulse with excess bandwidth of 50%.

(c) What is the fractional increase in bandwidth efficiency if we use these eight waveforms as building blocks for biorthogonal signaling with coherent reception?

Problem 2.27 (Bandwidth occupancy as a function of modulation format) We wish to send at a rate of 10 Mbit/s over a passband channel. Assuming that an excess bandwidth of 50% is used, how much bandwidth is needed for each of the following schemes: QPSK, 64-QAM, and 64-ary noncoherent orthogonal modulation using a Walsh–Hadamard code?

Problem 2.28 (Binary DPSK) Consider binary DPSK with encoding as described in Section 2.7. Assume that we fix $b[0] = -1$, and that the stream of information bits $\{i[n], n = 1, \ldots, 10\}$ to be sent is 0110001011.

(a) Find the transmitted symbol sequence $\{b[n]\}$ corresponding to the preceding bit sequence.

(b) Assuming that we use a rectangular timelimited pulse, draw the corresponding complex baseband transmitted waveform. Is the Q component being used?

(c) Now, suppose that the channel imposes a phase shift of $-\pi/6$. Draw the I and Q components of the noiseless received complex baseband signal.

(d) Suppose that the complex baseband signal is sent through a matched filter to the rectangular timelimited pulse, and is sampled at the peaks. What are the received samples $\{r[n]\}$ that are obtained corresponding to the transmitted symbol sequence $\{b[n]\}$?

(e) Find $r[2]r^*[1]$. How do you figure out the information bit $i[2]$ based on this complex number?

Problem 2.29 (Differential QPSK) Consider differential QPSK as shown in Figure 2.22. Suppose that $b[0] = e^{-j\pi/4}$, and that $b[1], b[2], \ldots, b[10]$ are determined by using the mapping shown in the figure, where the information bit sequence to be sent is given by 00, 11, 01, 10, 10, 01, 11, 00, 01, 10.

(a) Specify the phases $\arg(b[n])$, $n = 1, \ldots, 10$.

(b) If you received noisy samples $r[1] = 2 - j$ and $r[2] = 1 + j$, what would be a sensible decision for the pair of bits corresponding to the phase transition from $n = 1$ to $n = 2$? Does this match the true value of these bits? (A systematic treatment of differential demodulation in the presence of noise is given in Chapter 4.)

CHAPTER 3
Demodulation

We now know that information is conveyed in a digital communication system by selecting one of a set of signals to transmit. The received signal is a distorted and noisy version of the transmitted signal. A fundamental problem in receiver design, therefore, is to decide, based on the received signal, which of the set of possible signals was actually sent. The task of the link designer is to make the probability of error in this decision as small as possible, given the system constraints. Here, we examine the problem of receiver design for a simple channel model, in which the received signal equals one of M possible deterministic signals, plus white Gaussian noise (WGN). This is called the additive white Gaussian noise (AWGN) channel model. An understanding of transceiver design principles for this channel is one of the first steps in learning digital communication theory. White Gaussian noise is an excellent model for thermal noise in receivers, whose PSD is typically flat over most signal bandwidths of interest.

In practice, when a transmitted signal goes through a channel, at the very least, it gets attenuated and delayed, and (if it is a passband signal) undergoes a change of carrier phase. Thus, the model considered here applies to a receiver that can estimate the effects of the channel, and produce a noiseless copy of the received signal corresponding to each possible transmitted signal. Such a receiver is termed a *coherent* receiver. Implementation of a coherent receiver involves synchronization in time, carrier frequency, and phase, which are all advanced receiver functions discussed in the next chapter. In this chapter, we assume that such synchronization functions have already been taken care of. Despite such idealization, the material in this chapter is perhaps the most important tool for the communication systems designer. For example, it is the performance estimates provided here that are used in practice for *link budget analysis*, which provides a methodology for quick link designs, allowing for nonidealities with a *link margin*.

Prerequisites for this chapter We assume a familiarity with the modulation schemes described in Chapter 2. We also assume familiarity with common terminology and important concepts in probability, random variables, and random processes. See Appendix A for a quick review, as well as for recommendations for further reading on these topics.

Map of this chapter In this chapter, we provide the classical derivation of optimal receivers for the AWGN channel using the framework of hypothesis testing, and describe techniques for obtaining quick performance estimates. Hypothesis testing is the process of deciding which of a fixed number of hypotheses best explains an observation. In our application, the observation is the received signal, while the hypotheses are the set of possible signals that could have been transmitted. We begin with a quick review of Gaussian random variables, vectors and processes in Section 3.1. The basic ingredients and concepts of hypothesis testing are developed in Section 3.2. We then show in Section 3.3 that, for M-ary signaling in AWGN, the receiver can restrict attention to the M-dimensional *signal space* spanned by the M signals without loss of optimality. The optimal receiver is then characterized in Section 3.4, with performance analysis discussed in Section 3.5. In addition to the classical discussion of hard decision demodulation, we also provide a quick introduction to soft decisions, as a preview to their extensive use in coded systems in Chapter 7. We end with an example of a *link budget* in Section 3.7, showing how the results in this chapter can be applied to get a quick characterization of the combination of system parameters (e.g., signaling scheme, transmit power, range, and antenna gains) required to obtain an operational link.

Notation This is the chapter in which we begin to deal more extensively with random variables, hence it is useful to clarify and simplify notation at this point. Given a random variable X, a common notation for probability density function or probability mass function is $p_X(x)$, with X denoting the random variable, and x being a dummy variable which we might integrate out when computing probabilities. However, when there is no scope for confusion, we use the less cumbersome (albeit incomplete) notation $p(x)$, using the dummy variable x not only as the argument of the density, but also to indicate that the density corresponds to the random variable X. (Similarly, we would use $p(y)$ to denote the density for a random variable Y.) The same convention is used for joint and conditional densities as well. For random variables X and Y, we use the notation $p(x, y)$ instead of $p_{X,Y}(x, y)$, and $p(y|x)$ instead of $p_{Y|X}(y|x)$, to denote the joint and conditional densities, respectively.

3.1 Gaussian basics

The key reason why Gaussian random variables crop up so often in both natural and manmade systems is the central limit theorem (CLT). In its elementary form, the CLT states that the sum of a number of independent and identically distributed random variables is well approximated as a Gaussian random variable. However, the CLT holds in far more general settings: without going into

Demodulation

technical detail, it holds as long as dependencies or correlations among the random variables involved in the sum die off rapidly enough, and no one random variable contributes too greatly to the sum. The Gaussianity of receiver thermal noise can be attributed to its arising from the movement of a large number of electrons. However, because the CLT kicks in with a relatively small number of random variables, we shall see the CLT invoked in a number of other contexts, including performance analysis of equalizers in the presence of ISI as well as AWGN, and the modeling of multipath wireless channels.

Gaussian random variable The random variable X is said to follow a *Gaussian*, or *normal* distribution if its density is of the form:

$$p(x) = \frac{1}{\sqrt{2\pi v^2}} \exp\left(-\frac{(x-m)^2}{2v^2}\right), \quad -\infty < x < \infty, \qquad (3.1)$$

where $m = \mathbb{E}[X]$ is the mean of X, and $v^2 = \text{var}(X)$ is the variance of X. The Gaussian density is therefore completely characterized by its mean and variance. Figure 3.1 shows an $N(-5, 4)$ Gaussian density.

Notation for Gaussian distribution We use $N(m, v^2)$ to denote a Gaussian distribution with mean m and variance v^2, and use the shorthand $X \sim N(m, v^2)$ to denote that a random variable X follows this distribution.

Standard Gaussian random variable A zero mean, unit variance Gaussian random variable, $X \sim N(0, 1)$, is termed a standard Gaussian random variable.

An extremely important property of Gaussian random variables is that they remain Gaussian when we scale them or add constants to them (i.e., when we put them through an *affine* transformation).

Gaussianity is preserved under affine transformations If X is Gaussian, then $aX + b$ is Gaussian for any constants a and b.

In particular, probabilities involving Gaussian random variables can be expressed compactly by normalizing them into standard Gaussian form.

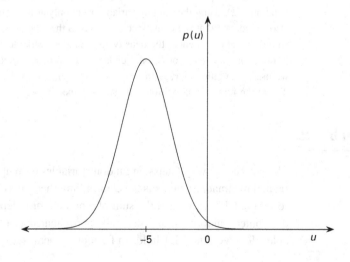

Figure 3.1 The shape of an $N(-5, 4)$ density.

3.1 Gaussian basics

Figure 3.2 The Φ and Q functions are obtained by integrating the N(0, 1) density over appropriate intervals.

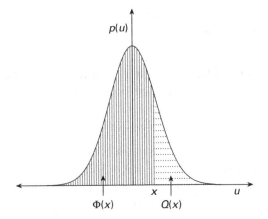

Conversion of a Gaussian random variable into standard form If $X \sim N(m, v^2)$, then $(X - m)/v \sim N(0, 1)$.

We set aside special notation for the cumulative distribution function (CDF) $\Phi(x)$ and complementary cumulative distribution function (CCDF) $Q(x)$ of a standard Gaussian random variable. By virtue of the standard form conversion, we can easily express probabilities involving any Gaussian random variable in terms of the Φ or Q functions. The definitions of these functions are illustrated in Figure 3.2, and the corresponding formulas are specified below.

$$\Phi(x) = P[N(0, 1) \leq x] = \int_{-\infty}^{x} \frac{1}{\sqrt{2\pi}} \exp\left(-\frac{t^2}{2}\right) dt, \qquad (3.2)$$

$$Q(x) = P[N(0, 1) > x] = \int_{x}^{\infty} \frac{1}{\sqrt{2\pi}} \exp\left(-\frac{t^2}{2}\right) dt. \qquad (3.3)$$

See Figure 3.3 for a plot of these functions. By definition, $\Phi(x) + Q(x) = 1$. Furthermore, by the symmetry of the Gaussian density around zero, $Q(-x) = \Phi(x)$. Combining these observations, we note that $Q(-x) = 1 - Q(x)$, so that

Figure 3.3 The Φ and Q functions.

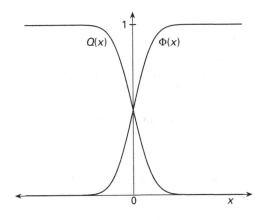

it suffices to consider only positive arguments for the Q function in order to compute probabilities of interest.

> **Example 3.1.1** X is a Gaussian random variable with mean $m = -3$ and variance $v^2 = 4$. Find expressions in terms of the Q function with positive arguments for the following probabilities: $P[X > 5]$, $P[X < -1]$, $P[1 < X < 4]$, $P[X^2 + X > 2]$.
>
> *Solution* We solve this problem by normalizing X to a standard Gaussian random variable $X - mv = X + 3/2$:
>
> $$P[X > 5] = P\left[\frac{X+3}{2} > \frac{5+3}{2} = 4\right] = Q(4),$$
>
> $$P[X < -1] = P\left[\frac{X+3}{2} < \frac{-1+3}{2} = 1\right] = \Phi(1) = 1 - Q(1),$$
>
> $$P[1 < X < 4] = P\left[2 = \frac{1+3}{2} < \frac{X+3}{2} < \frac{4+3}{2} = 3.5\right] = \Phi(3.5) - \Phi(2)$$
>
> $$= Q(2) - Q(3.5).$$
>
> Computation of the last probability needs a little more work to characterize the event of interest in terms of simpler events:
>
> $$P[X^2 + X > 2] = P[X^2 + X - 2 > 0] = P[(X+2)(X-1) > 0].$$
>
> The factorization shows that $X^2 + X > 2$ if and only if $X + 2 > 0$ and $X - 1 > 0$, or $X + 2 < 0$ and $X - 1 < 0$. This simplifies to the disjoint union (i.e., "or") of the mutually exclusive events $X > 1$ and $X < -2$. We therefore obtain
>
> $$P[X^2 + X > 2] = P[X > 1] + P[X < -2] = Q\left(\frac{1+3}{2}\right) + \Phi\left(\frac{-2+3}{2}\right)$$
>
> $$= Q(2) + \Phi\left(\frac{1}{2}\right) = Q(2) + 1 - Q\left(\frac{1}{2}\right).$$
>
> The Q function is ubiquitous in communication systems design, hence it is worth exploring its properties in some detail. The following bounds on the Q function are derived in Problem 3.3.

Bounds on $Q(x)$ for large arguments

$$\left(1 - \frac{1}{x^2}\right)\frac{e^{-x^2/2}}{x\sqrt{2\pi}} \leq Q(x) \leq \frac{e^{-x^2/2}}{x\sqrt{2\pi}}, \qquad x \geq 0. \tag{3.4}$$

These bounds are tight (the upper and lower bounds converge) for large values of x.

3.1 Gaussian basics

Upper bound on $Q(x)$ useful for small arguments and for analysis

$$Q(x) \leq \frac{1}{2}e^{-x^2/2}, \quad x \geq 0. \tag{3.5}$$

This bound is tight for small x, and gives the correct exponent of decay for large x. It is also useful for simplifying expressions involving a large number of Q functions, as we see when we derive transfer function bounds for the performance of optimal channel equalization and decoding in Chapters 5 and 7, respectively.

Figure 3.4 plots $Q(x)$ and its bounds for positive x. A logarithmic scale is used for the values of the function to demonstrate the rapid decay with x. The bounds (3.4) are seen to be tight even at moderate values of x (say $x \geq 2$).

Notation for asymptotic equivalence Since we are often concerned with exponential rates of decay (e.g., as SNR gets large), it is useful to introduce the notation $P \doteq Q$ (as we take some limit), which means that $\log P / \log Q \to 1$. An analogous notation $p \sim q$ denotes, on the other hand, that $p/q \to 1$. Thus, $P \doteq Q$ and $\log P \sim \log Q$ are two equivalent ways of expressing the same relationship.

Asymptotics of $Q(x)$ for large arguments For large $x > 0$, the exponential decay of the Q function dominates. We denote this by

$$Q(x) \doteq e^{-x^2/2}, \quad x \to \infty, \tag{3.6}$$

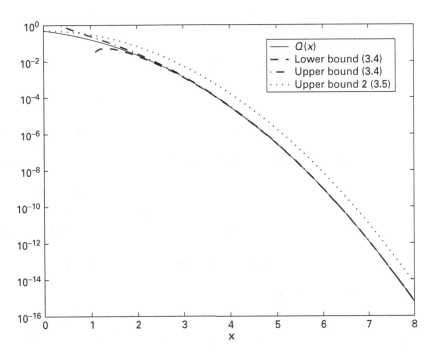

Figure 3.4 The Q function and bounds.

which is shorthand for the following limiting result:

$$\lim_{x \to \infty} \frac{\log Q(x)}{-x^2/2} = 1. \tag{3.7}$$

This can be proved by application of the upper and lower bounds in (3.4). The asymptotics of the Q function play a key role in design of communication systems. Events that cause bit errors have probabilities involving terms such as $Q(\sqrt{a\ \text{SNR}}) \doteq e^{-a\ \text{SNR}/2}$ as a function of the signal-to-noise ratio (SNR). When there are several events that can cause bit errors, the ones with the smallest rates of decay a dominate performance, and we often focus on these worst-case events in our designs for moderate and high SNR. This simplistic view does not quite hold in heavily coded systems operating at low SNR, but is still an excellent perspective for arriving at a coarse link design.

Often, we need to deal with multiple Gaussian random variables defined on the same probability space. These might arise, for example, when we sample filtered WGN. In many situations of interest, not only are such random variables individually Gaussian, but they satisfy a stronger *joint* Gaussianity property. Before discussing joint Gaussianity, however, we review mean and covariance for arbitrary random variables defined on the same probability space.

Mean vector and covariance matrix   Consider an arbitrary m-dimensional random vector $\mathbf{X} = (X_1, \ldots, X_m)^T$. The $m \times 1$ mean vector of $\mathbf{X}$ is defined as $\mathbf{m}_X = \mathbb{E}[\mathbf{X}] = (\mathbb{E}[X_1], \ldots, \mathbb{E}[X_m])^T$. The $m \times m$ covariance matrix $\mathbf{C}_X$ has its (i, j)th entry given by

$$\mathbf{C}_X(i, j) = \text{cov}(X_i, X_j) = \mathbb{E}[(X_i - \mathbb{E}[X_i])(X_j - \mathbb{E}[X_j])]$$
$$= \mathbb{E}[X_i X_j] - \mathbb{E}[X_i]\mathbb{E}[X_j].$$

More compactly,

$$\mathbf{C}_X = \mathbb{E}[(\mathbf{X} - \mathbb{E}[\mathbf{X}])(\mathbf{X} - \mathbb{E}[\mathbf{X}])^T] = \mathbb{E}[\mathbf{X}\mathbf{X}^T] - \mathbb{E}[\mathbf{X}](\mathbb{E}[\mathbf{X}])^T.$$

Some properties of covariance matrices are explored in Problem 3.31.

Variance   Variance is the covariance of a random variable with itself.

$$\text{var}(X) = \text{cov}(X, X).$$

We can also define a normalized version of covariance, as a scale-independent measure of the correlation between two random variables.

Correlation coefficient   The correlation coefficient $\rho(X_1, X_2)$ between random variables X_1 and X_2 is defined as the following normalized version of their covariance:

$$\rho(X_1, X_2) = \frac{\text{cov}(X_1, X_2)}{\sqrt{\text{var}(X_1)\text{var}(X_2)}}.$$

3.1 Gaussian basics

Using the Cauchy–Schwartz inequality for random variables, it can be shown that $|\rho(X_1, X_2)| \leq 1$, with equality if and only if $X_2 = aX_1 + b$ with probability one, for some constants a, b.

Notes on covariance computation Computations of variance and covariance come up often when we deal with Gaussian random variables, hence it is useful to note the following properties of covariance.

Property 1 Covariance is unaffected by adding constants.

$$\text{cov}(X+a, Y+b) = \text{cov}(X, Y) \quad \text{for any constants } a, b.$$

Covariance provides a measure of the correlation between random variables after subtracting out their means, hence adding constants to the random variables (which changes their means) does not affect covariance.

Property 2 Covariance is a bilinear function.

$$\text{cov}(a_1 X_1 + a_2 X_2, a_3 X_3 + a_4 X_4) = a_1 a_3 \text{cov}(X_1, X_3) + a_1 a_4 \text{cov}(X_1, X_4) \\ + a_2 a_3 \text{cov}(X_2, X_3) + a_2 a_4 \text{cov}(X_2, X_4).$$

By Property 1, it is clear that we can always consider zero mean versions of random variables when computing the covariance. An example that frequently arises in performance analysis of communication systems is a random variable which is a sum of a deterministic term (e.g., due to a signal), and a zero mean random term (e.g., due to noise). In this case, dropping the signal term is often convenient when computing variance or covariance.

Mean and covariance evolution under affine transformations Consider an $m \times 1$ random vector $\mathbf{X}$ with mean vector $\mathbf{m}_X$ and covariance matrix $\mathbf{C}_X$. Define $\mathbf{Y} = \mathbf{AX} + \mathbf{b}$, where $\mathbf{A}$ is an $n \times m$ matrix, and $\mathbf{b}$ is an $n \times 1$ vector. Then the random vector $\mathbf{Y}$ has mean vector $\mathbf{m}_Y = \mathbf{Am}_X + \mathbf{b}$ and covariance matrix $\mathbf{C}_Y = \mathbf{AC}_X \mathbf{A}^T$. To see this, first compute the mean vector of $\mathbf{Y}$ using the linearity of the expectation operator:

$$\mathbf{m}_Y = \mathbb{E}[\mathbf{Y}] = \mathbb{E}[\mathbf{AX} + \mathbf{b}] = \mathbf{A}\mathbb{E}[\mathbf{X}] + \mathbf{b} = \mathbf{Am}_X + \mathbf{b}. \tag{3.8}$$

This also implies that the "zero mean" version of $\mathbf{Y}$ is given by

$$\mathbf{Y} - \mathbb{E}[\mathbf{Y}] = (\mathbf{AX} + \mathbf{b}) - (\mathbf{Am}_X + \mathbf{b}) = \mathbf{A}(\mathbf{X} - \mathbf{m}_X),$$

so that the covariance matrix of $\mathbf{Y}$ is given by

$$\mathbf{C}_Y = \mathbb{E}[(\mathbf{Y} - \mathbb{E}[\mathbf{Y}])(\mathbf{Y} - \mathbb{E}[\mathbf{Y}])^T] = \mathbb{E}[\mathbf{A}(\mathbf{X} - \mathbf{m}_X)(\mathbf{X} - \mathbf{m}_X)^T \mathbf{A}^T] = \mathbf{AC}_X \mathbf{A}^T. \tag{3.9}$$

Mean and covariance evolve separately under affine transformations The mean of $\mathbf{Y}$ depends only on the mean of $\mathbf{X}$, and the covariance of $\mathbf{Y}$ depends only on the covariance of $\mathbf{X}$. Furthermore, the additive constant $\mathbf{b}$ in the transformation does not affect the covariance, since it influences only the mean of $\mathbf{Y}$.

Jointly Gaussian random variables, or Gaussian random vectors Random variables $X_1, \ldots, X_m$ defined on a common probability space are said to be *jointly Gaussian*, or the $m \times 1$ random vector $\mathbf{X} = (X_1, \ldots, X_m)^T$ is termed a *Gaussian random vector*, if any linear combination of these random variables is a Gaussian random variable. That is, for any scalar constants $a_1, \ldots, a_m$, the random variable $a_1 X_1 + \cdots + a_m X_m$ is Gaussian.

A Gaussian random vector is completely characterized by its mean vector and covariance matrix The definition of joint Gaussianity only requires us to characterize the distribution of an arbitrarily chosen linear combination of $X_1, \ldots, X_m$. For a Gaussian random vector $\mathbf{X} = (X_1, \ldots, X_m)^T$, consider $Y = a_1 X_1 + \cdots + a_m X_m$, where $a_1, \ldots, a_m$ can be any scalar constants. By definition, Y is a Gaussian random variable, and is completely characterized by its mean and variance. We can compute these in terms of $\mathbf{m}_X$ and $\mathbf{C}_X$ using (3.8) and (3.9) by noting that $Y = \mathbf{a}^T \mathbf{X}$, where $\mathbf{a} = (a_1, \ldots, a_m)^T$. Thus,

$$m_Y = \mathbf{a}^T \mathbf{m}_X,$$

$$C_Y = \text{var}(Y) = \mathbf{a}^T \mathbf{C}_X \mathbf{a}.$$

We have, therefore, shown that we can characterize the mean and variance, and hence the density, of an arbitrarily chosen linear combination Y if and only if we know the mean vector $\mathbf{m}_X$ and covariance matrix $\mathbf{C}_X$. This implies the desired result that the distribution of Gaussian random vector $\mathbf{X}$ is completely characterized by $\mathbf{m}_X$ and $\mathbf{C}_X$.

Notation for joint Gaussianity We use the notation $\mathbf{X} \sim N(\mathbf{m}, \mathbf{C})$ to denote a Gaussian random vector $\mathbf{X}$ with mean vector $\mathbf{m}$ and covariance matrix $\mathbf{C}$.

The preceding definitions and observations regarding joint Gaussianity apply even when the random variables involved do not have a joint density. For example, it is easy to check that, according to this definition, X_1 and $X_2 = 2X_1 - 3$ are jointly Gaussian. However, the joint density of X_1 and X_2 is not well defined (unless we allow delta functions), since all of the probability mass in the two-dimensional (x_1, x_2) plane is collapsed onto the line $x_2 = 2x_1 - 3$. Of course, since X_2 is completely determined by X_1, any probability involving X_1, X_2 can be expressed in terms of X_1 alone. In general, when the m-dimensional joint density does not exist, probabilities involving $X_1, \ldots, X_m$ can be expressed in terms of a smaller number of random variables, and can be evaluated using a joint density over a lower-dimensional space. A simple necessary and sufficient condition for the joint density to exist is as follows:

Joint Gaussian density exists if and only if the covariance matrix is invertible The proof of this result is sketched in Problem 3.32.

3.1 Gaussian basics

Joint Gaussian density For $\mathbf{X} = (X_1, \ldots, X_m) \sim N(\mathbf{m}, \mathbf{C})$, if $\mathbf{C}$ is invertible, the joint density exists and takes the following form:

$$p(x_1, \ldots, x_m) = p(\mathbf{x}) = \frac{1}{\sqrt{(2\pi)^m |\mathbf{C}|}} \exp\left(-\frac{1}{2}(\mathbf{x}-\mathbf{m})^T \mathbf{C}^{-1}(\mathbf{x}-\mathbf{m})\right). \tag{3.10}$$

In Problem 3.32, we derive the joint density above, starting from the definition that any linear combination of jointly Gaussian random variables is a Gaussian random variable.

Uncorrelatedness X_1 and X_2 are said to be uncorrelated if $\text{cov}(X_1, X_2) = 0$.

Independent random variables are uncorrelated If X_1 and X_2 are independent, then

$$\text{cov}(X_1, X_2) = \mathbb{E}[X_1 X_2] - \mathbb{E}[X_1]\mathbb{E}[X_2] = \mathbb{E}[X_1]\mathbb{E}[X_2] - \mathbb{E}[X_1]\mathbb{E}[X_2] = 0.$$

The converse is not true in general, but does hold when the random variables are jointly Gaussian.

Uncorrelated jointly Gaussian random variables are independent This follows from the form of the joint Gaussian density (3.10). If $X_1, \ldots, X_m$ are pairwise uncorrelated and joint Gaussian, then the covariance matrix $\mathbf{C}$ is diagonal, and the joint density decomposes into a product of marginal densities.

Example 3.1.2 (Variance of a sum of random variables) For random variables $X_1, \ldots, X_m$,

$$\text{var}(X_1 + \cdots + X_m) = \text{cov}(X_1 + \cdots + X_m, X_1 + \cdots + X_m)$$

$$= \sum_{i=1}^{m} \sum_{j=1}^{m} \text{cov}(X_i, X_j)$$

$$= \sum_{i=1}^{m} \text{var}(X_i) + \sum_{\substack{i,j=1 \\ i \neq j}}^{m} \text{cov}(X_i, X_j).$$

Thus, for uncorrelated random variables, the variance of the sum equals the sum of the variances:

$$\text{var}(X_1 + \cdots + X_m) = \text{var}(X_1) + \cdots + \text{var}(X_m) \quad \text{for uncorrelated random variables.}$$

We now characterize the distribution of affine transformations of jointly Gaussian random variables.

Joint Gaussianity is preserved under affine transformations If **X** above is a Gaussian random vector, then $\mathbf{Y} = \mathbf{AX} + \mathbf{b}$ is also Gaussian. To see this, note that any linear combination of $Y_1, \ldots, Y_n$ equals a linear combination of $X_1, \ldots, X_m$ (plus a constant), which is a Gaussian random variable by the Gaussianity of **X**. Since **Y** is Gaussian, its distribution is completely characterized by its mean vector and covariance matrix, which we have just computed. We can now state the following result:

If $\mathbf{X} \sim N(\mathbf{m}, \mathbf{C})$, then

$$\mathbf{AX} + \mathbf{b} \sim N(\mathbf{Am} + \mathbf{b}, \mathbf{A}^T \mathbf{C} \mathbf{A}). \tag{3.11}$$

Example 3.1.3 (Computations with jointly Gaussian random variables)
The random variables X_1 and X_2 are jointly Gaussian, with $\mathbb{E}[X_1] = 1$, $\mathbb{E}[X_2] = -2$, $\text{var}(X_1) = 4$, $\text{var}(X_2) = 1$, and correlation coefficient $\rho(X_1, X_2) = -1$.

(a) Write down the mean vector and covariance matrix for the random vector $\mathbf{X} = (X_1, X_2)^T$.
(b) Evaluate the probability $P[2X_1 - 3X_2 < 6]$ in terms of the Q function with positive arguments.
(c) Suppose that $Z = X_1 - aX_2$. Find the constant a such that Z is independent of X_1.

Let us solve this problem in detail in order to provide a concrete illustration of the properties we have discussed.

Solution to (a) The mean vector is given by

$$\mathbf{m}_X = \begin{pmatrix} \mathbb{E}[X_1] \\ \mathbb{E}[X_2] \end{pmatrix} = \begin{pmatrix} 1 \\ -2 \end{pmatrix}.$$

We know the diagonal entries of the covariance matrix, which are simply the variances of X_1 and X_2. The cross terms

$$\mathbf{C}_X(1,2) = \mathbf{C}_X(2,1) = \rho(X_1, X_2)\sqrt{\text{var}(X_1)\text{var}(X_2)} = -1\sqrt{4} = -2,$$

so that

$$\mathbf{C}_X = \begin{pmatrix} 4 & -2 \\ -2 & 1 \end{pmatrix}.$$

Solution to (b) The random variable $Y = 2X_1 - 3X_2$ is Gaussian, by the joint Gaussianity of X_1 and X_2. To compute the desired probability, we need to compute

$$\mathbb{E}[Y] = \mathbb{E}[2X_1 - 3X_2] = 2\mathbb{E}[X_1] - 3\mathbb{E}[X_2] = 2(1) - 3(-2) = 8;$$

$$\begin{aligned}\text{var}(Y) &= \text{cov}(Y, Y) = \text{cov}(2X_1 - 3X_2, 2X_1 - 3X_2) \\ &= 4\,\text{cov}(X_1, X_1) - 6\,\text{cov}(X_1, X_2) \\ &\quad - 6\,\text{cov}(X_2, X_1) + 9\,\text{cov}(X_2, X_2) \\ &= 4(4) - 6(-2) - 6(-2) + 9(1) = 49.\end{aligned}$$

Thus,

$$P[2X_1 - 3X_2 < 6] = P[Y < 6] = \Phi\left(\frac{6-8}{\sqrt{49}}\right) = \Phi\left(-\frac{2}{7}\right) = Q\left(\frac{2}{7}\right).$$

When using software such as MATLAB, which is good at handling vectors and matrices, it is convenient to use vector-based computations. To do this, we note that $Y = \mathbf{A}\mathbf{X}$, where $\mathbf{A} = (2, -3)$ is a row vector, and apply (3.11) to conclude that

$$\mathbb{E}[Y] = \mathbf{A}\mathbf{m}_X = (2\ \ -3)\binom{1}{-2} = 8$$

and

$$\text{var}(Y) = \text{cov}(Y, Y) = \mathbf{A}^T \mathbf{C}_X \mathbf{A} = \binom{2}{-3}\begin{pmatrix}4 & -2\\ -2 & 1\end{pmatrix}(2\ \ -3) = 49.$$

Solution to (c) Since $Z = X_1 - aX_2$ and X_1 are jointly Gaussian (why?), they are independent if they are uncorrelated. The covariance is given by

$$\text{cov}(Z, X_1) = \text{cov}(X_1 - aX_2, X_1) = \text{cov}(X_1, X_1) - a\,\text{cov}(X_2, X_1) = 4 + 2a,$$

so that we need $a = -2$ for Z and X_1 to be independent.

We are now ready to move on to Gaussian random processes, which are just generalizations of Gaussian random vectors to an arbitrary number of components (countable or uncountable).

Gaussian random process A random process $X = \{X(t), t \in T\}$ is said to be Gaussian if any linear combination of samples is a Gaussian random variable. That is, for any number n of samples, any sampling times $t_1, \ldots, t_n$, and any scalar constants $a_1, \ldots, a_n$, the linear combination $a_1 X(t_1) + \cdots + a_n X(t_n)$ is a Gaussian random variable. Equivalently, the samples $X(t_1), \ldots, X(t_n)$ are jointly Gaussian.

A linear combination of samples from a Gaussian random process is completely characterized by its mean and variance. To compute the latter quantities for an arbitrary linear combination, we can show, as we did for random vectors, that all we need to know are the mean function and the autocovariance function of the random process. These functions therefore provide a complete statistical characterization of a Gaussian random process, since the definition of a Gaussian random process requires only that we be able to characterize the distribution of an arbitrary linear combination of samples.

Characterizing a Gaussian random process The statistics of a Gaussian random process are completely specified by its mean function $m_X(t) = \mathbb{E}[X(t)]$ and its autocovariance function $C_X(t_1, t_2) = \mathbb{E}[X(t_1)X(t_2)]$. Since the autocorrelation function $R_X(t_1, t_2)$ can be computed from $C_X(t_1, t_2)$, and vice versa, given the mean function $m_X(t)$, it also follows that a Gaussian random process is completely specified by its mean and autocorrelation functions.

Wide sense stationary Gaussian random processes are stationary We know that a stationary random process is WSS. The converse is not true in general, but Gaussian WSS processes are indeed stationary. This is because the statistics of a Gaussian random process are characterized by its first and second order statistics, and if these are shift invariant (as they are for WSS processes), the random process is statistically indistinguishable under a time shift.

As in the previous chapter, we use the notation $R_X(\tau)$ and $C_X(\tau)$ to denote the autocorrelation and autocovariance functions, respectively, for a WSS process. The PSD $S_X(f) = \mathcal{F}(R_X)$. We are now ready to define WGN.

White Gaussian noise Real-valued WGN $n(t)$ is a zero mean, WSS, Gaussian random process with $S_n(f) \equiv N_0/2 = \sigma^2$. Equivalently, $R_n(\tau) = \frac{N_0}{2}\delta(\tau) = \sigma^2\delta(\tau)$. The quantity $N_0/2 = \sigma^2$ is often termed the two-sided PSD of WGN, since we must integrate over both positive and negative frequencies in order to compute power using this PSD. The quantity N_0 is therefore referred to as the one-sided PSD, and has the dimension of watt/hertz, or joules. Complex-valued WGN has real and imaginary components modeled as i.i.d. real WGN processes, and has two-sided PSD N_0 which is the sum of the two-sided PSDs of its components. Figure 3.5 shows the role played by WGN in modeling receiver noise in bandlimited systems.

WGN as model for receiver noise in bandlimited systems White Gaussian noise has infinite power, whereas receiver noise power in any practical system is always finite. However, since receiver processing always involves some form of bandlimiting, it is convenient to assume that the input to the system is infinite-power WGN. After filtering, the noise statistics obtained with this simplified description are the same as those obtained by bandlimiting the noise upfront. Figure 3.5 shows that real-valued WGN can serve as a model for bandlimited receiver noise in a passband system, as well as for each of the I and Q noise components after downconversion. It can also model the receiver noise in a physical baseband system, which is analogous to using

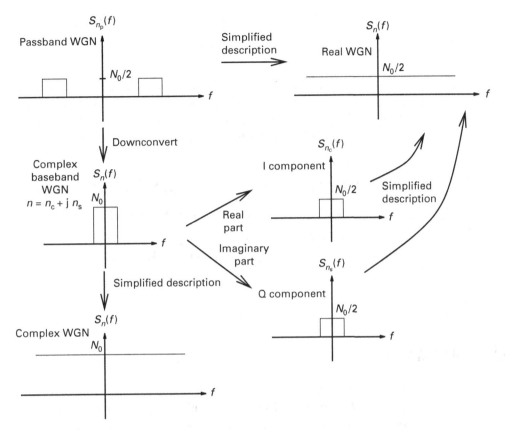

Figure 3.5 Since receiver processing always involves some form of band limitation, it is not necessary to impose band limitation on the WGN model. Real-valued infinite-power WGN provides a simplified description for both passband WGN, and for each of the I and Q components for complex baseband WGN. Complex-valued infinite-power WGN provides a simplified description for bandlimited complex baseband WGN.

only the I component in a passband system. Complex-valued WGN, on the other hand, models the complex envelope of passband WGN. Its PSD is double that of real-valued WGN because the PSDs of the real and imaginary parts of the noise, modeled as i.i.d. real-valued WGN, add up. The PSD is also double that of the noise model for passband noise; this is consistent with the relations developed in Chapter 2 between the PSD of a passband random process and its complex envelope.

Numerical value of noise PSD For an ideal receiver at room temperature, we have

$$N_0 = kT_0,$$

where $k = 1.38 \times 10^{-23}$ joule/kelvin is Boltzmann's constant, and T_0 is a reference temperature, usually set to $290\,\text{K}$ ("room temperature") by convention.

A receiver with a noise figure of F dB has a higher noise PSD, given by

$$N_0 = kT 10^{F/10}.$$

> **Example 3.1.4 (Noise power computation)** A 5 GHz wireless local area network (WLAN) link has a receiver bandwidth B of 20 MHz. If the receiver has a noise figure of 6 dB, what is the receiver noise power P_n?
>
> *Solution* The noise power
>
> $$P_n = N_0 B = kT_0 10^{F/10} B = (1.38 \times 10^{-23})(290)(10^{6/10})(20 \times 10^6)$$
> $$= 3.2 \times 10^{-13} \text{ watt} = 3.2 \times 10^{-10} \text{ milliwatts (mW)}.$$
>
> The noise power is often expressed in dBm, which is obtained by converting the raw number in milliwatts (mW) into dB. We therefore get
>
> $$P_{n,\text{dBm}} = 10 \log_{10} P_n(\text{mW}) = -95 \, \text{dBm}.$$

3.2 Hypothesis testing basics

Hypothesis testing is a framework for deciding which of M possible hypotheses, $H_1, \ldots, H_M$, "best" explains an observation Y. We assume that the observation Y takes values in a finite-dimensional observation space Γ; that is, Y is a scalar or vector. (It is possible to consider a more general observation space Γ, but that is not necessary for our purpose.) The observation is related to the hypotheses using a statistical model: given the hypothesis H_i, the conditional density of the observation, $p(y|i)$, is known, for $i = 1, \ldots, M$. In Bayesian hypothesis testing, the prior probabilities for the hypotheses, $\pi(i) = P[H_i]$, $i = 1, \ldots, M$, are known ($\sum_{i=1}^M \pi(i) = 1$). We often (but not always) consider the special case of equal priors, which corresponds to $\pi(i) = 1/M$ for all $i = 1, \ldots, M$.

> **Example 3.2.1 (Basic Gaussian example)** Consider binary hypothesis testing, in which H_0 corresponds to 0 being sent, H_1 corresponds to 1 being sent, and Y is a scalar decision statistic (e.g., generated by sampling the output of a receive filter or an equalizer). The conditional distributions for the observation given the hypotheses are $H_0 : Y \sim N(0, v^2)$ and $H_1 : Y \sim N(m, v^2)$, so that
>
> $$p(y|0) = \frac{\exp\left(-\frac{y^2}{2v^2}\right)}{\sqrt{2\pi v^2}}; \quad p(y|1) = \frac{\exp\left(-\frac{(y-m)^2}{2v^2}\right)}{\sqrt{2\pi v^2}}. \quad (3.12)$$

3.2 Hypothesis testing basics

Decision rule A decision rule $\delta : \Gamma \to \{1, \ldots, M\}$ is a mapping from the observation space to the set of hypotheses. Alternatively, a decision rule can be described in terms of a partition of the observation space Γ into disjoint decision regions $\{\Gamma_i, i = 1, \ldots, M\}$, where

$$\Gamma_i = \{y \in \Gamma : \delta(y) = i\}.$$

That is, when $y \in \Gamma_i$, the decision rule says that H_i is true.

Example 3.2.2 A "sensible" decision rule for the basic Gaussian example (assuming that $m > 0$) is

$$\delta(y) = \begin{cases} 1, & y > \frac{m}{2}, \\ 0, & y \leq \frac{m}{2}. \end{cases} \quad (3.13)$$

This corresponds to the decision regions $\Gamma_1 = (\frac{m}{2}, \infty)$, and $\Gamma_0 = (-\infty, \frac{m}{2})$.

The conditional densities and the "sensible" rule for the basic Gaussian example are illustrated in Figure 3.6.

We would like to quantify our intuition that the preceding sensible rule, which splits the difference between the means under the two hypotheses, is a good one. Indeed, this rule need not always be the best choice: for example, if we knew for sure that 0 was sent, then clearly a better rule is to say that H_0 is true, regardless of the observation. Thus, a systematic framework is needed to devise good decision rules, and the first step toward doing this is to define

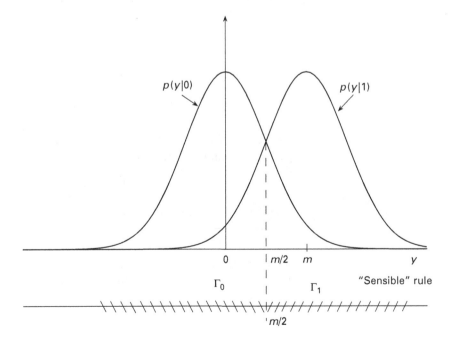

Figure 3.6 The conditional densities and "sensible" decision rule for the basic Gaussian example.

criteria for evaluating the goodness of a decision rule. Central to such criteria is the notion of conditional error probability, defined as follows.

Conditional error probability For an M-ary hypothesis testing problem, the conditional error probability, conditioned on H_i, for a decision rule δ is defined as

$$P_{e|i} = P[\text{say } H_j \text{ for some } j \neq i | H_i \text{ is true}] = \sum_{j \neq i} P[Y \in \Gamma_j | H_i]$$
$$= 1 - P[Y \in \Gamma_i | H_i], \quad (3.14)$$

where we have used the equivalent specification of the decision rule in terms of the decision regions it defines. We denote by $P_{c|i} = P[Y \in \Gamma_i | H_i]$, the conditional probability of correct decision, given H_i.

If the prior probabilities are known, then we can define the (average) error probability as

$$P_e = \sum_{i=1}^{M} \pi(i) P_{e|i}. \quad (3.15)$$

Similarly, the average probability of a correct decision is given by

$$P_c = \sum_{i=1}^{M} \pi(i) P_{c|i} = 1 - P_e. \quad (3.16)$$

Example 3.2.3 The conditional error probabilities for the "sensible" decision rule (3.13) for the basic Gaussian example (Example 3.2.2) are

$$P_{e|0} = P\left[Y > \frac{m}{2} \Big| H_0\right] = Q\left(\frac{m}{2v}\right),$$

since $Y \sim N(0, v^2)$ under H_0, and

$$P_{e|1} = P\left[Y \leq \frac{m}{2} \Big| H_1\right] = \Phi\left(\frac{\frac{m}{2} - m}{v}\right) = Q\left(\frac{m}{2v}\right),$$

since $Y \sim N(m, v^2)$ under H_1. Furthermore, since $P_{e|1} = P_{e|0}$, the average error probability is also given by

$$P_e = Q\left(\frac{m}{2v}\right),$$

regardless of the prior probabilities.

Notation Let us denote by "arg max" the argument of the maximum. That is, for a function $f(x)$ with maximum occurring at x_0, we have

$$\max_x f(x) = f(x_0), \quad \arg\max_x f(x) = x_0.$$

3.2 Hypothesis testing basics

Maximum likelihood decision rule The maximum likelihood (ML) decision rule is defined as

$$\delta_{ML}(y) = \arg \max_{1 \leq i \leq M} p(y|i) = \arg \max_{1 \leq i \leq M} \log p(y|i). \quad (3.17)$$

The ML rule chooses the hypothesis for which the conditional density of the observation is maximized. In rather general settings, it can be proven to be asymptotically optimal as the quality of the observation improves (e.g., as the number of samples gets large, or the signal-to-noise ratio gets large). It can be checked that the sensible rule in Example 3.2.2 is the ML rule for the basic Gaussian example.

Another popular decision rule is the minimum probability of error (MPE) rule, which seeks to minimize the average probability of error. It is assumed that the prior probabilities $\{\pi(i)\}$ are known. We now derive the form of the MPE decision rule.

Derivation of MPE rule Consider the equivalent problem of maximizing the probability of a correct decision. For a decision rule δ corresponding to decision regions $\{\Gamma_i\}$, the conditional probabilities of making a correct decision are given by

$$P_{c|i} = \int_{\Gamma_i} p(y|i) dy, \quad i = 1, \ldots, M$$

and the average probability of a correct decision is given by

$$P_c = \sum_{i=1}^{M} \pi(i) P_{c|i} = \sum_{i=1}^{M} \pi(i) \int_{\Gamma_i} p(y|i) dy.$$

Now, pick a point $y \in \Gamma$. If we see $Y = y$ and decide H_i (i.e., $y \in \Gamma_i$), the contribution to the integrand in the expression for P_c is $\pi(i) p(y|i)$. Thus, to maximize the contribution to P_c for that potential observation value y, we should put $y \in \Gamma_i$ such that $\pi(i) p(y|i)$ is the largest. Doing this for each possible y leads to the MPE decision rule. We summarize and state this as a theorem below.

Theorem 3.2.1 (MPE decision rule) *For M-ary hypothesis testing, the MPE rule is given by*

$$\delta_{\text{MPE}}(y) = \arg \max_{1 \leq i \leq M} \pi(i) p(y|i) = \arg \max_{1 \leq i \leq M} \log \pi(i) + \log p(y|i). \quad (3.18)$$

A number of important observations related to the characterization of the MPE rule are now stated below.

Remark 3.2.1 (MPE rule maximizes posterior probabilities) By Bayes' rule, the conditional probability of hypothesis H_i given the observation is $Y = y$ is given by

$$P(H_i|y) = \frac{\pi(i) p(y|i)}{p(y)},$$

where $p(y)$ is the unconditional density of Y, given by $p(y) = \sum_j \pi(j) p(y|j)$. The MPE rule (3.18) is therefore equivalent to the maximum a posteriori probability (MAP) rule, as follows:

$$\delta_{\text{MAP}}(y) = \arg \max_{1 \leq i \leq M} P(H_i|y). \qquad (3.19)$$

This has a nice intuitive interpretation: the error probability is minimized by choosing the hypothesis that is most likely, given the observation.

Remark 3.2.2 (ML rule is MPE for equal priors) By setting $\pi(i) = 1/M$ in the MPE rule (3.18), we see that it specializes to the ML rule (3.17). For example, the rule in Example 3.2.2 minimizes the error probability in the basic Gaussian example, if 0 and 1 are equally likely to be sent. While the ML rule minimizes the error probability for equal priors, it may also be used as a matter of convenience when the hypotheses are not equally likely.

We now introduce the notion of a *likelihood ratio*, a fundamental notion in hypothesis testing.

Likelihood ratio test for binary hypothesis testing For binary hypothesis testing, the MPE rule specializes to

$$\delta_{\text{MPE}}(y) = \begin{cases} 1, & \pi(1) p(y|1) > \pi(0) p(y|0), \\ 0, & \pi(1) p(y|1) < \pi(0) p(y|0), \\ \text{don't care}, & \pi(1) p(y|1) = \pi(0) p(y|0), \end{cases} \qquad (3.20)$$

which can be rewritten as

$$L(y) = \frac{p(y|1)}{p(y|0)} \underset{H_0}{\overset{H_1}{\gtrless}} \frac{\pi(0)}{\pi(1)}, \qquad (3.21)$$

where $L(y)$ is called the **likelihood ratio** (**LR**). A test that compares the likelihood ratio with a threshold is called a **likelihood ratio test** (**LRT**). We have just shown that the MPE rule is an LRT with threshold $\pi(0)/\pi(1)$. Similarly, the ML rule is an LRT with threshold one. Often, it is convenient (and equivalent) to employ the log likelihood ratio test (LLRT), which consists of comparing $\log L(y)$ with a threshold.

Example 3.2.4 (Likelihood ratio for the basic Gaussian example) Substituting (3.12) into (3.21), we obtain the likelihood ratio for the basic Gaussian example as

$$L(y) = \exp\left(\frac{1}{v^2}\left(my - \frac{m^2}{2}\right)\right). \qquad (3.22)$$

We shall encounter likelihood ratios of similar form when considering the more complicated scenario of a continuous-time signal in WGN. Comparing $\log L(y)$ with zero gives the ML rule, which reduces to the decision rule (3.13) for $m > 0$. For $m < 0$, the inequalities in (3.13) are reversed.

Irrelevant statistics In many settings, the observation Y to be used for hypothesis testing is complicated to process. For example, over the AWGN channel to be considered in the next section, the observation is a continuous-time waveform. In such scenarios, it is useful to identify simpler decision statistics that we can use for hypothesis testing, without any loss in performance. To this end, we introduce the concept of irrelevance, which is used to derive optimal receivers for signaling over the AWGN channel in the next section. Suppose that we can decompose the observation into two components: $Y = (Y_1, Y_2)$. We say that Y_2 is *irrelevant* for the hypothesis testing problem if we can throw it away (i.e., use only Y_1 instead of Y) without any performance degradation.

As an example, consider binary hypothesis testing with observation (Y_1, Y_2) as follows:
$$\begin{aligned} H_1 &: Y_1 = m + N_1, \quad Y_2 = N_2, \\ H_0 &: Y_1 = N_1, \quad\quad\;\; Y_2 = N_2, \end{aligned} \quad (3.23)$$

where $N_1 \sim N(0, v^2)$, $N_2 \sim N(0, v^2)$ are jointly Gaussian "noise" random variables. Note that only Y_1 contains the "signal" component m. However, does this automatically imply that the component Y_2, which contains only noise, is irrelevant? Intuitively, we feel that if N_2 is independent of N_1, then Y_2 will carry no information relevant to the decision. On the other hand, if N_2 is highly correlated with N_1, then Y_2 contains valuable information that we could exploit. As an extreme example, if $N_2 \equiv N_1$, then we could obtain perfect detection by constructing a noiseless observation $\hat{Y} = Y_1 - Y_2$, which takes value m under H_1 and value 0 under H_0. Thus, a systematic criterion for recognizing irrelevance is useful, and we provide this in the following theorem.

Theorem 3.2.2 (Characterizing an irrelevant statistic) *For M-ary hypothesis testing using an observation $Y = (Y_1, Y_2)$, the statistic Y_2 is irrelevant if the conditional distribution of Y_2, given Y_1 and H_i, is independent of i. In terms of densities, we can state the condition for irrelevance as $p(y_2|y_1, i) = p(y_2|y_1)$ for all i.*

Proof If $p(y_2|y_1, i)$ does not depend on i, then it is easy to see that $p(y_2|y_1, i) \equiv p(y_2|y_1)$ for all $i = 1, \ldots, M$. The statistical relationship between the observation Y and the hypotheses $\{H_i\}$ is through the conditional densities $\{p(y|i)\}$. We have

$$p(y|i) = p(y_1, y_2|i) = p(y_2|y_1, i)p(y_1|i) = p(y_2|y_1)p(y_1|i).$$

From the form of the MPE rule (3.18), we know that terms independent of i can be discarded, which means that we can restrict attention to the conditional densities $p(y_1|i)$ for the purpose of hypothesis testing. That is, Y_2 is irrelevant for hypothesis testing. □

> **Example 3.2.5 (Application of irrelevance criterion)** In (3.23), suppose that N_2 is independent of N_1. Then $Y_2 = N_2$ is independent of H_i and N_1, and hence of H_i and Y_1 and
>
> $$p(y_2|y_1, i) = p(y_2),$$
>
> which is a stronger version of the irrelevance condition in Theorem 3.2.2. In the next section, we use exactly this argument when deriving optimal receivers over AWGN channels.

We note in passing that the concept of *sufficient* statistic, which plays a key role in detection and estimation theory, is closely related to that of an irrelevant statistic. Consider a hypothesis testing problem with observation Y. Consider the augmented observation $\tilde{Y} = (Y_1 = f(Y), Y_2 = Y)$, where f is a function. Then $f(Y)$ is a sufficient statistic if $Y_2 = Y$ is irrelevant for hypothesis testing using $\tilde{Y}$. That is, once we know $Y_1 = f(Y)$, we have all the information we need to make our decision, and no longer need the original observation $Y_2 = Y$.

3.3 Signal space concepts

We are now ready to take the first step in deriving optimal receivers for M-ary signaling in AWGN. We restrict attention to real-valued signals and noise to start with (this model applies to passband and real baseband systems). Consider a communication system in which one of M continuous-time signals, $s_1(t), \ldots, s_M(t)$ is sent. The received signal equals the transmitted signal corrupted by AWGN. Of course, when we say "transmitted signal," we actually mean the noiseless copy produced by the coherent receiver of each possible transmitted signal, accounting for the effects of the channel.

In the language of hypothesis testing, we have M hypotheses for explaining the received signal, with

$$H_i : y(t) = s_i(t) + n(t), \quad i = 1, \ldots, M, \tag{3.24}$$

where $n(t)$ is WGN with PSD $\sigma^2 = N_0/2$. We show in this section that, without any loss of detection performance, we can reduce the continuous-time received signal to a finite-dimensional received vector.

3.3 Signal space concepts

A note on signal and noise scaling Even before we investigate this model in detail, we can make the following simple but important observation. If we scale the signal and the noise by the same factor, the performance of an optimal receiver remains the same (assuming that the receiver knows the scaling). Consider a scaled observation $\tilde{y}$ satisfying

$$H_i : \tilde{y}(t) = As_i(t) + An(t), \quad i = 1, \ldots, M. \tag{3.25}$$

We can now argue, without knowing anything about the structure of the optimal receiver, that the performance of optimal reception for models (3.24) and (3.25) is identical. An optimal receiver designed for model (3.24) provides exactly the same performance with model (3.25), by operating on $\tilde{y}/A$. Similarly, an optimal receiver designed for model (3.25) would provide exactly the same performance with model (3.25) by operating on $Ay(t)$. Hence, the performance of these two optimal receivers must be the same, otherwise we could improve the performance of one of the optimal receivers simply by scaling and using an optimal receiver for the scaled received signal. A consequence of this observation is that system performance is determined by the *ratio* of signal and noise strengths (in a sense to be made precise later), rather than individually on the signal and noise strengths. Therefore, when we discuss the structure of a given set of signals, our primary concern is with the relative geometry of the signal set, rather than with scale factors that are common to the entire signal set.

Next, we derive a fundamental property of WGN related to its distribution when linearly transformed. Any number obtained by linear processing of WGN can be expressed as the output of a correlation operation of the form

$$Z = \int_{-\infty}^{\infty} n(t)u(t)\mathrm{d}t = \langle n, u \rangle,$$

where $u(t)$ is a deterministic, finite-energy, signal. Since WGN is a Gaussian random process, we know that Z is a Gaussian random variable. To characterize its distribution, therefore, we need only compute its mean and variance. Since n has zero mean, the mean of Z is seen to be zero by the following simple computation:

$$\mathbb{E}[Z] = \int_{-\infty}^{\infty} \mathbb{E}[n(t)]u(t)\mathrm{d}t = 0,$$

where expectation and integral can be interchanged, both being linear operations. Instead of computing the variance of Z, however, we state a more general result below on covariance, from which the result on variance can be inferred. This result is important enough to state formally as a proposition.

Proposition 3.3.1 (WGN through correlators) *Let $u_1(t)$ and $u_2(t)$ denote finite-energy signals, and let $n(t)$ denote WGN with PSD $\sigma^2 = N_0/2$. Then $\langle n, u_1 \rangle$ and $\langle n, u_2 \rangle$ are jointly Gaussian with covariance*

$$\mathrm{cov}(\langle n, u_1 \rangle, \langle n, u_2 \rangle) = \sigma^2 \langle u_1, u_2 \rangle.$$

In particular, setting $u_1 = u_2 = u$, we obtain that

$$\text{var}(\langle n, u \rangle) = \text{cov}(\langle n, u \rangle, \langle n, u \rangle) = \sigma^2 \|u\|^2.$$

Proof of Proposition 3.3.1 The random variables $\langle n, u_1 \rangle$ and $\langle n, u_2 \rangle$ are zero mean and jointly Gaussian, since n is zero mean and Gaussian. Their covariance is computed as

$$\begin{aligned}
\text{cov}(\langle n, u_1 \rangle, \langle n, u_2 \rangle) &= \mathbb{E}[\langle n, u_1 \rangle \langle n, u_2 \rangle] = \mathbb{E}[\int n(t) u_1(t) dt \int n(s) u_2(s) ds] \\
&= \int \int u_1(t) u_2(s) \mathbb{E}[n(t) n(s)] dt\, ds \\
&= \int \int u_1(t) u_2(s) \sigma^2 \delta(t-s) dt\, ds \\
&= \sigma^2 \int u_1(t) u_2(t) dt = \sigma^2 \langle u_1, u_2 \rangle.
\end{aligned}$$

This completes the proof. □

The preceding result is simple but powerful, leading to the following geometric interpretation for white Gaussian noise.

Remark 3.3.1 (Geometric interpretation of WGN) Proposition 3.3.1 implies that the projection of WGN along any "direction" in the space of signals (i.e., the result of correlating WGN with a unit energy signal) has variance $\sigma^2 = N_0/2$. Also, its projections in orthogonal directions are jointly Gaussian and uncorrelated, and hence independent.

Armed with this geometric understanding of white Gaussian noise, we plan to argue as follows:

(1) The *signal space* spanned by the M possible received signals is finite-dimensional, of dimension at most M. There is no signal energy outside this signal space, regardless of which signal is transmitted.
(2) The component of WGN orthogonal to the signal space is independent of the component in the signal space, and its distribution does not depend on which signal was sent. It is therefore irrelevant to our hypothesis testing problem (it satisfies the condition of Theorem 3.2.2).
(3) We can therefore restrict attention to the signal and noise components lying in the signal space. These can be represented by finite-dimensional vectors, thus simplifying the problem immensely relative to our original problem of detection in continuous time.

Let us now flesh out the details of the preceding chain of reasoning. We begin by indicating how to construct a vector representation of the signal space. The signal space $\mathcal{S}$ is the finite-dimensional subspace (of dimension $n \leq M$) spanned by $s_1(t), \ldots, s_M(t)$. That is, $\mathcal{S}$ consists of all signals of the form $a_1 s_1(t) + \cdots + a_M s_M(t)$, where $a_1, \ldots, a_M$ are arbitrary scalars. Let

3.3 Signal space concepts

$\psi_1(t), \ldots, \psi_n(t)$ denote an orthonormal basis for S. Such a basis can be constructed systematically by Gramm–Schmidt orthogonalization (described below) of the set of signals $s_1(t), \ldots, s_M(t)$, or may be evident from inspection in some settings.

Example 3.3.1 (Developing a signal space representation for a 4-ary signal set) Consider the example depicted in Figure 3.7, where there are four possible received signals, $s_1, \ldots, s_4$. It is clear from inspection that these span a three-dimensional signal space, with a convenient choice of basis signals,

$$\psi_1(t) = I_{[-1,0]}(t), \quad \psi_2(t) = I_{[0,1]}(t), \quad \psi_3(t) = I_{[1,2]}(t),$$

as shown in Figure 3.8. Let $\mathbf{s}_i = (s_i[1], s_i[2], s_i[3])^T$ denote the vector representation of the signal s_i with respect to the basis, for $i = 1, \ldots 4$. That is, the coefficients of the vector $\mathbf{s}_i$ are such that

$$s_i(t) = \sum_{k=1}^{3} s_i[k] \psi_k(t).$$

We obtain, again by inspection,

$$\mathbf{s}_1 = \begin{pmatrix} 0 \\ 1 \\ 1 \end{pmatrix}, \quad \mathbf{s}_2 = \begin{pmatrix} 1 \\ 1 \\ 0 \end{pmatrix}, \quad \mathbf{s}_3 = \begin{pmatrix} 0 \\ 2 \\ 0 \end{pmatrix}, \quad \mathbf{s}_4 = \begin{pmatrix} -1 \\ 1 \\ -1 \end{pmatrix}.$$

In general, for any signal set with M signals $\{s_i(t), i = 1, \ldots, M\}$, we can find an orthonormal basis $\{\psi_k, k = 1, \ldots, n\}$, where the dimension of the signal space, n, is at most equal to the number of signals, M. The vector

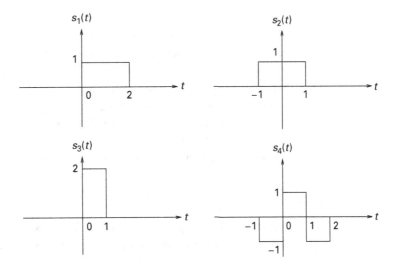

Figure 3.7 Four signals spanning a three-dimensional signal space.

Figure 3.8 An orthonormal basis for the signal set in Figure 3.7, obtained by inspection.

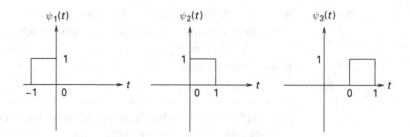

representation of signal $s_i(t)$ with respect to the basis is given by $\mathbf{s}_i = (s_i[1], \ldots, s_i[n])^T$, where

$$s_i[k] = \langle s_i, \psi_k \rangle, \quad i = 1, \ldots, M, \quad k = 1, \ldots, n.$$

Finding a basis by inspection is not always feasible. A systematic procedure for finding a basis is Gramm–Schmidt orthogonalization, described next.

Gramm–Schmidt orthogonalization Letting $\mathcal{S}_k$ denote the subspace spanned by $s_1, \ldots, s_k$, the Gramm–Schmidt algorithm proceeds iteratively: given an orthonormal basis for $\mathcal{S}_k$, it finds an orthonormal basis for $\mathcal{S}_{k+1}$. The procedure stops when $k = M$. The method is identical to that used for finite-dimensional vectors, except that the definition of the inner product involves an integral, rather than a sum, for the continuous-time signals considered here.

Step 1 (Initialization) Let $\phi_1 = s_1$. If $\phi_1 \neq 0$, then set $\psi_1 = \phi_1 / \|\phi_1\|$. Note that ψ_1 provides a basis function for $\mathcal{S}_1$.

Step $k+1$ Suppose that we have constructed an orthonormal basis $\mathcal{B}_k = \{\psi_1, \ldots \psi_m\}$ for the subspace $\mathcal{S}_k$ spanned by the first k signals (note that $m \leq k$). Define

$$\phi_{k+1}(t) = s_{k+1}(t) - \sum_{i=1}^{m} \langle s_{k+1}, \psi_i \rangle \psi_i(t).$$

The signal $\phi_{k+1}(t)$ is the component of $s_{k+1}(t)$ orthogonal to the subspace $\mathcal{S}_k$. If $\phi_{k+1} \neq 0$, define a new basis function $\psi_{m+1}(t) = \phi_{k+1}(t) / \|\phi_{k+1}\|$, and update the basis as $\mathcal{B}_{k+1} = \{\psi_1, \ldots, \psi_m, \psi_{m+1}\}$. If $\phi_{k+1} = 0$, then $s_{k+1} \in \mathcal{S}_k$, and it is not necessary to update the basis; in this case, we set $\mathcal{B}_{k+1} = \mathcal{B}_k = \{\psi_1, \ldots, \psi_m\}$.

The procedure terminates at step M, which yields a basis $\mathcal{B} = \{\psi_1, \ldots, \psi_n\}$ for the signal space $\mathcal{S} = \mathcal{S}_M$. The basis is not unique, and may depend (and typically does depend) on the order in which we go through the signals in the set. We use the Gramm–Schmidt procedure here mainly as a conceptual tool, in assuring us that there is indeed a finite-dimensional vector representation for a finite set of continuous-time signals.

Exercise 3.3.1 (Application of the Gramm–Schmidt procedure) Apply the Gramm–Schmidt procedure to the signal set in Figure 3.7. When the

3.3 Signal space concepts

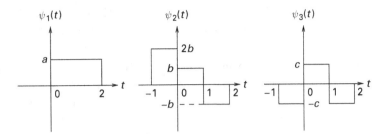

Figure 3.9 An orthonormal basis for the signal set in Figure 3.7, obtained by applying the Gramm–Schmidt procedure. The unknowns a, b, and c are to be determined in Exercise 3.3.1.

signals are considered in increasing order of index in the Gramm–Schmidt procedure, verify that the basis signals are as in Figure 3.9, and fill in the missing numbers. While the basis thus obtained is not as "nice" as the one obtained by inspection in Figure 3.8, the Gramm–Schmidt procedure has the advantage of general applicability.

Projection onto signal space We now project the received signal $y(t)$ onto the signal space to obtain an n-dimensional vector $\mathbf{Y}$. Specifically, set $\mathbf{Y} = (\langle y, \psi_1 \rangle, \ldots, \langle y, \psi_n \rangle)^T$. Under hypothesis H_i ($i = 1, \ldots, M$), we have $\mathbf{Y} = \mathbf{s}_i + \mathbf{N}$, where $\mathbf{s}_i = (\langle s_i, \psi_1 \rangle, \ldots, \langle s_i, \psi_n \rangle)^T$, $i = 1, \ldots, M$, and $\mathbf{N} = (\langle n, \psi_1 \rangle, \ldots, \langle n, \psi_n \rangle)^T$ are obtained by projecting the signals and noise onto the signal space. Note that the vector $\mathbf{Y} = (y[1], \ldots, y[n])^T$ completely describes the component of the received signal $y(t)$ in the signal space, given by

$$y_S(t) = \sum_{j=1}^{n} \langle y, \psi_j \rangle \psi_j(t) = \sum_{j=1}^{n} y[j] \psi_j(t).$$

The component of $y(t)$ orthogonal to the signal space is given by

$$y^{\perp}(t) = y(t) - y_S(t) = y(t) - \sum_{j=1}^{n} y_j \psi_j(t).$$

We now explore the structure of the signal space representation further.

Inner products are preserved We will soon show that performance of optimal reception of M-ary signaling on an AWGN channel depends only on the inner products between the signal, once the noise PSD is fixed. It is therefore important to check that the inner products of the continuous-time signals and their signal space counterparts remain the same. Specifically, plugging in the representation of the signals in terms of the basis functions, we get ($s_i[k]$ denotes $\langle s_i, \psi_k \rangle$, for $1 \le i \le M$, $1 \le k \le n$)

$$\langle s_i, s_j \rangle = \langle \sum_{k=1}^{n} s_i[k] \psi_k, \sum_{l=1}^{n} s_j[l] \psi_l \rangle = \sum_{k=1}^{n} \sum_{l=1}^{n} s_i[k] s_j[l] \langle \psi_k, \psi_l \rangle$$
$$= \sum_{k=1}^{n} \sum_{l=1}^{n} s_i[k] s_j[l] \delta_{kl} = \sum_{k=1}^{n} s_i[k] s_j[k] = \langle \mathbf{s}_i, \mathbf{s}_j \rangle.$$

Recall that δ_{kl} denotes the Kronecker delta function, defined as

$$\delta_{kl} = \begin{cases} 1 & k = l, \\ 0 & k \ne l. \end{cases}$$

In the above, we have used the orthonormality of the basis functions $\{\psi_k, k = 1, \ldots, n\}$ in collapsing the two summations into one.

Noise vector is discrete WGN The noise vector $\mathbf{N} = (N[1], \ldots, N[n])^T$ corrupting the observation within the signal space is discrete-time WGN. That is, it is a zero mean Gaussian random vector with covariance matrix $\sigma^2 \mathbf{I}$, so that its components $\{N[j]\}$ are i.i.d. $N(0, \sigma^2)$ random variables. This follows immediately from Proposition 3.3.1 and Remark 3.3.1.

Now that we understand the signal and noise structure within the signal space, we state and prove the fundamental result that the component of the received signal *orthogonal* to the signal space, $y^\perp(t)$, is *irrelevant* for detection in AWGN. Thus, it suffices to restrict attention to the finite-dimensional vector $\mathbf{Y}$ in the signal space for the purpose of optimal reception in AWGN.

Theorem 3.3.1 (Restriction to signal space is optimal) *For the model (3.24), there is no loss in detection performance in ignoring the component $y^\perp(t)$ of the received signal orthogonal to the signal space. Thus, it suffices to consider the equivalent hypothesis testing model given by*

$$H_i: \mathbf{Y} = \mathbf{s}_i + \mathbf{N} \quad i = 1, \ldots, M.$$

Proof of Theorem 3.3.1 Conditioning on hypothesis H_i, we first note that $y^\perp$ does not have any signal contribution, since all of the M possible transmitted signals are in the signal space. That is, for $y(t) = s_i(t) + n(t)$, we have

$$y^\perp(t) = y(t) - \sum_{j=1}^{n} \langle y, \psi_j \rangle \psi_j(t) = s_i(t) + n(t) - \sum_{j=1}^{n} \langle s_i + n, \psi_j \rangle \psi_j(t)$$

$$= n(t) - \sum_{j=1}^{n} \langle n, \psi_j \rangle \psi_j(t) = n^\perp(t),$$

where $n^\perp$ is the noise contribution orthogonal to the signal space. Next, we show that $n^\perp$ is independent of $\mathbf{N}$, the noise contribution in the signal space. Since $n^\perp$ and $\mathbf{N}$ are jointly Gaussian, it suffices to demonstrate that they are uncorrelated. Specifically, for any t and k, we have

$$\operatorname{cov}(n^\perp(t), N[k]) = \mathbb{E}[n^\perp(t)N[k]] = \mathbb{E}[\{n(t) - \sum_{j=1}^{n} N[j]\psi_j(t)\}N[k]]$$

$$= \mathbb{E}[n(t)N[k]] - \sum_{j=1}^{n} \mathbb{E}[N[j]N[k]]\psi_j(t). \quad (3.26)$$

The first term on the extreme right-hand side can be simplified as

$$\mathbb{E}[n(t)\langle n, \psi_k \rangle] = \mathbb{E}[n(t) \int n(s)\psi_k(s)ds]$$

$$= \int \mathbb{E}[n(t)n(s)]\psi_k(s)ds = \int \sigma^2 \delta(s-t)\psi_k(s)ds = \sigma^2 \psi_k(t).$$

$$(3.27)$$

3.3 Signal space concepts

Plugging (3.27) into (3.26), and noting that $\mathbb{E}[N[j]N[k]] = \sigma^2 \delta_{jk}$, we obtain

$$\text{cov}(n^\perp(t), N[j]) = \sigma^2 \psi_k(t) - \sigma^2 \psi_k(t) = 0.$$

Thus, conditioned on H_i, $y^\perp = n^\perp$ does not contain any signal contribution, and is independent of the noise vector $\mathbf{N}$ in the signal space. It is therefore irrelevant to the detection problem; applying Theorem 3.2.2 in a manner exactly analogous to the observation Y_2 in Example 3.2.5. (We have not discussed how to define densities for infinite-dimensional random processes such as $y^\perp$, but let us assume this can be done. Then $y^\perp$ plays exactly the role of Y_2 in the example.) □

Example 3.3.2 (Application to two-dimensional linear modulation)
Consider linear modulation in passband, for which the transmitted signal corresponding to a given symbol is of the form

$$s_{b_c, b_s}(t) = A b_c p(t)(\sqrt{2} \cos 2\pi f_c t) - A b_s p(t)(\sqrt{2} \sin 2\pi f_c t),$$

where the information is encoded in the pair of real numbers (b_c, b_s), and where $p(t)$ is a baseband pulse whose bandwidth is smaller than the carrier frequency f_c. We assume that there is no intersymbol interference, hence it suffices to consider each symbol separately. In this case, the signal space is two-dimensional, and a natural choice of basis functions for the signal space is $\psi_c(t) = \alpha p(t) \cos 2\pi f_c t$ and $\psi_s(t) = \alpha p(t) \sin 2\pi f_c t$, where α is a normalization constant. From Chapter 2, we know that ψ_c and ψ_s are indeed orthogonal. The signal space representation for $s_{b_c, b_s}(t)$ is therefore (a possibly scaled version of) $(b_c, b_s)^T$. The absolute scaling of the signal constellation can be chosen arbitrarily, since, as we have already observed, it is the signal-to-noise *ratio* that determines the performance. The two-dimensional received signal vector (the first dimension is the I component, and the second the Q component) can therefore be written as

$$\mathbf{y} = \begin{pmatrix} y_c \\ y_s \end{pmatrix} = \begin{pmatrix} b_c \\ b_s \end{pmatrix} + \begin{pmatrix} N_c \\ N_s \end{pmatrix}, \quad (3.28)$$

where N_c, N_s are i.i.d. $N(0, \sigma^2)$ random variables. While the received vector $\mathbf{y}$ is written as a column vector above, we reuse the same notation (y or $\mathbf{y}$) to denote the corresponding row vector (y_c, y_s) when convenient. Figure 3.10 shows the signal space representations of some PSK and QAM constellations (which we have just observed is just the symbol alphabet). We have not specified the scale for the constellations, since it is the constellation geometry, rather than the scaling, that determines performance.

Now that we have reduced the detection problem to finite dimensions, we can write down the density of the observation $\mathbf{Y}$, conditioned on the hypotheses, and infer the optimal decision rules using the detection theory basics described earlier. This is done in the next section.

Figure 3.10 For linear modulation with no intersymbol interference, the complex symbols themselves provide a two-dimensional signal space representation. Three different constellations are shown here.

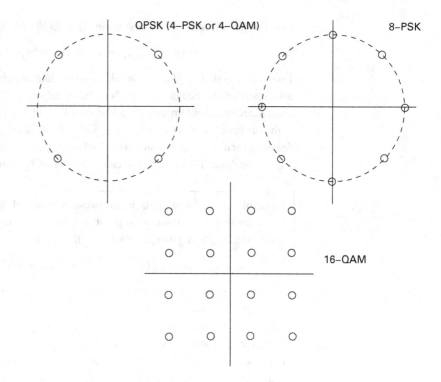

3.4 Optimal reception in AWGN

We begin with a theorem characterizing the optimal receiver when the received signal is a finite-dimensional vector. Using this, we infer the optimal receiver for continuous-time received signals.

Theorem 3.4.1 (Optimal detection in discrete-time AWGN) *Consider the finite-dimensional M-ary hypothesis testing problem where the observation is a random vector $\mathbf{Y}$ modeled as*

$$H_i : \mathbf{Y} = \mathbf{s}_i + \mathbf{N} \quad i = 1, \ldots, M, \tag{3.29}$$

where $\mathbf{N} \sim N(0, \sigma^2 \mathbf{I})$ is discrete-time WGN.

(a) When we observe $\mathbf{Y} = \mathbf{y}$, the ML decision rule is a "minimum distance rule," given by

$$\delta_{\mathrm{ML}}(\mathbf{y}) = \arg \min_{1 \leq i \leq M} \|\mathbf{y} - \mathbf{s}_i\|^2 = \arg \max_{1 \leq i \leq M} \langle \mathbf{y}, \mathbf{s}_i \rangle - \frac{\|\mathbf{s}_i\|^2}{2}. \tag{3.30}$$

(b) If hypothesis H_i has prior probability $\pi(i)$, $i = 1, \ldots, M$ ($\sum_{i=1}^{M} \pi(i) = 1$), then the MPE decision rule is given by

$$\delta_{\mathrm{MPE}}(\mathbf{y}) = \arg \min_{1 \leq i \leq M} \|\mathbf{y} - \mathbf{s}_i\|^2 - 2\sigma^2 \log \pi(i)$$

$$= \arg \max_{1 \leq i \leq M} \langle \mathbf{y}, \mathbf{s}_i \rangle - \frac{\|\mathbf{s}_i\|^2}{2} + \sigma^2 \log \pi(i). \tag{3.31}$$

3.4 Optimal reception in AWGN

Proof of Theorem 3.4.1 Under hypothesis H_i, **Y** is a Gaussian random vector with mean $\mathbf{s}_i$ and covariance matrix $\sigma^2 \mathbf{I}$ (the translation of the noise vector **N** by the deterministic signal vector $\mathbf{s}_i$ does not change the covariance matrix), so that

$$p_{\mathbf{Y}|i}(\mathbf{y}|H_i) = \frac{1}{(2\pi\sigma^2)^{n/2}} \exp\left(-\frac{\|\mathbf{y}-\mathbf{s}_i\|^2}{2\sigma^2}\right). \tag{3.32}$$

Plugging (3.32) into the ML rule (3.17), we obtain the rule (3.30) upon simplification. Similarly, we obtain (3.31) by substituting (3.32) in the MPE rule (3.18). □

We now provide the final step in deriving the optimal detector for the original continuous-time model (3.24), by mapping the optimal decision rules in Theorem 3.4.1 back to continuous time via Theorem 3.3.1.

Theorem 3.4.2 (Optimal coherent demodulation with real-valued signals) *For the continuous-time model (3.24), the optimal detectors are given as follows:*

(a) The ML decision rule is

$$\delta_{\text{ML}}(y) = \arg\max_{1\leq i\leq M} \langle y, s_i\rangle - \frac{\|s_i\|^2}{2}. \tag{3.33}$$

(b) If hypothesis H_i has prior probability $\pi(i)$, $i=1,\ldots,M$ ($\sum_{i=1}^{M}\pi(i)=1$), then the MPE decision rule is given by

$$\delta_{\text{MPE}}(\mathbf{y}) = \arg\max_{1\leq i\leq M} \langle y, s_i\rangle - \frac{\|s_i\|^2}{2} + \sigma^2 \log \pi(i). \tag{3.34}$$

Proof of Theorem 3.4.2 From Theorem 3.3.1, we know that the continuous-time model (3.24) is equivalent to the discrete-time model (3.29) in Theorem 3.4.1. It remains to map the optimal decision rules (3.30) and (3.31) back to continuous time. These rules involve correlation between the received and transmitted signals and the transmitted signal energies. It suffices to show that these quantities are the same for both the continuous-time model and the equivalent discrete-time model. We know now that signal inner products are preserved, so that

$$\|\mathbf{s}_i\|^2 = \|s_i\|^2.$$

Further, the continuous-time correlator output can be written as

$$\langle y, s_i\rangle = \langle y_s + y^\perp, s_i\rangle = \langle y_s, s_i\rangle + \langle y^\perp, s_i\rangle,$$
$$= \langle y_s, s_i\rangle \quad\quad = \langle \mathbf{y}, \mathbf{s}_i\rangle,$$

where the last equality follows because the inner product between the signals y_s and s_i (which both lie in the signal space) is the same as the inner product between their vector representations. □

Remark 3.4.1 (A technical remark of the form of optimal rules in continuous time) Notice that Theorem 3.4.2 does not contain the continuous-time version of the minimum distance rule in Theorem 3.4.1. This is because of a technical subtlety. In continuous time, the squares of the distances would be

$$||y - s_i||^2 = ||y_s - s_i||^2 + ||y^\perp||^2 = ||y_s - s_i||^2 + ||n^\perp||^2.$$

Under the AWGN model, the noise power orthogonal to the signal space is infinite, hence from a purely mathematical point of view, the preceding quantities are infinite for each i (so that we cannot minimize over i). Hence, it only makes sense to talk about the minimum distance rule in a finite-dimensional space in which the noise power is finite. The correlator-based form of the optimal detector, on the other hand, automatically achieves the projection onto the finite-dimensional signal space, and hence does not suffer from this technical difficulty. Of course, in practice, even the continuous-time received signal may be limited to a finite-dimensional space by filtering and time-limiting, but correlator-based detection still has the practical advantage that only components of the received signal that are truly useful appear in the decision statistics.

Correlators and matched filters The decision statistics for optimal detection can be computed using a bank of M correlators or matched filters as follows:

$$\langle y, s_i \rangle = \int y(t) s_i(t) \mathrm{d}t = (y * s_{i,\mathrm{MF}})(0),$$

where $s_{i,\mathrm{MF}}(t) = s_i(-t)$ is the impulse response of the matched filter for $s_i(t)$.

Coherent demodulation in complex baseband We can now infer the form of the optimal receiver for complex baseband signals by applying Theorem 3.4.2 to real-valued passband signals, and then expressing the decision rule in terms of their complex envelopes. Specifically, suppose that $s_{i,\mathrm{p}}(t)$, $i = 1, \ldots, M$, are M possible real passband transmitted signals, $y_\mathrm{p}(t)$ is the noisy received signal, and $n_\mathrm{p}(t)$ is real-valued AWGN with PSD $N_0/2$ (see Figure 3.5). Let $s_i(t)$ denote the complex envelope of $s_{i,\mathrm{p}}(t)$, $i = 1, \ldots, M$, and let $y(t)$ denote the complex envelope of $y_\mathrm{p}(t)$. Then the passband model

$$H_i : y_\mathrm{p}(t) = s_{i,\mathrm{p}}(t) + n_\mathrm{p}(t), \quad i = 1, \ldots, M \tag{3.35}$$

translates to the complex baseband model

$$H_i : y(t) = s_i(t) + n(t), \quad i = 1, \ldots, M, \tag{3.36}$$

where $n(t)$ is complex WGN with PSD N_0, as shown in Figure 3.5.

Applying Theorem 3.4.2, we know that the decision statistics based on the real passband received signal are given by

$$\langle y_p, s_{i,p} \rangle - \frac{||s_{i,p}||^2}{2} = \text{Re}\left(\langle y, s_i \rangle\right) - \frac{||s_i||^2}{2},$$

where we have translated passband inner products to complex baseband inner products as in Chapter 2. We therefore obtain the following theorem.

Theorem 3.4.3 (Optimal coherent demodulation in complex baseband)
For the passband model (3.35), and its equivalent complex baseband model (3.36), the optimal coherent demodulator is specified in complex baseband as follows:

(a) The ML decision rule is

$$\delta_{\text{ML}}(y) = \arg \max_{1 \leq i \leq M} \text{Re}\left(\langle y, s_i \rangle\right) - \frac{||s_i||^2}{2}. \quad (3.37)$$

(b) If hypothesis H_i has prior probability $\pi(i)$, $i = 1, \ldots, M$ ($\sum_{i=1}^{M} \pi(i) = 1$), then the MPE decision rule is given by

$$\delta_{\text{MPE}}(\mathbf{y}) = \arg \max_{1 \leq i \leq M} \text{Re}\left(\langle y, s_i \rangle\right) - \frac{||s_i||^2}{2} + \sigma^2 \log \pi(i). \quad (3.38)$$

Coherent reception can be understood in terms of real-valued vector spaces In Theorem 3.4.3, even though we are dealing with complex baseband signals, the decision statistics can be evaluated by interpreting each complex signal as a pair of real-valued signals. Specifically, the coherent correlation

$$\text{Re}(\langle y, s_i \rangle) = \langle y_c, s_{i,c} \rangle + \langle y_s, s_{i,s} \rangle$$

corresponds to separate correlation of the I and Q components, followed by addition, and the signal energy

$$||s_i||^2 = ||s_{i,c}||^2 + ||s_{i,s}||^2$$

is the sum of the energies of the I and Q components. Thus, there is no cross coupling between the I and Q components in a coherent receiver, because the receiver can keep the components separate. We can therefore develop signal space concepts for coherent receivers in real-valued vector spaces, as done for the example of two-dimensional modulation in Example 3.3.2.

When do we really need statistical models for complex-valued signals?
We have seen in Example 2.2.5 in Chapter 2 that, for noncoherent receivers that are not synchronized in carrier phase to the incoming signal, the I and Q components cannot be processed separately. We explore this observation in

far more detail in Chapter 4, which considers estimation of parameters such as delay, carrier frequency, and phase (which typically occur prior to carrier phase synchronization), as well as optimal noncoherent reception. At that point, it becomes advantageous to understand complex WGN on its own terms, rather than thinking of it as a pair of real-valued WGN processes, and to develop geometric notions specifically tailored to complex-valued vector spaces.

3.4.1 Geometry of the ML decision rule

The minimum distance interpretation for the ML decision rule implies that the decision regions (in signal space) for M-ary signaling in AWGN are constructed as follows. Interpret the signal vectors $\{\mathbf{s}_i\}$, and the received vector $\mathbf{y}$, as points in n-dimensional Euclidean space. It is easiest to think about this in two dimensions ($n = 2$). For any given i, draw a line between $\mathbf{s}_i$ and $\mathbf{s}_j$ for all $j \neq i$. The perpendicular bisector of the line between $\mathbf{s}_i$ and $\mathbf{s}_j$ defines two half planes, one in which we choose $\mathbf{s}_i$ over $\mathbf{s}_j$, the other in which we choose $\mathbf{s}_j$ over $\mathbf{s}_i$. The intersection of the half planes in which $\mathbf{s}_i$ is chosen over $\mathbf{s}_j$, for $j \neq i$, defines the decision region Γ_i. This procedure is illustrated for a two-dimensional signal space in Figure 3.11. The line L_{1i} is the perpendicular bisector of the line between $\mathbf{s}_1$ and $\mathbf{s}_i$. The intersection of these lines defines Γ_1 as shown. Note that L_{16} plays no role in determining Γ_1, since signal $\mathbf{s}_6$ is "too far" from $\mathbf{s}_1$, in the following sense: if the received signal is closer to $\mathbf{s}_6$ than to $\mathbf{s}_1$, then it is also closer to $\mathbf{s}_i$ than to $\mathbf{s}_1$ for some $i = 2, 3, 4, 5$. This kind of observation plays an important role in the performance analysis of ML reception in Section 3.5.

The preceding procedure can now be applied to the simpler scenario of the two-dimensional constellations depicted in Figure 2.16. The resulting ML decision regions are shown in Figure 3.12. For QPSK, the ML regions are simply the four quadrants. For 8-PSK, the ML regions are sectors of a circle. For 16-QAM, the ML regions take a rectangular form.

Figure 3.11 Maximum likelihood (ML) decision region Γ_1 for signal s_1.

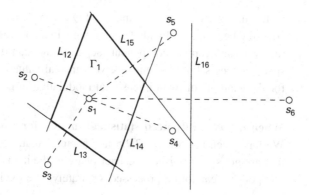

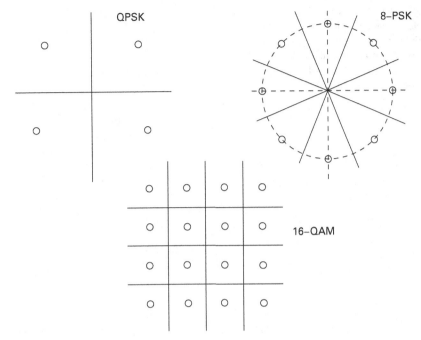

Figure 3.12 Maximum likelihood (ML) decision regions for some two-dimensional constellations.

3.4.2 Soft decisions

Maximum likelihood and MPE demodulation correspond to "hard" decisions regarding which of M signals have been sent. Each such M-ary "symbol" corresponds to $\log_2 M$ bits. Often, however, we send many such symbols (and hence many more than $\log_2 M$ bits), and may employ an error-correcting code over the entire sequence of transmitted symbols or bits. In such a situation, the decisions from the demodulator, which performs M-ary hypothesis testing for each symbol, must be fed to a decoder which accounts for the structure of the error-correcting code to produce more reliable decisions. It becomes advantageous in such a situation to feed the decoder more information than that provided by hard decisions. Consider the model (3.29), where the receiver is processing a finite-dimensional observation vector $\mathbf{Y}$. Two possible values of the observation (dark circles) are shown in Figure 3.13 for a QPSK constellation.

Clearly, we would have more confidence in the decision for the observed value $(1.5, -2)$, which lies further away from the edge of the decision region in which it falls. "Soft" decisions are a means of quantifying our estimate of the reliability of our decisions. While there are many mechanisms that could be devised for conveying more information than hard decisions, the maximal amount of information that the demodulator can provide is the posterior probabilities

$$\pi(i|\mathbf{y}) = P[\mathbf{s}_i \text{ sent}|\mathbf{y}] = P[H_i|\mathbf{y}],$$

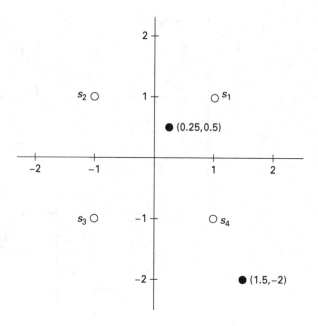

Figure 3.13 Two possible observations (shown in black circles) for QPSK signaling, with signal points denoted by $\{s_i, i = 1, \ldots, 4\}$. The signal space is two-dimensional.

where $\mathbf{y}$ is the value taken by the observation $\mathbf{Y}$. These posterior probabilities can be computed using Bayes' rule, as follows:

$$\pi(i|\mathbf{y}) = P[H_i|\mathbf{y}] = \frac{p(\mathbf{y}|i)P[H_i]}{p(\mathbf{y})} = \frac{p(\mathbf{y}|i)P[H_i]}{\sum_{j=1}^{M} p(\mathbf{y}|j)P[H_j]}.$$

Plugging in the expression (3.32) for the conditional densities $p(\mathbf{y}|j)$ and setting $\pi(i) = P[H_i]$, we obtain

$$\pi(i|\mathbf{y}) = \frac{\pi(i)\exp\left(-\frac{\|\mathbf{y}-\mathbf{s}_i\|^2}{2\sigma^2}\right)}{\sum_{j=1}^{M} \pi(j)\exp\left(-\frac{\|\mathbf{y}-\mathbf{s}_j\|^2}{2\sigma^2}\right)}. \quad (3.39)$$

For the example in Figure 3.13, suppose that we set $\sigma^2 = 1$ and $\pi(i) \equiv 1/4$. Then we can use (3.39) to compute the values shown in Table 3.1 for the posterior probabilities:

The observation $\mathbf{y} = (0.25, 0.5)$ falls in the decision region for $\mathbf{s}_1$, but is close to the decision boundary. The posterior probabilities in Table 3.1 reflect the resulting uncertainty, with significant probabilities assigned to all symbols. On the other hand, the observation $\mathbf{y} = (1.5, -2)$, which falls within the decision region for $\mathbf{s}_4$, is far away from the decision boundaries, hence we would expect it to provide a reliable decision. The posterior probabilities reflect this: the posterior probability for $\mathbf{s}_4$ is significantly larger than that of the other possible symbol values. In particular, the posterior probability for $\mathbf{s}_2$, which is furthest away from the received signal, is very small (and equals zero when rounded to three decimal places as in the table).

Unlike ML hard decisions, which depend only on the distances between the observation and the signal points, the posterior probabilities also depend

Table 3.1 Posterior probabilities for the QPSK constellation in Figure 3.13, assuming equal priors and $\sigma^2 = 1$.

$\pi(i\|\mathbf{y})$	$\mathbf{y} = (0.25, 0.5)$	$\mathbf{y} = (1.5, -2)$
1	0.455	0.017
2	0.276	0
3	0.102	0.047
4	0.167	0.935

Table 3.2 Posterior probabilities for the QPSK constellation in Figure 3.13, assuming equal priors and $\sigma^2 = 4$.

$\pi(i\|\mathbf{y})$	$\mathbf{y} = (0.25, 0.5)$	$\mathbf{y} = (1.5, -2)$
1	0.299	0.183
2	0.264	0.086
3	0.205	0.235
4	0.233	0.497

on the noise variance. If the noise variance is higher, then the decision becomes more unreliable. Table 3.2 illustrates what happens when the noise variance is increased to $\sigma^2 = 4$ for the scenario depicted in Figure 3.13. The posteriors for $\mathbf{y} = (0.25, 0.5)$, which is close to the decision boundaries, are close to uniform, which indicates that the observation is highly unreliable. Even for $\mathbf{y} = (1.5, -2)$, the posterior probabilities for symbols other than $\mathbf{s}_3$ are significant.

In Chapter 7, I consider the role of posterior probabilities in far greater detail for systems with error-correction coding.

3.5 Performance analysis of ML reception

We focus on performance analysis for the ML decision rule, assuming equal priors (for which the ML rule minimizes the error probability). The analysis for MPE reception with unequal priors is similar, and is sketched in one of the problems.

Now that we have firmly established the equivalence between continuous-time signals and signal space vectors, we can become sloppy about the distinction between them in our notation, using the notation y, s_i and n to denote the received signal, the transmitted signal, and the noise, respectively, in both settings.

3.5.1 Performance with binary signaling

The basic building block for performance analysis is binary signaling. Specifically, consider on–off signaling with

$$\begin{aligned} H_1 &: y(t) = s(t) + n(t), \\ H_0 &: y(t) = n(t). \end{aligned} \qquad (3.40)$$

Applying Theorem 3.4.2, we find that the ML rule reduces to

$$\langle y, s \rangle \underset{H_0}{\overset{H_1}{\gtrless}} \frac{||s||^2}{2}. \qquad (3.41)$$

Setting $Z = \langle y, s \rangle$, we wish to compute the conditional error probabilities given by

$$P_{e|1} = P\left[Z < \frac{||s||^2}{2} \Big| H_1\right] \quad P_{e|0} = P\left[Z > \frac{||s||^2}{2} \Big| H_0\right]. \qquad (3.42)$$

To this end, note that, conditioned on either hypothesis, Z is a Gaussian random variable. The conditional mean and variance of Z under H_0 are given by

$$\begin{aligned} \mathbb{E}[Z|H_0] &= \mathbb{E}[\langle n, s \rangle] &= 0, \\ \mathrm{var}(Z|H_0) &= \mathrm{cov}(\langle n, s \rangle, \langle n, s \rangle) &= \sigma^2 ||s||^2, \end{aligned}$$

where we have used Proposition 3.3.1, and the fact that $n(t)$ has zero mean. The corresponding computation under H_1 is as follows:

$$\mathbb{E}[Z|H_1] = \mathbb{E}[\langle s+n, s \rangle] = ||s||^2$$

$$\mathrm{var}(Z|H_1) = \mathrm{cov}(\langle s+n, s \rangle, \langle s+n, s \rangle) = \mathrm{cov}(\langle n, s \rangle, \langle n, s \rangle) = \sigma^2 ||s||^2,$$

noting that covariances do not change upon adding constants. Thus, $Z \sim N(0, v^2)$ under H_0 and $Z \sim N(m, v^2)$ under H_1, where $m = ||s||^2$ and $v^2 = \sigma^2 ||s||^2$. Substituting in (3.42), it is easy to check that

$$P_{e,\mathrm{ML}} = P_{e|1} = P_{e|0} = Q\left(\frac{||s||}{2\sigma}\right). \qquad (3.43)$$

In the language of detection theory, the correlation decision statistic Z is a sufficient statistic for the decision, in that it contains all the statistical information relevant to the decision. Thus, the ML or MPE decision rules based on Z must be equivalent (in form as well as performance) to the corresponding rules based on the original observation $y(t)$. This is easy to check as follows. The statistics of Z are exactly as in the basic scalar Gaussian example in Example 3.2.1, so that the ML rule is given by

$$Z \underset{H_0}{\overset{H_1}{\gtrless}} \frac{m}{2}$$

3.5 Performance analysis of ML reception

and its performance is given by

$$P_{e,\text{ML}} = Q\left(\frac{m}{2v}\right),$$

as discussed previously: see Examples 3.2.1, 3.2.2, and 3.2.3. It is easy to see that these results are identical to (3.41) and (3.43) by plugging in the values of m and v^2.

Next, consider binary signaling in general, with

$$H_1 : y(t) = s_1(t) + n(t),$$
$$H_0 : y(t) = s_0(t) + n(t).$$

The ML rule for this can be inferred from Theorem 3.4.2 as

$$\langle y, s_1 \rangle - \frac{\|s_1\|^2}{2} \underset{H_0}{\overset{H_1}{\gtrless}} \langle y, s_0 \rangle - \frac{\|s_0\|^2}{2}.$$

We can analyze this system by considering the joint distribution of the correlator statistics $Z_i = \langle y, s_i \rangle$, $i = 0, 1$, conditioned on the hypotheses. Alternatively, we can rewrite the ML decision rule as

$$\langle y, s_1 - s_0 \rangle \underset{H_0}{\overset{H_1}{\gtrless}} \frac{\|s_1\|^2}{2} - \frac{\|s_0\|^2}{2},$$

which corresponds to an implementation using a single correlator. The analysis now involves the conditional distributions of the single decision statistic $Z = \langle y, s_1 - s_0 \rangle$. Analyzing the performance of the ML rule using these approaches is left as an exercise for the reader.

Yet another alternative is to consider a transformed system, where the received signal is $\tilde{y}(t) = y(t) - s_0(t)$. Since this transformation is invertible, the performance of an optimal rule is unchanged under it. But the transformed received signal $\tilde{y}(t)$ falls under the on–off signaling model (3.40), with $s(t) = s_1(t) - s_0(t)$. The ML error probability therefore follows from the formula (3.43), and is given by

$$P_{e,\text{ML}} = P_{e|1} = P_{e|0} = Q\left(\frac{\|s_1 - s_0\|}{2\sigma}\right) = Q\left(\frac{d}{2\sigma}\right), \tag{3.44}$$

where $d = \|s_1 - s_0\|$ is the distance between the two possible received signals.

Before investigating the performance of some commonly used binary signaling schemes, let us establish some standard measures of signal and noise strength.

Energy per bit, E_b This is a measure of the signal strength that is universally employed to compare different communication system designs. A design is more power efficient if it gives the same performance with a smaller E_b, if we fix the noise strength. Since binary signaling conveys one bit of information, E_b is given by the formula

$$E_b = \frac{1}{2}(||s_0||^2 + ||s_1||^2),$$

assuming that 0 and 1 are equally likely to be sent.

Performance scaling with signal and noise strengths If we scale up both s_1 and s_0 by a factor A, E_b scales up by a factor A^2, while the distance d scales up by a factor A. We therefore define the scale-invariant parameter

$$\eta_P = \frac{d^2}{E_b}. \qquad (3.45)$$

Now, substituting, $d = \sqrt{\eta_P E_b}$ and $\sigma = \sqrt{N_0/2}$ into (3.44), we find that the ML performance is given by

$$P_{e,ML} = Q\left(\sqrt{\frac{\eta_P E_b}{2N_0}}\right) = Q\left(\sqrt{\frac{d^2}{E_b}}\sqrt{\frac{E_b}{2N_0}}\right). \qquad (3.46)$$

Two important observations follow.

Performance depends on signal-to-noise ratio We observe from (3.46) that the performance depends on the *ratio* E_b/N_0, rather than separately on the signal and noise strengths.

Concept of power efficiency For fixed E_b/N_0, the performance is better for a signaling scheme that has a higher value of η_P. We therefore use the term *power efficiency* for $\eta_P = d^2/E_b$.

Let us now compute the performance of some common binary signaling schemes in terms of E_b/N_0, using (3.46). Since inner products (and hence energies and distances) are preserved in signal space, we can compute η_P for each scheme using the signal space representations depicted in Figure 3.14. The absolute scale of the signals is irrelevant, since the performance depends on the signaling scheme only through the scale-invariant parameter η_P. We therefore choose a convenient scaling for the signal space representation.

Figure 3.14 Signal space representations with conveniently chosen scaling for three binary signaling schemes.

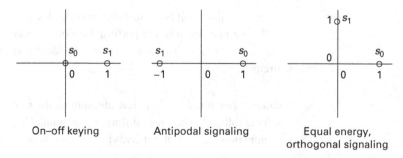

3.5 Performance analysis of ML reception

On–off keying Here $s_1(t) = s(t)$ and $s_0(t) = 0$. As shown in Figure 3.14, the signal space is one-dimensional. For the scaling in the figure, we have $d = 1$ and $E_b = 1/2(1^2 + 0^2) = 1/2$, so that $\eta_P = d^2/E_b = 2$. Substituting into (3.46), we obtain $P_{e,ML} = Q(\sqrt{E_b/N_0})$.

Antipodal signaling Here $s_1(t) = -s_0(t)$, leading again to a one-dimensional signal space representation. One possible realization of antipodal signaling is BPSK, discussed in the previous chapter. For the scaling chosen, $d = 2$ and $E_b = 1/2(1^2 + (-1)^2) = 1$, which gives $\eta_P = d^2/E_b = 4$. Substituting into (3.46), we obtain $P_{e,ML} = Q(\sqrt{2E_b/N_0})$.

Equal-energy orthogonal signaling Here s_1 and s_0 are orthogonal, with $||s_1||^2 = ||s_0||^2$. This is a two-dimensional signal space. Several possible realizations of orthogonal signaling were discussed in the previous chapter, including FSK and Walsh–Hadamard codes. From Figure 3.14, we have $d = \sqrt{2}$ and $E_b = 1$, so that $\eta_P = d^2/E_b = 2$. This gives $P_{e,ML} = Q(\sqrt{E_b/N_0})$.

Thus, on–off keying (which is orthogonal signaling with unequal energies) and equal-energy orthogonal signaling have the same power efficiency, while the power efficiency of antipodal signaling is a factor of two (i.e., 3 dB) better.

In plots of bit error rate (BER) versus SNR, we typically express BER on a log scale (to capture the rapid decay of error probability with SNR) and express SNR in decibels (to span a large range). Such a plot is provided for antipodal and orthogonal signaling in Figure 3.15.

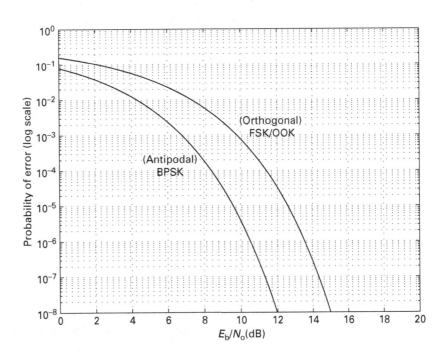

Figure 3.15 Bit error rate versus E_b/N_0 (dB) for antipodal and orthogonal signaling.

3.5.2 Performance with *M*-ary signaling

We turn now to M-ary signaling. Recall that the ML rule can be written as

$$\delta_{\mathrm{ML}}(y) = \arg \max_{1 \leq i \leq M} Z_i,$$

where, for $1 \leq i \leq M$, the decision statistics

$$Z_i = \langle y, s_i \rangle - \frac{1}{2} \|s_i\|^2.$$

For a finite-dimensional signal space, it is also convenient to use the minimum distance form of the decision rule:

$$\delta_{\mathrm{ML}}(y) = \arg \min_{1 \leq i \leq M} D_i,$$

where

$$D_i = \|y - s_i\|.$$

For convenience, we do not show the dependence of Z_i or D_i on the received signal y in our notation. However, it is worth noting that the ML decision regions can be written as

$$\Gamma_i = \{y : \delta_{\mathrm{ML}}(y) = i\} = \{y : Z_i \geq Z_j \text{ for all } j \neq i\} = \{y : D_i \leq D_j \text{ for all } j \neq i\}. \tag{3.47}$$

In the following, we first note the basic structural property that the performance is completely determined by the signal inner products and noise variance. We then observe that exact performance analysis is difficult in general. This leads into a discussion of performance bounds and approximations, and the kind of design tradeoffs we can infer from them.

Performance is determined by signal inner products normalized by noise strength The correlator decision statistics in the ML rule in Theorem 3.4.2 are jointly Gaussian, conditioned on a given hypothesis. To see this, condition on H_i. The conditional error probability is then given by

$$P_{e|i} = P[y \notin \Gamma_i | i \text{ sent}] = P[Z_i < Z_j \text{ for some } j \neq i | i \text{ sent}]. \tag{3.48}$$

To compute this probability, we need to know the joint distribution of the decision statistics $\{Z_j\}$, conditioned on H_i. Let us now examine the structure of this joint distribution. Conditioned on H_i, the received signal is given by

$$y(t) = s_i(t) + n(t).$$

The decision statistics $\{Z_j, 1 \leq j \leq M\}$ are now given by

$$Z_j = \langle y, s_j \rangle - \frac{1}{2} \|s_j\|^2 = \langle s_i, s_j \rangle + \langle n, s_j \rangle - \frac{1}{2} \|s_j\|^2. \tag{3.49}$$

The random variables $\{Z_j\}$ are jointly Gaussian, since n is a Gaussian random process, so that their joint distribution is completely determined by

3.5 Performance analysis of ML reception

means and covariances. Taking expectation in (3.49), we have (suppressing the conditioning on H_i from the notation)

$$\mathbb{E}[Z_j] = \langle s_i, s_j \rangle - \frac{1}{2}\|s_j\|^2.$$

Furthermore, using Proposition 3.3.1,

$$\text{cov}(Z_j, Z_k) = \sigma^2 \langle s_k, s_j \rangle;$$

note that only the noise terms in (3.49) contribute to the covariance. Thus, conditioned on H_i, the joint distribution of $\{Z_j\}$ depends only on the noise variance σ^2 and the signal inner products $\{\langle s_i, s_j \rangle, 1 \leq i, j \leq M\}$. Indeed, it is easy to show, replacing Z_j by Z_j/σ, that the joint distribution depends only on the normalized inner products $\{\langle s_i, s_j \rangle / \sigma^2, 1 \leq i, j \leq M\}$. We can now infer that $P_{e|i}$ for each i, and hence the unconditional error probability P_e, is completely determined by these normalized inner products.

Performance-invariant transformations Since performance depends only on normalized inner products, any transformation of the signal constellation that leaves these unchanged does not change the performance. Mapping finite-dimensional signal vectors $\{\mathbf{s}_i\}$ to continuous-time signals $\{s_i(t)\}$ using an orthonormal basis is one example of such a transformation that we have already seen. Another example is a transformation of the signal vectors $\{\mathbf{s}_i\}$ to another set of signal vectors $\tilde{\mathbf{s}}_i$ by using a different orthonormal basis for the vector space in which they lie. Such a transformation is called a *rotation*, and we can write $\tilde{\mathbf{s}}_i = \mathbf{Q}\mathbf{s}_i$, where $\mathbf{Q}$ is a rotation matrix containing as rows the new orthonormal basis vectors we wish to use. For this basis to be orthonormal, the rows must have unit energy and must be orthogonal, which we can write as $\mathbf{Q}\mathbf{Q}^T = \mathbf{I}$ (that is, the inverse of a rotation matrix is its transpose). We can now check explicitly that the inner products between signal vectors are unchanged by the rotation:

$$\langle \mathbf{Q}\mathbf{s}_i, \mathbf{Q}\mathbf{s}_j \rangle = \mathbf{s}_j^T \mathbf{Q}^T \mathbf{Q} \mathbf{s}_i = \mathbf{s}_j^T \mathbf{s}_i = \langle \mathbf{s}_i, \mathbf{s}_j \rangle.$$

Figure 3.16 provides a pictorial summary of these performance-invariant transformations.

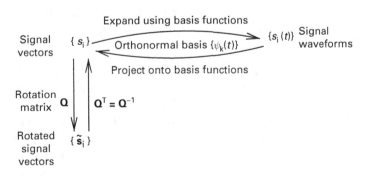

Figure 3.16 Transformations of the signal constellation that leave performance over the AWGN channel unchanged.

We have derived the preceding properties without having to explicitly compute any error probabilities. Building further on this, we can make some broad comments on how the performance depends on scale-invariant properties of the signal constellation and signal-to-noise ratio measures such as E_b/N_0. Let us first define the energy per symbol and energy per bit for M-ary signaling.

Energy per symbol, E_s For M-ary signaling with equal priors, the energy per symbol E_s is given by

$$E_s = \frac{1}{M} \sum_{i=1}^{M} ||s_i||^2.$$

Energy per bit, E_b Since M-ary signaling conveys $\log_2 M$ bit/symbol, the energy per bit is given by

$$E_b = \frac{E_s}{\log_2 M}.$$

If all signals in an M-ary constellation are scaled up by a factor A, then E_s and E_b get scaled up by A^2, as do all inner products $\{\langle s_i, s_j \rangle\}$. Thus, we can define scale-invariant inner products $\{(\langle s_i, s_j \rangle)/E_b\}$ that depend only on the shape of the signal constellation. Setting $\sigma^2 = N_0/2$, we can now write the normalized inner products determining performance as follows:

$$\frac{\langle s_i, s_j \rangle}{\sigma^2} = \frac{\langle s_i, s_j \rangle}{E_b} \frac{2E_b}{N_0}. \quad (3.50)$$

We can now infer the following statement.

Performance depends only on E_b/N_0 and constellation shape This follows from (3.50), which shows that the signal inner products normalized by noise strength (which we have already observed determine performance) depend only on E_b/N_0 and the *scale-invariant* inner products $\{\langle s_i, s_j \rangle / E_b\}$. The latter depend only on the shape of the signal constellation, and are completely independent of the signal and noise strengths.

Specialization to binary signaling Note that the preceding observations are consistent with our performance analysis of binary signaling in Section 3.5.1. We know from (3.46) that the performance depends only on E_b/N_0 and the power efficiency. As shown below, the power efficiency is a function of the scale-invariant inner products defined above.

$$\eta_P = \frac{d^2}{E_b} = \frac{||s_1 - s_0||^2}{E_b} = \frac{\langle s_1, s_1 \rangle + \langle s_0, s_0 \rangle - \langle s_1, s_0 \rangle - \langle s_0, s_1 \rangle}{E_b}.$$

The preceding scaling arguments yield insight into the factors determining the performance for M-ary signaling. We now discuss how to estimate the performance explicitly for a given M-ary signaling scheme. We have shown that there is a compact formula for ML performance with binary signaling. However, exact performance analysis of M-ary signaling for $M > 2$ requires the computation of $P_{e|i}$ (for each $i = 1, \ldots, M$) using the joint distribution of the $\{Z_j\}$ conditioned on H_i. This involves, in general, an integral of a multidimensional

3.5 Performance analysis of ML reception

Gaussian density over the decision regions defined by the ML rule. In many cases, computer simulation of the ML rule is a more straightforward means of computing error probabilities than multidimensional Gaussian integrals. Either method is computationally intensive for large constellations, but is important for accurately evaluating performance for, say, a completed design. However, during the design process, simple formulas that can be quickly computed, and can provide analytical insight, are often indispensable. We therefore proceed to develop bounds and approximations for ML performance, building on the simple analysis for binary signaling in Section 3.5.1.

We employ performance analysis for QPSK signaling as a running example, since it is possible to perform an exact analysis of ML performance in this case, and to compare it with the bounds and approximations that we develop. The ML decision regions (boundaries coincide with the axes) and the distances between the signal points for QPSK are depicted in Figure 3.17.

Exact analysis for QPSK Let us find $P_{e|1}$, the conditional error probability for the ML rule conditioned on s_1 being sent. For the scaling shown in the figure, $s_1 = (d/2, d/2)$, and the two-dimensional observation y is given by

$$y = s_1 + (N_c, N_s) = \left(N_c + \frac{d}{2}, N_s + \frac{d}{2}\right),$$

where N_c, N_s are i.i.d. $N(0, \sigma^2)$ random variables, using the geometric interpretation of WGN after Proposition 3.3.1. An error occurs if the noise moves the observation out of the positive quadrant, which is the decision region for s_1. This happens if $N_c + d/2 < 0$ *or* $N_s + d/2 < 0$. We can therefore write

$$P_{e|1} = P\left[N_c + \frac{d}{2} < 0 \text{ or } N_s + \frac{d}{2} < 0\right] = P\left[N_c + \frac{d}{2} < 0\right] + P\left[N_s + \frac{d}{2} < 0\right]$$
$$- P\left[N_c + \frac{d}{2} < 0 \text{ and } N_s + \frac{d}{2} < 0\right].$$

Figure 3.17 Distances between signal points for QPSK.

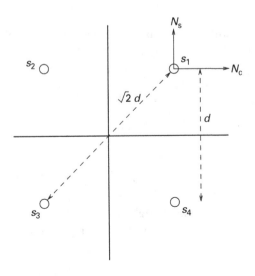

It is easy to see that

$$P\left[N_c + \frac{d}{2} < 0\right] = P\left[N_s + \frac{d}{2} < 0\right] = Q\left(\frac{d}{2\sigma}\right).$$

Using the independence of N_c, N_s, we obtain

$$P_{e|1} = 2Q\left(\frac{d}{2\sigma}\right) - \left[Q\left(\frac{d}{2\sigma}\right)\right]^2. \qquad (3.51)$$

By symmetry, the preceding equals $P_{e|i}$ for all i, which implies that the average error probability is also given by the expression above. To express the error probability in terms of E_b/N_0, we compute the scale-invariant parameter d^2/E_b, and use the relation

$$\frac{d}{2\sigma} = \sqrt{\frac{d^2}{E_b}}\sqrt{\frac{E_b}{2N_0}},$$

as we did for binary signaling. The energy per symbol is given by

$$E_s = \frac{1}{M}\sum_{i=1}^{M}||s_i||^2 = ||s_1||^2 = \left(\frac{d}{2}\right)^2 + \left(\frac{d}{2}\right)^2 = \frac{d^2}{2},$$

which implies that the energy per bit is

$$E_b = \frac{E_s}{\log_2 M} = \frac{E_s}{\log_2 4} = \frac{d^2}{4}.$$

This yields $d^2/E_b = 4$, and hence $d/2\sigma = \sqrt{2E_b/N_0}$. Substituting into (3.51), we obtain

$$P_e = P_{e|1} = 2Q\left(\sqrt{\frac{2E_b}{N_0}}\right) - Q^2\left(\sqrt{\frac{2E_b}{N_0}}\right) \qquad (3.52)$$

as the exact error probability for QPSK.

Union bound and variants We now discuss the union bound on the performance of M-ary signaling in AWGN. We can rewrite (3.48), the conditional error probability, conditioned on H_i, as a union of $M-1$ events, as follows:

$$P_{e|i} = P[\cup_{j \neq i}\{Z_i < Z_j\}|i \text{ sent}].$$

Since the probability of the union of events is upper bounded by the sum of their probabilities, we obtain

$$P_{e|i} \leq \sum_{j \neq i} P[Z_i < Z_j|i \text{ sent}]. \qquad (3.53)$$

But the jth term on the right-hand side above is simply the error probability of ML reception for binary hypothesis testing between the signals s_i and s_j.

3.5 Performance analysis of ML reception

From the results of Section 3.5.1, we therefore obtain the following *pairwise error probability*:

$$P[Z_i < Z_j | i \text{ sent}] = Q\left(\frac{\|s_j - s_i\|}{2\sigma}\right).$$

Substituting into (3.53), we obtain the following union bound.

Union bound The conditional error probabilities for the ML rule are bounded as

$$P_{e|i} \leq \sum_{j \neq i} Q\left(\frac{\|s_j - s_i\|}{2\sigma}\right) = \sum_{j \neq i} Q\left(\frac{d_{ij}}{2\sigma}\right), \quad (3.54)$$

introducing the notation d_{ij} for the distance between signals s_i and s_j. This can be averaged using the prior probabilities to obtain a bound on the average error probability as follows:

$$P_e = \sum_i \pi(i) P_{e|i} \leq \sum_i \pi(i) \sum_{j \neq i} Q\left(\frac{\|s_j - s_i\|}{2\sigma}\right) = \sum_i \pi(i) \sum_{j \neq i} Q\left(\frac{d_{ij}}{2\sigma}\right). \quad (3.55)$$

We can now rewrite the union bound in terms of E_b/N_0 and the scale-invariant squared distances d_{ij}^2/E_b as follows:

$$P_{e|i} \leq \sum_{j \neq i} Q\left(\sqrt{d_{ij}^2/E_b}\sqrt{E_b/2N_0}\right), \quad (3.56)$$

$$P_e = \sum_i \pi(i) P_{e|i} \leq \sum_i \pi(i) \sum_{j \neq i} Q\left(\sqrt{d_{ij}^2/E_b}\sqrt{E_b/2N_0}\right). \quad (3.57)$$

Union bound for QPSK For QPSK, we infer from Figure 3.17 that the union bound for $P_{e|1}$ is given by

$$P_e = P_{e|1} \leq Q\left(\frac{d_{12}}{2\sigma}\right) + Q\left(\frac{d_{13}}{2\sigma}\right) + Q\left(\frac{d_{14}}{2\sigma}\right) = 2Q\left(\frac{d}{2\sigma}\right) + Q\left(\frac{\sqrt{2}d}{2\sigma}\right).$$

Using $d^2/E_b = 4$, we obtain the union bound in terms of E_b/N_0 to be

$$P_e \leq 2Q\left(\sqrt{\frac{2E_b}{N_0}}\right) + Q\left(\sqrt{\frac{4E_b}{N_0}}\right) \quad \textbf{QPSK union bound}. \quad (3.58)$$

For moderately large E_b/N_0, the dominant term in terms of the decay of the error probability is the first one, since $Q(x)$ falls off rapidly as x gets large. Thus, while the union bound (3.58) is larger than the exact error probability (3.52), as it must be, it gets the multiplicity and argument of the dominant term correct.

The union bound can be quite loose for large signal constellations. However, if we understand the geometry of the constellation well enough, we can tighten this bound by pruning a number of terms from (3.54). Let us first discuss this in the context of QPSK. Condition again on s_1 being sent. Let E_1 denote the event that y falls outside the first quadrant, the decision region for s_1. We see from Figure 3.17 that this implies that event E_2 holds, where E_2 is the event that either y is closer to s_2 than to s_1 (if y lies in the left half plane), or y is closer to s_4 than to s_1 (if it lies in the bottom half plane). Since E_1 implies E_2, it is contained in E_2, and its (conditional) probability is bounded by that of E_2. In terms of the decision statistics Z_i, we can bound the conditional error probability (i.e., the conditional probability of E_1) as follows:

$$P_{e|1} \leq P[Z_2 > Z_1 \text{ or } Z_4 > Z_1 | s_1 \text{ sent}] \leq P[Z_2 > Z_1 | s_1 \text{ sent}]$$
$$+ P[Z_4 > Z_1 | s_1 \text{ sent}] = 2Q\left(\frac{d}{2\sigma}\right).$$

In terms of E_b/N_0, we obtain the "intelligent" union bound;

$$P_e = P_{e|1} \leq 2Q\left(\sqrt{\frac{2E_b}{N_0}}\right) \quad \textbf{QPSK intelligent union bound.} \quad (3.59)$$

This corresponds to dropping the term corresponding to s_3 from the union bound for $P_{e|1}$. We term the preceding bound an "intelligent" union bound because we have used our knowledge of the geometry of the signal constellation to prune the terms in the union bound, while still obtaining an upper bound for the error probability.

We now provide a characterization of the intelligent union bound for M-ary signaling in general. Denote by $N_{\text{ML}}(i)$ the indices of the set of neighbors of signal s_i (we exclude i from $N_{\text{ML}}(i)$ by definition) that characterize the ML decision region Γ_i. That is, the half planes that we intersect to obtain Γ_i correspond to the perpendicular bisectors of lines joining s_i and s_j, $j \in N_{\text{ML}}(i)$. In particular, we can express the decision region in (3.47) as

$$\Gamma_i = \{y : \delta_{\text{ML}}(y) = i\} = \{y : Z_i \geq Z_j \text{ for all } j \in N_{\text{ML}}(i)\}. \quad (3.60)$$

We can now say the following: y falls outside Γ_i if and only if $Z_i < Z_j$ for some $j \in N_{\text{ML}}(i)$. We can therefore write

$$P_{e|i} = P[y \notin \Gamma_i | i \text{ sent}] = P[Z_i < Z_j \text{ for some } j \in N_{\text{ML}}(i) | i \text{ sent}] \quad (3.61)$$

and from there, following the same steps as in the union bound, get a tighter bound, which we express as follows.

Intelligent union bound A better bound on $P_{e|i}$ is obtained by considering only the neighbors of s_i that determine its ML decision region, as follows:

$$P_{e|i} \leq \sum_{j \in N_{\text{ML}}(i)} Q\left(\frac{\|s_j - s_i\|}{2\sigma}\right). \quad (3.62)$$

In terms of E_b/N_0, we get

$$P_{e|i} \leq \sum_{j \in N_{\text{ML}}(i)} Q\left(\sqrt{\frac{d_{ij}^2}{E_b}}\sqrt{\frac{E_b}{2N_0}}\right) \quad (3.63)$$

3.5 Performance analysis of ML reception

(the bound on the unconditional error probability P_e is computed as before by averaging the bounds on $P_{e|i}$).

For QPSK, we see from Figure 3.17 that $N_{ML}(1) = \{2, 4\}$, which means that we need only consider terms corresponding to s_2 and s_4 in the union bound for $P_{e|1}$, yielding the result (3.59).

As another example, consider the signal constellation depicted in Figure 3.11. The union bound is given by

$$P_{e|1} \leq Q\left(\frac{d_{12}}{2\sigma}\right) + Q\left(\frac{d_{13}}{2\sigma}\right) + Q\left(\frac{d_{14}}{2\sigma}\right) + Q\left(\frac{d_{15}}{2\sigma}\right) + Q\left(\frac{d_{16}}{2\sigma}\right).$$

However, since $N_{ML}(1) = \{2, 3, 4, 5\}$, the last term above can be dropped to get the following intelligent union bound:

$$P_{e|1} \leq Q\left(\frac{d_{12}}{2\sigma}\right) + Q\left(\frac{d_{13}}{2\sigma}\right) + Q\left(\frac{d_{14}}{2\sigma}\right) + Q\left(\frac{d_{15}}{2\sigma}\right).$$

The gains from employing intelligent pruning of the union bound are larger for larger signal constellations. In Chapter 5, for example, we apply more sophisticated versions of these pruning techniques when discussing the performance of ML demodulation for channels with intersymbol interference.

Another common approach for getting a better (and quicker to compute) estimate than the original union bound is the *nearest neighbors approximation*. This is a loose term employed to describe a number of different methods for pruning the terms in the summation (3.54). Most commonly, it refers to regular signal sets in which each signal point has a number of nearest neighbors at distance $d_{\min}$ from it, where $d_{\min} = \min_{i \neq j} ||s_i - s_j||$. Letting $N_{d_{\min}}(i)$ denote the number of nearest neighbors of s_i, we obtain the following approximation.

Nearest neighbors approximation

$$P_{e|i} \approx N_{d_{\min}}(i) Q\left(\frac{d_{\min}}{2\sigma}\right). \qquad (3.64)$$

Averaging over i, we obtain

$$P_e \approx \bar{N}_{d_{\min}} Q\left(\frac{d_{\min}}{2\sigma}\right), \qquad (3.65)$$

where $\bar{N}_{d_{\min}}$ denotes the average number of nearest neighbors for a signal point. The rationale for the nearest neighbors approximation is that, since $Q(x)$ decays rapidly, $Q(x) \sim e^{-x^2/2}$, as x gets large, the terms in the union bound corresponding to the smallest arguments for the Q function dominate at high SNR.

The corresponding formulas as a function of scale-invariant quantities and E_b/N_0 are:

$$P_{e|i} \approx N_{d_{min}}(i) Q\left(\sqrt{\frac{d_{min}^2}{E_b}}\sqrt{\frac{E_b}{2N_0}}\right). \tag{3.66}$$

It is also worth explicitly writing down an expression for the average error probability, averaging the preceding over i:

$$P_e \approx \bar{N}_{d_{min}} Q\left(\sqrt{\frac{d_{min}^2}{E_b}}\sqrt{\frac{E_b}{2N_0}}\right), \tag{3.67}$$

where

$$\bar{N}_{d_{min}} = \frac{1}{M}\sum_{i=1}^{M} N_{d_{min}}(i)$$

is the *average* number of nearest neighbors for the signal points in the constellation.

For QPSK, we have from Figure 3.17 that

$$N_{d_{min}}(i) \equiv 2 = \bar{N}_{d_{min}}$$

and

$$\sqrt{\frac{d_{min}^2}{E_b}} = \sqrt{\frac{d^2}{E_b}} = 4,$$

yielding

$$P_e \approx 2Q\left(\sqrt{\frac{2E_b}{N_0}}\right).$$

In this case, the nearest neighbors approximation coincides with the intelligent union bound (3.59). This happens because the ML decision region for each signal point is determined by its nearest neighbors for QPSK. Indeed, the latter property holds for many regular constellations, including all of the PSK and QAM constellations whose ML decision regions are depicted in Figure 3.12.

Power efficiency While the performance analysis for M-ary signaling is difficult, we have now obtained simple enough estimates that we can define concepts such as power efficiency, analogous to the development for binary signaling. In particular, comparing the nearest neighbors approximation (3.65) with the error probability for binary signaling (3.46), we define in analogy the power efficiency of an M-ary signaling scheme as

$$\eta_P = \frac{d_{min}^2}{E_b}. \tag{3.68}$$

3.5 Performance analysis of ML reception

We can rewrite the nearest neighbors approximation as

$$P_e \approx \bar{N}_{d_{\min}} Q\left(\sqrt{\frac{\eta_P E_b}{2N_0}}\right). \quad (3.69)$$

Since the argument of the Q function in (3.69) plays a bigger role than the multiplicity $\bar{N}_{d_{\min}}$ for moderately large SNR, η_P offers a means of quickly comparing the power efficiency of different signaling constellations, as well as for determining the dependence of performance on E_b/N_0.

Performance analysis for 16-QAM We now apply the preceding performance analysis to the 16-QAM constellation depicted in Figure 3.18, where we have chosen a convenient scale for the constellation. We now compute the nearest neighbors approximation, which coincides with the intelligent union bound, since the ML decision regions are determined by the nearest neighbors. Noting that the number of nearest neighbors is four for the four innermost signal points, two for the four outermost signal points, and three for the remaining eight signal points, we obtain upon averaging

$$\bar{N}_{d_{\min}} = 3. \quad (3.70)$$

It remains to compute the power efficiency η_P and apply (3.69). For the scaling shown, we have $d_{\min} = 2$. The energy per symbol is obtained as follows:

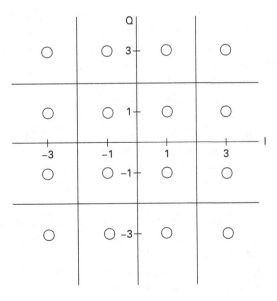

Figure 3.18 ML decision regions for 16-QAM with scaling chosen for convenience in computing power efficiency.

E_s = average energy of I component + average energy of Q component
     = 2(average energy of I component)

by symmetry. Since the I component is equally likely to take the four values ± 1 and ± 3, we have:

$$\text{average energy of I component} = \frac{1}{2}(1^2 + 3^2) = 5$$

and

$$E_s = 10.$$

We therefore obtain

$$E_b = \frac{E_s}{\log_2 M} = \frac{10}{\log_2 16} = \frac{5}{2}.$$

The power efficiency is therefore given by

$$\eta_P = \frac{d_{\min}^2}{E_b} = \frac{2^2}{\frac{5}{2}} = \frac{8}{5}. \tag{3.71}$$

Substituting (3.70) and (3.71) into (3.69), we obtain

$$P_e(\text{16-QAM}) \approx 3Q\left(\sqrt{\frac{4E_b}{5N_0}}\right) \tag{3.72}$$

as the nearest neighbors approximation and intelligent union bound for 16-QAM. The bandwidth efficiency for 16-QAM is 4 bit/2 dimensions, which is twice that of QPSK, whose bandwidth efficiency is 2 bit/2 dimensions. It is not surprising, therefore, that the power efficiency of 16-QAM ($\eta_P = 1.6$) is smaller than that of QPSK ($\eta_P = 4$). We often encounter such tradeoffs between power and bandwidth efficiency in the design of communication systems, including when the signaling waveforms considered are sophisticated codes that are constructed from multiple symbols drawn from constellations such as PSK and QAM.

Figure 3.19 shows the symbol error probabilities for QPSK, 16-QAM, and 16PSK, comparing the intelligent union bounds (which coincide with nearest neighbors approximations) with exact (up to, of course, the numerical accuracy of the computer programs used) results. The exact computations for 16-QAM and 16PSK use expressions (3.72) and (3.92), as derived in the problems. It can be checked that the power efficiencies of the constellations accurately predict the distance between the curves. For example, $\eta_P(QPSK)/\eta_P(16-QAM) = 4/1.6$, which equals about 4 dB. From Figure 3.19, we see that the distance between the QPSK and 16-QAM curves at small error probabilities is about 4 dB.

3.5 Performance analysis of ML reception

Figure 3.19 Symbol error probabilities for QPSK, 16-QAM, and 16PSK.

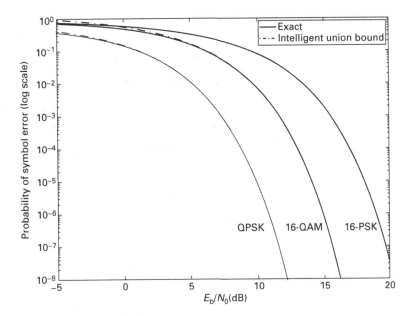

Performance analysis for equal-energy M-ary orthogonal signaling This is a signaling technique that lies at an extreme of the power–bandwidth tradeoff space. The signal space is M-dimensional, hence it is convenient to take the M orthogonal signals as unit vectors along the M axes. With this scaling, we have $E_s = 1$, so that $E_b = 1/(\log_2 M)$. All signals are equidistant from each other, so that the union bound, the intelligent union bound, and the nearest neighbors approximation all coincide, with $d_{\min}^2 = 2$ for the chosen scaling. We therefore get the power efficiency

$$\eta_P = \frac{d_{\min}^2}{E_b} = 2\log_2 M.$$

Note that the power efficiency gets better as M gets large. On the other hand, the bandwidth efficiency, or the number of bits per dimension, is given by

$$\eta_B = \frac{\log_2 M}{M}$$

and goes to zero as $M \to \infty$. We now examine the behavior of the error probability as M gets large, using the union bound.

Expressions for the probabilities of correct detection and error are derived in Problem 3.25. Note here one of these expressions:

$$P_e = (M-1) \int_{-\infty}^{\infty} [\Phi(x)]^{M-2} \Phi(x-m) \frac{1}{\sqrt{2\pi}} e^{-x^2/2} \, dx \tag{3.73}$$

Exact error probability for orthogonal signaling,

where

$$m = \sqrt{\frac{2E_s}{N_0}} = \sqrt{\frac{2E_b \log_2 M}{N_0}}.$$

The union bound is given by

$$P_e \leq (M-1)Q\left(\sqrt{\frac{E_s}{N_0}}\right) = (M-1)Q\left(\sqrt{\frac{E_b \log_2 M}{N_0}}\right) \quad (3.74)$$

Union bound for orthogonal signaling.

Let us now examine the behavior of this bound as M gets large. Noting that the Q function goes to zero, and the term $M-1$ goes to infinity, we employ L'Hôpital's rule to evaluate the limit of the right-hand side above, interpreting M as a real variable rather than an integer. Specifically, let

$$f_1(M) = Q\left(\sqrt{\frac{E_b}{N_0}\log_2 M}\right) = Q\left(\sqrt{\frac{E_b}{N_0}\frac{\ln M}{\ln 2}}\right), \quad f_2(M) = \frac{1}{M-1}.$$

Since

$$\frac{dQ(x)}{dx} = -\frac{1}{\sqrt{2\pi}}e^{-\frac{x^2}{2}},$$

we have

$$\frac{df_1(M)}{dM} = \left[\frac{d}{dM}\left(\sqrt{\frac{E_b}{N_0}\frac{\ln M}{\ln 2}}\right)\right]\left[-\frac{1}{\sqrt{2\pi}}e^{-\left(\sqrt{\frac{E_b}{N_0}\frac{\ln M}{\ln 2}}\right)^2/2}\right]$$

and

$$\frac{df_2(M)}{dM} = -(M-1)^{-2} \approx -M^{-2}.$$

We obtain upon simplification that

$$\lim_{M\to\infty} P_e \leq \lim_{M\to\infty} \frac{\frac{df_1(M)}{dM}}{\frac{df_2(M)}{dM}} = \lim_{M\to\infty} A(\ln M)^{-\frac{1}{2}} M^{1-\frac{E_b}{2\ln 2}}, \quad (3.75)$$

where A is a constant independent of M. The asymptotics as $M \to \infty$ are dominated by the power of M on the right-hand side. If $E_b/N_0 < 2\ln 2$, the right-hand side of (3.75) tends to infinity; that is, the union bound becomes useless, since P_e is bounded above by one. However, if $E_b/N_0 > 2\ln 2$, the right-hand side of (3.75) tends to zero, which implies that P_e tends to zero. The union bound has quickly revealed a remarkable thresholding effect: *M-ary orthogonal signaling can be made arbitrarily reliable by increasing M, as long as E_b/N_0 is above a threshold.*

A more detailed analysis shows that the union bound threshold is off by 3 dB. One can actually show the following result (see Problem 3.26):

$$\lim_{M\to\infty} P_e = \begin{cases} 0, & \frac{E_b}{N_0} > \ln 2, \\ 1, & \frac{E_b}{N_0} < \ln 2. \end{cases} \quad (3.76)$$

That is, by letting M get large, we can get arbitrarily reliable performance as long as E_b/N_0 exceeds -1.6 dB ($\ln 2$ expressed in dB). Using the tools of information theory, we observe in a later chapter that it is not possible to do any better than this in the limit of communication over AWGN channels, as

Figure 3.20 Symbol error probabilities for *M*-ary orthogonal signaling.

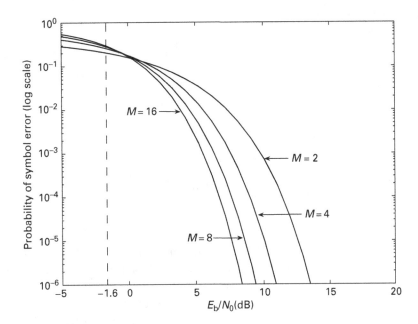

the bandwidth efficiency is allowed to go to zero. That is, M-ary orthogonal signaling is asymptotically optimum in terms of power efficiency.

Figure 3.20 shows the probability of symbol error as a function of E_b/N_0 for several values of M. We see that the performance is quite far away from the asymptotic limit of -1.6 dB (also marked on the plot) for the moderate values of M considered. For example, the E_b/N_0 required for achieving an error probability of 10^{-6} for $M = 16$ is more than 9 dB away from the asymptotic limit.

3.6 Bit-level demodulation

So far, we have discussed how to decide which of M signals have been sent, and how to estimate the performance of decision rules we employ. In practice, however, the information to be sent is encoded in terms of binary digits, or bits, taking value 0 or 1. Sending one of M signals conveys $\log_2 M$ bits of information. Thus, an ML decision rule that picks one of these M signals is actually making a decision on $\log_2 M$ bits. For hard decisions, we wish to compute the probability of bit error, also termed the bit error rate (BER), as a function of E_b/N_0.

For a given SNR, the symbol error probability is only a function of the constellation geometry, but the BER depends also on the manner in which bits are mapped to signals. Let me illustrate this using the example of QPSK. Figure 3.21 shows two possible bitmaps for QPSK, along with the ML decision regions demarcated by bold lines. The first is a *Gray code*, in which the bit mapping is such that the labels for nearest neighbors differ in exactly

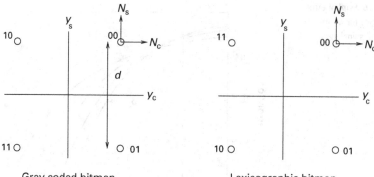

Figure 3.21 QPSK with Gray and lexicographic bitmaps.

one bit. The second corresponds to a lexicographic binary representation of 0, 1, 2, 3, numbering the signals in counterclockwise order. Let us denote the symbol labels as $b[1]b[2]$ for the transmitted symbol, where $b[1]$ and $b[2]$ each take values 0 and 1. Letting $\hat{b}[1]\hat{b}[2]$ denote the label for the ML symbol decision, the probabilities of bit error are given by $p_1 = P[\hat{b}[1] \neq b[1]]$ and $p_2 = P[\hat{b}[2] \neq b[2]]$. The average probability of bit error, which we wish to estimate, is given by $p_b = 1/2\,(p_1 + p_2)$.

Bit error probability for QPSK with Gray bitmap Conditioned on 00 being sent, the probability of making an error on $b[1]$ is as follows:

$$P[\hat{b}[1] = 1 | 00 \text{ sent}] = P[\text{ML decision is 10 or 11} | 00 \text{ sent}]$$

$$= P\left[N_c < -\frac{d}{2}\right] = Q\left(\frac{d}{2\sigma}\right) = Q\left(\sqrt{\frac{2E_b}{N_0}}\right),$$

where, as before, we have expressed the result in terms of E_b/N_0 using the power efficiency $d^2/E_b = 4$. Also note, by the symmetry of the constellation and the bitmap, that the conditional probability of error of $b[1]$ is the same, regardless of which symbol we condition on. Moreover, exactly the same analysis holds for $b[2]$, except that errors are caused by the noise random variable N_s. We therefore obtain

$$p_b = p_1 = p_2 = Q\left(\sqrt{\frac{2E_b}{N_0}}\right). \qquad (3.77)$$

The fact that this expression is identical to the bit error probability for binary antipodal signaling is not a coincidence; QPSK with Gray coding can be thought of as two independent BPSK (or binary antipodal signaling) systems, one signaling along the I (or "cosine") component, and the other along the Q (or "sine") component.

3.6 Bit-level demodulation

Bit error probability for QPSK with lexicographic bitmap Conditioned on 00 being sent, it is easy to see that the error probability for $b[1]$ is as with the Gray code. That is,

$$p_1 = Q\left(\sqrt{\frac{2E_b}{N_0}}\right).$$

However, the conditional error probability for $b[2]$ is different: to make an error in $b[2]$, the noise must move the received signal from the first quadrant into the second or fourth quadrants, the probability of which is given as follows:

$$P[\hat{b}[2] \neq b[2]|00 \text{ sent}] = P[\hat{b}[2] = 1|00 \text{ sent}]$$
$$= P[\text{ML decision is 01 or 11}|00 \text{ sent}]$$
$$= P\left[N_c < -\frac{d}{2}, N_s > -\frac{d}{2}\right] + P\left[N_c > -\frac{d}{2}, N_s < -\frac{d}{2}\right].$$

We have a similar situation regardless of which symbol we condition on. An error in $b[2]$ occurs if the noise manages to move the received signal into either one of the two quadrants adjacent to the one in which the transmitted signal lies. We obtain, therefore,

$$p_2 = 2Q\left(\frac{d}{2\sigma}\right)\left[1 - Q\left(\frac{d}{2\sigma}\right)\right] = 2Q\left(\sqrt{\frac{2E_b}{N_0}}\right)\left[1 - Q\left(\sqrt{\frac{2E_b}{N_0}}\right)\right] \approx 2Q\left(\sqrt{\frac{2E_b}{N_0}}\right)$$

for moderately large E_b/N_0. Thus, p_2 is approximately two times larger than the corresponding quantity for Gray coding, and the average bit error probability is about 1.5 times larger than for Gray coding.

While we have invoked large SNR to discuss the impact of bitmaps, the superiority of Gray coding over an arbitrary bitmap, such as a lexicographic map, plays a bigger role for coded systems operating at low SNR. Gray coding also has the advantage of simplifying the specification of the bit-level demodulation rules for regular constellations. Gray coding is an important enough concept to merit a systematic definition, as follows.

Gray coding Consider a 2^n-ary constellation in which each point is represented by a binary string $\mathbf{b} = (b_1, \ldots, b_n)$. The bit assigment is said to be *Gray coded* if, for any two constellation points $\mathbf{b}$ and $\mathbf{b}'$ which are nearest neighbors, the bit representations $\mathbf{b}$ and $\mathbf{b}'$ differ in exactly one bit location.

Fortunately, QPSK is a simple enough constellation to allow for an exact analysis. A similar analysis can be carried out for larger rectangular constellations such as 16-QAM. It is not possible to obtain simple analytical expressions for BER for nonrectangular constellations such as 8-PSK. In general, it is useful to develop quick estimates for the bit error probability, analogous to the results derived earlier for symbol error probability. Finding bounds on the bit error probability is difficult, hence the discussion here is restricted to the nearest neighbors approximation.

Nearest neighbors approximation to the bit error probability Consider a 2^n-ary constellation in which each constellation point is represented by a binary string of length n as above. Define $d(\mathbf{b}, \mathbf{b}')$ as the distance between constellation points labeled by $\mathbf{b}$ and $\mathbf{b}'$. Define $d_{\min}(\mathbf{b}) = \min_{\mathbf{b}' \neq \mathbf{b}} d(\mathbf{b}, \mathbf{b}')$ as the distance of $\mathbf{b}$ from its nearest neighbors. Let $N_{d_{\min}}(\mathbf{b}, i)$ denote the number of nearest neighbors of $\mathbf{b}$ that differ in the ith bit location, i.e., $N_{d_{\min}}(\mathbf{b}, i) =$ card$\{\mathbf{b}' : d(\mathbf{b}, \mathbf{b}') = d_{\min}(\mathbf{b}), b_i \neq b_i'\}$. Given that $\mathbf{b}$ is sent, the conditional probability of error for the ith bit can be approximated by

$$P(b_i \text{ wrong}|\mathbf{b} \text{ sent}) \approx N_{d_{\min}}(\mathbf{b}, i) Q\left(\frac{d_{\min}(\mathbf{b})}{2\sigma}\right)$$

so that, for equiprobable signaling, the unconditional probability of the ith bit being in error is

$$P(b_i \text{ wrong}) \approx \frac{1}{2^n} \sum_{\mathbf{b}} N_{d_{\min}}(\mathbf{b}, i) Q\left(\frac{d_{\min}(\mathbf{b})}{2\sigma}\right).$$

For an arbitrary constellation or an arbitrary bit mapping, the probability of error for different bit locations may be different. This may indeed be desirable for certain applications in which we wish to provide *unequal error protection* among the bits. Usually, however, we attempt to protect each bit equally. In this case, we are interested in the average bit error probability

$$P(\text{bit error}) = \frac{1}{n} \sum_{i=1}^{n} P(b_i \text{ wrong}).$$

BER with Gray coding For a Gray coded constellation, $N_{d_{\min}}(\mathbf{b}, i) \leq 1$ for all $\mathbf{b}$ and all i. It follows that the value of the nearest neighbors approximation for bit error probability is at most $Q(d_{\min}/2\sigma) = Q\left(\sqrt{\eta_P E_b/2N_0}\right)$, where $\eta_P = d_{\min}^2/E_b$ is the power efficiency.

$$P(\text{bit error}) \approx Q\left(\sqrt{\frac{\eta_P E_b}{2N_0}}\right) \quad \text{with Gray coding.} \tag{3.78}$$

Figure 3.22 shows the BER of 16-QAM and 16PSK with Gray coding, comparing the nearest neighbors approximation with exact results (obtained analytically for 16-QAM, and by simulation for 16PSK). The slight pessimism and ease of computation of the nearest neighbors approximation implies that it is an excellent tool for link design.

Note that Gray coding may not always be possible. Indeed, for an arbitrary set of $M = 2^n$ signals, we may not understand the geometry well enough to assign a Gray code. In general, a necessary (but not sufficient) condition for an n-bit Gray code to exist is that the number of nearest neighbors for any signal point should be at most n.

Figure 3.22 BER for 16-QAM and 16PSK with Gray coding.

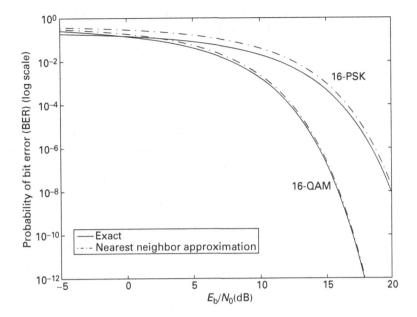

Bit error rate for orthogonal modulation For $M = 2^m$-ary equal energy, orthogonal modulation, each of the m bits split the signal set into half. By the symmetric geometry of the signal set, any of the $M - 1$ wrong symbols is equally likely to be chosen, given a symbol error, and $M/2$ of these will correspond to error in a given bit. We therefore have

$$P(\text{bit error}) = \frac{\frac{M}{2}}{M-1} P(\text{symbol error}), \quad \textbf{BER for } M\textbf{-ary orthogonal signaling.} \quad (3.79)$$

Note that Gray coding is out of the question here, since there are only m bits and $2^m - 1$ neighbors, all at the same distance.

Alternative bit-to-symbol maps Gray coding tries to minimize the number of bit errors due to likely symbol error events. It therefore works well for uncoded systems, or for coded systems in which the bits sent to the modulator are all of equal importance. However, there are coded modulation strategies that can be built on the philosophy of assigning different levels of importance to different bits, for which alternatives to the Gray map are more appropriate. We discuss this in the context of trellis coded modulation in Chapter 7.

3.6.1 Bit-level soft decisions

Bit-level soft decisions can be computed from symbol level soft decisions. Consider the posterior probabilities computed for the scenario depicted in

Figure 3.13 in Section 3.4.2. If we now assume Gray coding as in Figure 3.21, we have

$$P[b[1] = 0|y] = P[s_1 \text{ or } s_4 \text{ sent}|y] = \pi_1(y) + \pi_4(y).$$

Similarly,

$$P[b[2] = 0|y] = P[s_1 \text{ or } s_2 \text{ sent}|y] = \pi_1(y) + \pi_2(y).$$

We can now read off the bit-level soft decisions from Table 3.1. For example, for $\mathbf{y} = (0.25, 0.5)$, we have $P[b[1] = 0|\mathbf{y}] = 0.455 + 0.167 = 0.622$ and $P[b[2] = 0|\mathbf{y}] = 0.455 + 0.276 = 0.731$. As shall be seen in Chapter 7, it is often convenient to express the bit-level soft decisions as log likelihood ratios (LLRs), where we define the LLR for a bit b as

$$\text{LLR}(b) = \log \frac{P[b=0]}{P[b=1]}.$$

We therefore obtain the LLR for $b[1]$, conditioned on the observation, as

$$\text{LLR}(b[1]|\mathbf{y}) = \log \frac{0.622}{1 - 0.622} = 0.498.$$

For Gray coded QPSK, it is easier to compute the bit-level soft decisions directly, using the fact that $b[1]$ and $b[2]$ may be interpreted as being transmitted using BPSK on two parallel channels. We now outline how to compute soft decisions for BPSK signaling.

Soft decisions for BPSK Suppose that a bit $b \in \{0, 1\}$ is sent by mapping it to the symbol $(-1)^b \in \{-1, +1\}$. Then the decision statistic Y for BPSK follows the model:

$$Y = \begin{cases} A + N, & b = 0 \\ -A + N, & b = 1, \end{cases}$$

where $A > 0$ is the amplitude and $N \sim N(0, \sigma^2)$. Suppose that the prior probability of 0 being sent is π_0. (While 0 and 1 are usually equally likely, we shall see the benefit of this general formulation when we discuss iterative decoding in Chapter 7: decoding modules exchange information, with the output of one module in the decoder providing priors to be used for LLR computation in another.) Using Bayes' rule, $P[b|y] = P[b]p(y|b)/p(y)$, $b = 0, 1$, so that the LLR is given by

$$\text{LLR}(b|y) = \log \frac{P[b=0|y]}{P[b=1|y]} = \log \frac{\pi_0 p(y|0)}{\pi_1 p(y|1)}.$$

Plugging in the Gaussian densities $p(y|0)$ and $p(y|1)$ and simplifying, we obtain

$$\text{LLR}(b|y) = \text{LLR}_{\text{prior}}(b) + \frac{2Ay}{\sigma^2}, \tag{3.80}$$

where $\text{LLR}_{\text{prior}} = \log \pi_0/\pi_1$ is the LLR based on the priors. The formula reveals a key advantage of using LLR (as opposed to, say, the conditional probability $P[0|y]$) to express soft decisions: the LLR is simply a sum of information from the priors and from the observation.

3.7 Elements of link budget analysis

Communication link design corresponds to making choices regarding transmit power, antenna gains at transmitter and receiver, quality of receiver circuitry, and range, which have not been mentioned so far in either this chapter or the previous one. We now discuss how these physical parameters are related to the quantities we have been working with. Before doing this, let us summarize what we do know:

(a) Given the bit rate R_b and the signal constellation, we know the symbol rate (or more generally, the number of modulation degrees of freedom required per unit time), and hence the minimum Nyquist bandwidth $B_{\min}$. We can then factor in the excess bandwidth a dictated by implementation considerations to find the bandwidth $B = (1+a)B_{\min}$ required.

(b) Given the constellation and a desired bit error probability, we can infer the E_b/N_0 we need to operate at. Since the SNR satisfies $\text{SNR} = E_b R_b / N_0 B$, we have

$$\text{SNR}_{\text{reqd}} = \left(\frac{E_b}{N_0}\right)_{\text{reqd}} \frac{R_b}{B}. \tag{3.81}$$

(c) Given the receiver noise figure F (dB), we can infer the noise power $P_n = N_0 B = kT_0 10^{F/10} B$, and hence the minimum required received signal power is given by

$$P_{\text{RX}}(\min) = \text{SNR}_{\text{reqd}} P_n = \left(\frac{E_b}{N_0}\right)_{\text{reqd}} \frac{R_b}{B} P_n. \tag{3.82}$$

This is called the required *receiver sensitivity*, and is usually quoted in dBm, as $P_{\text{RX,dBm}}(\min) = 10 \log_{10} P_{\text{RX}}(\min)(\text{mW})$.

Now that we know the required received power for "closing" the link, all we need to do is to figure out link parameters such that the receiver actually gets at least that much power, plus a link margin (typically expressed in dB). Let us do this using the example of an idealized line-of-sight wireless link. In this case, if P_{TX} is the transmitted power, then the received power is obtained using Friis' formula for propagation loss in free space:

$$P_{\text{RX}} = P_{\text{TX}} G_{\text{TX}} G_{\text{RX}} \frac{\lambda^2}{16\pi^2 R^2}, \tag{3.83}$$

where

- G_{TX} is the gain of the transmit antenna,
- G_{RX} is the gain of the receive antenna,

- $\lambda = c/f_c$ is the carrier wavelength ($c = 3 \times 10^8$ m/s, is the speed of light, f_c the carrier frequency),
- R is the range (line-of-sight distance between transmitter and receiver).

Note that the antenna gains are with respect to an isotropic radiator. As with most measures related to power in communication systems, antenna gains are typically expressed on the logarithmic scale, in dBi, where $G_{\text{dBi}} = 10 \log_{10} G$ for an antenna with raw gain G.

It is convenient to express the preceding equation in the logarithmic scale to convert the multiplicative factors into addition. For example, expressing the powers in dBm and the gains in dB, we have

$$P_{\text{RX,dBm}} = P_{\text{TX,dBm}} + G_{\text{TX,dBi}} + G_{\text{RX,dBi}} + 10 \log_{10} \frac{\lambda^2}{16\pi^2 R^2} \tag{3.84}$$

where the antenna gains are expressed in dBi (referenced to the 0 dB gain of an isotropic antenna). More generally, we have the link budget equation

$$P_{\text{RX,dBm}} = P_{\text{TX,dBm}} + G_{\text{TX,dB}} + G_{\text{RX,dB}} - L_{\text{path,dB}}(R) \tag{3.85}$$

where $L_{\text{path,dB}}(R)$ is the path loss in dB. For free space propagation, we have from Friis' formula (3.84) that

$$L_{\text{path,dB}}(R) = -10 \log_{10} \frac{\lambda^2}{16\pi^2 R^2} \quad \text{path loss in dB for free space propagation.} \tag{3.86}$$

However, we can substitute any other expression for path loss in (3.85), depending on the propagation environment we are operating under. For example, for wireless communication in a cluttered environment, the signal power may decay as $1/R^4$ rather than the free space decay of $1/R^2$. Propagation measurements, along with statistical analysis, are typically used to characterize the path loss as a function of range for the system being designed. Once we decide on the path loss formula ($L_{\text{path,dB}}(R)$) to be used in the design, the transmit power required to attain a given receiver sensitivity can be determined as a function of range R. Such a path loss formula typically characterizes an "average" operating environment, around which there might be significant statistical variations that are not captured by the model used to arrive at the receiver sensitivity. For example, the receiver sensitivity for a wireless link may be calculated based on the AWGN channel model, whereas the link may exhibit rapid amplitude variations due to multipath fading, and slower variations due to shadowing (e.g., due to buildings and other obstacles). Even if fading or shadowing effects are factored into the channel model used to compute the BER, and the model for path loss, the actual environment encountered may be worse than that assumed in the model. In general, therefore, we add a link margin $L_{\text{margin,dB}}$, again expressed in dB, in an attempt to budget for potential performance losses due to unmodeled or unforeseen

3.7 Elements of link budget analysis

impairments. The size of the link margin depends, of course, on the confidence of the system designer in the models used to arrive at the rest of the link budget.

Putting this all together, if $P_{\text{RX,dBm}}(\min)$ is the desired receiver sensitivity (i.e., the minimum required received power), then we compute the transmit power for the link to be

$$P_{\text{TX,dBm}} = P_{\text{RX,dBm}}(\min) - G_{\text{TX,dB}} - G_{\text{RX,dB}} + L_{\text{path,dB}}(R) + L_{\text{margin,dB}}. \quad (3.87)$$

Let me now illustrate these concepts using an example.

Example 3.7.1 Consider again the 5 GHz WLAN link of Example 3.1.4. We wish to utilize a 20 MHz channel, using QPSK and an excess bandwidth of 33%. The receiver has a noise figure of 6 dB.

(a) What is the bit rate?
(b) What is the receiver sensitivity required to achieve a bit error rate (BER) of 10^{-6}?
(c) Assuming transmit and receive antenna gains of 2 dBi each, what is the range achieved for 100 mW transmit power, using a link margin of 20 dB? Use link budget analysis based on free space path loss.

Solution to (a) For bandwidth B and fractional excess bandwidth a, the symbol rate

$$R_s = \frac{1}{T} = \frac{B}{1+a} = \frac{20}{1+0.33} = 15 \text{ Msymbol/s}$$

and the bit rate for an M-ary constellation is

$$R_b = R_s \log_2 M = 15 \text{ Msymbol/s} \times 2 \text{ bit/symbol} = 30 \text{ Mbit/s}.$$

Solution to (b) The BER for QPSK with Gray coding is $Q\left(\sqrt{2E_b/N_0}\right)$. For a desired BER of 10^{-6}, we obtain that $(E_b/N_0)_{\text{reqd}}$ is about 10.2 dB. From (3.81), we obtain

$$\text{SNR}_{\text{reqd}} = 10.2 + 10\log_{10}\frac{30}{20} \approx 12 \text{ dB}.$$

We know from Example 3.1.4 that the noise power is -95 dBm. Thus, the desired receiver sensitivity is

$$P_{\text{RX,dBm}}(\min) = P_{n,\text{dBm}} + \text{SNR}_{\text{reqd,dB}} = -95 + 12 = -83 \text{ dBm}.$$

Solution to (c) The transmit power is 100 mW, or 20 dBm. Rewriting (3.87), the allowed path loss to attain the desired sensitivity at the desired link margin is

$$L_{\text{path,dB}}(R) = P_{\text{TX,dBm}} - P_{\text{RX,dBm}}(\min) + G_{\text{TX,dBi}} + G_{\text{RX,dBi}} - L_{\text{margin,dB}}$$
$$= 20 - (-83) + 2 + 2 - 20 = 87 \text{ dB}. \quad (3.88)$$

> We can now invert the formula for free space loss, (3.86), to get a range R of 107 meters, which is of the order of the advertised ranges for WLANs under nominal operating conditions. The range decreases, of course, for higher bit rates using larger constellations. What happens, for example, when we use 16-QAM or 64-QAM?

3.8 Further reading

Most communication theory texts, including those mentioned in Section 1.3, cover signal space concepts in some form. These concepts were first presented in a cohesive manner in the classic text by Wozencraft and Jacobs [10], which remains recommended reading. The fundamentals of detection and estimation can be explored further using the text by Poor [19].

3.9 Problems

3.9.1 Gaussian basics

Problem 3.1 Two random variables X and Y have joint density

$$p_{X,Y}(x, y) = \begin{cases} Ke^{-\frac{2x^2+y^2}{2}} & xy \geq 0, \\ 0 & xy < 0. \end{cases}$$

(a) Find K.
(b) Show that X and Y are each Gaussian random variables.
(c) Express the probability $P[X^2 + X > 2]$ in terms of the Q function.
(d) Are X and Y jointly Gaussian?
(e) Are X and Y independent?
(f) Are X and Y uncorrelated?
(g) Find the conditional density $p_{X|Y}(x|y)$. Is it Gaussian?

Problem 3.2 (Computations for Gaussian random vectors) The random vector $\mathbf{X} = (X_1 X_2)^T$ is Gaussian with mean vector $\mathbf{m} = (2, 1)^T$ and covariance matrix $\mathbf{C}$ given by

$$\mathbf{C} = \begin{pmatrix} 1 & -1 \\ -1 & 4 \end{pmatrix}.$$

(a) Let $Y_1 = X_1 + 2X_2$, $Y_2 = -X_1 + X_2$. Find $\text{cov}(Y_1, Y_2)$.
(b) Write down the joint density of Y_1 and Y_2.
(c) Express the probability $P[Y_1 > 2Y_2 + 1]$ in terms of the Q function.

3.9 Problems

Problem 3.3 (Bounds on the Q function) We derive the bounds (3.5) and (3.4) for

$$Q(x) = \int_x^\infty \frac{1}{\sqrt{2\pi}} e^{-t^2/2} dt. \tag{3.89}$$

(a) Show that, for $x \geq 0$, the following upper bound holds:

$$Q(x) \leq \frac{1}{2} e^{-x^2/2}.$$

Hint Try pulling out a factor of e^{-x^2} from (3.89), and then bounding the resulting integrand. Observe that $t \geq x \geq 0$ in the integration interval.

(b) For $x \geq 0$, derive the following upper and lower bounds for the Q function:

$$\left(1 - \frac{1}{x^2}\right) \frac{e^{-x^2/2}}{\sqrt{2\pi} x} \leq Q(x) \leq \frac{e^{-x^2/2}}{\sqrt{2\pi} x}.$$

Hint Write the integrand in (3.89) as a product of $1/t$ and $te^{-t^2/2}$ and then integrate by parts to get the upper bound. Integrate by parts once more using a similar trick to get the lower bound. Note that you can keep integrating by parts to get increasingly refined upper and lower bounds.

Problem 3.4 (From Gaussian to Rayleigh, Rician, and exponential random variables) Let X_1, X_2 be i.i.d. Gaussian random variables, each with mean zero and variance v^2. Define (R, Φ) as the polar representation of the point (X_1, X_2), i.e.,

$$X_1 = R \cos \Phi, \quad X_2 = R \sin \Phi,$$

where $R \geq 0$ and $\Phi \in [0, 2\pi]$.

(a) Find the joint density of R and Φ.
(b) Observe from (a) that R, Φ are independent. Show that Φ is uniformly distributed in $[0, 2\pi]$, and find the marginal density of R.
(c) Find the marginal density of R^2.
(d) What is the probability that R^2 is at least 20 dB below its mean value? Does your answer depend on the value of v^2?

Remark The random variable R is said to have a Rayleigh distribution. Further, you should recognize that R^2 has an exponential distribution. We use these results when we discuss noncoherent detection and Rayleigh fading in Chapters 4 and 8.

(e) Now, assume that $X_1 \sim N(m_1, v^2)$, $X_2 \sim N(m_2, v^2)$ are independent, where m_1 and m_2 may be nonzero. Find the joint density of R and Φ, and the marginal density of R. Express the latter in terms of the modified Bessel function

$$I_0(x) = \frac{1}{2\pi}\int_0^{2\pi} \exp(x\cos\theta)\, d\theta.$$

Remark The random variable R is said to have a Rician distribution in this case. This specializes to a Rayleigh distribution when $m_1 = m_2 = 0$.

Problem 3.5 (Geometric derivation of Q function bound) Let X_1 and X_2 denote independent standard Gaussian random variables.

(a) For $a > 0$, express $P[|X_1| > a, |X_2| > a]$ in terms of the Q function.
(b) Find $P[X_1^2 + X_2^2 > 2a^2]$.

Hint Transform to polar coordinates. Or use the results of Problem 3.4.

(c) Sketch the regions in the (x_1, x_2) plane corresponding to the events considered in (a) and (b).
(d) Use (a)–(c) to obtain an alternative derivation of the bound $Q(x) \leq \frac{1}{2}e^{-x^2/2}$ for $x \geq 0$ (i.e., the bound in Problem 3.3(a)).

3.9.2 Hypothesis testing basics

Problem 3.6 The received signal in a digital communication system is given by

$$y(t) = \begin{cases} s(t) + n(t) & 1 \text{ sent}, \\ n(t) & 0 \text{ sent}, \end{cases}$$

where n is AWGN with PSD $\sigma^2 = N_0/2$ and $s(t)$ is as shown below. The received signal is passed through a filter, and the output is sampled to yield a decision statistic. An ML decision rule is employed based on the decision statistic. The set-up is shown in Figure 3.23.

Figure 3.23 Set-up for Problem 3.6.

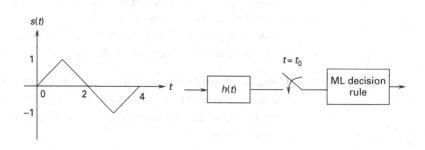

(a) For $h(t) = s(-t)$, find the error probability as a function of E_b/N_0 if $t_0 = 1$.
(b) Can the error probability in (a) be improved by choosing the sampling time t_0 differently?
(c) Now find the error probability as a function of E_b/N_0 for $h(t) = I_{[0,2]}$ and the best possible choice of sampling time.
(d) Finally, comment on whether you can improve the performance in (c) by using a linear combination of two samples as a decision statistic, rather than just using one sample.

Problem 3.7 Find and sketch the decision regions for a binary hypothesis testing problem with observation Z, where the hypotheses are equally likely, and the conditional distributions are given by
H_0: Z is uniform over $[-2, 2]$,
H_1: Z is Gaussian with mean 0 and variance 1.

Problem 3.8 The receiver in a binary communication system employs a decision statistic Z which behaves as follows:
$Z = N$ if 0 is sent,
$Z = 4 + N$ if 1 is sent,
where N is modeled as Laplacian with density
$$p_N(x) = \frac{1}{2}e^{-|x|}, \quad -\infty < x < \infty.$$
Note Parts (a) and (b) can be done independently.

(a) Find and sketch, as a function of z, the log likelihood ratio
$$K(z) = \log L(z) = \log \frac{p(z|1)}{p(z|0)},$$
where $p(z|i)$ denotes the conditional density of Z given that i is sent ($i = 0, 1$).
(b) Find $P_{e|1}$, the conditional error probability given that 1 is sent, for the decision rule
$$\delta(z) = \begin{cases} 0, & z < 1, \\ 1, & z \geq 1. \end{cases}$$
(c) Is the rule in (b) the MPE rule for any choice of prior probabilities? If so, specify the prior probability $\pi_0 = P[0 \text{ sent}]$ for which it is the MPE rule. If not, say why not.

Problem 3.9 The output of the receiver in an optical on–off keyed system is a photon count Y, where Y is a Poisson random variable with mean m_1 if 1 is sent, and mean m_0 if 0 is sent (assume $m_1 > m_0$). Assume that 0 and 1 are equally likely to be sent.

(a) Find the form of the ML rule. Simplify as much as possible, and explicitly specify it for $m_1 = 100$, $m_0 = 10$.

(b) Find expressions for the conditional error probabilities $P_{e|i}$, $i = 0, 1$ for the ML rule, and give numerical values for $m_1 = 100$, $m_0 = 10$.

Problem 3.10 Consider hypothesis testing based on the decision statistic Y, where $Y \sim N(1, 4)$ under H_1 and $Y \sim N(-1, 1)$ under H_0.

(a) Show that the optimal (ML or MPE) decision rule is equivalent to comparing a function of the form $ay^2 + by$ with a threshold.

(b) Specify the rule explicitly (i.e., specify a, b and the threshold) for the MPE rule when $\pi_0 = 1/3$.

3.9.3 Receiver design and performance analysis for the AWGN channel

Problem 3.11 Let $p_1(t) = I_{[0,1]}(t)$ denote a rectangular pulse of unit duration. Consider two 4-ary signal sets as follows:

Signal set A: $s_i(t) = p_1(t-i)$, $i = 0, 1, 2, 3$.
Signal set B: $s_0(t) = p_1(t) + p_1(t-3)$, $s_1(t) = p_1(t-1) + p_1(t-2)$, $s_2(t) = p_1(t) + p_1(t-2)$, $s_3(t) = p_1(t-1) + p_1(t-3)$.

(a) Find signal space representations for each signal set with respect to the orthonormal basis $\{p_1(t-i), i = 0, 1, 2, 3\}$.

(b) Find union bounds on the average error probabilities for both signal sets as a function of E_b/N_0. At high SNR, what is the penalty in dB for using signal set B?

(c) Find an exact expression for the average error probability for signal set B as a function of E_b/N_0.

Problem 3.12 Three 8-ary signal constellations are shown in Figure 3.24.

(a) Express R and $d_{min}^{(2)}$ in terms of $d_{min}^{(1)}$ so that all three constellations have the same E_b.

(b) For a given E_b/N_0, which constellation do you expect to have the smallest bit error probability over a high SNR AWGN channel?

(c) For each constellation, determine whether you can label signal points using three bits so that the label for nearest neighbors differs by at most

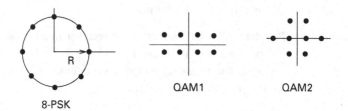

Figure 3.24 Signal constellations for Problem 3.12.

one bit. If so, find such a labeling. If not, say why not and find some "good" labeling.

(d) For the labelings found in part (c), compute nearest neighbors approximations for the average bit error probability as a function of E_b/N_0 for each constellation. Evaluate these approximations for $E_b/N_0 = 15$ dB.

Problem 3.13 Consider the signal constellation shown in Figure 3.25, which consists of two QPSK constellations of different radii, offset from each other by $\pi/4$. The constellation is to be used to communicate over a passband AWGN channel.

(a) Carefully redraw the constellation (roughly to scale, to the extent possible) for $r = 1$ and $R = \sqrt{2}$. Sketch the ML decision regions.
(b) For $r = 1$ and $R = \sqrt{2}$, find an intelligent union bound for the conditional error probability, given that a signal point from the inner circle is sent, as a function of E_b/N_0.
(c) How would you choose the parameters r and R so as to optimize the power efficiency of the constellation (at high SNR)?

Problem 3.14 (Exact symbol error probabilities for rectangular constellations) Assuming each symbol is equally likely, derive the following expressions for the average error probability for 4-PAM and 16-QAM:

$$P_e = \frac{3}{2} Q\left(\sqrt{\frac{4E_b}{5N_0}}\right), \quad \text{symbol error probability for 4-PAM} \quad (3.90)$$

$$P_e = 3Q\left(\sqrt{\frac{4E_b}{5N_0}}\right) - \frac{9}{4} Q^2\left(\sqrt{\frac{4E_b}{5N_0}}\right), \quad \text{symbol error probability for 16-QAM.} \quad (3.91)$$

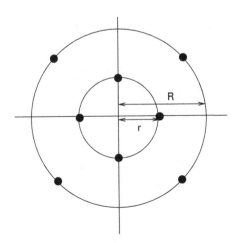

Figure 3.25 Constellation for Problem 3.13.

(Assume 4-PAM with equally spaced levels symmetric about the origin, and rectangular 16-QAM equivalent to two 4-PAM constellations independently modulating the I and Q components.)

Problem 3.15 (Symbol error probability for PSK) In this problem, we derive an expression for the symbol error probability for M-ary PSK that requires numerical evaluation of a single integral over a finite interval. Figure 3.26 shows the decision boundaries corresponding to a point in a PSK constellation. A two-dimensional noise vector originating from the signal point must reach beyond the boundaries to cause an error. A direct approach to evaluating error probability requires integration of the two-dimensional Gaussian density over an infinite region. We avoid this by switching to polar coordinates, with the noise vector having radius L and angle θ as shown.

(a) Owing to symmetry, the error probability equals twice the probability of the noise vector crossing the top decision boundary. Argue that this happens if $L > d(\theta)$ for some $\theta \in (0, \pi - \pi/M)$.

(b) Show that the probability of error is given by

$$P_e = 2 \int_0^{\pi - \frac{\pi}{M}} P[L > d(\theta)|\theta] p(\theta) \, d\theta.$$

(c) Use Problem 3.4 to show that $P[L > d] = e^{-\frac{d^2}{2\sigma^2}}$, that L is independent of θ, and that θ is uniform over $[0, 2\pi]$.

(d) Show that $d(\theta) = (R \sin \pi/M)/(\sin(\theta + \pi/M))$.

(e) Conclude that the error probability is given by

$$P_e = \frac{1}{\pi} \int_0^{\pi - \frac{\pi}{M}} e^{-\frac{R^2 \sin^2 \frac{\pi}{M}}{2\sigma^2 \sin^2(\theta + \frac{\pi}{M})}} \, d\theta.$$

(f) Use the change of variable $\phi = \pi - (\theta + \pi/M)$ (or alternatively, realize that $\theta + \pi/M \pmod{2\pi}$ is also uniform over $[0, 2\pi]$) to conclude that

$$P_e = \frac{1}{\pi} \int_0^{\pi - \frac{\pi}{M}} e^{-\frac{R^2 \sin^2 \frac{\pi}{M}}{2\sigma^2 \sin^2 \phi}} \, d\phi = \frac{1}{\pi} \int_0^{\pi - \frac{\pi}{M}} e^{-\frac{E_b \log_2 M \sin^2 \frac{\pi}{M}}{N_0 \sin^2 \phi}} \, d\phi, \qquad (3.92)$$

symbol error probability for M-ary PSK.

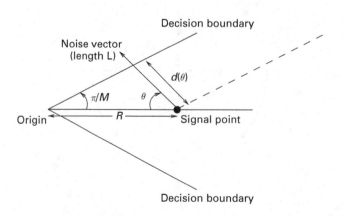

Figure 3.26 Figure for Problem 3.15.

Figure 3.27 Constellation for Problem 3.16.

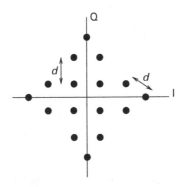

Problem 3.16 The signal constellation shown in Figure 3.27 is obtained by moving the outer corner points in rectangular 16-QAM to the I and Q axes.

(a) Sketch the ML decision regions.
(b) Is the constellation more or less power efficient than rectangular 16-QAM?

Problem 3.17 A 16-ary signal constellation consists of four signals with coordinates $(\pm-1, \pm-1)$, four others with coordinates $(\pm 3, \pm 3)$, and two each having coordinates $(\pm 3, 0)$, $(\pm 5, 0)$, $(0, \pm 3)$, and $(0, \pm 5)$, respectively.

(a) Sketch the signal constellation and indicate the ML decision regions.
(b) Find an intelligent union bound on the average symbol error probability as a function of E_b/N_0.
(c) Find the nearest neighbors approximation to the average symbol error probability as a function of E_b/N_0.
(d) Find the nearest neighbors approximation to the average symbol error probability for 16-QAM as a function of E_b/N_0.
(e) Comparing (c) and (d) (i.e., comparing the performance at high SNR), which signal set is more power efficient?

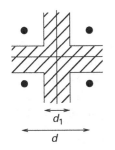

Figure 3.28 QPSK with erasures.

Problem 3.18 (adapted from [13]) A QPSK demodulator is designed to put out an *erasure* when the decision is ambivalent. Thus, the decision regions are modified as shown in Figure 3.28, where the crosshatched region corresponds to an erasure. Set $\alpha = d_1/d$, where $0 \leq \alpha \leq 1$.

(a) Use the intelligent union bound to find approximations to the probability p of symbol error and the probability q of symbol erasure in terms of E_b/N_0 and α.
(b) Find exact expressions for p and q as functions of E_b/N_0 and α.
(c) Using the approximations in (a), find an approximate value for α such that $q = 2p$ for $E_b/N_0 = 4$ dB.

Figure 3.29 Signal constellation with unequal error protection (Problem 3.19).

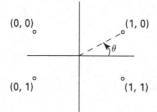

Remark The motivation for (c) is that a typical error-correcting code can correct twice as many erasures as errors.

Problem 3.19 Consider the constant modulus constellation shown in Figure 3.29 where $\theta \leq \pi/4$. Each symbol is labeled by two bits (b_1, b_2), as shown. Assume that the constellation is used over a complex baseband AWGN channel with noise power spectral density (PSD) $N_0/2$ in each dimension. Let $(\hat{b}_1, \hat{b}_2)$ denote the maximum likelihood (ML) estimates of (b_1, b_2).

(a) Find $P_{e1} = P[\hat{b}_1 \neq b_1]$ and $P_{e2} = P[\hat{b}_2 \neq b_2]$ as a function of E_s/N_0, where E_s denotes the signal energy.

(b) Assume now that the transmitter is being heard by two receivers, $R1$ and $R2$, and that $R2$ is twice as far away from the transmitter as $R1$. Assume that the received signal energy falls off as $1/r^4$, where r is the distance from the transmitter, and that the noise PSD for both receivers is identical. Suppose that $R1$ can demodulate both bits $b1$ and $b2$ with error probability at least as good as 10^{-3}, i.e., so that $\max\{P_{e1}(R1), P_{e2}(R1)\} = 10^{-3}$. Design the signal constellation (i.e., specify θ) so that $R2$ can demodulate at least one of the bits with the same error probability, i.e., such that $\min\{P_{e1}(R2), P_{e2}(R2)\} = 10^{-3}$.

Remark You have designed an unequal error protection scheme in which the receiver that sees a poorer channel can still extract part of the information sent.

Problem 3.20 (Demodulation with amplitude mismatch) Consider a 4-PAM system using the constellation points $\{\pm 1, \pm 3\}$. The receiver has an accurate estimate of its noise level. An automatic gain control (AGC) circuit is supposed to scale the decision statistics so that the noiseless constellation points are in $\{\pm 1, \pm 3\}$. The ML decision boundaries are set according to this nominal scaling.

(a) Suppose that the AGC scaling is faulty, and the *actual* noiseless signal points are at $\{\pm 0.9, \pm 2.7\}$. Sketch the points and the mismatched decision regions. Find an intelligent union bound for the symbol error probability in terms of the Q function and E_b/N_0.

(b) Repeat (a), assuming that faulty AGC scaling puts the noiseless signal points at $\{\pm 1.1, \pm 3.3\}$.

(c) The AGC circuits try to maintain a constant output power as the input power varies, and can be viewed as imposing a scale factor on the input inversely proportional to the square root of the input power. In (a), does the AGC circuit overestimate or underestimate the input power?

Problem 3.21 (Demodulation with phase mismatch) Consider a BPSK system in which the receiver's estimate of the carrier phase is off by θ.

(a) Sketch the I and Q components of the decision statistic, showing the noiseless signal points and the decision region.

(b) Derive the BER as a function of θ and E_b/N_0 (assume that $\theta < \pi/2$).

(c) Assuming now that θ is a random variable taking values uniformly in $[-\pi/4, \pi/4]$, numerically compute the BER averaged over θ, and plot it as a function of E_b/N_0. Plot the BER without phase mismatch as well, and estimate the dB degradation due to the phase mismatch.

Problem 3.22 (Soft decisions for BPSK) Consider a BPSK system in which 0 and 1 are equally likely to be sent, with 0 mapped to $+1$ and 1 to -1 as usual.

(a) Show that the LLR is conditionally Gaussian given the transmitted bit, and that the conditional distribution is scale-invariant, depending only on the SNR.

(b) If the BER for hard decisions is 10%, specify the conditional distribution of the LLR, given that 0 is sent.

Problem 3.23 (Soft decisions for PAM) Consider a 4-PAM constellation in which the signal levels at the receiver have been scaled to $\pm 1, \pm 3$. The system is operating at E_b/N_0 of 6 dB. Bits $b_1, b_2 \in \{0, 1\}$ are mapped to the symbols using Gray coding. Assume that $(b_1, b_2) = (0, 0)$ for symbol -3, and $(1, 0)$ for symbol $+3$.

(a) Sketch the constellation, along with the bitmaps. Indicate the ML hard decision boundaries.

(b) Find the posterior symbol probability $P[-3|y]$ as a function of the noisy observation y. Plot it as a function of y.

Hint The noise variance can be inferred from the signal levels and SNR.

(c) Find $P[b_1 = 1|y]$ and $P[b_2 = 1|y]$, and plot each as a function of y.

(d) Display the results of part (c) in terms of LLRs.

$$LLR_1(y) = \log \frac{P[b_1 = 0|y]}{P[b_1 = 1|y]}, \quad LLR_2(y) = \log \frac{P[b_2 = 0|y]}{P[b_2 = 1|y]}.$$

Plot the LLRs as a function of y, saturating the values as ± 50.

(e) Try other values of E_b/N_0 (e.g., 0 dB, 10 dB). Comment on any trends you notice. How do the LLRs vary as a function of distance from the noiseless signal points? How do they vary as you change E_b/N_0?

(f) Simulate the system over multiple symbols at E_b/N_0 such that the BER is about 5%. Plot the histograms of the LLRs for each of the two bits, and comment on whether they look Gaussian. What happens as you increase or decrease E_b/N_0?

Problem 3.24 (Soft decisions for PSK) Consider the Gray coded 8-PSK constellation shown in Figure 3.30, labeled with bits (b_1, b_2, b_3). The received samples are ISI-free, with the noise contribution modeled as discrete-time WGN with variance 0.1 per dimension. The system operates at an E_b/N_0 of 8 dB.

(a) Use the nearest neighbors approximation to estimate the BER for hard decisions.
(b) For a received sample $y = 2e^{-j2\pi/3}$, find the hard decisions on the bits.
(c) Find the LLRs for each of the three bits for the received sample in (b).
(d) Now, simulate the system over multiple symbols at E_b/N_0 such that the BER for hard decisions is approximately 5%. Plot the histograms of the LLRs of each of the three bits, and comment on their shapes. What happens as you increase or decrease E_b/N_0?

Problem 3.25 (Exact performance analysis for M-ary orthogonal signaling) Consider an M-ary orthogonal equal-energy signal set

Figure 3.30 Gray coded 8-PSK constellation for Problem 3.24.

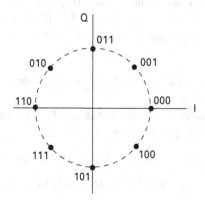

$\{s_i, i = 1, \ldots, M\}$ with $\langle s_i, s_j \rangle = E_s \delta_{ij}$, for $1 \leq i, j \leq M$. Condition on s_1 being sent, so that the received signal $y = s_1 + n$, where n is WGN with PSD $\sigma^2 = N_0/2$. The ML decision rule in this case is given by

$$\delta_{ML}(y) = \arg \max_{1 \leq i \leq M} Z_i,$$

where $Z_i = \langle y, s_i \rangle$, $i = 1, \ldots, M$. Let $\mathbf{Z} = (Z_1, \ldots, Z_M)^T$.

(a) Show that $\{Z_i\}$ are (conditionally) independent, with $Z_1 \sim N(E_s, \sigma^2 E_s)$ and $Z_i \sim N(0, \sigma^2 E_s)$.

(b) Show that the conditional probability of correct reception (given s_1 sent) is given by

$$P_{c|1} = P[Z_1 = \max_i Z_i] = P[Z_1 \geq Z_2, Z_1 \geq Z_3, \ldots, Z_1 \geq Z_M]$$

$$= \int_{-\infty}^{\infty} [\Phi(x)]^{M-1} \frac{1}{\sqrt{2\pi}} e^{-(x-m)^2/2} \, dx, \tag{3.93}$$

where

$$m = \sqrt{\frac{2E_s}{N_0}}.$$

Hint Scale the $\{Z_i\}$ so that they have unit variance (this does not change the outcome of the decision, since they all have the same variance). Condition on the value of Z_1.

(c) Show that the conditional probability of error (given s_1 sent) is given by

$$P_{e|1} = P[Z_1 < \max_i Z_i] = 1 - P[Z_1 = \max_i Z_i]$$

$$= (M-1) \int_{-\infty}^{\infty} [\Phi(x)]^{M-2} \Phi(x-m) \frac{1}{\sqrt{2\pi}} e^{-x^2/2} \, dx. \tag{3.94}$$

Hint One approach is to use (3.93) and integrate by parts. Alternatively, decompose the event of getting an error $\{Z_1 < \max_i Z_i\}$ into $M-1$ disjoint events and evaluate their probabilities. Note that events such as $Z_i = Z_j$ for $i \neq j$ have zero probability.

Remark The probabilities (3.93) and (3.94) sum up to one, but (3.94) is better for numerical evaluation when the error probability is small.

(d) Compute and plot the probability of error (log scale) versus E_b/N_0 (dB), for $M = 4, 8, 16, 32$. Comment on what happens with increasing M.

Problem 3.26 (M-ary orthogonal signaling performance as $M \to \infty$) We wish to derive the result that

$$\lim_{M \to \infty} P(\text{correct}) = \begin{cases} 1, & \frac{E_b}{N_0} > \ln 2, \\ 0, & \frac{E_b}{N_0} < \ln 2. \end{cases} \tag{3.95}$$

(a) Show that

$$P(\text{correct}) = \int_{-\infty}^{\infty} \left[\Phi\left(x + \sqrt{\frac{2E_b \log_2 M}{N_0}}\right) \right]^{M-1} \frac{1}{\sqrt{2\pi}} e^{-x^2/2} \, dx.$$

Hint Use Problem 3.25(b).

(b) Show that, for any x,

$$\lim_{M \to \infty} \left[\Phi\left(x + \sqrt{\frac{2E_b \log_2 M}{N_0}}\right) \right]^{M-1} = \begin{cases} 0 & \frac{E_b}{N_0} < \ln 2, \\ 1 & \frac{E_b}{N_0} > \ln 2. \end{cases}$$

Hint Use L'Hôpital's rule on the log of the expression whose limit is to be evaluated.

(c) Substitute (b) into the integral in (a) to infer the desired result.

Problem 3.27 (Preview of Rayleigh fading) We shall show in Chapter 8 that constructive and destructive interference between multiple paths in wireless systems lead to large fluctuations in received amplitude, modeled as a Rayleigh random variable A (see Problem 3.4). The energy per bit is therefore proportional to A^2. Thus, using Problem 3.4(c), we can model E_b as an exponential random variable with mean $\bar{E}_b$, where $\bar{E}_b$ is the *average* energy per bit.

(a) Show that the BER for BPSK over a Rayleigh fading channel is given by

$$P_e = \frac{1}{2}\left(1 - \left(1 + \frac{N_0}{\bar{E}_b}\right)^{-\frac{1}{2}}\right).$$

How does the BER decay with $\bar{E}_b/N_0$ at high SNR?

Hint Compute $\mathbb{E}\left[Q\left(\sqrt{2E_b/N_0}\right)\right]$ using the distribution of E_b/N_0. Integrate by parts to evaluate.

(b) Plot BER versus $\bar{E}_b/N_0$ for BPSK over the AWGN and Rayleigh fading channels (BER on log scale, $\bar{E}/N_0$ in dB). Note that $\bar{E}_b = E_b$ for the AWGN channel. At BER of 10^{-4}, what is the degradation in dB due to Rayleigh fading?

Problem 3.28 (ML decision rule for multiuser systems) Consider 2-user BPSK signaling in AWGN, with received signal

$$\mathbf{y} = b_1 \mathbf{u}_1 + b_2 \mathbf{u}_2 + \mathbf{n}, \tag{3.96}$$

where $\mathbf{u}_1 = (-1, -1)^T$, $\mathbf{u}_2 = (2, 1)^T$, b_1, b_2 take values ± 1 with equal probability, and $\mathbf{n}$ is AWGN of variance σ^2 per dimension.

(a) Draw the decision regions in the (y_1, y_2) plane for making joint ML decisions for b_1, b_2, and specify the decision if $\mathbf{y} = (2.5, 1)^T$.

Hint Think of this as M-ary signaling in WGN, with $M = 4$.

(b) Find an intelligent union bound on the conditional error probability, conditioned on $b_1 = +1, b_2 = +1$, for the joint ML rule (an error occurs if either of the bits are wrong). Repeat for $b_1 = +1, b_2 = -1$.

(c) Find the average error probability $P_{e,1}$ that the joint ML rule makes an error in b_1.

Hint Use the results of part (b).

(d) If the second user were not transmitting (remove the term $b_2\mathbf{u}_2$ from (3.96)), sketch the ML decision region for b_1 in the (y_1, y_2) plane and evaluate the error probability $P_{e,1}^{su}$ for b_1, where the superscript "su" denotes "single user."

(e) Find the rates of exponential decay as $\sigma^2 \to 0$ for the error probabilities in (c) and (d). That is, find $a, b \geq 0$ such that $P_{e,1} \doteq e^{-\frac{a^2}{\sigma^2}}$ and $P_{e,1}^{su} \doteq e^{-\frac{b^2}{\sigma^2}}$.

Remark The ratio a/b is the *asymptotic efficiency* (of the joint ML decision rule) for user 1. It measures the fractional degradation in the exponential rate of decay (as SNR increases) of error probability for user 1 due to the presence of user 2.

(f) Redo parts (d) and (e) for user 2.

3.9.4 Link budget analysis

Problem 3.29 You are given an AWGN channel of bandwidth 3 MHz. Assume that implementation constraints dictate an excess bandwidth of 50%. Find the achievable bit rate, the E_b/N_0 required for a BER of 10^{-8}, and the receiver sensitivity (assuming a receiver noise figure of 7 dB) for the following modulation schemes, assuming that the bit-to-symbol map is optimized to minimize the BER whenever possible:

(a) QPSK;
(b) 8-PSK;
(c) 64QAM;
(d) Coherent 16-ary orthogonal signaling.

Remark Use nearest neighbors approximations for the BER.

Problem 3.30 Consider the setting of Example 3.7.1.

(a) For all parameters remaining the same, find the range and bit rate when using a 64QAM constellation.

(b) Suppose now that the channel model is changed from AWGN to Rayleigh fading (see Problem 3.27). Find the receiver sensitivity required for QPSK at BER of 10^{-6}. What is the range, assuming all other parameters are as in Example 3.7.1?

3.9.5 Some mathematical derivations

Problem 3.31 (Properties of covariance matrices) Let $\mathbf{C}$ denote the covariance matrix of a random vector $\mathbf{X}$ of dimension m. Let $\{\lambda_i, i = 1, \ldots, m\}$ denote its eigenvalues, and let $\{\mathbf{v}_i, i = 1, \ldots, m\}$ denote the corresponding eigenvectors, chosen to form an orthonormal basis for $\mathbb{R}^m$ (let us take it for granted that this can always be done). That is, we have $\mathbf{C}\mathbf{v}_i = \lambda_i \mathbf{v}_i$ and $\mathbf{v}_i^T \mathbf{v}_j = \delta_{ij}$.

(a) Show that $\mathbf{C}$ is nonnegative definite. That is, for any vector $\mathbf{a}$, we have $\mathbf{a}^T \mathbf{C} \mathbf{a} \geq 0$.

Hint Show that you can write $\mathbf{a}^T \mathbf{C} \mathbf{a} = E[Y^2]$ for some random variable Y.

(b) Show that any eigenvalue $\lambda_i \geq 0$.

(c) Show that $\mathbf{C}$ can be written in terms of its eigenvectors as follows:

$$\mathbf{C} = \sum_{i=1}^{m} \lambda_i \mathbf{v}_i \mathbf{v}_i^T. \tag{3.97}$$

Hint The matrix equality $\mathbf{A} = \mathbf{B}$ is equivalent to saying that $\mathbf{A}\mathbf{x} = \mathbf{B}\mathbf{x}$ for any vector $\mathbf{x}$. We use this to show that the two sides of (3.97) are equal. For any vector $\mathbf{x}$, consider its expansion $\mathbf{x} = \sum_i x_i \mathbf{v}_i$ with respect to the basis $\{\mathbf{v}_i\}$. Now, show that applying the matrices on each side of (3.97) gives the same result.

The expression (3.97) is called the *spectral factorization* of $\mathbf{C}$, with the eigenvalues $\{\lambda_i\}$ playing the role of a discrete spectrum. The advantage of this view is that, as shown in the succeeding parts of this problem, algebraic operations on the eigenvalues, such as taking their inverse or square root, correspond to analogous operations on the matrix $\mathbf{C}$.

(d) Show that, for $\mathbf{C}$ invertible, the inverse is given by

$$\mathbf{C}^{-1} = \sum_{i=1}^{m} \frac{1}{\lambda_i} \mathbf{v}_i \mathbf{v}_i^T. \tag{3.98}$$

Hint Check this by directly multiplying the right-hand sides of (3.97) and (3.98), and using the orthonormality of the eigenvectors.

(e) Show that the matrix

$$\mathbf{A} = \sum_{i=1}^{m} \sqrt{\lambda_i} \mathbf{v}_i \mathbf{v}_i^T \tag{3.99}$$

3.9 Problems

can be thought of as a square root of $\mathbf{C}$, in that $\mathbf{A}^2 = \mathbf{C}$. We denote this as $\mathbf{C}^{\frac{1}{2}}$.

(f) Suppose now that $\mathbf{C}$ is not invertible. Show that there is a nonzero vector $\mathbf{a}$ such that the entire probability mass of $\mathbf{X}$ lies along the $(m-1)$-dimensional plane $\mathbf{a}^T(\mathbf{X}-\mathbf{m}) = 0$. That is, the m-dimensional joint density of $\mathbf{X}$ does not exist.

Hint If $\mathbf{C}$ is not invertible, then there is a nonzero $\mathbf{a}$ such that $\mathbf{Ca} = 0$. Now left multiply by $\mathbf{a}^T$ and write out $\mathbf{C}$ as an expectation.

Remark In this case, it is possible to write one of the components of $\mathbf{X}$ as a linear combination of the others, and work in a lower-dimensional space for computing probabilities involving $\mathbf{X}$. Note that this result does not require Gaussianity.

Problem 3.32 (Derivation of joint Gaussian density) We wish to derive the density of a real-valued Gaussian random vector $\mathbf{X} = (X_1, \ldots, X_m)^T \sim N(0, \mathbf{C})$, starting from the assumption that any linear combination of the elements of $\mathbf{X}$ is a Gaussian random variable. This can then be translated by $\mathbf{m}$ to get any desired mean vector. To this end, we employ the characteristic function of $\mathbf{X}$, defined as

$$\phi_{\mathbf{X}}(\mathbf{w}) = E[e^{j\mathbf{w}^T\mathbf{X}}] = E[e^{j(w_1 X_1 + \cdots + w_m X_m)}] = \int e^{j\mathbf{w}^T\mathbf{x}} p_{\mathbf{X}}(\mathbf{x}) \, d\mathbf{x}, \quad (3.100)$$

as a multidimensional Fourier transform of $\mathbf{X}$. The density $p_{\mathbf{X}}(\mathbf{x})$ is therefore given by a multidimensional inverse Fourier transform,

$$p_{\mathbf{X}}(\mathbf{x}) = \frac{1}{(2\pi)^m} \int e^{-j\mathbf{w}^T\mathbf{x}} \phi_{\mathbf{X}}(\mathbf{w}) \, d\mathbf{w}. \quad (3.101)$$

(a) Show that the characteristic function of a standard Gaussian random variable $Z \sim N(0,1)$ is given by $\phi_Z(w) = e^{-w^2/2}$.

(b) Set $Y = \mathbf{w}^T \mathbf{X}$. Show that $Y \sim N(0, v^2)$, where $v^2 = \mathbf{w}^T \mathbf{C} \mathbf{w}$.

(c) Use (a) and (b) to show that

$$\phi_{\mathbf{X}}(\mathbf{w}) = e^{-\frac{1}{2}\mathbf{w}^T\mathbf{C}\mathbf{w}}. \quad (3.102)$$

(d) To obtain the density using the integral in (3.101), make the change of variable $\mathbf{u} = \mathbf{C}^{\frac{1}{2}}\mathbf{w}$. Show that you get

$$p_{\mathbf{X}}(\mathbf{x}) = \frac{1}{(2\pi)^m} \frac{1}{|\mathbf{C}|^{\frac{1}{2}}} \int e^{-j\mathbf{u}^T \mathbf{C}^{-\frac{1}{2}}\mathbf{x}} e^{-\frac{1}{2}\mathbf{u}^T\mathbf{u}} \, d\mathbf{u},$$

where $|\mathbf{A}|$ denotes the determinant of a matrix $\mathbf{A}$.

Hint We have $\phi_{\mathbf{X}}(\mathbf{w}) = e^{-\frac{1}{2}\mathbf{u}^T\mathbf{u}}$ and $d\mathbf{u} = |\mathbf{C}^{\frac{1}{2}}| d\mathbf{w}$.

(e) Now, set $\mathbf{C}^{-\frac{1}{2}}\mathbf{x} = \mathbf{z}$, with $p_\mathbf{X}(\mathbf{x}) = f(\mathbf{z})$. Show that

$$f(\mathbf{z}) = \frac{1}{(2\pi)^m} \frac{1}{|\mathbf{C}|^{\frac{1}{2}}} \int e^{-j\mathbf{u}^T\mathbf{z}} e^{-\frac{1}{2}\mathbf{u}^T\mathbf{u}} \, d\mathbf{u} = \frac{1}{|\mathbf{C}|^{\frac{1}{2}}} \prod_{i=1}^{m} \left(\frac{1}{2\pi} \int e^{-ju_i z_i} e^{-u_i^2/2} \, du_i \right).$$

(f) Using (a) to evaluate $f(\mathbf{z})$ in (e), show that

$$f(\mathbf{z}) = \frac{1}{|\mathbf{C}|^{\frac{1}{2}}} \frac{1}{(2\pi)^{\frac{m}{2}}} e^{-\frac{1}{2}\mathbf{z}^T\mathbf{z}}.$$

Now substitute $\mathbf{C}^{-\frac{1}{2}}\mathbf{x} = \mathbf{z}$ to get the density $p_\mathbf{X}(\mathbf{x})$.

CHAPTER 4
Synchronization and noncoherent communication

In Chapter 3, we established a framework for demodulation over AWGN channels under the assumption that the receiver knows and can reproduce the noiseless received signal for each possible transmitted signal. These provide "templates" against which we can compare the noisy received signal (using correlation), and thereby make inferences about the likelihood of each possible transmitted signal. Before the receiver can arrive at these templates, however, it must estimate unknown parameters such as the amplitude, frequency and phase shifts induced by the channel. We discuss *synchronization* techniques for obtaining such estimates in this chapter. Alternatively, the receiver might fold in implicit estimation of these parameters, or average over the possible values taken by these parameters, in the design of the demodulator. *Noncoherent demodulation*, discussed in detail in this chapter, is an example of such an approach to dealing with unknown channel phase. Noncoherent communication is employed when carrier synchronization is not available (e.g., because of considerations of implementation cost or complexity, or because the channel induces a difficult-to-track time-varying phase, as for wireless mobile channels). Noncoherent processing is also an important component of many synchronization algorithms (e.g., for timing synchronization, which often takes place prior to carrier synchronization).

Since there are many variations in individual (and typically proprietary) implementations of synchronization and demodulation algorithms, the focus here is on developing basic principles, and on providing some simple examples of how these principles might be applied. Good transceiver designs are often based on a sound understanding of such principles, together with a willingness to make approximations guided by intuition, driven by implementation constraints, and verified by simulations.

The framework for demodulation developed in Chapter 3 exploited signal space techniques to project the continuous-time signal to a finite-dimensional vector space, and then applied detection theory to characterize optimal receivers. We now wish to apply a similar strategy for the more general problem of parameter estimation, where the parameter may be continuous-valued, e.g., an unknown delay, phase or amplitude. The resulting framework also enables us to recover, as a special case, the results derived earlier for optimal detection for M-ary signaling in AWGN, since this problem can be interpreted as that of estimating an M-valued parameter. The model for the received signal is

$$y(t) = s_\theta(t) + n(t), \tag{4.1}$$

where $\theta \in \Lambda$ indexes the set of possible noiseless received signals, and $n(t)$ is WGN with PSD $\sigma^2 = N_0/2$ *per dimension*. Note that this description captures both real-valued and complex-valued WGN; for the latter, the real part n_c and the imaginary part n_s each has PSD $N_0/2$, so that the sum $n_c + jn_s$ has PSD N_0. The parameter θ may be vector-valued (e.g., when we wish to obtain joint estimates of the amplitude, delay and phase). We develop a framework for optimal parameter estimation that applies to both real-valued and complex-valued signals. We then apply this framework to some canonical problems of synchronization, and to the problem of noncoherent communication.

Map of this chapter We begin by providing a qualitative discussion of the issues facing the receiver designer in Section 4.1, with a focus on the problem of synchronization, which involves estimation and tracking of parameters such as delay, phase, and frequency. We then summarize some basic concepts of parameter estimation in Section 4.2. Estimation of a parameter θ using an observation Y requires knowledge of the conditional density of Y, conditioned on each possible value of θ. In the context of receiver design, the observation is actually a continuous-time analog signal. Thus, an important result is the establishment of the concept of (conditional) density for such signals. To this end, we develop the concept of a *likelihood function*, which is an appropriately defined likelihood ratio playing the role of density for a signal corrupted by AWGN. We then apply this to receiver design in the subsequent sections. Section 4.3 discusses application of parameter estimation to some canonical synchronization problems. Section 4.4 derives optimal noncoherent receivers using the framework of composite hypothesis testing, where we choose between multiple hypotheses (i.e., the possible transmitted signals) when there are some unknown "nuisance" parameters in the statistical relationship between the observation and the hypotheses. In the case of noncoherent communication, the unknown parameter is the carrier phase. Classical examples of modulation formats amenable to noncoherent demodulation, including orthogonal modulation and differential PSK (DPSK), are discussed. Finally, Section 4.5 is devoted to techniques for analyzing the performance of noncoherent systems. An important tool is the concept of proper complex Gaussianity, discussed in Section 4.5.1. Binary noncoherent communication is analyzed in Section 4.5.2; in addition to exact analysis for orthogonal modulation, we also develop geometric insight analogous to the signal space concepts developed for coherent communication in Chapter 3. These concepts provide the building block for the rest of Section 4.5, which discusses M-ary orthogonal signaling, DPSK, and block differential demodulation.

4.1 Receiver design requirements

In this section, we discuss the synchronization tasks underlying a typical receiver design. For concreteness, the discussion is set in the context of linear modulation over a passband channel. Some key transceiver blocks are shown in Figure 4.1.

The transmitted complex baseband signal is given by

$$u(t) = \sum_n b[n] g_{\text{TX}}(t - nT),$$

and is upconverted to passband using a local oscillator (LO) at carrier frequency f_c. Both the local oscillator and the sampling clock are often integer or rational multiples of the natural frequency f_{XO} of a crystal oscillator, and can be generated using a phase locked loop, as shown in Figure 4.2. Detailed description of the operation of a PLL does not fall within our agenda (of developing optimal estimators) in this chapter, but we briefly interpret the PLL as an ML estimator in Example 4.3.3.

Effect of delay The passband signal $u_p(t) = \text{Re}\left(u(t) e^{j2\pi f_c t}\right)$ goes through the channel. For simplicity, we consider a nondispersive channel which causes

Figure 4.1 Block diagram of key transceiver blocks for synchronization and demodulation.

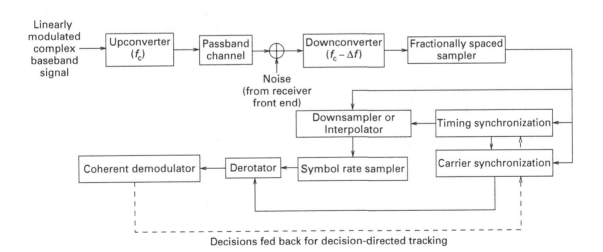

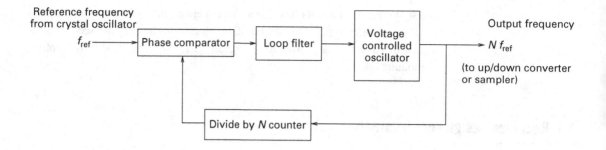

Figure 4.2 Generation of LOs and clocks from a crystal oscillator reference using a PLL.

only amplitude scaling and delay (dispersive channels are considered in the next chapter). Thus, the passband received signal is given by

$$y_p(t) = A u_p(t - \tau) + n_p(t),$$

where A is an unknown amplitude, τ is an unknown delay, and n_p is passband noise. Let us consider the effect of the delay τ in complex baseband. We can write the passband signal as

$$u_p(t - \tau) = \mathrm{Re}\left(u(t-\tau)e^{j2\pi f_c(t-\tau)}\right) = \mathrm{Re}\left(u(t-\tau)e^{j\theta}e^{j2\pi f_c t}\right),$$

where the phase $\theta = -2\pi f_c \tau$ mod 2π is very sensitive to the delay τ, since the carrier frequency f_c is typically very large. We can therefore safely model θ as uniformly distributed over $[0, 2\pi]$, and read off the complex baseband representation of $A u_p(t - \tau)$ with respect to f_c as $A u(t-\tau)e^{j\theta}$, where τ, θ are unknown parameters.

Effect of LO offset The passband received signal y_p is downconverted to complex baseband using a local oscillator, again typically synthesized from a crystal oscillator using a PLL. Crystal oscillators typically have tolerances of the order of 10–100 parts per million (ppm), so that the frequency of the local oscillator at the receiver typically differs from that of the transmitter. Assuming that the frequency of the receiver's LO is $f_c - \Delta f$, the output y of the downconverter is the complex baseband representation of y_p with respect to $f_c - \Delta f$. We therefore obtain the following complex baseband model including unknown delay, frequency offset, and phase.

Complex baseband model prior to synchronization

$$\begin{aligned} y(t) &= A e^{j\theta} u(t - \tau) e^{j2\pi \Delta f t} + n(t) \\ &= A e^{j(2\pi \Delta f t + \theta)} \sum_n b[n] g_{TX}(t - nT - \tau) + n(t), \end{aligned} \quad (4.2)$$

where n is complex WGN.

4.1 Receiver design requirements

Sampling In many modern receiver architectures, the operations on the downconverted complex baseband signal are made using DSP, which can, in principle, implement any desired operation on the original analog signal with arbitrarily high fidelity, as long as the sampling rate is high enough and the analog-to-digital converter has high enough precision. The sampling rate is usually chosen to be an integer multiple of the symbol rate; this is referred to as *fractionally spaced sampling*. For signaling with moderate excess bandwidth (less than 100%), the signal bandwidth is at most $2/T$, hence sampling at rate $2/T$ preserves all the information in the analog received signal. Recall from Chapter 2, however, that reconstruction of an analog signal from its sample requires complicated interpolation using sinc functions, when sampling at the minimum rate required to avoid aliasing. Such interpolation can be simplified (or even eliminated) by sampling even faster, so that sampling at four or eight times the symbol rate is not uncommon in modern receiver implementations. For example, consider the problem of timing synchronization for Nyquist signaling over an ideal communication channel. When working with the original analog signal, our task is to choose sampling points spaced by T which have no ISI. If we sample at rate $8/T$, we have eight symbol-spaced substreams, at least one of which is within at most $T/8$ of the best sampling instant. In this case, we may be willing to live with the performance loss incurred by sampling slightly away from the optimum point, and simply choose the best among the eight substreams. On the other hand, if we sample at rate $2/T$, then there are only two symbol-spaced substreams, and the worst-case offset of $T/2$ yields too high a performance degradation. In this case, we need to interpolate the samples in order to generate a T-spaced stream of samples that we can use for symbol decisions.

The two major synchronization blocks shown in Figure 4.1 are timing synchronization and carrier synchronization.

Timing synchronization The first important task of the timing synchronization block is to estimate the delay τ in (4.2). If the symbols $\{b[n]\}$ are stationary, then the delay τ can only be estimated modulo T, since shifts in the symbol stream are undistinguishable from each other. Thus, to estimate the absolute value of τ, we typically require a subset of the symbols $\{b[n]\}$ to be known, so that we can match what we receive against the expected signal corresponding to these known symbols. These *training symbols* are usually provided in a *preamble* at the beginning of the transmission. This part of timing synchronization usually occurs before carrier synchronization.

We have already observed in (4.2) the consequences of the offset between the transmitter and receiver LOs. A similar observation also applies to the nominal symbol rate at the transmitter and receiver. That is, the symbol time T in the model (4.2) corresponds to the symbol rate clock at the transmitter. The (fractionally spaced) sampler at the receiver operates at $(1+\delta)m/T$, where δ

is of the order of 10–100 ppm (and can be positive or negative), and where m is a positive integer. The relative timing of the T-spaced "ticks" generated by the transmitter and receiver clocks therefore drifts apart significantly over a period of tens of thousands of symbols. If the number of transmitted symbols is significantly smaller than this, which is the case for some packetized systems, then this drift can be ignored. However, when a large number of symbols are sent, the timing synchronization block must track the drift in symbol timing. Training symbols are no longer available at this point, hence the algorithms must either operate in *decision-directed* mode, with the decisions from the demodulator being fed back to the synchronization block, or they must be *blind*, or insensitive to the specific values of the symbols transmitted. Blind algorithms are generally derived by averaging over all possible values of the transmitted symbols, but often turn out to have a form similar to decision-directed algorithms, with hard decisions replaced by soft decisions. See Example 4.2.2 for a simple instance of this observation.

Carrier synchronization This corresponds to estimation of Δf and θ in (4.2). These estimates would then be used to undo the rotations induced by the frequency and phase offsets before coherent demodulation. Initial estimates of the frequency and phase offset are often obtained using a training sequence, with subsequent tracking in decision-directed mode. Another classical approach is first to remove the data modulation by nonlinear operations (e.g., squaring for BPSK modulation), and then to use a PLL for carrier frequency and phase acquisition.

As evident from the preceding discussion, synchronization typically involves two stages: obtaining an *initial estimate* of the unknown parameters (often using a block of known training symbols sent as a preamble at the beginning of transmission), and then *tracking* these parameters as they vary slowly over time (typically after the training phase, so that the $\{b[n]\}$ are unknown). For packetized communication systems, which are increasingly common in both wireless and wireline communication, the variations of the synchronization parameters over a packet are often negligible, and the tracking phase can often be eliminated. The estimation framework developed in this chapter consists of choosing parameter values that optimize an appropriately chosen cost function. Typically, initial estimates from a synchronization algorithm can be viewed as directly optimizing the cost function, while feedback loops for subsequent parameter tracking can be interpreted as using the derivative of the cost function to drive recursive updates of the estimate. Many classical synchronization algorithms, originally obtained using intuitive reasoning, can be interpreted as approximations to optimal estimators derived in this fashion. More importantly, the optimal estimation framework in this chapter gives us a systematic method to approach *new* receiver design scenarios, with the understanding that creative approximations may be needed when computation of optimal estimators is too difficult.

4.2 Parameter estimation basics

We begin by outlining a basic framework for parameter estimation. Given an observation Y, we wish to estimate a parameter θ. The relation between Y and θ is statistical: we know $p(y|\theta)$, the conditional density of Y given θ.

The maximum likelihood estimate (MLE) of θ is given by

$$\hat{\theta}_{\text{ML}}(y) = \arg\max_\theta \; p(y|\theta) = \arg\max_\theta \; \log p(y|\theta), \qquad (4.3)$$

where it is sometimes more convenient to maximize a monotonic increasing function, such as the logarithm, of $p(y|\theta)$.

If prior information about the parameter is available, that is, if the density $p(\theta)$ is known, then it is possible to apply *Bayesian* estimation, wherein we optimize the value of an appropriate cost function, averaged using the joint distribution of Y and Θ. It turns out that the key to such minimization is the a posteriori density of θ (i.e., the conditional density of Θ given Y)

$$p(\theta|y) = \frac{p(y|\theta)p(\theta)}{p(y)}. \qquad (4.4)$$

For our purpose, we only define the maximum a posteriori probability (MAP) estimator, which maximizes the posterior density (4.4) over θ. The denominator of (4.4) does not depend on θ, and can therefore be dropped. Furthermore, we can maximize any monotonic increasing function, such as the logarithm, of the cost function. We therefore obtain several equivalent forms of the MAP rule, as follows:

$$\begin{aligned}\hat{\theta}_{\text{MAP}}(y) &= \arg\max_\theta \; p(\theta|y) = \arg\max_\theta \; p(y|\theta)p(\theta) \\ &= \arg\max_\theta \; \log p(y|\theta) + \log p(\theta).\end{aligned} \qquad (4.5)$$

Example 4.2.1 (Density conditioned on amplitude) As an example of the kinds of conditional densities used in parameter estimation, consider a single received sample in a linearly modulated BPSK system, of the form:

$$Y = A\,b + N, \qquad (4.6)$$

where A is an amplitude, b is a BPSK symbol taking values ± 1 with equal probability, and $N \sim N(0, \sigma^2)$ is noise. If b is known (e.g., because it is part of a training sequence), then, conditioned on $A = a$, the received sample Y is Gaussian: $Y \sim N(a, \sigma^2)$ for $b = +1$, and $Y \sim N(-a, \sigma^2)$ for $b = -1$. That is,

$$p(y|a, b = +1) = \frac{e^{-\frac{(y-a)^2}{2\sigma^2}}}{\sqrt{2\pi\sigma^2}}, \qquad p(y|a, b = -1) = \frac{e^{-\frac{(y+a)^2}{2\sigma^2}}}{\sqrt{2\pi\sigma^2}}. \qquad (4.7)$$

However, if b is not known, then we must average over the possible values it can take in order to compute the conditional density $p(y|a)$. For $b = \pm 1$ with equal probability, we obtain

$$p(y|a) = P[b = +1]p(y|a, b = +1) + P[b = +-1]p(y|a, b = -1)$$

$$= \frac{1}{2} \frac{e^{-\frac{(y-a)^2}{2\sigma^2}}}{\sqrt{2\pi\sigma^2}} + \frac{1}{2} \frac{e^{-\frac{(y+a)^2}{2\sigma^2}}}{\sqrt{2\pi\sigma^2}} = e^{-\frac{a^2}{2\sigma^2}} \cosh\left(\frac{ay}{\sigma^2}\right) \frac{e^{-\frac{y^2}{2\sigma^2}}}{\sqrt{2\pi\sigma^2}}.$$
$$(4.8)$$

We can now maximize (4.6) or (4.8) over a to obtain an ML estimate, depending on whether the transmitted symbol is known or not. Of course, amplitude estimation based on a single symbol is unreliable at typical SNRs, hence we use the results of this example as a building block for developing an amplitude estimator for a block of symbols.

Example 4.2.2 (ML amplitude estimation using multiple symbols) Consider a linearly modulated system in which the samples at the receive filter output are given by

$$Y[k] = A\, b[k] + N[k], \quad k = 1, \ldots, K, \qquad (4.9)$$

where A is an unknown amplitude, $b[k]$ are transmitted symbols taking values ± 1, and $N[k]$ are i.i.d. $N(0, \sigma^2)$ noise samples. We wish to find an ML estimate for the amplitude **A**, using the vector observation $\mathbf{Y} = (Y[1], \ldots, Y[K])^T$. The vector of K symbols is denoted by $\mathbf{b} = (b[1], \ldots, b[K])^T$. We consider two cases separately: first, when the symbols $\{b[k]\}$ are part of a known training sequence, and second, when the symbols $\{b[k]\}$ are unknown, and modeled as i.i.d., taking values ± 1 with equal probability.

Case 1 (Training-based estimation) The ML estimate is given by

$$\hat{A}_{\text{ML}} = \arg\max_A \log p(\mathbf{y}|A, \mathbf{b}) = \arg\max_A \sum_{k=1}^{K} \log p(Y[k]|A, b[k]),$$

where the logarithm of the joint density decomposes into a sum of the logarithms of the marginals because of the conditional independence of the $Y[k]$, given **A** and **b**. Substituting from (4.6), we can show that (see Problem 4.1)

4.2 Parameter estimation basics

$$\hat{A}_{\text{ML}} = \frac{1}{K} \sum_{k=1}^{K} Y[k]b[k]. \quad (4.10)$$

That is, the ML estimate is obtained by correlating the received samples against the training sequence, which is an intuitively pleasing result. The generalization to complex-valued symbols is straightforward, and is left as an exercise.

Case 2 (Blind estimation) "Blind" estimation refers to estimation without the use of a training sequence. In this case, we model the $\{b[k]\}$ as i.i.d., taking values ± 1 with equal probability. Conditioned on A, the $\{Y[k]\}$ are independent, with marginals given by (4.7). The ML estimate is therefore given by

$$\hat{A}_{\text{ML}} = \arg\max_{A} \log p(\mathbf{y}|A) = \arg\max_{A} \sum_{k=1}^{K} \log p(Y[k]|A).$$

Substituting from (4.8) and setting the derivative of the cost function with respect to A to zero, we can show that (see Problem 4.1) the ML estimate $\hat{A}_{\text{ML}} = a$ satisfies the transcendental equation

$$a = \frac{1}{K} \sum_{k=1}^{K} Y[k] \tanh\left(\frac{aY[k]}{\sigma^2}\right) = \frac{1}{K} \sum_{k=1}^{K} Y[k]\hat{b}[k], \quad (4.11)$$

where the analogy with correlation in the training-based estimator (4.10) is evident, interpreting $\hat{b}[k] = \tanh((aY[k])/\sigma^2)$ as a "soft" estimate of the symbol $b[k]$, $k = 1,\ldots,K$. How would the preceding estimators need to be modified if we wished to implement the MAP rule, assuming that the prior distribution of A is $N(0, \sigma_A^2)$?

We see that the key ingredient of parameter estimation is the conditional density of the observation, given the parameter. To apply this framework to a continuous-time observation as in (4.1), therefore, we must be able to define a conditional "density" for the infinite-dimensional observation $y(t)$, conditioned on θ. To do this, let us first reexamine the notion of density for scalar random variables more closely.

Example 4.2.3 (There are many ways to define a density) Consider the Gaussian random variable $Y \sim N(\theta, \sigma^2)$, where θ is an unknown parameter. The conditional density of Y, given θ, is given by

$$p(y|\theta) = \frac{1}{\sqrt{2\pi\sigma^2}} \exp\left(-\frac{(y-\theta)^2}{2\sigma^2}\right).$$

The conditional probability that Y takes values in a subset A of real numbers is given by

$$P(Y \in A|\theta) = \int_A p(y|\theta)\,dy. \tag{4.12}$$

For any arbitrary function $q(y)$ satisfying the property that $q(y) > 0$ wherever $p(y) > 0$, we may rewrite the above probability as

$$P(Y \in A) = \int_A \frac{p(y|\theta)}{q(y)} q(y)dy = \int_A L(y|\theta)\, q(y)dy. \tag{4.13}$$

An example of such a function $q(y)$ is a Gaussian $N(0, \sigma^2)$ density, given by

$$q(y) = \frac{1}{\sqrt{2\pi\sigma^2}} \exp\left(-\frac{y^2}{2\sigma^2}\right).$$

In this case, we obtain

$$L(y|\theta) = \frac{p(y|\theta)}{q(y)} = \exp\left(\frac{1}{\sigma^2}\left(y\theta - \frac{\theta^2}{2}\right)\right).$$

Comparing (4.12) and (4.13), we observe the following. The probability of an infinitesimal interval $(y, y+dy)$ is given by the product of the density and the size of the interval. Thus, $p(y|\theta)$ is the (conditional) probability density of Y when the *measure* of the size of an infinitesimal interval $(y, y+dy)$ is its length dy. However, if we redefine the measure of the interval size as $q(y)dy$ (this measure now depends on the location of the interval as well as its length), then the (conditional) density of Y with respect to this new measure of length is $L(y|\theta)$. The two notions of density are equivalent, since the probability of the infinitesimal interval is the same in both cases. In this particular example, the new density $L(y|\theta)$ can be interpreted as a likelihood ratio, since $p(y|\theta)$ and $q(y)$ are both probability densities.

Suppose, now, that we wish to estimate the parameter θ based on Y. Noting that $q(y)$ does not depend on θ, dividing $p(y|\theta)$ by $q(y)$ does not affect the MLE for θ based on Y: check that $\hat{\theta}_{ML}(y) = y$ in both cases. In general, we can choose to define the density $p(y|\theta)$ with respect to any convenient measure, to get a form that is easy to manipulate. This is the idea we use to define the notion of a density for a continuous-time signal in WGN: we define the density as the likelihood function of a hypothesis corresponding to the model (4.1), with respect to a *dummy* hypothesis that is independent of the signal $s_\theta(t)$.

4.2.1 Likelihood function of a signal in AWGN

Let H_s be the hypothesis corresponding to the signal model of (4.1), dropping the subscript θ for convenience:

$$H_s : y(t) = s(t) + n(t), \tag{4.14}$$

where $n(t)$ is WGN and $s(t)$ has finite energy. Define a noise-only dummy hypothesis as follows:

$$H_n : y(t) = n(t). \tag{4.15}$$

We now use signal space concepts in order to compute the likelihood ratio for the hypothesis testing problem H_s versus H_n. Define

$$Z = \langle y, s \rangle \tag{4.16}$$

as the component of the received signal along the signal s. Let

$$y^\perp(t) = y(t) - \langle y, s \rangle \frac{s(t)}{||s||^2}$$

denote the component of y orthogonal to the signal space spanned by s. Since Z and $y^\perp$ provide a complete description of the received signal y, it suffices to compute the likelihood ratio for the pair $(Z, y^\perp(t))$. We can now argue as in Chapter 3. First, note that $y^\perp = n^\perp$, where $n^\perp(t) = n(t) - \langle n, s \rangle s(t)/||s||^2$ is the noise component orthogonal to the signal space. Thus, $y^\perp$ is unaffected by s. Second, $n^\perp$ is independent of the noise component in Z, since components of WGN in orthogonal directions are uncorrelated and jointly Gaussian, and hence independent. This implies that Z and $y^\perp$ are conditionally independent, conditioned on each hypothesis, and that $y^\perp$ is identically distributed under each hypothesis. Thus, it is irrelevant to the decision and does not appear in the likelihood ratio. We can interpret this informally as follows: when taking the ratio of the conditional densities of $(Z, y^\perp(t))$ under the two hypotheses, the conditional density of $y^\perp(t)$ cancels out. We therefore obtain $L(y) = L(z)$. The random variable Z is conditionally Gaussian under each hypothesis, and its mean and variance can be computed in the usual fashion. The problem has now reduced to computing the likelihood ratio for the scalar random variable Z under the following two hypotheses:

$$H_s : Z \sim N(||s||^2, \sigma^2 ||s||^2),$$
$$H_n : Z \sim N(0, \sigma^2 ||s||^2).$$

Taking the ratio of the densities yields

$$L(z) = \exp\left(\frac{1}{\sigma^2 ||s||^2} \left(||s||^2 z - (||s||^2)^2/2\right)\right) = \exp\left(\frac{1}{\sigma^2}\left(z - ||s||^2/2\right)\right).$$

Expressing the result in terms of y, using (4.16), we obtain the following result.

Likelihood function for a signal in real AWGN

$$L(y|s) = \exp\left(\frac{1}{\sigma^2}\left(\langle y, s \rangle - ||s||^2/2\right)\right), \tag{4.17}$$

where we have made the dependence on s explicit in the notation for the likelihood function. If $s(t) = s_\theta(t)$, the likelihood function may be denoted by $L(y|\theta)$.

We can now immediately extend this result to complex baseband signals, by applying (4.17) to a real passband signal, and then translating the results to complex baseband. To this end, consider the hypotheses

$$H_s : y_p(t) = s_p(t) + n_p(t),$$
$$H_n : y_p(t) = n_p(t),$$

where y_p is the passband received signal, s_p is the noiseless passband signal, and n_p is passband WGN. The equivalent model in complex baseband is

$$H_s : y(t) = s(t) + n(t),$$
$$H_n : y(t) = n(t),$$

where s is the complex envelope of s_p, and n is complex WGN. The likelihood functions computed in passband and complex baseband must be the same, since the information in the two domains is identical. Thus,

$$L(y|s) = \exp\left(\frac{1}{\sigma^2}\left(\langle y_p, s_p \rangle - ||s_p||^2/2\right)\right).$$

We can now replace the passband inner products by the equivalent computations in complex baseband, noting that $\langle y_p, s_p \rangle = \text{Re}(\langle y, s \rangle)$ and that $||s_p||^2 = ||s||^2$. We therefore obtain the following generalization of (4.17) to complex-valued signals in complex AWGN, which we can state as a theorem.

Theorem 4.2.1 (Likelihood function for a signal in complex AWGN) *For a signal $s(t)$ corrupted by complex AWGN $n(t)$, modeled as*

$$y(t) = s(t) + n(t),$$

the likelihood function (i.e., the likelihood ratio with respect to a noise-only dummy hypothesis) is given by

$$L(y|s) = \exp\left(\frac{1}{\sigma^2}\left(\text{Re}(\langle y, s \rangle) - ||s||^2/2\right)\right). \quad (4.18)$$

We can use (4.18) for both complex-valued and real-valued received signals from now on, since the prior formula (4.17) for real-valued received signals reduces to a special case of (4.18).

Discrete-time likelihood functions The preceding formulas also hold for the analogous scenario of discrete-time signals in AWGN. Consider the signal model

$$y[k] = s[k] + n[k], \quad (4.19)$$

where $\text{Re}(n[k])$, $\text{Im}(n[k])$ are i.i.d. $N(0, \sigma^2)$ random variables for all k. We say that $n[k]$ is complex WGN with variance $\sigma^2 = N_0/2$ per dimension, and discuss this model in more detail in Section 4.5.1 in the context of performance analysis. For now, however, it is easy to show that a formula entirely analogous to (4.18) holds for this model (taking the likelihood ratio with respect to a noise-only hypothesis) as follows:

$$L(\mathbf{y}|\mathbf{s}) = \exp\left(\frac{1}{\sigma^2}\left(\text{Re}(\langle \mathbf{y}, \mathbf{s}\rangle) - \|\mathbf{s}\|^2/2\right)\right), \tag{4.20}$$

where $\mathbf{y} = (y[1], \ldots, y[K])^T$ and $\mathbf{s} = (s[1], \ldots, s[K])^T$ are the received vector and the signal vector, respectively.

In the next two sections, we apply the results of this section to derive receiver design principles for synchronization and noncoherent communication. Before doing that, however, let us use the framework of parameter estimation to quickly rederive the optimal receiver structures in Chapter 3 as follows.

Example 4.2.4 (M-ary signaling in AWGN revisited) The problem of testing among M hypotheses of the form

$$H_i : y(t) = s_i(t) + n(t), \quad i = 1, \ldots, M$$

is a special case of parameter estimation, where the parameter takes one of M possible values. For a complex baseband received signal y, the conditional density, or likelihood function, of y follows from setting $s(t) = s_i(t)$ in (4.18):

$$L(y|H_i) = \exp\left(\frac{1}{\sigma^2}\left(\text{Re}(\langle y, s_i \rangle) - \|s_i\|^2/2\right)\right).$$

The ML decision rule can now be interpreted as the MLE of an M-valued parameter, and is given by

$$\hat{i}_{\text{ML}}(y) = \arg\max_i L(y|H_i) = \arg\max_i \text{Re}(\langle y, s_i\rangle) - \|s_i\|^2/2,$$

thus recovering our earlier result on the optimality of a bank of correlators.

4.3 Parameter estimation for synchronization

We now discuss several canonical examples of parameter estimation in AWGN, beginning with phase estimation. The model (4.2) includes the effect of frequency offset between the local oscillators at the transmitter and receiver. Such a phase offset is of the order of 10–100 ppm, relative to the carrier frequency. In addition, certain channels, such as the wireless mobile channel, can induce a Doppler shift in the signal of the order of vf_c/c, where v is the relative velocity between the transmitter and receiver, and c is the speed of light. For a velocity of 200 km/hr, the ratio v/c of the Doppler shift to the carrier frequency is about 0.2 ppm. On the other hand, typical baseband signal bandwidths are about 1–10% of the carrier frequency. Thus, the time variations of the modulated signal are typically much faster than those induced by the frequency offsets due to LO mismatch and Doppler. Consider, for example, a linearly modulated system, in which the signal bandwidth is of the order of the symbol rate $1/T$. Thus, for a frequency shift Δf which is small compared with the signal bandwidth, the change in phase $2\pi\Delta f\, T$ over a symbol interval T is small. Thus, the phase can often be taken to be constant over multiple symbol intervals. This can be exploited for both explicit phase estimation, as in the following example, and for implicit phase estimation in noncoherent and differentially coherent reception, as discussed in Section 4.4.

Example 4.3.1 (ML phase estimation) Consider a noisy signal with unknown phase, modeled in complex baseband as

$$y(t) = s(t)e^{j\theta} + n(t), \qquad (4.21)$$

where θ is an unknown phase, s is a known complex-valued signal, and n is complex WGN with PSD N_0. To find the ML estimate of θ, we write down the likelihood function of y conditioned on θ, replacing s with $se^{j\theta}$ in (4.18) to get

$$L(y|\theta) = \exp\left(\frac{1}{\sigma^2}\left(\mathrm{Re}\left(\langle y, se^{j\theta}\rangle\right) - \|se^{j\theta}\|^2/2\right)\right). \qquad (4.22)$$

Setting $\langle y, s\rangle = |Z|e^{j\phi} = Z_c + jZ_s$, we have

$$\langle y, se^{j\theta}\rangle = e^{-j\theta}Z = |Z|e^{j(\phi-\theta)},$$

so that

$$\mathrm{Re}\left(\langle y, se^{j\theta}\rangle\right) = |Z|\cos(\phi-\theta).$$

Further, $\|se^{j\theta}\|^2 = \|s\|^2$. The conditional likelihood function, therefore, can be rewritten as

$$L(y|\theta) = \exp\left(\frac{1}{\sigma^2}\left(|Z|\cos(\phi-\theta) - \|s\|^2/2\right)\right). \qquad (4.23)$$

The ML estimate of θ is obtained by maximizing the exponent in (4.23), which corresponds to

$$\hat{\theta}_{\mathrm{ML}} = \phi = \arg(\langle y, s\rangle) = \tan^{-1}\frac{Z_s}{Z_c}.$$

Note that this is also the MAP estimate if the prior distribution of θ is uniform over $[0, 2\pi]$. The ML phase estimate, therefore, equals the phase of the complex inner product between the received signal y and the template

4.3 Parameter estimation for synchronization

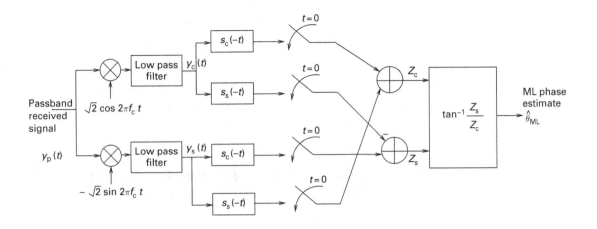

Figure 4.3 Maximum likelihood phase estimation: the complex baseband operations in Example 4.3.1 are implemented after downconverting the passband received signal.

signal s. The implementation of the phase estimate, starting from the passband received signal y_p is shown in Figure 4.3. The four real-valued correlations involved in computing the complex inner product $\langle y, s \rangle$ are implemented using matched filters.

Example 4.3.2 (ML delay estimation) Let us now consider the problem of estimating delay in the presence of an unknown phase (recall that timing synchronization is typically done prior to carrier synchronization). The received signal is given by

$$y(t) = As(t - \tau)e^{j\theta} + n(t), \tag{4.24}$$

where n is complex WGN with PSD N_0, with unknown vector parameter $\Gamma = (\tau, A, \theta)$. We can now apply (4.18), replacing $s(t)$ by $s_\Gamma(t) = As(t - \tau)e^{j\theta}$, to obtain

$$L(y|\Gamma) = \exp\left(\frac{1}{\sigma^2}\left(\text{Re}\left(\langle y, s_\Gamma\rangle\right) - ||s_\Gamma||^2/2\right)\right).$$

Defining the filter matched to s as $s_{\text{MF}}(t) = s^*(-t)$, we obtain

$$\langle y, s_\Gamma \rangle = Ae^{-j\theta} \int y(t) s^*(t - \tau) dt = Ae^{-j\theta} \int y(t) s_{\text{MF}}(\tau - t) dt$$

$$= Ae^{-j\theta}(y * s_{\text{MF}})(\tau).$$

Note also that, assuming a large enough observation interval, the signal energy does not depend on the delay, so that $||s_\Gamma||^2 = A^2 ||s||^2$. Thus, we obtain

$$L(y|\Gamma) = \exp\left(\frac{1}{\sigma^2}\left(\text{Re}(Ae^{-j\theta}(y * s_{\text{MF}})(\tau)) - A^2||s||^2/2\right)\right).$$

The MLE of the vector parameter Γ is now given by

$$\hat{\Gamma}_{\text{ML}}(y) = \arg\max_{\Gamma} L(y|\Gamma).$$

This is equivalent to maximizing the cost function

$$J(\tau, A, \theta) = \text{Re}\left(Ae^{-j\theta}(y * s_{\text{MF}})(\tau)\right) - A^2 ||s||^2/2 \quad (4.25)$$

over τ, A, and θ. Since our primary objective is to maximize over τ, let us first eliminate the dependence on A and θ. We first maximize over θ for τ, A fixed, proceeding exactly as in Example 4.3.1. We can write $(y * s_{\text{MF}})(\tau) = Z(\tau) = |Z(\tau)|e^{j\phi(\tau)}$, and realize that

$$\text{Re}\left(Ae^{-j\theta}(y * s_{\text{MF}})(\tau)\right) = A|Z(\tau)|\cos\left(\phi(\tau) - \theta\right).$$

Thus, the maximizing value of $\theta = \phi(\tau)$. Substituting into (4.25), we get a cost function which is now a function of only two arguments:

$$J(\tau, A) = \max_{\theta} J(\tau, A, \theta) = A|(y * s_{\text{MF}})(\tau)| - A^2 ||s||^2/2.$$

For any fixed value of A, the preceding is maximized by maximizing $|(y * s_{\text{MF}})(\tau)|$. We can conclude, therefore, that the ML estimate of the delay is

$$\hat{\tau}_{\text{ML}} = \arg \max_{\tau} |(y * s_{\text{MF}})(\tau)|. \quad (4.26)$$

That is, the ML delay estimate corresponds to the intuitively pleasing strategy of picking the peak of the magnitude of the matched filter output in complex baseband. As shown in Figure 4.4, this requires noncoherent processing, with building blocks similar to those used for phase estimation.

We have implicitly assumed in Examples 4.3.1 and 4.3.2 that the data sequence $\{b[n]\}$ is known. This *data-aided* approach can be realized either by using a training sequence, or in decision-directed mode, assuming that the symbol decisions fed to the synchronization algorithms are reliable enough. An alternative nondata-aided (NDA) approach, illustrated in the blind amplitude estimator in Example 4.2.2 is to average over the unknown symbols, typically assuming that they are i.i.d., drawn equiprobably from a fixed constellation. The resulting receiver structure in Case 2 of Example 4.2.2 is

Figure 4.4 Maximum likelihood delay estimation.

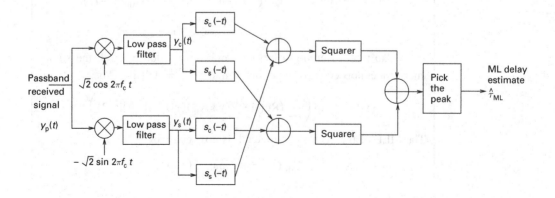

4.3 Parameter estimation for synchronization

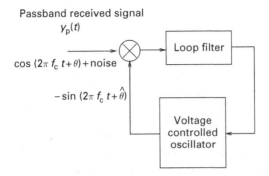

Figure 4.5 Passband implementation of PLL approximating ML phase tracking.

actually quite typical of NDA estimators, which have a structure similar to the data-aided, or decision-directed setting, except that "soft" rather than "hard" decisions are fed back.

Finally, we consider tracking time variations in a synchronization parameter once we are close enough to the true value. For example, we may wish to update a delay estimate to track the offset between the clocks of the transmitter and the receiver, or to update the carrier frequency or phase to track a wireless mobile channel. Most tracking loops are based on the following basic idea. Consider a cost function $J(\theta)$, typically proportional to the log likelihood function, to be maximized over a parameter θ. The tracking algorithm consists of a feedback loop that performs "steepest ascent," perturbing the parameter so as to go in the direction in which the cost function is increasing:

$$\frac{d\theta}{dt} = a \frac{dJ(\theta)}{d\theta}\Big|_{\theta=\hat{\theta}}. \tag{4.27}$$

The success of this strategy of following the derivative depends on our being close enough to the global maximum so that the cost function is locally concave.

We now illustrate this ML tracking framework by deriving the classical PLL structure for tracking carrier phase. Similar interpretations can also be given to commonly used feedback structures such as the Costas loop for phase tracking, and the delay locked loop for delay tracking, and are explored in the problems.

Example 4.3.3 (ML interpretation of phase locked loop) Consider a noisy unmodulated sinusoid with complex envelope

$$y(t) = e^{j\theta} + n(t), \tag{4.28}$$

where θ is the phase to be tracked (its dependence on t has been suppressed from the notation), and $n(t)$ is complex WGN. Writing down the likelihood function over an observation interval of length T_o, we have

$$L(y|\theta) = \exp\left(\frac{1}{\sigma^2}\left(\text{Re}\langle y, e^{-j\theta}\rangle - \frac{T_o}{2}\right)\right),$$

so that the cost function to be maximized is

$$J(\theta) = \text{Re}\langle y, e^{j\theta}\rangle = \int_0^{T_0} (y_c(t)\cos\theta(t) + y_s(t)\sin\theta(t))\,dt.$$

Applying the steepest ascent (4.27), we obtain

$$\frac{d\theta}{dt} = a\int_0^{T_0} \left(-y_c(t)\sin\hat{\theta}(t) + y_s(t)\cos\hat{\theta}(t)\right)dt. \qquad (4.29)$$

Since $1/2\pi\,d\theta/dt$ equals the frequency, we can implement steepest ascent by applying the right-hand side of (4.29) to a voltage controlled oscillator. Furthermore, this expression can be recognized to be the real part of the complex inner product between $y(t)$ and $v(t) = -\sin\hat{\theta} + j\cos\hat{\theta} = je^{j\hat{\theta}}$. The corresponding passband signals are $y_p(t)$ and $v_p(t) = -\sin(2\pi f_c t + \hat{\theta})$. Recognizing that $\text{Re}\langle y, v\rangle = \langle y_p, s_p\rangle$, we can rewrite the right-hand side of (4.29) as a passband inner product to get:

$$\frac{d\theta}{dt} = -a\int_0^{T_0} y_p(t)\sin(2\pi f_c t + \hat{\theta})\,dt. \qquad (4.30)$$

In both (4.29) and (4.30), the integral can be replaced by a low pass filter for continuous tracking. Doing this for (4.30) gives us the well-known structure of a passband PLL, as shown in Figure 4.5.

Further examples of amplitude, phase, and delay estimation, including block-based estimators, as well as classical structures such as the Costas loop for phase tracking in linearly modulated systems, are explored in the problems.

4.4 Noncoherent communication

We have shown that the frequency offsets due to LO mismatch at the transmitter and receiver, and the Doppler induced by relative mobility between the transmitter and receiver, are typically small compared with the bandwidth of the transmitted signal. Noncoherent communication exploits this observation to eliminate the necessity for carrier synchronization, modeling the phase over the duration of a demodulation interval as unknown, but constant. The mathematical model for a noncoherent system is as follows.

Model for M-ary noncoherent communication The complex baseband received signal under the ith hypothesis is as follows:

$$H_i : y(t) = s_i(t)e^{j\theta} + n(t), \quad i = 1,\ldots,M, \qquad (4.31)$$

where θ is an unknown phase, and n is complex AWGN with PSD $\sigma^2 = N_0/2$ per dimension.

Before deriving the receiver structure in this setting, we need some background on *composite* hypothesis testing, or hypothesis testing with one or more unknown parameters.

4.4.1 Composite hypothesis testing

As in a standard detection theory framework, we have an observation Y taking values in Γ, and M hypotheses $H_1, \ldots, H_M$. However, the conditional density of the observation given the hypothesis is not known, but is parameterized by an unknown parameter. That is, we know the conditional density $p(y|i, \theta)$ of the observation Y given H_i and an unknown parameter θ taking values in Λ. The unknown parameter θ may not have the same interpretation under all hypotheses, in which case the set Λ may actually depend on i. However, we do not introduce this complication into the notation, since it is not required for our intended application of these concepts to noncoherent demodulation (where the unknown parameter for each hypothesis is the carrier phase).

Generalized likelihood ratio test (GLRT) approach This corresponds to joint ML estimation of the hypothesis (treated as an M-valued parameter) and θ, so that

$$(\hat{i}, \hat{\theta})(y) = \arg \max_{1 \leq i \leq M, \theta \in \Lambda} p(y|i, \theta).$$

This can be interpreted as maximizing first with respect to θ, for each i, getting

$$\hat{\theta}_i(y) = \arg \max_{\theta \in \Lambda} p(y|i, \theta)$$

then plugging into the conditional density $p(y|i, \theta)$ to get the "generalized density,"

$$q_i(y) = p(y|i, \hat{\theta}_i(y)) = \max_{\theta \in \Lambda} p(y|i, \theta)$$

(note that q_i is not a true density, in that it does not integrate to one). The GLRT decision rule can be then expressed as

$$\delta_{\text{GLRT}}(y) = \arg \max_{1 \leq i \leq M} q_i(y).$$

This is of similar form to the ML decision rule for simple hypothesis testing, hence the term GLRT.

Bayesian approach If $p(\theta|i)$, the conditional density of θ given H_i, is known, then the unknown parameter θ can be integrated out, yielding a simple hypothesis testing problem that we know how to solve. That is, we can compute the conditional density of Y given H_i as follows:

$$p(y|i) = \int_\Lambda p(y|i, \theta) p(\theta|i) \, d\theta.$$

4.4.2 Optimal noncoherent demodulation

We now apply the GLRT and Bayesian approaches to derive receiver structures for noncoherent communication. For simplicity, we consider equal-energy signaling.

Equal energy M-ary noncoherent communication: receiver structure
The model is as in (4.31) with equal signal energies under all hypotheses, $||s_i||^2 \equiv E_s$. From (4.22), we find that

$$L(y|i, \theta) = \exp\left(\frac{1}{\sigma^2}[|Z_i|\cos(\theta - \phi_i) - ||s_i||^2/2]\right), \quad (4.32)$$

where $Z_i = \langle y, s_i \rangle$ is the result of complex correlation with s_i, and $\phi_i = \arg(Z_i)$.

Applying the GLRT approach, we note that the preceding is maximized at $\theta = \phi_i$ to get the generalized density

$$q_i(y) = \exp\left(\frac{1}{\sigma^2}[|Z_i| - E_s/2]\right),$$

where we have used the equal energy assumption. Maximizing over i, we get the GLRT rule

$$\delta_{\text{GLRT}}(y) = \arg\max_{1 \leq i \leq M} |Z_i| = \arg\max_{1 \leq i \leq M} Z_{i,c}^2 + Z_{i,s}^2,$$

where $Z_{i,c} = \text{Re}(Z_i)$ and $Z_{i,s} = \text{Im}(Z_i)$.

Figure 4.6 shows the computation of the noncoherent decision statistic $|Z|^2 = |\langle y, s \rangle|^2$ for a signal s. The noncoherent receiver chooses the maximum among the outputs of a bank of such processors, for $s = s_i$, $i = 1, \ldots, M$.

Now, let us apply the Bayesian approach, modeling the unknown phase under each hypothesis as a random variable uniformly distributed over $[0, 2\pi]$. We can now average out θ in (4.32) as follows:

$$L(y|i) = \frac{1}{2\pi}\int_0^{2\pi} L(y|i, \theta)\, d\theta. \quad (4.33)$$

Figure 4.6 Computation of the noncoherent decision statistic for a complex baseband signal s. The optimal noncoherent receiver employs a bank of such processors, one for each of M signals, and picks the maximum of M noncoherent decision statistics.

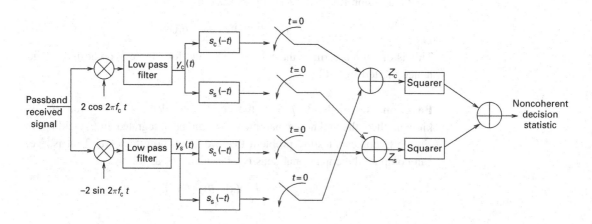

It is now useful to introduce the following *modified Bessel function of the first kind, of order zero*:
$$I_0(x) = \frac{1}{2\pi} \int_0^{2\pi} e^{x \cos \theta} \, d\theta. \tag{4.34}$$

Noting that the integrand is a periodic function of θ, we note that
$$I_0(x) = \frac{1}{2\pi} \int_0^{2\pi} e^{x \cos(\theta - \phi)} \, d\theta,$$

for any fixed phase offset ϕ. Using (4.32) and (4.33), we obtain
$$L(y|i) = e^{-\frac{\|s_i\|^2}{2\sigma^2}} I_0\left(\frac{|Z_i|}{\sigma^2}\right). \tag{4.35}$$

For equal-energy signaling, the first term above is independent of i. Noting that $I_0(x)$ is increasing in $|x|$, we obtain the following ML decision rule (which is also MPE for equal priors) by maximizing (4.35) over i:
$$\delta_{\text{ML}}(y) = \arg \max_{1 \leq i \leq M} |Z_i|,$$

which is the same receiver structure that we derived using the GLRT rule. The equivalence of the GLRT and Bayesian rules is a consequence of the specific models that we use; in general, the two approaches can lead to different receivers.

4.4.3 Differential modulation and demodulation

A drawback of noncoherent communication is the inability to encode information in the signal phase, since the channel produces an unknown phase shift that would destroy such information. However, if this unknown phase can be modeled as approximately constant over more than one symbol, then we can get around this problem by encoding information in the phase differences between successive symbols. This enables recovery of the encoded information even if the absolute phase is unknown. This method is known as differential phase shift keying (DPSK), and is robust against unknown channel amplitude as well as phase. We have already introduced this concept in Section 2.7, and are now able to discuss it in greater depth as an instance of noncoherent communication, where the signal of interest now spans several symbols.

In principle, differential modulation can also be employed with QAM alphabets, by encoding information in amplitude and phase transitions, assuming that the channel is roughly constant over several symbols, but there are technical issues with both encoding (unrestricted amplitude transitions may lead to poor performance) and decoding (handling an unknown amplitude is trickier) that are still a subject of research. We therefore restrict attention to DPSK here.

We explain the ideas in the context of the following discrete-time model, in which the nth sample at the receiver is given by
$$y[n] = h[n]b[n] + w[n], \tag{4.36}$$

where $\{b[n]\}$ is the sequence of complex-valued transmitted symbols, $\{h[n]\}$ is the effective channel gain seen by the nth symbol, and $\{w[n]\}$ is discrete-time complex AWGN with variance $\sigma^2 = N_0/2$ per dimension. The sequence $\{y[n]\}$ would typically be generated by downconversion of a continuous-time passband signal, followed by baseband filtering, and sampling at the symbol rate. We have assumed that there is no ISI.

Suppose that the complex-valued channel gains $h[n] = A[n]e^{j\theta[n]}$ are unknown. This could occur, for example, as a result of inherent channel time variations (e.g., in a wireless mobile environment), or of imperfect carrier phase recovery (e.g., due to free running local oscillators at the receiver). If the $\{h[n]\}$ can vary arbitrarily, then there is no hope of recovering the information encoded in $\{b[n]\}$. However, now suppose that $h[n] \approx h[n-1]$ (i.e., the rate of variation of the channel gain is slower than the symbol rate). Consider the vector of two successive received samples, given by $\mathbf{y}[n] = (y[n-1], y[n])^T$. Setting $h[n] = h[n-1] = h$, we have

$$\begin{pmatrix} y[n-1] \\ y[n] \end{pmatrix} = h \begin{pmatrix} b[n-1] \\ b[n] \end{pmatrix} + \begin{pmatrix} w[n-1] \\ w[n] \end{pmatrix}$$

$$= hb[n-1] \begin{pmatrix} 1 \\ b[n]/b[n-1] \end{pmatrix} + \begin{pmatrix} w[n-1] \\ w[n] \end{pmatrix}. \quad (4.37)$$

The point of the above manipulations is that we can now think of $\tilde{h} = hb[n-1]$ as an unknown gain, and treat $(1, b[n]/b[n-1])^T$ as a signal to be demodulated noncoherently. The problem now reduces to one of noncoherent demodulation for M-ary signaling: the set of signals is given by $\mathbf{s}_{a[n]} = (1, a[n])^T$, where M is the number of possible values of $a[n] = b[n]/b[n-1]$. That is, we can rewrite (4.37) as

$$\mathbf{y}[n] = \tilde{h}\mathbf{s}_{a[n]} + \mathbf{w}[n], \quad (4.38)$$

where $\mathbf{y}[n] = (y[n-1], y[n])^T$ and $\mathbf{w}[n] = (w[n-1], w[n])^T$. In DPSK, we choose $a[n] \in A$ from a standard PSK alphabet, and set $b[n] = b[n-1]a[n]$ (the initial condition can be set arbitrarily, say, as $b[0] = 1$). Thus, the transmitted symbols $\{b[n]\}$ are also drawn from a PSK constellation. The information bits are mapped to $a[n]$ in standard fashion, and then are recovered via noncoherent demodulation based on the model (4.38).

Example 4.4.1 (Binary DPSK) Suppose that we wish to transmit a sequence $\{a[n]\}$ of ± 1 bits. Instead of sending these directly over the channel, we send the ± 1 sequence $b[n]$, defined by $b[n] = a[n]b[n-1]$. Thus, we are encoding information in the phase transitions of successive symbols drawn from a BPSK constellation: $b[n] = b[n-1]$ (no phase transition) if $a[n] = 1$, and $b[n] = -b[n-1]$ (phase transition of π) if $a[n] = -1$. The signaling set $\mathbf{s}_{a[n]}$, $a[n] = \pm 1$ is $(1, 1)^T$ and $(1, -1)$, which corresponds to equal-energy, binary, orthogonal signaling.

For the DPSK model (4.38), the noncoherent decision rule for equal-energy signaling becomes

$$\hat{a}[n] = \arg \max_{a \in A} |\langle \mathbf{y}[n], \mathbf{s}_a \rangle|^2. \quad (4.39)$$

For binary DPSK, this reduces to taking the sign of

$$|\langle \mathbf{y}, \mathbf{s}_{+1}\rangle|^2 - |\langle \mathbf{y}, \mathbf{s}_{-1}\rangle|^2 = |y[n] + y[n-1]|^2 - |y[n] - y[n-1]|^2$$
$$= 2\text{Re}(y[n]y^*[n-1]).$$

That is, we have

$$\hat{a}_{\text{binary}}[n] = \text{sign}[\text{Re}(y[n]y^*[n-1])]. \quad (4.40)$$

That is, we take the phase difference between the two samples, and check whether it falls into the right half plane or the left half plane.

For M-ary DPSK, a similar rule is easily derived by examining the decision statistics in (4.39) in more detail:

$$|\langle \mathbf{y}[n], \mathbf{s}_a \rangle|^2 = |y[n-1] + y[n]a^*[n]|^2$$
$$= |y[n-1]|^2 + |y[n]|^2|a[n]|^2 + 2\text{Re}(y[n]y^*[n-1]a^*[n])$$
$$= |y[n-1]|^2 + |y[n]|^2 + 2\text{Re}(y[n]y^*[n-1]a^*[n]),$$

where we have used $|a[n]| = 1$ for PSK signaling. Since only the last term depends on a, the decision rule can be simplified to

$$\hat{a}_{M\text{-ary}}[n] = \arg \max_{a \in A} \text{Re}(y[n]y^*[n-1]a^*). \quad (4.41)$$

This corresponds to taking the phase difference between two successive received samples, and mapping it to the closest constellation point in the PSK alphabet from which $a[n]$ is drawn.

4.5 Performance of noncoherent communication

Performance analysis for noncoherent receivers is typically more complicated than for coherent receivers, since we need to handle complex-valued decision statistics going through nonlinear operations. As a motivating example, consider noncoherent detection for equal-energy, binary signaling, with complex baseband received signal under hypothesis H_i, $i = 0, 1$, given by

$$H_i : y(t) = s_i(t)e^{j\theta} + n(t),$$

where θ is an unknown phase shift induced by the channel. For equal priors, and assuming that θ is uniformly distributed over $[0, 2\pi]$, the MPE rule has been shown to be

$$\delta_{\text{MPE}}(y) = \arg \max_{i=0,1} |\langle y, s_i \rangle|.$$

We are interested in evaluating the error probability for this decision rule. As usual, we condition on one of the hypotheses, say H_0, so that $y = s_0 e^{j\theta} + n$. The conditional error probability is then given by

$$P_{e|0} = P(|Z_1| > |Z_0| | H_0),$$

where $Z_i = \langle y, s_i \rangle$, $i = 0, 1$. Conditioned on H_0, we obtain

$$Z_0 = \langle s_0 e^{j\theta} + n, s_0 \rangle = ||s_0||^2 e^{j\theta} + \langle n, s_0 \rangle,$$
$$Z_1 = \langle s_0 e^{j\theta} + n, s_1 \rangle = \langle s_0, s_1 \rangle e^{j\theta} + \langle n, s_1 \rangle.$$

Each of the preceding statistics contains a complex-valued noise contribution obtained by correlating complex WGN with a (possibly complex) signal. Since our prior experience has been with real random variables, before proceeding further, we devote the next section to developing a machinery for handling complex-valued random variables generated in this fashion.

4.5.1 Proper complex Gaussianity

For real-valued signals, performance analysis in a Gaussian setting is made particularly easy by the fact that (joint) Gaussianity is preserved under linear transformations, and that probabilities are completely characterized by means and covariances. For complex AWGN models, joint Gaussianity is preserved for the real and imaginary parts under operations such as filtering, sampling, and correlation, and probabilities can be computed by keeping track of the covariance of the real part, the covariance of the imaginary part, and the crosscovariance of the real and imaginary parts. However, we describe below a simpler and more elegant approach based purely on complex-valued covariances. This approach works when the complex-valued random processes involved are *proper complex Gaussian* (to be defined shortly), as is the case for the random processes of interest to us, which are obtained from complex WGN through linear transformations.

Definition 4.5.1 (Covariance and pseudocovariance for complex-valued random vectors) *Let* **U** *denote an* $m \times 1$ *complex-valued random vector, and* **V** *an* $n \times 1$ *complex-valued random vector, defined on a common probability space. The* $m \times n$ *covariance matrix is defined as*

$$\mathbf{C}_{\mathbf{U},\mathbf{V}} = \mathbb{E}[(\mathbf{U} - \mathbb{E}[\mathbf{U}])(\mathbf{V} - \mathbb{E}[\mathbf{V}])^H] = \mathbb{E}[\mathbf{U}\mathbf{V}^H] - \mathbb{E}[\mathbf{U}](\mathbb{E}[\mathbf{V}])^H.$$

The $m \times n$ *pseudocovariance matrix is defined as*

$$\tilde{\mathbf{C}}_{\mathbf{U},\mathbf{V}} = \mathbb{E}[(\mathbf{U} - \mathbb{E}[\mathbf{U}])(\mathbf{V} - \mathbb{E}[\mathbf{V}])^T] = \mathbb{E}[\mathbf{U}\mathbf{V}^T] - \mathbb{E}[\mathbf{U}](\mathbb{E}[\mathbf{V}])^T.$$

Note that covariance and pseudocovariance are the same for real random vectors.

Definition 4.5.2 (Complex Gaussian random vector) *The $n \times 1$ complex random vector $\mathbf{X} = \mathbf{X}_c + j\mathbf{X}_s$ is Gaussian if the real random vectors $\mathbf{X}_c$ and $\mathbf{X}_s$ are Gaussian, and $\mathbf{X}_c$, $\mathbf{X}_s$ are jointly Gaussian.*

To characterize probabilities involving an $n \times 1$ complex Gaussian random vector $\mathbf{X}$, one general approach is to use the statistics of a real random vector formed by concatenating the real and imaginary parts of $\mathbf{X}$ into a single $2n \times 1$ random vector

$$\mathbf{X}_r = \begin{pmatrix} \mathbf{X}_c \\ \mathbf{X}_s \end{pmatrix}.$$

Since $\mathbf{X}_r$ is a Gaussian random vector, it can be described completely in terms of its $2n \times 1$ mean vector, and its $2n \times 2n$ covariance matrix, given by

$$\mathbf{C}_r = \begin{pmatrix} \mathbf{C}_c & \mathbf{C}_{cs} \\ \mathbf{C}_{sc} & \mathbf{C}_s \end{pmatrix},$$

where $\mathbf{C}_{cc} = \mathrm{cov}(\mathbf{X}_c, \mathbf{X}_c)$, $\mathbf{C}_{ss} = \mathrm{cov}(\mathbf{X}_s, \mathbf{X}_s)$ and $\mathbf{C}_{cs} = \mathrm{cov}(\mathbf{X}_c, \mathbf{X}_s) = \mathbf{C}_{sc}^T$.

The preceding approach is cumbersome, requiring us to keep track of three $n \times n$ covariance matrices, and can be simplified if $\mathbf{X}$ satisfies some special properties.

Definition 4.5.3 (Proper complex random vector) *The complex random vector $\mathbf{X} = \mathbf{X}_c + j\mathbf{X}_s$ is proper if its pseudocovariance matrix, given by*

$$\hat{\mathbf{C}}_X = \mathbb{E}[(\mathbf{X} - \mathbb{E}[\mathbf{X}])(\mathbf{X} - \mathbb{E}[\mathbf{X}])^T] = 0. \tag{4.42}$$

In terms of the real covariance matrices defined above, $\mathbf{X}$ is proper if

$$\mathbf{C}_{cc} = \mathbf{C}_{ss} \quad \text{and} \quad \mathbf{C}_{cs} = -\mathbf{C}_{sc} = -\mathbf{C}_{cs}^T. \tag{4.43}$$

We now state a very important result: a proper complex Gaussian random vector is characterized completely by its mean vector and covariance matrix.

Characterizing a proper complex Gaussian random vector Suppose that the complex random vector $\mathbf{X} = \mathbf{X}_c + j\mathbf{X}_s$ is proper (i.e., it has zero pseudocovariance) and Gaussian (i.e., X_c, X_s are jointly Gaussian real random vectors). In this case, $\mathbf{X}$ is completely characterized by its mean vector $\mathbf{m}_X = E[\mathbf{X}]$ and its complex covariance matrix

$$\mathbf{C}_X = \mathbb{E}[(\mathbf{X} - \mathbb{E}[\mathbf{X}])(\mathbf{X} - \mathbb{E}[\mathbf{X}])^H] = 2\mathbf{C}_{cc} + 2j\mathbf{C}_{sc}. \tag{4.44}$$

The probability density function of $\mathbf{X}$ is given by

$$p(\mathbf{x}) = \frac{1}{\pi^n \det(\mathbf{C}_X)} \exp\left(-(\mathbf{x} - \mathbf{m}_X)^H \mathbf{C}_X^{-1} (\mathbf{x} - \mathbf{m}_X)\right). \tag{4.45}$$

We denote the distribution of $\mathbf{X}$ as $CN(\mathbf{m}_X, \mathbf{C}_X)$.

Remark 4.5.1 (Loss of generality due to insisting on properness) In general,
$$\mathbf{C}_X = \mathbf{C}_{cc} + \mathbf{C}_{ss} + j(\mathbf{C}_{sc} - \mathbf{C}_{cs}).$$
Thus, knowledge of $\mathbf{C}_X$ is not enough to infer knowledge of $\mathbf{C}_{cc}$, $\mathbf{C}_{ss}$, and $\mathbf{C}_{sc}$, which are needed, in general, to characterize an n-dimensional complex Gaussian random vector in terms of a $2n$-dimensional real Gaussian random vector $\mathbf{X}_r$. However, under the properness condition (4.43), $\mathbf{C}_X$ contains all the information needed to infer $\mathbf{C}_{cc}$, $\mathbf{C}_{ss}$, and $\mathbf{C}_{sc}$, which is why $\mathbf{C}_X$ (together with the mean) provides a complete statistical characterization of $\mathbf{X}$.

Remark 4.5.2 (Proper complex Gaussian density) The form of the density (4.45) is similar to that of a real Gaussian random vector, but the constants are a little different, because the density integrates to one over complex n-dimensional space. As with real Gaussian random vectors, we can infer from the form of the density (4.45) that two jointly proper complex Gaussian random variables are independent if their complex covariance vanishes.

Proposition 4.5.1 (Scalar proper complex Gaussian random variable) *If $X = X_c + jX_s$ is a scalar complex Gaussian random variable, then its covariance C_X must be real and nonnegative, and its real and imaginary parts, X_c and X_s, are i.i.d. $N(0, C_X/2)$.*

Proof of Proposition 4.5.1 The covariance matrices C_{cc}, C_{ss}, and C_{sc} are now scalars. Using (4.44), the condition $C_{sc} = -C_{sc}^T$ implies that $C_{sc} = 0$. Since X_c, X_s are jointly Gaussian, their uncorrelatedness implies their independence. It remains to note that $C_X = 2C_{cc} = 2C_{ss}$ to complete the proof. □

Remark 4.5.3 (Functions of a scalar proper complex Gaussian random variable) Proposition 4.5.1 and Problem 3.4 imply that for scalar proper complex Gaussian X, the magnitude $|X|$ is Rayleigh, the phase $\arg(X)$ is uniform over $[0, 2\pi]$ (and independent of the magnitude), the magnitude squared $|X|^2$ is exponential with mean C_X, and the magnitude $|m + X|$, where m is a complex constant, is Rician.

Proposition 4.5.2 (Preservation of properness and Gaussianity under linear transformations) *If $\mathbf{X}$ is proper, so is $\mathbf{Y} = \mathbf{AX} + \mathbf{b}$, where $\mathbf{A}$, $\mathbf{b}$ are arbitrary complex matrices. If $\mathbf{X}$ is proper Gaussian, so is $\mathbf{Y} = \mathbf{AX} + \mathbf{b}$. The mean and covariance of $\mathbf{X}$ and $\mathbf{Y}$ are related as follows:*
$$\mathbf{m}_Y = \mathbf{A}\mathbf{m}_X + \mathbf{b}, \quad \mathbf{C}_Y = \mathbf{A}\mathbf{C}_X\mathbf{A}^H. \tag{4.46}$$

Proof of Proposition 4.5.2 To check the properness of $\mathbf{Y}$, we compute
$$\mathbb{E}\left[(\mathbf{AX}+\mathbf{b}-\mathbb{E}[\mathbf{AX}+\mathbf{b}])(\mathbf{AX}+\mathbf{b}-\mathbb{E}[\mathbf{AX}+\mathbf{b}])^T\right]$$
$$= \mathbf{A}\mathbb{E}[(\mathbf{X}-\mathbb{E}[\mathbf{X}])(\mathbf{X}-\mathbb{E}[\mathbf{X}])^T]\mathbf{A}^T = 0$$

4.5 Performance of noncoherent communication

by the properness of **X**. The expressions for mean and covariance follow from similar computations. To check the Gaussianity of **Y**, note that any linear combination of real and imaginary components of **Y** can be expressed as a linear combination of real and imaginary components of **X**, which is Gaussian by the Gaussianity of **X**. □

We can now extend the definition of properness to random processes.

Definition 4.5.4 (Proper complex random process) *A random process $X(t) = X_c(t) + jX_s(t)$ is proper if any set of samples forms a proper complex random vector. Since the sampling times and number of samples are arbitrary, X is proper if*

$$\mathbb{E}[(X(t_1) - \mathbb{E}[X(t_1)])(X(t_2) - \mathbb{E}[X(t_2)])] = 0$$

for all times, t_1, t_2. Equivalently, X is proper if

$$C_{X_c,X_c}(t_1,t_2) = C_{X_s,X_s}(t_1,t_2) \quad \text{and} \quad C_{X_s,X_c}(t_1,t_2) = -C_{X_s,X_c}(t_2,t_1) \quad (4.47)$$

for all t_1, t_2.

Definition 4.5.5 (Proper complex Gaussian random processes) *A random process X is proper complex Gaussian if any set of samples is a proper complex Gaussian random vector. Since a proper complex Gaussian random vector is completely characterized by its mean vector and covariance matrix, a proper complex Gaussian random process X is completely characterized by its mean function $m_X(t) = \mathbb{E}[X(t)]$ and its autocovariance function $C_X(t_1,t_2) = \mathbb{E}[X(t_1)X^*(t_2)]$ (which can be used to compute mean and covariance for an arbitrary set of samples).*

Proposition 4.5.3 (Complex WGN is proper) *Complex WGN $n(t)$ is a zero mean, proper complex Gaussian random process with autocorrelation and autocovariance functions given by*

$$C_n(t_1,t_2) = R_n(t_1,t_2) = \mathbb{E}[n(t_1)n^*(t_2)]$$
$$= 2\sigma^2 \delta(t_1 - t_2).$$

Proof of Proposition 4.5.3 We have $n(t) = n_c(t) + jn_s(t)$, where n_c, n_s are i.i.d. zero mean real WGN, so that

$$C_{n_c}(t_1,t_2) = C_{n_s}(t_1,t_2) = \sigma^2 \delta(t_1 - t_2) \quad \text{and} \quad C_{n_s,n_c}(t_1,t_2) \equiv 0,$$

which satisfies the definition of properness in (4.47). Since n is zero mean, all that remains to specify its statistics completely is its autocovariance function. We compute this as

$$C_n(t_1,t_2) = R_n(t_1,t_2) = \mathbb{E}[n(t_1)n^*(t_2)] = \mathbb{E}[(n_c(t_1) + jn_s(t_1))(n_c(t_2) - jn_s(t_2))]$$
$$= R_{n_c}(t_1,t_2) + R_{n_s}(t_1,t_2) = 2\sigma^2 \delta(t_1 - t_2),$$

where cross terms such as $\mathbb{E}[n_c(t_1)n_s(t_2)] = 0$ because of the independence, and hence uncorrelatedness, of n_c and n_s. □

Notation Since the autocovariance and autocorrelation functions for complex WGN depend only on the time difference $\tau = t_1 - t_2$, it is often convenient to denote them as functions of one variable, as follows:

$$C_n(\tau) = R_n(\tau) = \mathbb{E}[n(t+\tau)n^*(t)] = 2\sigma^2 \delta(\tau).$$

Proposition 4.5.4 (Complex WGN through a correlator) *Let $n(t) = n_c(t) + jn_s(t)$ denote complex WGN, and let $s(t) = s_c(t) + js_s(t)$ denote a finite-energy complex-valued signal. Let*

$$Z = \langle n, s \rangle = \int n(t)s^*(t)dt$$

denote the result of correlating n against s. Denoting $Z = Z_c + jZ_s$ (Z_c, Z_s real), we have the following equivalent statements:

(a) Z is zero mean, proper complex Gaussian with variance $2\sigma^2 ||s||^2$.
(b) Z_c, Z_s are i.i.d. $N(0, \sigma^2 ||s||^2)$ real random variables.

Proof of Proposition 4.5.4 (The "proper" way) The proof is now simple, since the hard work has already been done in developing the machinery of proper Gaussianity. Since n is zero mean, proper complex Gaussian, so is Z, since it is obtained via a linear transformation from n. It remains to characterize the covariance of Z, given by

$$C_Z = \mathbb{E}[\langle n, s \rangle \langle n, s \rangle^*] = \mathbb{E}\left[\int n(t)s^*(t)dt \int n^*(u)s(u)du\right]$$

$$= \int\int \mathbb{E}[n(t)n^*(u)]s^*(t)s(u)dtdu$$

$$= \int\int 2\sigma^2 \delta(t-u)s^*(t)s(u)dtdu$$

$$= 2\sigma^2 \int |s(t)|^2 dt = 2\sigma^2 ||s||^2.$$

The equivalence of (a) and (b) follows from Proposition 4.5.1, since Z is a scalar proper complex Gaussian random variable. □

Proof of Proposition 4.5.4 (Without invoking properness) We can also infer these results, using only what we know about real WGN. We provide this alternative proof to illustrate that the computations get somewhat messy (and do not scale well when we would like to consider the outputs of multiple complex correlators), compared with the prior proof exploiting properness. First, recall that for real WGN (n_c for example), if u_1 and u_2 are two finite-energy real-valued signals, then $\langle n_c, u_1 \rangle$ and $\langle n_c, u_2 \rangle$ are jointly Gaussian with covariance

$$\text{cov}(\langle n_c, u_1 \rangle, \langle n_c, u_2 \rangle) = \sigma^2 \langle u_1, u_2 \rangle. \tag{4.48}$$

Setting $u_1 = u_2 = u$, specialize to the result that $\text{Var}(\langle n_c, u \rangle) = \sigma^2 ||u||^2$. The preceding results also hold if n_c is replaced by n_s. Now, note that

$$Z_c = \langle n_c, s_c \rangle + \langle n_s, s_s \rangle \quad Z_s = \langle n_s, s_c \rangle - \langle n_c, s_s \rangle.$$

Since n_c, n_s are independent Gaussian random processes, the two terms in the equation for Z_c above are independent Gaussian random variables. Using (4.48) to compute the variances of these terms, and then adding these variances up, we obtain

$$\mathrm{var}(Z_c) = \mathrm{var}(\langle n_c, s_c \rangle) + \mathrm{var}(\langle n_s, s_s \rangle) = \sigma^2 ||s_c||^2 + \sigma^2 ||s_s||^2 = \sigma^2 ||s||^2.$$

A similar computation yields the same result for $\mathrm{Var}(Z_s)$. Finally, the covariance of Z_c and Z_s is given by

$$\begin{aligned}
\mathrm{cov}(Z_c, Z_s) &= \mathrm{cov}(\langle n_c, s_c \rangle + \langle n_s, s_s \rangle, \langle n_s, s_c \rangle - \langle n_c, s_s \rangle) \\
&= \mathrm{cov}(\langle n_c, s_c \rangle, \langle n_s, s_c \rangle) + \mathrm{cov}(\langle n_s, s_s \rangle, \langle n_s, s_c \rangle) \\
&\quad - \mathrm{cov}(\langle n_c, s_c \rangle, \langle n_c, s_s \rangle) - \mathrm{cov}(\langle n_s, s_s \rangle, \langle n_c, s_s \rangle) \\
&= 0 + \sigma^2 \langle s_s, s_c \rangle - \sigma^2 \langle s_c, s_s \rangle - 0 = 0,
\end{aligned}$$

where we have used (4.48), and the fact that the contribution of cross terms involving n_c and n_s is zero because of their independence. □

Remark 4.5.4 (Complex WGN through multiple correlators) Using the same arguments as in the proof of Proposition 4.5.4, we can characterize the joint distribution of complex WGN through multiple correlators. Specifically, for finite-energy signals $s_1(t)$ and $s_0(t)$, it is left as an exercise to show that $\langle n, s_1 \rangle$ and $\langle n, s_0 \rangle$ are jointly proper complex Gaussian with covariance

$$\mathrm{cov}(\langle n, s_1 \rangle, \langle n, s_0 \rangle) = 2\sigma^2 \langle s_0, s_1 \rangle. \tag{4.49}$$

4.5.2 Performance of binary noncoherent communication

We now return to noncoherent detection for equal-energy, equiprobable, binary signaling, with the complex baseband received signal under hypothesis H_i, $i = 0, 1$, given by

$$H_i : y(t) = s_i(t) e^{j\theta} + n(t).$$

We assume that the phase shift θ induced by the channel is uniformly distributed over $[0, 2\pi]$. Under these conditions, the MPE rule has been shown to be as follows:

$$\delta_{\mathrm{MPE}}(y) = \arg \max_{i=0,1} |\langle y, s_i \rangle|.$$

We denote the signal energies by $E_s = ||s_1||^2 = ||s_0||^2$, and define the complex correlation coefficient $\rho = (\langle s_0, s_1 \rangle)/(||s_0|| ||s_1||)$, so that $\langle s_0, s_1 \rangle = \rho E_s = \langle s_1, s_0 \rangle^*$.

Conditioned on H_0, the received signal $y = s_0 e^{j\theta} + n$. The conditional error probability is then given by

$$P_{e|0} = P(|Z_1| > |Z_0| | H_0),$$

where $Z_i = \langle y, s_i \rangle$, $i = 0, 1$ are given by

$$Z_0 = \langle s_0 e^{j\theta} + n, s_0 \rangle = E_s e^{j\theta} + \langle n, s_0 \rangle,$$
$$Z_1 = \langle s_0 e^{j\theta} + n, s_1 \rangle = \rho E_s e^{j\theta} + \langle n, s_1 \rangle.$$

Conditioned on H_0 and θ (we soon show that the conditional error probability does not depend on θ), $\mathbf{Z} = (Z_0, Z_1)^T$ is proper complex Gaussian, because n is proper complex Gaussian. Using Proposition 4.5.4 and Remark 4.5.4, we find that the covariance matrix for $\mathbf{Z}$ is

$$\mathbf{C_Z} = 2\sigma^2 E_s \begin{pmatrix} 1 & \rho^* \\ \rho & 1 \end{pmatrix} \tag{4.50}$$

and the mean vector is

$$\mathbf{m_Z} = E_s e^{j\theta} \begin{pmatrix} 1 \\ \rho \end{pmatrix}. \tag{4.51}$$

In general, developing an expression for the exact error probability involves the painful process of integration over contours in the complex plane, and does not give insight into how, for example, the error probability varies with SNR. We therefore restrict ourselves here to broader observations on the dependence of the error probability on system parameters, including high SNR asymptotics. We do, however, derive the exact error probability for the special case of orthogonal signaling ($\rho = 0$). We state these results as propositions, discuss their implications, and then provide proofs (in the case of Proposition 4.5.5 below, we only sketch the proof, providing a reference for the details).

Proposition 4.5.5 (Dependence on $|\rho|$ and SNR) *The error probability depends only on $|\rho|$ and E_s/N_0, and its high SNR asymptotics are given by*

$$P_e(\text{noncoh}) \sim \exp\left(-\frac{E_s}{2N_0}(1 - |\rho|)\right), \quad \frac{E_s}{2N_0} \to \infty. \tag{4.52}$$

Remark 4.5.5 (Contrast with coherent demodulation) For coherent detection, we know that the error probability is given by $Q(||s_1 - s_0||/2\sigma)$. Noting that $||s_1 - s_0||^2 = 2E_s(1 - \text{Re}(\rho))$ for equal-energy signals, and setting $\sigma^2 = N_0/2$, we have $P_e(\text{coh}) = Q(\sqrt{E_s(1 - \text{Re}(\rho))/N_0})$. Using $Q(x) \sim e^{-x^2/2}$ for large x, the high SNR asymptotics for coherent detection of equal-energy signaling are given by

$$P_e(\text{coh}) \sim \exp\left(-\frac{E_s}{2N_0}(1 - \text{Re}(\rho))\right), \quad \frac{E_s}{2N_0} \to \infty. \tag{4.53}$$

Proposition 4.5.6 (Error probability for orthogonal signaling) *For noncoherent demodulation of equal-energy orthogonal signals ($\rho = 0$), the error probability is given by*

$$P_e = \frac{1}{2} \exp\left(-\frac{E_s}{2N_0}\right) \quad \text{Binary equal-energy, noncoherent signaling.}$$
$$\tag{4.54}$$

4.5 Performance of noncoherent communication

Remark 4.5.6 (Orthogonal signaling with coherent and noncoherent detection) Comparing (4.52) and (4.53), we see that the high SNR asymptotics with orthogonal signaling are the same for both coherent and noncoherent demodulation. However, there are hidden costs associated with noncoherent demodulation. First, if coherent detection were possible, then we could design the signals such that $\text{Re}(\rho) < 0$ (e.g., $\rho = -1$ for antipodal signaling) in order to obtain better performance than with orthogonal signaling. Second, orthogonal signaling with coherent demodulation requires only that $\text{Re}(\rho) = 0$, while orthogonal signaling with noncoherent demodulation requires that $|\rho| = 0$. As shown in Chapter 2, this implies that noncoherent orthogonal signaling requires twice as many degrees of freedom than coherent signaling. For example, orthogonal FSK requires a tone spacing of $1/T$ for noncoherent demodulation, and only $1/2T$ for coherent demodulation, where T is the symbol interval.

We now proceed with the proofs.

Proof of Proposition 4.5.5 We condition on H_0 and θ, and our starting points are (4.50) and (4.51).

First, we show that the performance depends only on $|\rho|$. Suppose that $\rho = |\rho|e^{j\phi}$. We can now rotate one of the signals such that the correlation coefficient becomes positive. Specifically, set $\hat{s}_0(t) = s_0(t)e^{-j\phi}$, and replace Z_0 by $\hat{Z}_0 = \langle y, \hat{s}_0 \rangle$. The decision rule depends only on $|Z_i|$, $i = 0, 1$, and $|Z_0| = |\hat{Z}_0|$, so that the outcome, and hence performance, of the decision rule is unchanged. Conditioned on H_0 and θ, the statistics of $\hat{\mathbf{Z}} = (\hat{Z}_0, Z_1)^T$ are as in (4.50) and (4.51), except that ρ is now replaced by $\hat{\rho} = (\langle \hat{s}_0, s_1 \rangle)/(\|\hat{s}_0\| \|s_1\|) = |\rho|$.

A related point worth noting is that the performance is independent of θ; that is, from the point of performance analysis, we may set $\theta = 0$. To see this, replace Z_i by $Z_i e^{-j\theta}$; this does not change $|Z_i|$, and hence it does not change the decision. Now, write

$$Z_i e^{-j\theta} = \langle s_0, s_i \rangle + \langle n e^{-j\theta}, s_i \rangle$$

and note that the statistics of the proper Gaussian random process $ne^{-j\theta}$ are the same as those of n (check that the mean and autocovariance functions are unchanged).

Thus, we can replace ρ by $|\rho|$ and $\theta = 0$ in (4.50) and (4.51). Furthermore, let us normalize Z_0 and Z_1 to obtain the scale-invariant $U_i = Z_i/\sigma\sqrt{E_s}$, $i = 0, 1$. The conditional mean and covariance matrix (conditioned on 0 being sent) for the proper complex Gaussian vector $\mathbf{U} = (U_0, U_1)^T$ is now given by

$$\mathbf{m}_\mathbf{U} = \sqrt{\frac{E_s}{\sigma^2}} \begin{pmatrix} 1 \\ |\rho| \end{pmatrix} \qquad \mathbf{C}_\mathbf{U} = 2 \begin{pmatrix} 1 & |\rho| \\ |\rho| & 1 \end{pmatrix}. \qquad (4.55)$$

Since the decision based on comparing $|U_0|$ and $|U_1|$ is identical to those provided by the original decision rule, and the conditional distribution of these decision statistics depends on $|\rho|$ and E_s/N_0 alone, so does the conditional error probability (and hence also the unconditional error probability).

We now sketch a plausibility argument for the high SNR asymptotics given in (4.52). The noncoherent rule may be viewed as comparing the magnitudes of the projections of the received signal onto the one-dimensional complex subspaces S_0 and S_1 spanned by s_0 and s_1, respectively (each subspace has two real dimensions). High SNR asymptotics are determined by the most likely way to make an error. If s_0 is sent, then the most likely way for the noise to induce a wrong decision is to move the signal a distance d along the two-dimensional plane defined by the minimum angle between the subspaces S_0 and S_1, as shown in Figure 4.7.

The angle between two complex signals is given by
$$\cos\theta = \frac{\text{Re}(\langle u, v\rangle)}{||u||\,||v||}.$$

To determine the minimum angle between S_0 and S_1, we need to maximize $\cos\theta$ for $u(t) = \alpha s_0(t)$ and $v(t) = \beta s_1(t)$, where α, β are scalars. It is easy to see that the answer is $\cos\theta_{\min} = |\rho|$: this corresponds to rotating one of the signals so that the inner product becomes nonnegative ($u(t) = s_0(t)$ and $v(t) = s_1(t)e^{j\phi}$ works, where $\phi = \arg(\rho)$). We therefore find that the minimum angle is given by
$$\cos\theta_{\min} = |\rho|. \tag{4.56}$$

The minimum distance that the noise needs to move the signal is seen from Figure 4.7 to be
$$d = ||s_0||\sin\left(\frac{\theta_{\min}}{2}\right).$$

Since
$$\sin^2\left(\frac{\theta_{\min}}{2}\right) = \frac{1-\cos\theta_{\min}}{2} = \frac{1-|\rho|}{2},$$
we obtain
$$d^2 = \frac{E_s}{2}(1-|\rho|). \tag{4.57}$$

This yields the high SNR asymptotics
$$P_e \sim Q\left(\frac{d}{\sigma}\right) \sim \exp\left(-\frac{d^2}{2\sigma^2}\right),$$
which yields the desired result (4.52) upon substituting from (4.57). □

Figure 4.7 Geometric view of noncoherent demodulation.

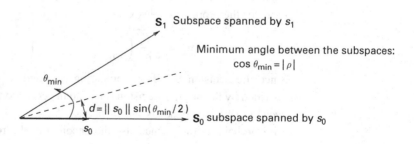

Proof of Proposition 4.5.6 We now consider the special case of orthogonal signaling, for which $\rho = 0$. Let us use the equivalent scaled decision statistics U_0 and U_1 as defined in the proof of Proposition 4.5.5, conditioning on H_0 and setting $\theta = 0$ without loss of generality, as before. Setting $\rho = 0$ in (4.55), we obtain $\mathbf{m_U} = m(1,0)^T$, and $\mathbf{C_U} = 2\mathbf{I}$, where $m = \sqrt{E_s/\sigma^2}$. Since U_0 and U_1 are uncorrelated, they are independent. Since U_0 is a scalar proper complex Gaussian random variable, its real and imaginary parts are independent, with $U_{0c} \sim N(m,1)$ and $U_{0s} \sim N(0,1)$. Similarly, $U_{1c} \sim N(0,1)$ and $U_{1s} \sim N(0,1)$ are independent Gaussian random variables. This implies that $R_0 = |U_0|$ is Rician (see Problem 3.4) with pdf

$$p_{R_0}(r) = r \exp\left(-\frac{m^2 + r^2}{2}\right) I_0(mr) \quad r \geq 0$$

and $R_1 = |U_1|$ is Rayleigh with pdf

$$p_{R_1}(r) = r \exp\left(-\frac{r^2}{2}\right) \quad r \geq 0,$$

where we have dropped the conditioning on H_0 in the notation. The conditional error probability is given by

$$P_{e|0} = P[R_1 > R_0 | H_0] = \int_0^\infty P[R_1 > r | R_0 = r] p_{R_0}(r) dr.$$

Noting that $P[R_1 > r | R_0 = r] = P[R_1 > r] = e^{-r^2/2}$, we obtain

$$P_{e|0} = \int_0^\infty \exp\left(-\frac{r^2}{2}\right) r \exp\left(-\frac{m^2 + r^2}{2}\right) I_0(mr) \, dr. \quad (4.58)$$

We can now massage the integrand above into the form of a new Rician density, multiplied by a constant factor. Since the density must integrate to one, the constant factor is our final answer. The general form of the Rician density is $r/v^2 e^{-\frac{a^2+r^2}{2v^2}} I_0(\frac{ar}{v^2})$. Comparing this with the terms involving r in (4.58), we obtain $r^2 = r^2/2v^2$ and $mr = ar/v^2$, which gives $v^2 = 1/2$ and $a = m/2$. It is left as an exercise to complete the proof by showing that the integral evaluates to $1/2 \exp(-m^2/4)$. Substituting $m = \sqrt{E_s/\sigma^2}$, we obtain the desired formula (4.54). □

4.5.3 Performance of *M*-ary noncoherent orthogonal signaling

An important class of noncoherent systems is M-ary orthogonal signaling. We have shown in Chapter 3 that coherent orthogonal signaling attains fundamental limits of power efficiency as $M \to \infty$. We now show that this property holds for noncoherent orthogonal signaling as well. We consider equal-energy M-ary orthogonal signaling with symbol energy $E_s = E_b \log_2 M$.

Exact error probability As shown in Problem 4.8, this is given by the expression

$$P_e = \sum_{k=1}^{M-1} \binom{M-1}{k} \frac{(-1)^{k+1}}{k+1} \exp\left(-\frac{k}{k+1}\frac{E_s}{N_0}\right). \quad (4.59)$$

Union bound For equal-energy orthogonal signaling with symbol energy E_s, Proposition 4.5.6 provides a formula for the pairwise error probability. We therefore obtain the following union bound:

$$P_e \leq \frac{M-1}{2} \exp\left(-\frac{E_s}{2N_0}\right). \quad (4.60)$$

Note that the union bound coincides with the first term in the summation (4.59) for the exact error probability.

As for coherent orthogonal signaling in Chapter 2, we can take the limit of the union bound as $M \to \infty$ to infer that $P_e \to 0$ if E_b/N_0 is larger than a threshold. However, as before, the threshold obtained from the union bound is off by 3 dB. As we show in Problem 4.9, the threshold for reliable communication for M-ary noncoherent orthogonal signaling is actually $E_b/N_0 > \ln 2$ (-1.6 dB). That is, coherent and noncoherent M-ary orthogonal signaling achieve the same asymptotically optimal power efficiency as M gets large.

Figure 4.8 shows the probability of symbol error as a function of E_b/N_0 for several values of M. As for coherent demodulation (see Figure 3.20), we see that the performance for the values of M considered is quite far from the asymptotic limit of -1.6 dB.

Figure 4.8 Symbol error probabilities for M-ary orthogonal signaling with noncoherent demodulation.

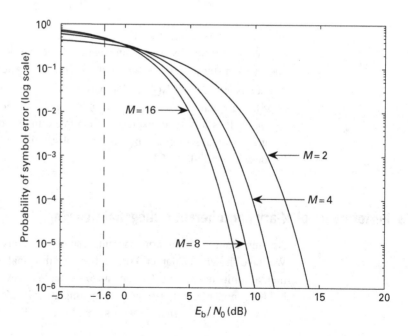

4.5.4 Performance of DPSK

Exact analysis of the performance of M-ary DPSK suffers from the same complication as the exact analysis of noncoherent demodulation of correlated signals. However, an exact result is available for the special case of binary DPSK, as follows.

Proposition 4.5.7 (Performance of binary DPSK) *For an AWGN channel with unknown phase, the error probability for demodulation of binary DPSK over a two-symbol window is given by*

$$P_e = \frac{1}{2} \exp\left(-\frac{E_b}{N_0}\right) \quad \text{Binary DPSK.} \qquad (4.61)$$

Proof of Proposition 4.5.7 Demodulation of binary DPSK over two symbols corresponds to noncoherent demodulation of binary, equal-energy, orthogonal signaling using the signals $\mathbf{s}_{+1} = (1, 1)^T$ and $\mathbf{s}_{-1} = (1, -1)^T$ in (4.38), so that the error probability is given by the formula $1/2 \exp(-E_s/2N_0)$. The result follows upon noting that $E_s = 2E_b$, since the signal $\mathbf{s}_a$ spans two bit intervals, $a = \pm 1$. $\square$

Remark 4.5.7 (Comparison of binary DPSK and coherent BPSK) The error probability for coherent BPSK, which is given by $Q\sqrt{(2E_b/N_0)} \sim \exp(-E_b/N_0)$. Comparing with (4.61), note that the high SNR asymptotics are not degraded due to differential demodulation in this case.

For M-ary DPSK, Proposition 4.5.5 implies that the high SNR asymptotics for the pairwise error probabilities are given by $\exp(-E_s/2N_0(1 - |\rho|))$, where ρ is the pairwise correlation coefficient between signals drawn from the set $\{\mathbf{s}_a, a \in A\}$, and $E_s = 2E_b \log_2 M$. The worst-case value of ρ dominates the high SNR asymptotics. For example, if $a[n]$ are drawn from a QPSK constellation $\{\pm 1, \pm j\}$, the largest value of $|\rho|$ can be obtained by correlating the signals $(1, 1)^T$ and $(1, j)^T$, which yields $|\rho| = 1/\sqrt{2}$. We therefore find that the high SNR asymptotics for DQPSK, demodulated over two successive symbols, are given by

$$P_e(\text{DQPSK}) \sim \exp\left(-\frac{E_b}{N_0}(2 - \sqrt{2})\right).$$

Comparing with the error probability for coherent QPSK, which is given by $Q\sqrt{(2E_b/N_0)} \doteq \exp(-E_b/N_0)$, we note that there is a degradation of 2.3 dB ($10 \log_{10}(2 - \sqrt{2}) = -2.3$). It can be checked using similar methods that the degradation relative to coherent demodulation gets worse with the size of the constellation.

4.5.5 Block noncoherent demodulation

Now that we have developed some insight into the performance of noncoherent communication, we can introduce some more advanced techniques in noncoherent and differential demodulation. If the channel is well approximated as constant over more than two symbols, the performance degradation of M-ary DPSK relative to coherent M-PSK can be alleviated by demodulating over a larger block of symbols. Specifically, suppose that $h[n] = h[n-1] = \cdots = h[n-L+1] = h$, where $L > 2$. Then we can group L received samples together, constructing a vector $\mathbf{y} = (y[n-L+1], \ldots, y[n])$, and obtain

$$\mathbf{y} = \tilde{h}\mathbf{s_a} + \mathbf{w},$$

where $\mathbf{w}$ is the vector of noise samples, $\tilde{h} = hb[n-L+1]$ is unknown, $\mathbf{a} = (a[n-L+2], \ldots, a[n])^T$ is the set of information symbols affecting the block of received samples, and, for DPSK,

$$\mathbf{s_a} = (1, a[n-L+2], a[n-L+2]a[n-L+3], \ldots, a[n])^T.$$

We can now make a joint decision on $\mathbf{a}$ by maximizing the noncoherent decision statistics $|\langle \mathbf{y}, \mathbf{s_a} \rangle|^2$ over all possible values of $\mathbf{a}$.

Remark 4.5.8 (Approaching coherent performance with large block lengths) It can be shown that, for an M-ary PSK constellation, as $L \to \infty$, the high SNR asymptotics for the error probability of block differential demodulation approach that of coherent demodulation. For binary DPSK, however, there is no point in increasing the block size beyond $L = 2$, since the high SNR asymptotics are already as good as those for coherent demodulation.

Remark 4.5.9 (Complexity considerations) For block demodulation of M-ary DPSK, the number of candidate vectors $\mathbf{a}$ is M^{L-1}, so that the complexity of direct block differential demodulation grows exponentially with the block length. Contrast this with coherent, symbol-by-symbol, demodulation, for which the complexity of demodulating a block of symbols is linear. However, near-optimal, linear-complexity, techniques for block differential demodulation are available. The idea is to quantize the unknown phase corresponding to the effective channel gain $\tilde{h}$ into Q hypotheses, to perform symbol-by-symbol coherent demodulation over the block for each hypothesized phase, and to choose the best of the Q candidate sequences $\mathbf{a}_1, \ldots, \mathbf{a}_Q$ thus generated by picking the maximum among the noncoherent decision statistics $|\langle \mathbf{y}, \mathbf{s}_{\mathbf{a}_i} \rangle|$, $i = 1, \ldots, Q$. The complexity is larger than that of coherent demodulation by a fixed factor Q, rather than the exponential complexity of brute force block differential demodulation.

Figure 4.9 Symbol error probabilities for block noncoherent demodulation of differential QPSK, compared with the performance of "absolute" modulation (or coherent QPSK).

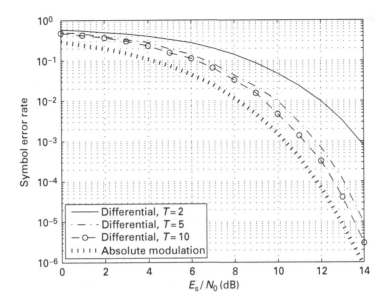

Figure 4.9 shows the effect of block length on block noncoherent demodulation of differential QPSK. Note the large performance improvement in going from a block length of $T = 2$ (standard differential demodulation) to $T = 5$; as we increase the block length further, the performance improves more slowly, and eventually approaches that of coherent QPSK or "absolute" modulation.

4.6 Further reading

For further reading on synchronization, we suggest the books by Mengali and D'Andrea [20], Meyr and Ascheid [21], and Meyr, Moenclaey, and Fechtel [22], and the references therein. We recommend the book by Poor [19] for a systematic treatment of estimation theory, including bounds on achievable performance such as the Cramer–Rao lower bound. An important classical reference, which includes a detailed analysis of the nonlinear dynamics of the PLL, is the text by Viterbi [11]. References related to synchronization for spread spectrum modulation formats are given in Chapter 8. The material on signal space concepts for noncoherent communication is drawn from the paper by Warrier and Madhow [23]. An earlier paper by Divsalar and Simon [24] was the first to point out that block noncoherent demodulation could approach the performance of coherent systems. For detailed analysis of noncoherent communication with correlated signals, we refer to Appendix B of the book by Proakis [3]. Finally, extensive tabulation of the properties of special functions such as Bessel functions can be found in Abramowitz and Stegun [25] and Gradshteyn *et al.* [26].

4.7 Problems

Problem 4.1 (Amplitude estimation) Fill in the details for the amplitude estimates in Example 4.2.2 by deriving (4.10) and (4.11).

Problem 4.2 (NDA amplitude and phase estimation for sampled QPSK system) The matched filter outputs in a linearly modulated system are modeled as

$$z[k] = Ae^{j\theta}b[k] + N[k], \quad k = 1, \ldots, K,$$

where $A > 0$, $\theta \in [0, 2\pi]$ are unknown, $b[k]$ are i.i.d. QPSK symbols taking values equiprobably in $\pm 1, \pm j$, and $N[k]$ are i.i.d. complex WGN samples with variance σ^2 per dimension.

(a) Find the likelihood function of $z[k]$ given A and θ, using the discrete-time likelihood function (4.20). Show that it can be written as a sum of two hyperbolic cosines.
(b) Use the result of (a) to write down the log likelihood function for $z[1], \ldots, z[K]$, given A and θ.
(c) Show that the likelihood function is unchanged when θ is replaced by $\theta + \pi/2$. Conclude that the phase θ can only be estimated modulo $\pi/2$ in NDA mode, so that we can restrict attention to $\theta \in [0, \pi/2)$, without loss of generality.
(d) Show that

$$\mathbb{E}[|z[k]|^2] = A^2 + 2\sigma^2.$$

Use this to motivate an ad hoc estimator for A based on averaging $|z[k]|^2$.
(e) Maximize the likelihood function in (c) numerically over A and θ for $K = 4$, with

$$z[1] = -0.1 + 0.9j, \ z[2] = 1.2 + 0.2j, \ z[3] = 0.3 - 1.1j, \ z[4] = -0.8 + 0.4j$$

and $\sigma^2 = 0.1$.

Hint Use (c) to restrict attention to $\theta \in [0, \pi/2)$. You can try an iterative approach in which you fix the value of one parameter, and maximize numerically over the other, and continue until the estimates "settle." The amplitude estimator in (d) can provide a good starting point.

Problem 4.3 (Costas loop for phase tracking in linearly modulated systems) Consider the complex baseband received signal

$$y(t) = \sum_{k=1}^{M} b[k]p(t - kT)e^{j\theta} + n(t),$$

where $\{b[k]\}$ are drawn from a complex-valued constellation and θ is an unknown phase. For data-aided systems, we assume that the symbol sequence

4.7 Problems

$\mathbf{b} = \{b[k]\}$ is known. For nondata-aided systems, we assume that the symbols $\{b[k]\}$ are i.i.d., selected equiprobably from the constellation. Let $z(t) = (y * p_{\text{MF}})(t)$ denote the output, at time t, of the matched filter with impulse response $p_{\text{MF}}(t) = p^*(-t)$.

(a) Show that the likelihood function conditioned on θ and $\mathbf{b}$ depends on the received signal only through the sampled matched filter outputs $\{z(kT)\}$.

(b) For known symbol sequence $\mathbf{b}$, find the ML estimate of θ. It should depend on y only through the sampled matched filter outputs $\{z[k] = z(kT)\}$.

(c) For tracking slowly varying θ in data-aided mode, assume that $\mathbf{b}$ is known, and define the log likelihood function cost function

$$J_k(\theta) = \log L(z[k]|\theta, \mathbf{b}),$$

where $L(z[k]|\theta, \mathbf{b})$ is proportional to the conditional density of $z[k] = z(kT)$, given θ (and the known symbol sequence $\mathbf{b}$). Show that

$$\frac{\partial J_k(\theta)}{\partial \theta} = a \, \text{Im}\left(b^*[k]z[k]e^{-j\theta}\right),$$

where a is a constant.

(d) Suppose, now, that we wish to operate in decision-directed mode. Specialize to BPSK signaling (i.e., $b[k] \in \{-1, +1\}$). Show that the optimum coherent decision on $b[k]$, assume ideal phase tracking, is

$$\hat{b}[k] = \text{sign}\left(\text{Re}(z[k]e^{-j\theta})\right).$$

Assuming that this bit estimate is correct, substitute $\hat{b}[k]$ in place of $b[k]$ into the result of (c). Show that a discrete-time ascent algorithm of the form

$$\hat{\theta}[k+1] = \theta[k] + b\frac{\partial J_k(\theta)}{\partial \theta}\bigg|_{\theta=\theta[k]}$$

reduces to

$$\hat{\theta}[k+1] = \theta[k] + \alpha \, \text{sign}\left(\text{Re}(z[k]e^{-j\theta}[k])\right) \, \text{Im}(z[k]e^{-j\theta}[k]),$$

where $\alpha > 0$ is a parameter that governs the reaction time of the tracking algorithm. The block diagram for the algorithm is shown in Figure 4.10.

(e) Now, consider nondata-aided estimation for i.i.d. BPSK symbols taking values ± 1 equiprobably. Find the log likelihood function averaged over $\mathbf{b}$:

$$\log L(y|\theta) = \log \mathbb{E}\left[L(y|\theta, \mathbf{b})\right],$$

where the expectation is over the symbol sequence $\mathbf{b}$. Assume that p is square root Nyquist at rate $1/T$ if needed.

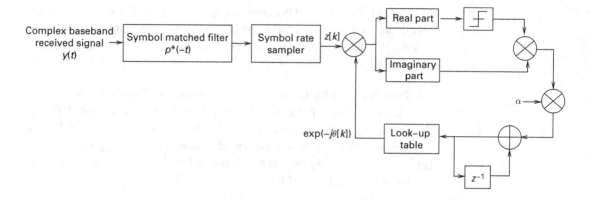

Figure 4.10 Discrete-time decision-directed Costas loop for BPSK modulation.

Hint Use techniques similar to those used to derive the NDA amplitude estimate in Example 4.2.2.

(f) Find an expression for a block-based ML estimate of θ using the NDA likelihood function in (d). Again, this should depend on y only through $\{z[k]\}$.

(g) Derive a tracking algorithm as in part (d), but this time within NDA mode. That is, use the cost function

$$J_k(\theta) = \log L(z[k]|\theta)$$

where $L(z[k]|\theta)$ is obtained by averaging over all possible values of the BPSK symbols. Show that the tracker can be implemented as in the block diagram in Figure 4.10 by replacing the hard decision by a hyperbolic tangent with appropriately scaled argument.

(h) Show that the tracker in (g) is approximated by the decision-directed tracker in (d) by using the high SNR approximation $\tanh x \approx \text{sign}(x)$. Can you think of a low SNR approximation?

Remark The low SNR approximation mentioned in (h) corresponds to what is generally known as a Costas loop. In this problem, we use the term for a broad class of phase trackers with similar structure.

Problem 4.4 (Frequency offset estimation using training sequence) Consider a linear modulated system with no ISI and perfect timing recovery, but with unknown frequency offset Δf and phase offset θ. The symbol rate samples are modeled as

$$y[k] = b[k]e^{j(2\pi\Delta f kT + \theta)} + N[k], \quad k = 1, \ldots, K,$$

where T is the symbol time, and $N[k]$ is discrete-time complex WGN with variance $\sigma^2 = N_0/2$ per dimension. Define $\Gamma = 2\pi\Delta fT$ as the normalized frequency offset. We wish to obtain ML estimates of Γ and θ, based on the observation $\mathbf{y} = (y[1], \ldots, y[K])^T$. Assume that the complex symbols $\{b[k]\}$ are part of a known training sequence.

(a) Find the log likelihood function conditioned on Γ and θ, simplifying as much as possible.

(b) Fixing Γ, maximize the log likelihood function over θ. Substitute the maximizing value of θ to derive a cost function $J(\Gamma)$ to be maximized over Γ.

(c) Discuss approximate computation of the ML estimate for Γ using a discrete Fourier transform (DFT).

Problem 4.5 (Example of one-shot timing estimation) The received signal in a real baseband system is given by

$$y(t) = p(t - \tau) + n(t),$$

where $p(t) = I_{[0,1]}(t)$, τ is an unknown delay taking values in $[0, 1]$, and n is real-valued WGN with PSD $\sigma^2 = N_0/2$. The received signal is passed through a filter matched to p to obtain $z(t) = (y * p_{MF})(t)$, where $p_{MF}(t) = p(-t)$, but the ML estimation algorithm only has access to the samples at times $0, 1/2, 1$.

(a) Specify the distribution of the sample vector $\mathbf{z} = (z(0), z(1/2), z(1))^T$, conditioned on $\tau \in [0, 1]$.

Hint Consider the cases $\tau \leq 1/2$ and $\tau \geq 1/2$ separately.

(b) Compute the ML estimate of τ if $\mathbf{z} = (0.7, 0.8, -0.1)^T$, assuming that $\sigma^2 = 0.1$. How does your answer change if $\sigma^2 = 0.01$?

Problem 4.6 (Block-based timing estimation for a linearly modulated signal) Consider the timing estimation problem in Example 4.3.2 for a linearly modulated signal. That is, the received signal y is given by

$$y(t) = As(t - \tau)e^{j\theta} + n(t),$$

for

$$s(t) = \sum_{k=1}^{K} b[k] p(t - kT),$$

where τ is to be estimated, A, θ are "nuisance" parameters which we eliminate by estimating (for fixed τ) and substituting into the cost function, as in Example 4.3.2, and n is complex WGN. Assume that $\{b[k], k = 1, \ldots, K\}$ are part of a known training sequence.

(a) Specializing the result in Example 4.3.2, show that the ML estimate of the delay τ can be implemented using the output $z(t) = (y * p_{MF})(t)$ of a filter with impulse response $p_{MF}(t) = p^*(-t)$ matched to the modulating pulse p.

(b) Now, suppose that we only have access to the matched filter outputs sampled at twice the symbol rate, at sample times $\ell T/2$. Discuss how you might try to approximate the delay estimate in (a), which has access to the matched filter outputs at all times.

Problem 4.7 Consider an on–off keyed system in which the receiver makes its decision based on a single complex number y, as follows:

$$y = hA + n, \quad 1 \text{ sent},$$
$$y = n, \quad 0 \text{ sent},$$

where $A > 0$, h is a random channel gain modeled as a zero mean, proper complex Gaussian random variable with $\mathbb{E}[|h|^2] = 1$, and n is zero mean, proper complex Gaussian noise with variance $\sigma^2 = N_0/2$ per dimension.

(a) Assume that h is unknown to the receiver, but that the receiver knows its distribution (given above). Show that the ML decision rule based on y is equivalent to comparing $|y|^2$ with a threshold. Find the value of the threshold in terms of the system parameters.

(b) Find the conditional probability of error, as a function of the average E_b/N_0 (averaged over all possible realizations of h), given that 0 is sent.

(c) Assume now that the channel gain is known to the receiver. What is the ML decision if $y = 1 + j$, $h = j$, $A = 3/2$, and $\sigma^2 = 0.01$ for the coherent receiver?

Problem 4.8 (Exact performance analysis for M-ary, equal-energy, noncoherent orthogonal signaling) Consider an M-ary orthogonal equal-energy signal set $\{s_i, i = 1, \ldots, M\}$ with $\langle s_i, s_j \rangle = E_s \delta_{ij}$, for $1 \leq i, j \leq M$. Condition on s_1 being sent, so that the received signal $y = s_1 e^{j\theta} + n$, where n is complex WGN with variance $\sigma^2 = N_0/2$ per dimension, and θ is an arbitrary unknown phase shift. The noncoherent decision rule is given by

$$\delta_{\mathrm{nc}}(y) = \arg \max_{1 \leq i \leq M} |Z_i|,$$

where we consider the normalized, scale-invariant decision statistics $Z_i = (\langle y, s_i \rangle)/(\sigma\sqrt{E_s})$, $i = 1, \ldots, M$. Let $\mathbf{Z} = (Z_1, \ldots, Z_M)^T$, and denote the magnitudes by $R_i = |Z_i|$, $i = 1, \ldots, M$.

(a) Show that the normalized decision statistics $\{Z_i\}$ are (conditionally) independent, with $Z_1 \sim CN(me^{j\theta}, 2)$ and $Z_i \sim CN(0, 2)$, where $m = \sqrt{(2E_s/N_0)}$.

(b) Conclude that, conditioned on s_1 sent, the magnitudes R_i, $i \neq 1$, obey a Rayleigh distribution (see Problem 3.4) satisfying

$$P[R_i \leq r] = 1 - e^{-\frac{r^2}{2}}, \quad r \geq 0.$$

(c) Show that, conditioned on s_1 sent, $R_1 = |Z_1|$ is Rician (see Problem 3.4) with conditional density

$$p_{R_1|1}(r|1) = r \exp\left(-\frac{m^2 + r^2}{2}\right) I_0(mr), \quad r \geq 0.$$

(d) Show that the conditional probability of correct reception (given s_1 sent), which also equals the unconditional probability of correct reception by symmetry, is given by

$$P_c = P_{c|1} = P[R_1 = \max_i R_i | H_1] = P[R_2 \leq R_1, R_3 \leq R_1, \ldots, R_M \leq R_1 | H_1]$$

$$= \int_0^\infty \left(1 - e^{-\frac{r^2}{2}}\right)^{M-1} p_{R_1|1}(r|1) \, dr \qquad (4.62)$$

$$= \int_0^\infty \left(1 - e^{-\frac{r^2}{2}}\right)^{M-1} r \exp\left(-\frac{m^2 + r^2}{2}\right) I_0(mr) \, dr$$

$(m = \sqrt{\frac{2E_s}{N_0}})$.

(e) Show that the error probability is given by

$$P_e = 1 - P_c = \int_0^\infty \left[1 - \left(1 - e^{-\frac{r^2}{2}}\right)^{M-1}\right] r \exp\left(-\frac{m^2 + r^2}{2}\right) I_0(mr) \, dr.$$

Using a binomial expansion within the integrand, conclude that

$$P_e = \sum_{k=1}^{M-1} \binom{M-1}{k} A_k,$$

where

$$A_k = (-1)^{k+1} \int_0^\infty r e^{-\frac{kr^2}{2}} \exp\left(-\frac{m^2 + r^2}{2}\right) I_0(mr) \, dr. \qquad (4.63)$$

(f) Now, massage the integrand into the form of a Rician density as we did when computing the error probability for binary orthogonal signaling. Use this to evaluate A_k and obtain the following final expression for error probability

$$P_e = \sum_{k=1}^{M-1} \binom{M-1}{k} \frac{(-1)^{k+1}}{k+1} \exp\left(\frac{k}{k+1} \frac{E_s}{N_0}\right).$$

Check that this specializes to the expression for binary orthogonal signaling by setting $M = 2$.

Problem 4.9 (Asymptotic performance of M-ary noncoherent orthogonal signaling) In the setting of Problem 4.8, we wish to derive the result that

$$\lim_{M \to \infty} P_c = \begin{cases} 1, & \frac{E_b}{N_0} > \ln 2, \\ 0, & \frac{E_b}{N_0} < \ln 2. \end{cases} \qquad (4.64)$$

Set

$$m = \sqrt{\frac{2E_s}{N_0}} = \sqrt{\frac{2E_b \log_2 M}{N_0}},$$

as in Problem 4.8.

(a) In (4.8), use a change of variables $U = R_1 - m$ to show that the probability of correct reception is given by

$$P_c = \int_0^\infty \left(1 - e^{-\frac{(u+m)^2}{2}}\right)^{M-1} p(u|1)\, du.$$

(b) Show that, for any $u \geq 0$,

$$\lim_{M \to \infty} \left(1 - e^{-\frac{(u+m)^2}{2}}\right)^{M-1} = \begin{cases} 0, & \frac{E_b}{N_0} < \ln 2, \\ 1, & \frac{E_b}{N_0} > \ln 2. \end{cases}$$

Hint Use L'Hôpital's rule on the log of the expression whose limit is to be evaluated.

(c) Show that, by a suitable change of coordinates, we can write

$$R_1 = \sqrt{(m + V_1)^2 + V_2^2},$$

where V_1, V_2 are i.i.d. $N(0, 1)$ random variables. Use this to show that, as $m \to \infty$, $U = R_1 - m$ converges to a random variable whose distribution does not depend on M (an intuitive argument rather than a rigorous proof is expected). What is the limiting distribution? (The specific form of the density is actually not required in the subsequent proof, which only uses the fact that there is some limiting distribution that does not depend on M.)

(d) Assume now that we can interchange limit and integral as we let $M \to \infty$, so that

$$\lim_{M \to \infty} P_c = \int_0^\infty \lim_{M \to \infty} \left(1 - e^{-\frac{(u+m)^2}{2}}\right)^{M-1} \lim_{M \to \infty} p(u|1) du.$$

Now use (b) and (c) to infer the desired result.

Problem 4.10 (Noncoherent orthogonal signaling over a Rayleigh fading channel) Binary orthogonal signaling over a Rayleigh fading channel can be modeled using the following hypothesis testing problem:

$$H_1: y(t) = As_1(t)e^{j\theta} + n(t), \quad 0 \leq t \leq T,$$
$$H_0: y(t) = As_0(t)e^{j\theta} + n(t), \quad 0 \leq t \leq T,$$

where $\langle s_1, s_0 \rangle = 0$, $||s_1||^2 = ||s_0||^2 = E_b$, n is complex AWGN with PSD $\sigma^2 = N_0/2$ per dimension. Conditioned on either hypothesis, the amplitude $A > 0$ is Rayleigh with $\mathbb{E}[A^2] = 1$, θ is uniformly distributed over $[0, 2\pi]$, and A, θ are independent of each other and of the noise n. Equivalently, $h = Ae^{j\theta} \sim CN(0, 1)$ is a proper complex Gaussian random variable. Define the complex-valued correlation decision statistics $Z_i = \langle y, s_i \rangle$, $i = 0, 1$.

(a) Show that the MPE decision rule is the noncoherent detector given by

$$\hat{i} = \arg\max\{|Z_1|, |Z_0|\}.$$

(b) Find the error probability as a function of E_b/N_0 by first conditioning on A and using Proposition 4.5.6, and then removing the conditioning.

(c) Now, find the error probability directly using the following reasoning. Condition throughout on H_0. Show that Z_1 and Z_0 are independent complex Gaussian random variables with i.i.d. real and imaginary parts. Infer that $|Z_1|^2$ and $|Z_0|^2$ are independent exponential random variables (see Problem 3.4), and use this fact to derive directly the error probability conditioned on H_0 (without conditioning on A or θ).

(d) Plot the error probability on a log scale as a function of E_b/N_0 in dB for the range 0–20 dB. Compare with the results for the AWGN channel (i.e., for $A \equiv 1$), and note the heavy penalty due to Rayleigh fading.

Problem 4.11 (Soft decisions with noncoherent demodulation) Consider noncoherent binary on–off keying over a Rayleigh fading channel, where the receiver decision statistic is modeled as:

$$Y = h + N, \quad 1 \text{ sent},$$
$$Y = N, \quad 0 \text{ sent},$$

where h is zero mean complex Gaussian with $\mathbb{E}[|h|^2] = 3$, N is zero mean complex Gaussian with $\mathbb{E}[|N|^2] = 1$, and h, N are independent. The receiver does not know the actual value of h, although it knows the distributions above. Find the posterior probability $P[1 \text{ sent}|Y = 1 - 2j]$, assuming the prior probability $P[1 \text{ sent}] = 1/3$.

Problem 4.12 (A toy model illustrating channel uncertainty and diversity) Consider binary, equiprobable signaling over a scalar channel in which the (real-valued) received sample is given by

$$y = hb + n, \tag{4.65}$$

where $b \in \{-1, +1\}$ is the transmitted symbol, $n \sim N(0, 1)$, and the channel gain h is a random variable taking one of two values, as follows:

$$P[h = 1] = \frac{1}{4}, \quad P[h = 2] = \frac{3}{4}. \tag{4.66}$$

(a) Find the probability of error, in terms of the Q function with positive arguments, for the decision rule $\hat{b} = \text{sign}(y)$. Express your answer in terms of E_b/N_0, where E_b denotes the average received energy per bit (averaged over channel realizations).

(b) **True or False** The decision rule in (a) is the minimum probability of error (MPE) rule. Justify your answer.

Now, suppose that we have two-channel diversity, with two received samples given by

$$y_1 = h_1 b + n_1, \qquad y_2 = h_2 b + n_2, \qquad (4.67)$$

where b is equally likely to be ± 1, n_1 and n_2 are independent and identically distributed (i.i.d.) $N(0, \sigma^2)$, and h_1 and h_2 are i.i.d., each with distribution given by (4.66).

(c) Find the probability of error, in terms of the Q function with positive arguments, for the decision rule $\hat{b} = \text{sign}(y_1 + y_2)$.

(d) **True or False** The decision rule in (b) is the MPE rule for the model (4.67), assuming that the receiver does not know h_1, h_2, but knows their joint distribution. Justify your answer.

Problem 4.13 (Preview of diversity for wireless channels) The performance degradation due to Rayleigh fading encountered in Problem 4.10 can be alleviated by the use of diversity, in which we see multiple Rayleigh fading channels (ideally independent), so that the probability of all channels having small amplitudes is small. We explore diversity in greater depth in Chapter 8, but this problem provides a quick preview. Consider, as in Problem 4.10, binary orthogonal signaling, except that we now have access to two copies of the noisy transmitted signal over independent Rayleigh fading channels. The resulting hypothesis testing problem can be written as follows:

$$H_1: y_1(t) = h_1 s_1(t) + n_1(t), \quad y_2(t) = h_2 s_1(t) + n_2(t), \quad 0 \le t \le T,$$
$$H_0: y_1(t) = h_1 s_0(t) + n_1(t), \quad y_2(t) = h_2 s_0(t) + n_2(t), \quad 0 \le t \le T,$$

where $\langle s_1, s_0 \rangle = 0$, $||s_1||^2 = ||s_0||^2 = E_b$, h_1, h_2 are i.i.d. $CN(0, 1/2)$ (normalizing so that the net average received energy per bit is still E_b), and n is complex AWGN with PSD $\sigma^2 = N_0/2$ per dimension.

(a) Assuming that h_1 and h_2 are known (i.e., coherent reception) to the receiver, find the ML decision rule based on y_1 and y_2.

(b) Find an expression for the error probability (averaged over the distribution of h_1 and h_2) for the decision rule in (a). Evaluate this expression for $E_b/N_0 = 15$ dB, either analytically or by simulation.

(c) Assuming now that the channel gains are unknown (i.e., noncoherent reception), find the ML decision rule based on y_1 and y_2.

(d) Find an expression for the error probability (averaged over the distribution of h_1 and h_2) for the decision rule in (c). Evaluate this expression for $E_b/N_0 = 15$ dB, either analytically or by simulation.

CHAPTER 5
Channel equalization

In this chapter, we develop *channel equalization* techniques for handling the intersymbol interference (ISI) incurred by a linearly modulated signal that goes through a dispersive channel. The principles behind these techniques also apply to dealing with interference from other users, which, depending on the application, may be referred to as co-channel interference, multiple-access interference, multiuser interference, or crosstalk. Indeed, we revisit some of these techniques in Chapter 8 when we briefly discuss multiuser detection. More generally, there is great commonality between receiver techniques for efficiently accounting for memory, whether it is introduced by nature, as considered in this chapter, or by design, as in the channel coding schemes considered in Chapter 7. Thus, the optimum receiver for ISI channels (in which the received signal is a convolution of the transmitted signal with the channel impulse response) uses the same Viterbi algorithm as the optimum receiver for convolutional codes (in which the encoded data are a convolution of the information stream with the code "impulse response") in Chapter 7.

The techniques developed in this chapter apply to single-carrier systems in which data are sent using linear modulation. An alternative technique for handling dispersive channels, discussed in Chapter 8, is the use of multi-carrier modulation, or orthogonal frequency division multiplexing (OFDM). Roughly speaking, OFDM, or multicarrier modulation, transforms a system with memory into a memoryless system in the frequency domain, by decomposing the channel into parallel narrowband subchannels, each of which sees a scalar channel gain.

Map of this chapter After introducing the channel model in Section 5.1, we discuss the choice of receiver front end in Section 5.2. We then briefly discuss the visualization of the effect of ISI using eye diagrams in Section 5.3. This is followed by a derivation of maximum likelihood sequence estimation (MLSE) for optimum equalization in Section 5.4. We introduce the Viterbi algorithm for efficient implementation of MLSE. Since the complexity of

MLSE is exponential in the channel memory, suboptimal equalizers with lower complexity are often used in practice. Section 5.5 describes a geometric model for design of such equalizers. The model is then used to design linear equalizers in Section 5.6, and decision feedback equalizers in Section 5.7. Techniques for evaluating the performance of these suboptimum equalizers are also discussed. Finally, Section 5.8 discusses the more complicated problem of estimating the performance of MLSE. The idea is to use the union bounds introduced in Chapter 3 for estimating the performance of M-ary signaling in AWGN, except that M can now be very large, since it equals the number of possible symbol sequences that could be sent. We therefore discuss "intelligent" union bounds to prune out unnecessary terms, as well as a transfer function bound for summing such bounds over infinitely many terms. Similar arguments are also used in performance analysis of ML decoding of coded systems (see Chapter 7).

5.1 The channel model

Consider the complex baseband model for linear modulation over a dispersive channel, as depicted in Figure 5.1.

The signal sent over the channel is given by

$$u(t) = \sum_{n=-\infty}^{\infty} b[n] g_{\text{TX}}(t - nT),$$

where $g_{\text{TX}}(t)$ is the impulse response of the transmit filter, and $\{b[n]\}$ is the symbol sequence, transmitted at rate $1/T$. The channel is modeled as a filter with impulse response $g_C(t)$, followed by AWGN. Thus, the received signal is given by

$$y(t) = \sum_{n=-\infty}^{\infty} b[n] p(t - nT) + n(t), \tag{5.1}$$

where

$$p(t) = (g_{\text{TX}} * g_C)(t)$$

is the impulse response of the cascade of the transmit and channel filters, and $n(t)$ is complex WGN with PSD $\sigma^2 = N_0/2$ per dimension. The task of the channel equalizer is to extract the transmitted sequence $\mathbf{b} = \{b[n]\}$ from the received signal $y(t)$.

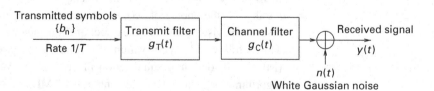

Figure 5.1 Linear modulation over a dispersive channel.

> **Running example** As a running example through this chapter, we consider the setting shown in Figure 5.2. The symbol rate is 1/2 (i.e., one symbol every two time units). The transmit pulse $g_{TX}(t) = I_{[0,2]}(t)$ is an ideal rectangular pulse in the time domain, while the channel response $g_C(t) = \delta(t-1) - 1/2\delta(t-2)$ corresponds to two discrete paths.
>
>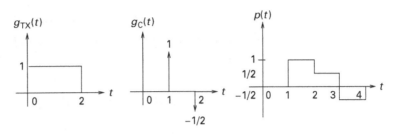
>
> **Figure 5.2** Transmit pulse $g_{TX}(t)$, channel impulse response $g_C(t)$, and overall pulse $p(t)$ for the running example. The symbol rate is 1/2 symbol per unit time.

5.2 Receiver front end

Most modern digital communication receivers are DSP-intensive. For example, for RF communication, relatively sloppy analog filters are used in the passband and at intermediate frequencies (for superheterodyne reception). The complex baseband version of these passband filtering operations corresponds to passing the complex envelope of the received signal through a sloppy analog complex baseband filter, which is a cascade of the complex baseband versions of the analog filters used in the receive chain. We would typically design this *equivalent* analog baseband filter to have a roughly flat transfer function over the band, say $[-W/2, W/2]$, occupied by the transmitted signal. Thus, there is no loss of information in the signal contribution to the output of the equivalent complex baseband receive filter if it is sampled at a rate faster than W. Typically, the sampling rate is chosen to be an integer multiple of $1/T$, the symbol rate. This provides an information-lossless front end which yields a discrete-time signal which we can now process in DSP. For example, we can implement the equivalent of a specific passband filtering operation on the passband received signal using DSP operations on the discrete-time complex baseband signal that implement the corresponding complex baseband filtering operations.

Now that we have assured ourselves that we can implement any analog operation in DSP using samples at the output of a sloppy wideband filter, let us now return to the analog complex baseband signal $y(t)$ and ask how to process it optimally if we did not have to worry about implementation details. The answer is given by the following theorem, which characterizes the optimal receiver front end.

Theorem 5.2.1 (Optimality of the matched filter) *The optimal receive filter is matched to the equivalent pulse $p(t)$, and is specified in the time and frequency domains as follows:*

$$\begin{aligned} g_{R,\text{opt}}(t) &= p_{\text{MF}}(t) = p^*(-t), \\ G_{R,\text{opt}}(f) &= P_{\text{MF}}(f) = P^*(f). \end{aligned} \quad (5.2)$$

In terms of a decision on the symbol sequence **b**, *there is no loss of relevant information by restricting attention to symbol rate samples of the matched filter output, given by*

$$z[n] = (y * p_{\text{MF}})(nT) = \int y(t) p_{\text{MF}}(nT - t) \, dt = \int y(t) p^*(t - nT) \, dt. \quad (5.3)$$

Proof of Theorem 5.2.1 We can prove this result using either the hypothesis testing framework of Chapter 3, or the broader parameter estimation framework of Chapter 4. Deciding on the sequence **b** is equivalent to testing between all possible hypothesized sequences **b**, with the hypothesis $H_\mathbf{b}$ corresponding to sequence **b** given by

$$H_\mathbf{b} : y(t) = s_\mathbf{b}(t) + n(t),$$

where

$$s_\mathbf{b}(t) = \sum_n b[n] p(t - nT)$$

is the noiseless received signal corresponding to transmitted sequence **b**. We know from Theorem 3.4.3 that the ML rule is given by

$$\delta_{\text{ML}}(y) = \arg \max_\mathbf{b} \text{Re}(\langle y, s_\mathbf{b} \rangle) - \frac{\|s_\mathbf{b}\|^2}{2}.$$

The MPE rule is similar, except for an additive correction term accounting for the priors. In both cases, the decision rule depends on the received signal only through the term $\langle y, s_\mathbf{b} \rangle$. The optimal front end, therefore, should capture enough information to be able to compute this inner product for all possible sequences **b**.

We can also use the more general framework of the likelihood function derived in Theorem 4.2.1 to infer the same result. For $y = s_\mathbf{b} + n$, the likelihood function (conditioned on **b**) is given by

$$L(y|\mathbf{b}) = \exp\left(\frac{1}{\sigma^2}[\text{Re}\langle y, s_\mathbf{b}\rangle - \frac{\|s_\mathbf{b}\|^2}{2}]\right).$$

We have sufficient information for deciding on **b** if we can compute the preceding likelihood function for any sequence **b**, and the observation-dependent part of this computation is the inner product $\langle y, s_\mathbf{b} \rangle$.

Figure 5.3 Typical implementation of optimal front end.

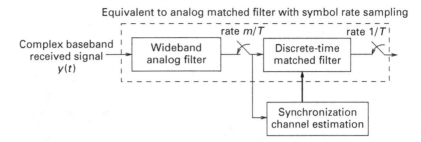

Let us now consider the structure of this inner product in more detail.

$$\langle y, s_{\mathbf{b}} \rangle = \langle y, \sum_n b[n] p(t-nT) \rangle = \sum_n b^*[n] \int y(t) p^*(t-nT) \, dt = \sum_n b^*[n] z[n],$$

where $\{z[n]\}$ are as in (5.3). Generation of $\{z[n]\}$ by sampling the outputs of the matched filter (5.2) at the symbol rate follows immediately from the definition of the matched filter. □

While the matched filter is an analog filter, as discussed earlier, it can be implemented in discrete time using samples at the output of a wideband analog filter. A typical implementation is shown in Figure 5.3. The matched filter is implemented in discrete time after estimating the effective discrete-time channel (typically using a sequence of known training symbols) from the input to the transmit filter to the output of the sampler after the analog filter.

For the suboptimal equalization techniques that we discuss, it is not necessary to implement the matched filter. Rather, the sampled outputs of the analog filter can be processed directly by an adaptive digital filter that is determined by the specific equalization algorithm employed.

5.3 Eye diagrams

An intuitive sense of the effect of ISI can be obtained using *eye diagrams*. Consider the noiseless signal $r(t) = \sum_n b[n] x(t-nT)$, where $\{b[n]\}$ is the transmitted symbol sequence. The waveform $x(t)$ is the effective symbol waveform: for an eye diagram at the input to the receive filter, it is the cascade of the transmit and channel filters; for an eye diagram at the output of the receive filter, it is the cascade of the transmit, channel, and receive filters. The effect of ISI seen by different symbols is different, depending on how the contributions due to neighboring symbols add up. The eye diagram superimposes the ISI patterns seen by different symbols into one plot, thus enabling us to see the variation between the best-case and worst-case effects of ISI. One way to generate such a plot is to generate $\{b[n]\}$ randomly, and then superimpose the waveforms $\{r(t-kT), k = 0, \pm 1, \pm 2, \ldots\}$, plotting the superposition over a basic interval of length chosen to be an integer multiple

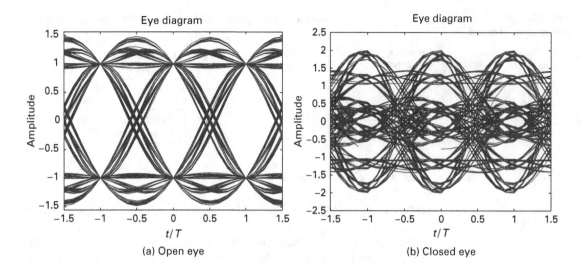

Figure 5.4 Eye diagrams for raised cosine pulse with 50% excess bandwidth for (a) an ideal channel and (b) a highly dispersive channel.

of T. The eye diagram for BPSK using a raised cosine pulse with 50% excess bandwidth is shown in Figure 5.4(a), where the interval chosen is of length $3T$. Note that, in every symbol interval, there is a sampling time at which we can clearly distinguish between symbol value of $+1$ and -1, for all possible ISI realizations. This desirable situation is termed an "open" eye. In contrast, Figure 5.4(b) shows the eye diagram when the raised cosine pulse is passed through a channel $\delta(t) - 0.6\delta(t - 0.5T) + 0.7\delta(t - 1.5T)$. Now there is no longer a sampling point where we can clearly distinguish the value $+1$ from the value -1 for all possible ISI realizations. That is, the eye is "closed," and simple symbol-by-symbol decisions based on samples at appropriately chosen times do not provide reliable performance. However, sophisticated channel equalization schemes such as the ones we discuss in this chapter can provide reliable performance even when the eye is closed.

5.4 Maximum likelihood sequence estimation

We develop a method for ML estimation of the entire sequence $\mathbf{b} = \{b[n]\}$ based on the received signal model (5.1). Theorem 5.2.1 tells us that the optimal front end is the filter matched to the cascade of the transmit and channel filters. We use the notation in Theorem 5.2.1 and its proof in the following.

We wish to maximize $L(y|\mathbf{b})$ over all possible sequences $\mathbf{b}$. Equivalently, we wish to maximize

$$\Lambda(\mathbf{b}) = \mathrm{Re}\langle y, s_\mathbf{b} \rangle - \frac{||s_\mathbf{b}||^2}{2}, \tag{5.4}$$

where the dependence on the MF outputs $\{z[n]\}$ has been suppressed from the notation. To see the computational infeasibility of a brute force approach to this problem, suppose that N symbols, each drawn from an M-ary alphabet,

5.4 Maximum likelihood sequence estimation

are sent. Then there are M^N possible sequences **b** that must be considered in the maximization, a number that quickly blows up for any reasonable sequence length (e.g., direct ML estimation for 1000 QPSK symbols incurs a complexity of 4^{1000}). We must therefore understand the structure of the preceding cost function in more detail, in order to develop efficient algorithms to maximize it. In particular, we would like to develop a form for the cost function that we can compute simply by adding terms as we increment the symbol time index n. We shall soon see that such a form is key to developing an efficient maximization algorithm.

It is easy to show that the first term in (5.4) has the desired additive form. From the proof of Theorem 5.2.1, we know that

$$\operatorname{Re}\langle y, s_\mathbf{b}\rangle = \sum_n \operatorname{Re}(b^*[n]z[n]). \qquad (5.5)$$

To simplify the term involving $||s_\mathbf{b}||^2$, it is convenient to introduce the sampled autocorrelation sequence of the pulse p as follows:

$$h[m] = \int p(t)p^*(t-mT)\,\mathrm{d}t = (p*p_{\mathrm{MF}})(mT). \qquad (5.6)$$

The sequence $\{h[m]\}$ is conjugate symmetric:

$$h[-m] = h^*[m]. \qquad (5.7)$$

This is proved as follows:

$$h[-m] = \int p(t)p^*(t+mT)\,\mathrm{d}t$$

$$= \int p(u-mT)p^*(u)\,\mathrm{d}u$$

$$= \left(\int p^*(u-mT)p(u)\,\mathrm{d}u\right)^* = h^*[m],$$

where we have used the change of variables $u = t + mT$.

Running example For our running example, it is easy to see from Figure 5.1 that $p(t)$ only has nontrivial overlap with $p(t-nT)$ for $n = 0, \pm 1$. In particular, we can compute that $h[0] = 3/2$, $h[1] = h[-1] = -1/2$, and $h[n] = 0$ for $|n| > 1$.

We can now write

$$||s_\mathbf{b}||^2 = \left\langle \sum_n b[n]p(t-nT), \sum_m b[m]p(t-mT)\right\rangle$$

$$= \sum_n \sum_m b[n]b^*[m]\int p(t-nT)p^*(t-mT)\,\mathrm{d}t \qquad (5.8)$$

$$= \sum_n \sum_m b[n]b^*[m]h[m-n].$$

This does not have the desired additive form, since, for each value of n, we must consider all possible values of m in the inner summation. To remedy this, rewrite the preceding as

$$||s_\mathbf{b}||^2 = \sum_n |b[n]|^2 h[0] + \sum_n \sum_{m<n} b[n]b^*[m]h[m-n]$$
$$+ \sum_n \sum_{m>n} b[n]b^*[m]h[m-n].$$

Interchanging the roles of m and n in the last summation, we obtain

$$||s_\mathbf{b}||^2 = h[0]\sum_n |b[n]|^2 + \sum_n \sum_{m<n} [b[n]b^*[m]h[m-n] + b^*[n]b[m]h[n-m]].$$

Using (5.7), we can rewrite the above as follows:

$$||s_\mathbf{b}||^2 = h[0]\sum_n |b[n]|^2 + \sum_n \sum_{m<n} 2\text{Re}(b^*[n]b[m]h[n-m]). \qquad (5.9)$$

Substituting (5.5) and (5.9) into (5.4), the cost function to be maximized becomes

$$\Lambda(\mathbf{b}) = \sum_n \left\{ \text{Re}(b^*[n]z[n]) - \frac{h[0]}{2}|b[n]|^2 \right.$$
$$\left. - \text{Re}\left(b^*[n]\sum_{m<n} b[m]h[n-m]\right) \right\}. \qquad (5.10)$$

Notice that the preceding cost function is additive in n, and that the term to be added at the nth step is a function of the "current" symbol $b[n]$ and the "past" symbols $\{b[m], m < n\}$.

In practice, the memory needed to compute the term that needs to be added at step n is truncated using the following *finite memory* condition:

$$h[n] = 0, \quad |n| > L. \qquad (5.11)$$

(For our running example, we have shown that $L = 1$.)

Under the condition (5.11), we can rewrite (5.10) as

$$\Lambda(\mathbf{b}) = \sum_n \{\text{Re}(b^*[n]z[n]) - \frac{h[0]}{2}|b[n]|^2 - \text{Re}[b^*[n]\sum_{m=n-L}^{n-1} b[m]h[n-m]]\}. \qquad (5.12)$$

Thus, to compute the term at time n for a candidate sequence $\mathbf{b}$, we need to keep track of the current symbol $b[n]$ and a *state* consisting of the past L symbols: $s[n] = (b[n-L], ..., b[n-1])$. This term is written as

$$\lambda_n(b[n], s[n]) = \lambda_n(s[n] \to s[n+1]) = \text{Re}(b^*[n]z[n]) - \frac{h[0]}{2}|b[n]|^2$$
$$- \text{Re}\left(b^*[n]\sum_{m=n-L}^{n-1} b[m]h[n-m]\right), \qquad (5.13)$$

where the two alternative notations reflect two useful interpretations for the metric: it is a function of the current symbol $b[n]$ and the current state $s[n]$,

5.4 Maximum likelihood sequence estimation

or it is a function of the transition between the current state $s[n]$ and the next state $s[n+1] = (b[n+1-L], \ldots, b[n])$. The cost function can therefore be written in the following additive form:

$$\Lambda(\mathbf{b}) = \sum_n \lambda_n(b[n], s[n]) = \sum_n \lambda_n(s[n] \to s[n+1]). \quad (5.14)$$

If $b[n]$ are drawn from an M-ary alphabet, the number of possible states at any time n is M^L. We can now define a trellis which consists of the set of states as we step through n: the set of states at n and $n+1$ are connected by edges. The label for an edge, or branch, between $s[n] = a$ and $s[n+1] = b$ is simply the "branch metric" $\lambda_n(a \to b)$. Note that, even when the states at either end of a branch are specified, the metric value depends on n through the matched filter output $z[n]$. A particular candidate sequence $\mathbf{b} = \{b[n], n = 1, 2, \ldots\}$ corresponds to a unique path through the trellis. The running sum up to time k of the metrics is defined as

$$\Lambda_k(\mathbf{b}) = \sum_{n=1}^{k} \lambda_n(b[n], s[n]) = \sum_{n=1}^{k} \lambda_n(s[n] \to s[n+1]). \quad (5.15)$$

We can update the running sum for any given sequence as we proceed through the trellis, since the metric at time n depends only on the states $s[n]$ and $s[n+1]$.

Branch metric for running example For our running example in Figure 5.1, suppose that we employ BPSK modulation, with $b[n] \in \{-1, +1\}$. The number of states is given by $M^L = 2^1 = 2$. The state at time n is $s[n] = b[n-1]$. The branch metric in going from state $s[n] = b[n-1]$ to state $s[n+1] = b[n]$ is given by specializing (5.13), to obtain

$$\lambda_n(b[n], s[n]) = \lambda_n(s[n] \to s[n+1])$$
$$= \text{Re}(b^*[n]z[n]) - \frac{h[0]}{2}|b[n]|^2 - \text{Re}[b^*[n]b[n-1]h[1]].$$

Since $\{b[k]\}$ are real-valued, we see that only $y[n] = \text{Re}(z[n])$ (i.e. the I component of the samples) affects the preceding metric. Furthermore, since $|b[n]|^2 \equiv 1$, the second term in the preceding equation does not depend on $b[n]$ (since $|b[n]|^2 \equiv 1$), and can be dropped from the branch metric. (This simplification applies more generally to PSK alphabets, but not to constellations with amplitude variations, such as 16-QAM.) We therefore obtain the modified metric

$$m_n(b[n], s[n]) = m_n(s[n] \to s[n+1])$$
$$= b[n]y[n] + \frac{1}{2}b[n]b[n-1]$$
$$= b[n]\left(y[n] + \frac{1}{2}b[n-1]\right). \quad (5.16)$$

Suppose now that we know that $b[0] = +1$, and that the first few samples at the output of the matched filter are given by $y[0] = -1$, $y[1] = 2$, $y[2] = -2$, and $y[3] = 1.5$. We can now use (5.16) to compute the branch metrics for the trellis. Figure 5.4 shows the corresponding trellis, with the branches labeled by the corresponding metrics. Note that, since we know that $s[1] = b[0] = +1$, we do not need the value of $y[0]$. The first branch metric we need is

$$m_1(s[1] \to s[2]) = b[1]y[1] + \frac{1}{2}b[0]b[1].$$

We compute this for $b[0] = +1$ and for $b[1] = \pm 1$. After this, we compute the branch metrics

$$m_n(s[n] \to s[n+1]) = b[n]y[n] - \frac{1}{2}b[n]b[n-1]$$
$$= b[n](y[n] - \frac{1}{2}b[n-1])$$

for $b[n] = \pm 1$ and $b[n+1] = \pm 1$ for $n = 1, 2, 3$.

Consider now a bit sequence $b[0] = +1$, $b[1] = +1$, $b[2] = +1$. From Figure 5.5, this has an accumulated metric of $1.5 - 2.5 = -1$. Compare it with the sequence $b[0] = +1$, $b[1] = -1$, $b[2] = +1$. This has an accumulated metric of $-1.5 - 1.5 = -3$. Both sequences start from the same state $s[1] = b[0] = +1$ and end in the same state $s[3] = b[2] = +1$. Thus, for each possible value of $b[3]$, we add the same branch metric $m_3(s[3] \to s[4])$ to the accumulated metric. Since the first sequence had a better accumulated metric coming into state $s[3]$, it continues to have a better accumulated metric for each possible value of state $s[4]$. This means that the second sequence cannot be part of the ML solution, since we can construct another sequence that has a better accumulated metric. Hence we can discard the second sequence from further consideration.

The preceding logic can be applied at any state. Two sequences meeting at a state $s[n]$ have a common starting state $s[1] = b[0] = +1$ and a common ending state $s[n]$ over the time $1, \ldots, n$. By the same reasoning as above, the sequence that has a worse accumulated metric at state $s[n]$ can be discarded from further consideration, since it cannot be part of the ML solution. Thus, at any given state, we only need to keep track of the sequence that has the

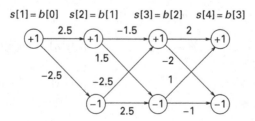

Figure 5.5 First few trellis branches for the running example.

5.4 Maximum likelihood sequence estimation

Figure 5.6 Example application of the Viterbi algorithm. Survivors are shown in bold.

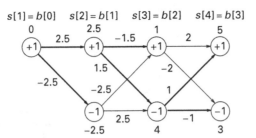

best accumulated metric up to that state. This sequence is called the *survivor* at state $s[n]$. Figure 5.6 shows this pruning procedure in action for the trellis in Figure 5.5, with the survivors at each state shown in bold. In the figure, the number labeling each state is the accumulated metric for the survivor at that state. We can now make the following observations:

- At each time, there are two survivors, since there are two states. We cannot compare across survivors ending at different states; a survivor with a poorer metric now could make up for it in the future. We do not know whether or not that happens until the survivors merge, at which point we can directly compare the accumulated metrics and choose the best.
- Based on the given information, we know that the ML solution will be an extension of the two survivors at any given state. Thus, if the survivors have merged in the past, then we know that the ML solution must contain this merged segment. Specifically, the two survivors at $s[4]$ have actually merged at $s[3]$. Thus, we know that the ML sequence must contain this merged segment, which implies that $\hat{b}[1] = +1$, $\hat{b}[2] = -1$ are the ML decisions for $b[1]$ and $b[2]$, respectively.
- Just as we have forced the starting state to be $s[1] = b[0] = +1$ to ensure a common starting point for all paths through the trellis, we can also force the ending state to be a predetermined state by specifying the final bit that is sent. For example, if we send a hundred bits $b[0], \ldots, b[99]$, we can specify $b[0] = b[99] = +1$ to ensure that all sequences have a common start and end state. The ML solution is then the survivor at state $s[100] = b[99] = +1$.

In the above discussion, we have invoked the *principle of optimality* to derive the Viterbi algorithm for efficient implementation of MLSE, in the context of our running example. Let us now state this principle formally, and in greater generality.

Principle of optimality Consider two sequences specified up to time k, $\mathbf{b} = \{b[1], \ldots, b[k]\}$ and $\mathbf{c} = \{c[1], \ldots, c[k]\}$. Suppose that the state at time k is the same for both sequences; that is, $s[k] = (b[k-L], \ldots, b[k-1]) = (c[k-L], \ldots, c[k-1])$. If $\mathbf{c}$ has a smaller running sum than $\mathbf{b}$ up to time k, then it cannot be the ML sequence, and can be eliminated from further

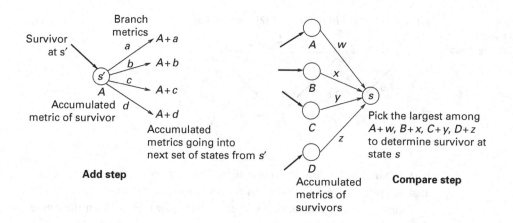

Figure 5.7 The add and compare steps in the Viterbi algorithm.

consideration. This follows from the following reasoning. Sequence **c** is worse than **b** up to time k. As we proceed further along the trellis, for any possible state $s[k+1]$, the term $\lambda_k(s[k] \to s[k+1])$ is the same for both sequences, so that the running sum for any extension of the sequence **c** will remain worse than that of the corresponding extension for sequence **b**.

Since any two sequences meeting at a state can be directly compared, we can define the surviving sequence, or survivor, at state $s[n] = a$ as the sequence entering the state with the largest running sum. By the principle of optimality, no other sequences entering state $s[n]$ need be considered further for the purpose of MLSE. Since there are $S = M^L$ possible states at any given time, we need to maintain a list of S survivors at each time. Given S survivors at time n, we extend each survivor in M possible ways, corresponding to the M possible values for $b[n]$, and update the corresponding running sums by adding $\lambda(s[n] \to s[n+1])$. At time $n+1$, for each possible state $s[n+1] = b$, we pick the sequence entering the state with the largest running sum, thus obtaining a new set of S survivors. If the first and last L symbols of the transmitted sequence are known, the start and end states are fixed, and the ML sequence is the only survivor at the end of this process. This is the Viterbi algorithm, which we state more formally below.

Viterbi algorithm Assume that the starting state of the encoder $s[1]$ is known. Now, all sequences through the trellis meeting at state $s[n]$ can be directly compared, using the principle of optimality between times 0 and n, and all sequences except the one with the best running sum can be discarded. If the trellis has $S = M^L$ states at any given time (the algorithm also applies to time-varying trellises where the number of states can depend on time), we have exactly S surviving sequences, or *survivors*, at any given time. We need to keep track of only these S sequences (i.e., the sequence of states through the trellis, or equivalently, the input sequence, that they correspond to) up to the current time. We apply this principle successively at times

5.4 Maximum likelihood sequence estimation

$n = 1, 2, 3, \ldots$. Consider the S survivors at time k. Let $F(s')$ denote the set of possible values of the next state $s[k+1]$, given that the current state is $s[k] = s'$. For an M-ary alphabet, there are M possible values of $s[k+1] = (b_k, b[k-1], \ldots, b[k-L+1])$ given that $s[k] = (b[k-1], \ldots, b[k-L+1], b[k-L])$ is fixed. Denote the running sum of metrics up to time k for the survivor at $s[k] = s'$ by $\Lambda^*(1:k, s')$. We now extend the survivors by one more time step as follows:

Add step: for each state s', extend the survivor at s' in all admissible ways, and add the corresponding branch metric to the current running sum to get

$$\Lambda_0(1:k+1, s' \to s) = \Lambda^*(1:k, s') + \lambda_{k+1}(s' \to s), \quad s \in F(s').$$

Compare step: after the "add" step, each possible state $s[k+1] = s$ has a number of candidate sequences coming into it, corresponding to different possible values of the prior state. We compare the metrics for these candidates and choose the best as the survivor at $s[k+1] = s$. Denote by $P(s)$ the set of possible values of $s[k] = s'$, given that $s[k+1] = s$. For an M-ary alphabet, $P(s)$ has M elements. We can now update the metric of the survivor at $s[k+1] = s$ as follows:

$$\Lambda^*(1:k+1, s) = \max_{s' \in P(s)} \Lambda_0(1:k+1, s' \to s)$$

and store the maximizing s' for each $s[k+1] = s$ (when we wish to minimize the metric, the maximization above is replaced by minimization).

At the end of the add and compare steps, we have extended the set of S survivors by one more time step. If the information sequence is chosen such that the terminating state is fixed, then we simply pick the survivor with the best metric at the terminal state as the ML sequence. The complexity of this algorithm is $O(S)$ per time step; that is, it is exponential in the channel memory, but linear in the (typically much larger) number of transmitted symbols. Contrast this with brute force ML estimation, which is exponential in the number of transmitted symbols.

The Viterbi algorithm is often simplified further in practical implementations. For true MLSE, we must wait until the terminal state to make bit decisions, which can be cumbersome in terms of both decoding delay and memory (we need to keep track of S surviving information sequences) for long information sequences. However, we can take advantage of the fact that the survivors at time k typically have merged at some point in the past, and make hard decisions on the bits corresponding to this common section with the confidence that this section must be part of the ML solution. For example, in Figure 5.4, the two survivors at time 4 have merged prior to $s[2]$, which means we can make the decision that $\hat{b}_{\text{ML}}[1] = +1$. In practice, we may impose a hard constraint on the decoding delay d and say that, if the Viterbi algorithm is at time step k, then we must make decisions on all information bits prior to time step $k - d$. If the survivors at time k have not merged by time step $k - d$, therefore, we must employ heuristic rules for making bit decisions: for example, we may make decisions prior to $k - d$ corresponding to the survivor with the best metric at time k. Alternatively, some form of majority logic, or weighted majority logic, may be used to combine the information contained in all survivors at time k.

5.4.1 Alternative MLSE formulation

The preceding MLSE formulation derived the cost function directly from the continuous-time model (5.1) of a signal depending on **b**, plus WGN. An alternative approach is to derive the cost function from a discrete-time WGN model. We briefly outline this approach here. Start with the matched filter outputs $\{z[n]\}$, which consist of a discrete-time signal depending on the transmitted sequence **b**, plus discrete-time colored noise. This noise is obtained by passing continuous-time WGN through the matched filter $p^*(-t)$, and sampling at the symbol rate. Knowing the matched filter impulse response, we know the noise correlation, and we can pass it through a discrete-time whitening filter to obtain discrete-time WGN. The whitening filter would also change the signal component, but the overall discrete-time system can be represented as the symbol sequence passed through a discrete-time filter (the cascade of the transmit filter, channel filter, receive filter, sampler, and whitening filter), plus discrete-time WGN. The received sequence $\mathbf{v} = \{v[k]\}$ is therefore given by

$$v[k] = \sum_{n=0}^{L} f[n]b[k-n] + \eta_k, \qquad (5.17)$$

where $\{\eta_k\}$ is discrete-time WGN with variance σ^2 per dimension, $\mathbf{f} = \{f[n]\}$ is the overall discrete-time channel impulse response, and $\mathbf{b} = \{b[n]\}$ is the transmitted symbol sequence. We have assumed that $f[n] = 0$ for $n < 0$ (causality) and $n > L$ (finite memory): the causality follows from an appropriate choice of the whitening filter. The details of this *whitened matched filter* approach are developed for the running example in Problem 5.7.

The MLSE for the discrete-time WGN model (5.17) is obtained simply by minimizing the distance between the received sequence **v** and the noiseless signal $\mathbf{s_b} = \{s(k, \mathbf{b})\}$, where the kth component of the signal is given by $s(k, \mathbf{b}) = \sum_{n=0}^{L} f[n]b[k-n]$. Thus, the cost function to be minimized is

$$g(\mathbf{b}) = \sum_k |v[k] - s(k, \mathbf{b})|^2 = \sum_k |v[k] - \sum_{n=0}^{L} f[n]b[k-n]|^2.$$

As before, the contribution at time k is a function of the current symbol $b[k]$ and the state $s[k] = (b[k-L], \ldots, b[k-1])$ consisting of the L past symbols, and can be written as

$$\Gamma_k(b[k], s[k]) = \Gamma(s[k] \to s[k+1]) = |v[k] - \sum_{n=0}^{L} f[n]b[k-n]|^2. \qquad (5.18)$$

Thus, the MLSE is defined as
$$\hat{\mathbf{b}}_{\mathrm{ML}} = \arg \min_{\mathbf{b}} \sum_k \Gamma(s[k] \to s[k+1]).$$

The principles of optimality and the Viterbi algorithm apply as before, except that maximization is replaced by minimization.

5.5 Geometric model for suboptimal equalizer design

The optimum MLSE receiver has complexity $O(M^L)$ per demodulated symbol, where M is the alphabet size and L is the channel memory. This may be excessive for large constellations or large channel memory. We now consider suboptimal equalization strategies whose complexity scales linearly with the channel memory. The schemes we describe are amenable to adaptive implementation, and do not require an optimal front end. While they can be developed in continuous time, we describe these equalization strategies in discrete time, which is almost invariably the setting in which they are implemented. Specifically, assume that the received signal is passed through an arbitrary receive filter $g_{\mathrm{RX}}(t)$, and being sampled at a rate $1/T_{\mathrm{s}} = m/T$, where m is a positive integer: $m = 1$ corresponds to *symbol spaced sampling*, while $m > 1$ corresponds to *fractionally spaced sampling*. The received signal is, as before, given by

$$y(t) = \sum_n b[n] p(t - nT) + n(t).$$

The output of the sampler is a discrete-time sequence $\{r[k]\}$, where

$$r[k] = (y * g_{\mathrm{RX}})(kT_{\mathrm{s}} + \delta),$$

where δ is a sampling offset. To understand the structure of $\{r[k]\}$, consider the signal and noise contributions to it separately. The signal contribution is best characterized by considering the response, at the output of the sampler, to a single symbol, say $b[0]$. This is given by the discrete-time impulse response

$$f[k] = (p * g_{\mathrm{RX}})(kT_{\mathrm{s}} + \delta), \quad k = \ldots, -1, 0, 1, 2, \ldots$$

The next symbol sees the same response, shifted by the symbol interval T, which corresponds to m samples, and so on. The noise sequence at the output of the sampler is given by

$$w[k] = (n * g_{\mathrm{RX}})(kT_{\mathrm{s}} + \delta).$$

If n is complex WGN, the noise at the receive filter output, $w(t) = (n * g_{\mathrm{RX}})(t)$, is zero mean, proper complex, Gaussian random process with autocorrelation/covariance function

$$2\sigma^2 \int g_{\mathrm{RX}}(t) g_{\mathrm{RX}}^*(t - \tau) \mathrm{d}t = 2\sigma^2 (g_{\mathrm{RX}} * g_{\mathrm{R,MF}})(\tau),$$

where $g_{R,MF}(t) = g_{RX}^*(-t)$. (derivation left to the reader). Thus, the sampled noise sequence $\{w[k] = w(kT_s + \delta)\}$ is zero mean, proper complex Gaussian, with autocovariance function

$$C_w(k) = \text{cov}(w_{n+k}, w_n) = 2\sigma^2 \int g_{RX}(t) g_{RX}^*(t - kT_s) dt. \quad (5.19)$$

In the following, we discuss equalization schemes which operate on a block of received samples for each symbol decision. The formula (5.19) can be used to determine the covariance matrix for the noise contribution to any such block of samples. Note that noise correlation depends on the autocorrelation function of $g_{RX}(t)$ evaluated at integer multiples of the sample spacing.

Running example Consider our running example of Figure 5.2, and consider a receive filter $g_{RX}(t) = I_{[0,1]}$. Note that this receive filter in this example is not matched to either the transmit filter or to the cascade of the transmit filter and the channel. The symbol interval $T = 2$, and we choose a sampling interval $T_s = 1$; that is, we sample twice as fast as the symbol rate. Note that the impulse response of the receive filter is of shorter duration than that of the transmit filter, which means that it has a higher bandwidth than the transmit filter. While we have chosen timelimited waveforms in the running example for convenience, this is consistent with the discussion in Section 5.2, in which a wideband filter followed by sampling, typically at a rate faster than the symbol rate, is employed to discretize the observation with no (or minimal) loss of information. The received samples are given by

$$r[k] = (y * g_{RX})(k) = \int_{k-1}^{k} y(t) dt.$$

The sampled response to the symbol $b[0]$ can be shown to be

$$(\ldots, 0, 1, \frac{1}{2}, -\frac{1}{2}, 0, \ldots). \quad (5.20)$$

The sampled response to successive symbols is shifted by two samples, since there are two samples per symbol. This defines the signal contribution to the output. To define the noise contribution, note that the autocovariance function of the complex Gaussian noise samples is given by

$$C_w[k] = 2\sigma^2 \delta_{k0}.$$

That is, the noise samples are complex WGN. Suppose, now, that we wish to make a decision on the symbol $b[n]$ based on a block of five samples $\mathbf{r}[n]$, chosen such that $b[n]$ makes a strong contribution to the block. The model for such a block can be written as

5.5 Geometric model for suboptimal equalizer design

$$\mathbf{r}[n] = b[n-1] \begin{pmatrix} \frac{1}{2} \\ -\frac{1}{2} \\ 0 \\ 0 \\ 0 \end{pmatrix} + b[n] \begin{pmatrix} 0 \\ 1 \\ \frac{1}{2} \\ -\frac{1}{2} \\ 0 \end{pmatrix} + b[n+1] \begin{pmatrix} 0 \\ 0 \\ 0 \\ 1 \\ \frac{1}{2} \end{pmatrix} + \mathbf{w}_n = \mathbf{U}\mathbf{b}[n] + \mathbf{w}[n],$$

(5.21)

where $\mathbf{w}[n]$ is discrete-time WGN,

$$\mathbf{b}[n] = \begin{pmatrix} b[n-1] \\ b[n] \\ b[n+1] \end{pmatrix} \tag{5.22}$$

is the block of symbols making a nonzero contribution to the block of samples, and

$$\mathbf{U} = \begin{pmatrix} \frac{1}{2} & 0 & 0 \\ -\frac{1}{2} & 1 & 0 \\ 0 & \frac{1}{2} & 0 \\ 0 & -\frac{1}{2} & 1 \\ 0 & 0 & \frac{1}{2} \end{pmatrix} \tag{5.23}$$

is a matrix whose columns equal the responses corresponding to the symbols contributing to $\mathbf{r}[n]$. The middle column corresponds to the *desired* symbol $b[n]$, while the other columns correspond to the interfering symbols $b[n-1]$ and $b[n+1]$. The columns are acyclic shifts of the basic discrete impulse response to a single symbol, with the entries shifting down by one symbol interval (two samples in this case) as the symbol index is incremented. We use $\mathbf{r}[n]$ to decide on $b[n]$ (using methods to be discussed shortly). For a decision on the next symbol, $b[n+1]$, we simply shift the window of samples to the right by a symbol interval (i.e., by two samples), to obtain a vector $\mathbf{r}[n+1]$. Now $b[n+1]$ becomes the desired symbol, and $b[n]$ and b_{n+2} the interfering symbols, but the basic model remains the same. Note that the blocks of samples used for successive symbol decisions overlap, in general.

Geometric model We are now ready to discuss a general model for finite-complexity, suboptimal equalizers. A block of L received samples $\mathbf{r}[n]$ is used to decide on $b[n]$, with successive blocks shifted with respect to each other by the symbol interval (m samples). The model for the received vector is

$$\mathbf{r}[n] = \mathbf{U}\,\mathbf{b}[n] + \mathbf{w}[n], \tag{5.24}$$

where $\mathbf{b}[n] = (b[n-k_1], \ldots, b[n-1], b[n], b[n+1], \ldots, b[n+k_2])^T$ is the $K \times 1$ vector of symbols making nonzero contributions to $\mathbf{r}[n]$, with $K = k_1 + k_2 + 1$. The $L \times K$ matrix $\mathbf{U}$ has as its columns the responses, or "signal vectors," corresponding to the individual symbols. All of these column vectors are acyclic shifts of the basic discrete-time impulse response to a single symbol, given by the samples of $g_{TX} * g_C * g_{RX}$. We denote the signal vector corresponding to symbol $b[n+i]$ as $\mathbf{u}_i$, $-k_1 \leq i \leq k_2$. The noise vector $\mathbf{w}[n]$ is zero mean, proper complex Gaussian with covariance matrix $\mathbf{C}_w$.

5.6 Linear equalization

Linear equalization corresponds to correlating $\mathbf{r}[n]$ with a vector $\mathbf{c}$ to produce a decision statistic $Z[n] = \langle \mathbf{r}[n], \mathbf{c} \rangle = \mathbf{c}^H \mathbf{r}[n]$. This decision statistic is then employed to generate either hard or soft decisions for $b[n]$. Rewriting $\mathbf{r}[n]$ as

$$\mathbf{r}[n] = b[n]\mathbf{u}_0 + \sum_{i \neq 0} b[n+i]\mathbf{u}_i + \mathbf{w}[n], \quad (5.25)$$

we obtain the correlator output as

$$Z[n] = \mathbf{c}^H \mathbf{r}[n] = b[n](\mathbf{c}^H \mathbf{u}_0) + \sum_{i \neq 0} b[n+i](\mathbf{c}^H \mathbf{u}_i) + \mathbf{c}^H \mathbf{w}[n]. \quad (5.26)$$

To make a reliable decision on $b[n]$ based on $Z[n]$, we must choose $\mathbf{c}$ such that the term $\mathbf{c}^H \mathbf{u}_0$ is significantly larger than the "residual ISI" terms $\mathbf{c}^H \mathbf{u}_i$, $i \neq 0$. We must also keep in mind the effects of the noise term $\mathbf{c}^H \mathbf{w}[n]$, which is zero mean proper Gaussian with covariance $\mathbf{c}^H \mathbf{C}_w \mathbf{c}$.

The correlator $\mathbf{c}$ can also be implemented as a discrete-time filter, whose outputs are sampled at the symbol rate to obtain the desired decision statistics $\{Z[n]\}$. Such an architecture is depicted in Figure 5.8.

Zero-forcing (ZF) equalizer The ZF equalizer addresses the preceding considerations by insisting that the ISI at the correlator output be set to zero. While doing this, we must constrain the desired term $\mathbf{c}^H \mathbf{u}_0$ so that it is not driven to zero. Thus, the ZF solution, if it exists, satisfies

$$\mathbf{c}^H \mathbf{u}_0 = 1, \quad (5.27)$$

and

$$\mathbf{c}^H \mathbf{u}_i = 0, \quad \text{for all } i \neq 0. \quad (5.28)$$

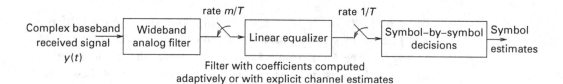

Figure 5.8 A typical architecture for implementing a linear equalizer.

5.6 Linear equalization

To obtain an expression for the ZF correlator, it is convenient to write (5.27) and (5.28) in matrix form:

$$\mathbf{c}^H \mathbf{U} = (0, \ldots, 0, 1, 0, \ldots, 0) = \mathbf{e}^T,$$

where the nonzero entry on the right-hand side corresponds to the column with the desired signal vector. It is more convenient to work with the conjugate transpose of the preceding equation:

$$\mathbf{U}^H \mathbf{c} = \mathbf{e}. \tag{5.29}$$

The solution to the preceding equation may not be unique (e.g., if the dimension L is larger than the number of signal vectors, as in the example considered earlier). Uniqueness is enforced by seeking a minimum norm solution to (5.29). To minimize $||\mathbf{c}||^2$ subject to (5.29), we realize that any component orthogonal to the subspace spanned by the signal vectors $\{\mathbf{u}_i\}$ must be set to zero, so that we may insist that $\mathbf{c}$ is a linear combination of the $\mathbf{u}_i$, given by

$$\mathbf{c} = \mathbf{U}\mathbf{a},$$

where the $K \times 1$ vector $\mathbf{a}$ contains the coefficients of the linear combination. Substituting in (5.29), we obtain

$$\mathbf{U}^H \mathbf{U} \mathbf{a} = \mathbf{e}.$$

We can now solve to obtain $\mathbf{a} = (\mathbf{U}^H \mathbf{U})^{-1} \mathbf{e}$, which yields

$$\mathbf{c}_{\text{ZF}} = \mathbf{U}(\mathbf{U}^H \mathbf{U})^{-1} \mathbf{e}. \tag{5.30}$$

Geometric view of the zero-forcing equalizer A linear correlator $\mathbf{c}$ must lie in the *signal space* spanned by the vectors $\{\mathbf{u}_i\}$, since any component of $\mathbf{c}$ orthogonal to this space only contributes noise to the correlator output. This signal space can be viewed as in Figure 5.9, which shows the desired vector $\mathbf{u}_0$, and the *interference subspace* $\mathbf{S}_I$ spanned by the interference vectors $\{\mathbf{u}_i, i \neq 0\}$. If there were no interference, then the best strategy is to point $\mathbf{c}$ along $\mathbf{u}_0$ to gather as much energy as possible from the desired vector: this is the matched filter receiver. However, if we wish to force the ISI to zero,

Figure 5.9 The geometry of zero-forcing equalization.

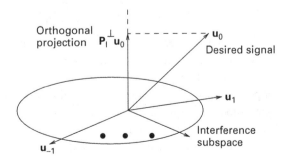

we must choose $\mathbf{c}$ orthogonal to the interference subspace $\mathbf{S}_I$. The correlator vector $\mathbf{c}$ that maximizes the contribution of the desired signal, while being orthogonal to the interference subspace, is simply (any scaled version of) the projection $\mathbf{P}_I^\perp \mathbf{u}_0$ of $\mathbf{u}_0$ orthogonal to $\mathbf{S}_I$. The ZF solution exists if and only if this projection is nonzero, which is the case if and only if the desired vector $\mathbf{u}_0$ is linearly independent of the interfering vectors. A rule of thumb for the existence of the ZF solution, therefore, is that the number of available dimensions L is greater than the number of interference vectors $K-1$.

How does the preceding geometric view relate to the algebraic specification (5.27), (5.28) of the ZF solution? Consider a correlator $\mathbf{c}$ which is a scalar multiple of the projection of $\mathbf{u}_0$ orthogonal to the interference subspace, $\mathbf{c} = \alpha \mathbf{P}_I^\perp \mathbf{u}_0$. By definition, this satisfies the zero ISI condition (5.28). The contribution of the desired signal at the output of the correlator is given by

$$\langle \mathbf{c}, \mathbf{u}_0 \rangle = \alpha \langle \mathbf{P}_I^\perp \mathbf{u}_0, \mathbf{u}_0 \rangle = a \|\mathbf{P}_I^\perp \mathbf{u}_0\|^2.$$

To obtain the normalization $\langle \mathbf{c}, \mathbf{u}_0 \rangle = 1$, we set

$$\alpha = \frac{1}{\|\mathbf{P}_I^\perp \mathbf{u}_0\|^2}.$$

Thus, the smaller the orthogonal projection $\mathbf{P}_I^\perp \mathbf{u}_0$, the larger the scale factor a required to obtain the normalization (5.27) for the contribution of the desired signal to the correlator output. As the scale factor increases, so does the noise at the correlator output: the variance v^2 (per dimension) of the output noise is given by

$$v_{ZF}^2 = \sigma^2 \|\mathbf{c}\|^2 = \sigma^2 \alpha^2 \|\mathbf{P}_I^\perp \mathbf{u}_0\|^2 = \frac{\sigma^2}{\|\mathbf{P}_I^\perp \mathbf{u}_0\|^2}. \quad (5.31)$$

The corresponding noise variance for matched filter reception (which is optimal if there is no ISI) is

$$v_{MF}^2 = \frac{\sigma^2}{\|\mathbf{u}_0\|^2}. \quad (5.32)$$

Thus, when we fix the desired signal contribution to the correlator output as in (5.27), the output noise variance for the ZF solution is larger than that of the matched filter receiver. The factor by which the noise variance increases is called the *noise enhancement* factor, and is given by

$$\frac{v_{ZF}^2}{v_{MF}^2} = \frac{\|\mathbf{u}_0\|^2}{\|\mathbf{P}_I^\perp \mathbf{u}_0\|^2}. \quad (5.33)$$

The noise enhancement factor is the price we pay for knocking out the ISI, and is often expressed in dB. Since there is no ISI at the output of the ZF equalizer, we obtain a scalar observation corrupted by Gaussian noise, so that performance is completely determined by SNR. Fixing the desired signal contribution at the output, the SNR scales inversely with the noise variance. Thus, the noise enhancement factor (5.33) is the factor by which the SNR must be increased for a system employing the ZF equalizer to combat ISI, in order to

5.6 Linear equalization

maintain the same performance as matched filter reception in a system with no ISI.

Running example Going back to our running example, we see that $L = 5$ and $K = 3$, so that the ZF solution is likely to exist. Applying (5.30) to the example, we obtain

$$\mathbf{c}_{\text{ZF}} = \frac{1}{8}(5, 5, 5, -1, 2)^T. \tag{5.34}$$

The output of the ZF equalizer for the example is therefore given by

$$Z[n] = \mathbf{c}_{\text{ZF}}^H \mathbf{r}[n] = b[n] + N[n],$$

where $N[n] \sim CN(0, 2v^2)$, where $2v^2 = \mathbf{c}_{\text{ZF}}^H \mathbf{C}_\mathbf{w} \mathbf{c}_{\text{ZF}} = 2\sigma^2 ||\mathbf{c}_{\text{ZF}}||^2 = 2.5\sigma^2$. For BPSK transmission $b[n] \in \{-1, 1\}$, our decision rule is

$$\hat{b}[n] = \text{sign}(\text{Re}(Z[n])).$$

Since $\text{Re}(N[n]) \sim N(0, v^2)$, the error probability is given by

$$Q\left(\frac{1}{v}\right).$$

Using scaling arguments as in Chapter 3, we know that this can be written as $Q(\sqrt{aE_b/N_0})$ for some constant a. We can now solve for a by noting that $v^2 = 5\sigma^2/4 = 5N_0/8$, and that the received energy per bit is $E_b = ||p||^2 = 3/2$. Setting

$$\frac{1}{v} = \sqrt{\frac{aE_b}{N_0}}$$

yields $a = 16/15$. Contrast this with $a = 2$ for ISI-free BPSK. The loss of $10 \log_{10} 2/16/15 = 2.73$ dB can be interpreted as "noise enhancement" due to the ZF solution.

In the preceding, we have enforced the constraint that the ZF equalizer must operate on a finite block of samples for each symbol. If this restriction is lifted (i.e., if each symbol decision can involve an arbitrary number of samples), then it is convenient to express the ZF solution in z-transform notation. For fractionally spaced sampling at rate m/T, think of the samples as m parallel symbol-spaced streams. The response to a single symbol for stream i is denoted as $\{h_i[n]\}$, $1 \leq i \leq m$, and has z-transform $H_i(z) = \sum_n h_i[n]z^{-n}$. In our example, we may set

$$H_1(z) = 1 - \frac{1}{2}z^{-1}, \quad H_2(z) = \frac{1}{2}.$$

A linear ZF equalizer can then be characterized as a set of parallel filters with z-transforms $\{G_i(z)\}$ such that, in the absence of noise, the sum of the parallel filter outputs reconstructs the original symbol stream up to a *decision delay d*, as follows:

$$\sum_{i=1}^{m} H_i(z) G_i(z) = z^{-d}. \quad (5.35)$$

The coefficients of the filters $\{G_i(z)\}$ are *time-reversed, subsampled* versions of the corresponding correlator operating on the fractionally spaced data. Thus, the ZF correlator (5.34) in our example corresponds to the following pair of parallel filters:

$$G_1(z) = \frac{1}{8}(-1 + 5z^{-1}), \quad G_2(z) = \frac{1}{8}(2 + 5z^{-1} + 5z^{-2}),$$

so that

$$H_1(z) G_1(z) + H_2(z) G_2(z) = z^{-1}.$$

For fractionally spaced equalization, it is known that finite-length $\{G_i(z)\}$ satisfying (5.35) exist, as long as the parallel channel z-transforms $\{H_i(z)\}$ do not have common zeros (although finite-length ZF solutions exist under milder conditions as well). On the other hand, for symbol-spaced samples, there is only one discrete-time channel, so that the ZF equalizer must take the form $(z^{-d})/(H_1(z))$. This has infinite length for a finite impulse response (FIR) channel $H_1(z)$, so that perfect ZF equalization using a finite-length equalizer is not possible for symbol-spaced sampling. This is one reason why fractionally spaced sampling is often preferred in practice, especially when the receive filter is suboptimal. Fractionally spaced sampling is also less sensitive to timing offsets. This is illustrated by Problem 5.8, which computes the ZF solution for the running example when the sampling times are shifted by a fraction of the symbol.

Even though perfect ZF equalization is not possible for symbol-spaced sampling using a finite window of samples, it can be realized approximately by choosing which of the ISI vectors to null out, and being reconciled to having residual ISI due to the other ISI vectors at the correlator output. In this case, we can compute the ZF solution as in (5.30), except that the matrix **U** contains as its columns the desired signal vector and the ISI vectors to be nulled out (the columns corresponding to the other ISI vectors are deleted).

Linear MMSE equalizer The design of the ZF equalizer ignores the effect of noise at the equalizer output. An alternative to this is the linear minimum mean squared error (MMSE) criterion, which trades off the effect of noise and ISI at the equalizer output. The mean squared error (MSE) at the output of a linear equalizer **c** is defined as

$$\text{MSE} = J(\mathbf{c}) = \mathbb{E}[|\mathbf{c}^H \mathbf{r}[n] - b[n]|^2], \quad (5.36)$$

where the expectation is taken over the symbol stream $\{b[n]\}$. The MMSE correlator is given by

$$\mathbf{c}_{\text{MMSE}} = \mathbf{R}^{-1} \mathbf{p}, \quad (5.37)$$

5.6 Linear equalization

where
$$\mathbf{R} = \mathbb{E}[\mathbf{r}[n](\mathbf{r}[n])^H], \quad \mathbf{p} = \mathbb{E}[b^*[n]\mathbf{r}[n]]. \tag{5.38}$$

The MMSE criterion is useful in many settings, not just equalization, and the preceding solution holds in great generality.

Direct Proof by differentiation For simplicity, consider real-valued $\mathbf{r}[n]$ and $\mathbf{c}$ first. The function $J(\mathbf{c})$ is quadratic in $\mathbf{c}$, so a global minimum exists, and can be found by setting the gradient with respect to $\mathbf{c}$ to zero, as follows:

$$\begin{aligned}\nabla_{\mathbf{c}} J(\mathbf{c}) &= \nabla_{\mathbf{c}} \mathbb{E}[(\mathbf{c}^T \mathbf{r}[n] - b[n])^2] \\ &= \mathbb{E}[\nabla_{\mathbf{c}} (\mathbf{c}^T \mathbf{r}[n] - b[n])^2] = \mathbb{E}[2(\mathbf{c}^T \mathbf{r}[n] - b[n])\mathbf{r}[n]] \\ &= 2(\mathbf{R}\mathbf{c} - \mathbf{p}).\end{aligned}$$

In addition to characterizing the optimal solution, the gradient can also be employed for descent algorithms for iterative computation of the optimal solution. For complex-valued $\mathbf{r}[n]$ and $\mathbf{c}$, there is a slight subtlety in computing the gradient. Letting $\mathbf{c} = \mathbf{c}_c + j\mathbf{c}_s$, where $\mathbf{c}_c$ and $\mathbf{c}_s$ are the real and imaginary parts of $\mathbf{c}$, respectively, note that the gradient to be used for descent is actually

$$\nabla_{\mathbf{c}_c} J + j \nabla_{\mathbf{c}_s} J.$$

While the preceding characterization treats the function J as a function of two independent real vector variables $\mathbf{c}_c$ and $\mathbf{c}_s$, a more compact characterization in the complex domain is obtained by interpreting it as a function of the independent complex vector variables $\mathbf{c}$ and $\mathbf{c}^*$. Since

$$\mathbf{c} = \mathbf{c}_c + j\mathbf{c}_s, \quad \mathbf{c}^* = \mathbf{c}_c - j\mathbf{c}_s,$$

we can show, using the chain rule, that

$$\begin{aligned}\nabla_{\mathbf{c}_c} J &= \nabla_{\mathbf{c}} J + \nabla_{\mathbf{c}^*} J, \\ \nabla_{\mathbf{c}_s} J &= j \nabla_{\mathbf{c}} J - j \nabla_{\mathbf{c}^*} J,\end{aligned}$$

so that

$$\nabla_{\mathbf{c}_c} J + j \nabla_{\mathbf{c}_s} J = 2 \nabla_{\mathbf{c}^*} J. \tag{5.39}$$

Thus, the right gradient to use for descent is $\nabla_{\mathbf{c}^*} J$. To compute this, rewrite the cost function as

$$J = \mathbb{E}[((\mathbf{c}^*)^T \mathbf{r}[n] - b[n])(\mathbf{r}[n]^H \mathbf{c} - b^*[n])],$$

so that

$$\nabla_{\mathbf{c}^*} J = \mathbb{E}[\mathbf{r}[n]((\mathbf{r}[n])^H \mathbf{c} - b^*[n])] = \mathbf{R}\mathbf{c} - \mathbf{p}. \tag{5.40}$$

Setting the gradient to zero proves the result. □

Alternative Proof using the orthogonality principle We do not prove the orthogonality principle here, but state a form of it convenient for our purpose. Suppose that we wish to find the best linear approximation for a complex random variable Y in terms of a sequence of complex random variables $\{X_i\}$. Thus, the approximation must be of the form $\sum_i a_i X_i$, where $\{a_i\}$ are complex scalars to be chosen to minimize the MSE:

$$\mathbb{E}\left[\left|Y - \sum_i a_i X_i\right|^2\right].$$

The preceding can be viewed as minimizing a distance in a space of random variables, in which the inner product is defined as

$$\langle U, V \rangle = \mathbb{E}[UV^*].$$

This satisfies all the usual properties of an inner product: $\langle aU, bV \rangle = ab^* \langle U, V \rangle$, and $\langle U, U \rangle = 0$ if and only if $U = 0$ (where equalities are to be interpreted as holding with probability one). The orthogonality principle holds for very general inner product spaces, and states that, for the optimal approximation, the approximation error is orthogonal to every element of the approximating space. Specifically, defining the error as

$$e = Y - \sum_i a_i X_i,$$

we must have

$$\langle X_i, e \rangle = 0 \quad \text{for all } i. \tag{5.41}$$

Applying it to our setting, we have $Y = b[n]$, X_i are the components of $\mathbf{r}[n]$, and

$$e = \mathbf{c}^H \mathbf{r}[n] - b[n].$$

In this setting, the orthogonality principle can be compactly stated as

$$0 = \mathbb{E}[\mathbf{r}[n]e^*] = \mathbb{E}[\mathbf{r}[n]((\mathbf{r}[n])^H \mathbf{c} - b^*[n])] = \mathbf{R}\mathbf{c} - \mathbf{p}.$$

This completes the proof. □

Let us now give an explicit formula for the MMSE correlator in terms of the model (5.24). Assuming that the symbols $\{b[n]\}$ are uncorrelated, with $\mathbb{E}[b[n]b^*[m]] = \sigma_b^2 \delta_{nm}$, (5.37) and (5.38) specialize to

$$\mathbf{c}_{\text{MMSE}} = \mathbf{R}^{-1}\mathbf{p}, \quad \text{where} \quad \mathbf{R} = \sigma_b^2 \mathbf{U}\mathbf{U}^H + \mathbf{C}_w = \sigma_b^2 \sum_j \mathbf{u}_j \mathbf{u}_j^H + \mathbf{C}_w, \quad \mathbf{p} = \sigma_b^2 \mathbf{u}_0. \tag{5.42}$$

While the ultimate performance measure for any equalizer is the error probability, a useful performance measure for the linear equalizer is the signal-to-interference ratio (SIR) at the equalizer output, given by

$$\text{SIR} = \frac{\sigma_b^2 |\langle \mathbf{c}, \mathbf{u}_0 \rangle|^2}{\sigma_b^2 \sum_{j \neq 0} |\langle \mathbf{c}, \mathbf{u}_j \rangle|^2 + \mathbf{c}^H \mathbf{C}_w \mathbf{c}}. \tag{5.43}$$

Two important properties of the MMSE equalizer are as follows:

- The MMSE equalizer maximizes the SIR (5.43) among all linear equalizers. Since scaling the correlator does not change the SIR, any scaled multiple of the MMSE equalizer also maximizes the SIR.
- In the limit of vanishing noise, the MMSE equalizer specializes to the ZF equalizer.

These, and other properties, are explored in Problem 5.9.

5.6.1 Adaptive implementations

Directly computing the expression (5.42) for the MMSE correlator requires knowledge of the matrix of signal vectors $\mathbf{U}$, which in turn requires an explicit channel estimate. This approach requires the use of the specific model (5.24) for the received vectors $\{\mathbf{r}[n]\}$, along with an explicit channel estimate for computing the matrix $\mathbf{U}$. An alternative, and more general, approach begins with the observation that the MSE cost function (5.36) and the solution (5.38) are based on expectations involving only the received vectors $\{\mathbf{r}[n]\}$ and the symbol sequence $\{b[n]\}$. At the receiver, we know the received vectors $\{\mathbf{r}[n]\}$, and, if we have a known training sequence, we know $\{b[n]\}$. Thus, we can compute estimates of (5.36) and (5.38) simply by replacing statistical expectation by empirical averages. This approach does not rely on a detailed model for the received vectors $\{\mathbf{r}[n]\}$, and is therefore quite general.

In the following, we derive the least squares (LS) and recursive least squares (RLS) implementations of the MMSE correlator by replacing the statistical expectations involved in the expression (5.38) by suitable empirical averages. An alternate approach is to employ a gradient descent on the MSE cost function: by replacing the gradient, which involves a statistical expectation, by its instantaneous empirical realization, we obtain the least mean squares (LMS) algorithm.

Training and decision-directed modes It is assumed that the symbol sequence $\{b[n]\}$ is known in the *training* phase of the adaptive algorithm. Once a correlator has been computed based on the training sequence, it can be used for making symbol decisions. These symbol decisions can then be used to further update the correlator, if necessary, in decision-directed mode, by replacing $\{b[n]\}$ by its estimates.

Least squares algorithm The LS implementation replaces the statistical expectations in (5.38) by empirical averages computed over a block of N received vectors, as follows:

$$\mathbf{c}_{\mathrm{LS}} = \hat{\mathbf{R}}^{-1}\hat{\mathbf{p}}, \\ \hat{\mathbf{R}} = \tfrac{1}{N}\sum_{n=1}^{N} \mathbf{r}[n](\mathbf{r}[n])^H, \quad \hat{\mathbf{p}} = \tfrac{1}{N}\sum_{n=1}^{N} b^*[n]\mathbf{r}[n], \tag{5.44}$$

where we would typically require an initial *training* sequence for $\{b[n]\}$ to compute $\hat{\mathbf{p}}$. Just as (5.38) is the solution that minimizes the MSE (5.36), the LS solution (5.44) is the solution that minimizes the empirical MSE

$$\text{Empirical MSE} = \hat{J}(\mathbf{c}) = \frac{1}{N} \sum_{n=1}^{N} |\mathbf{c}^H \mathbf{r}[n] - b[n]|^2, \qquad (5.45)$$

obtained by replacing the expectation in (5.36) by an empirical average. Note that the normalization factors of $1/N$ in (5.44) and (5.45) are included to reinforce the concept of an empirical average, but can be omitted without affecting the final result, since scaling a cost function does not change the optimizing solution.

Recursive least squares algorithm While the preceding empirical averages (or sums, if the normalizing factors of $1/N$ are omitted) are computed over a block of N received vectors, another approach is to sum over terms corresponding to all available received vectors $\{\mathbf{r}[n]\}$ (i.e., to use a potentially infinite number of received vectors) for computing the empirical MSE to be optimized, ensuring convergence of the cost function by putting in an exponential forget factor. This approach allows continual updating of the correlator, which is useful when we wish to adapt to a time-varying channel. The cost function evolves over time as follows:

$$\hat{J}_k(\mathbf{c}) = \sum_{n=0}^{k} \lambda^{k-n} |\mathbf{c}^H \mathbf{r}[n] - b[n]|^2, \qquad (5.46)$$

where $0 < \lambda < 1$ is an exponential forget factor, and $\mathbf{c}[k]$, the solution that minimizes the cost function $\hat{J}_k(\mathbf{c})$, computed based on all received vectors $\{\mathbf{r}[n], n \leq k\}$. In direct analogy with (5.38) and (5.44), we can write down the following formula for $\mathbf{c}[k]$:

$$\mathbf{c}[k] = \left(\hat{\mathbf{R}}[k]\right)^{-1} \hat{\mathbf{p}}[k],$$

$$\hat{\mathbf{R}}[k] = \sum_{n=0}^{k} \lambda^{k-n} \mathbf{r}[n](\mathbf{r}[n])^H, \quad \hat{\mathbf{p}}[k] = \sum_{n=0}^{k} \lambda^{k-n} b^*[n] \mathbf{r}[n]. \qquad (5.47)$$

At first sight, the RLS solution appears to be computationally inefficient, requiring a matrix inversion at every iteration, in contrast to the LS solution (5.44), which only requires one matrix inversion for the entire block of received vectors considered. However, the preceding computations can be simplified significantly by exploiting the special relationship between the sequence of matrices $\hat{\mathbf{R}}[k]$ to be inverted. Specifically, we have

$$\hat{\mathbf{R}}[k] = \lambda \hat{\mathbf{R}}[k-1] + \mathbf{r}[k](\mathbf{r}[k])^H, \qquad (5.48)$$

which says that the new matrix equals another matrix, plus an outer product. We can now invoke the matrix inversion lemma (see Problem 5.18 for a proof), which handles exactly this scenario.

5.6 Linear equalization

Matrix inversion Lemma If $\mathbf{A}$ is an $m \times m$ invertible conjugate symmetric matrix, and $\mathbf{x}$ is an $m \times 1$ vector, then

$$\left(\mathbf{A} + \mathbf{x}\mathbf{x}^H\right)^{-1} = \mathbf{A}^{-1} - \frac{\tilde{\mathbf{x}}\tilde{\mathbf{x}}^H}{1 + \mathbf{x}^H \tilde{\mathbf{x}}}, \quad \text{where } \tilde{\mathbf{x}} = \mathbf{A}^{-1}\mathbf{x}. \tag{5.49}$$

That is, if the matrix $\mathbf{A}$ is updated by adding the outer product of $\mathbf{x}$, then the inverse is updated by the scaled outer product of $\tilde{\mathbf{x}} = \mathbf{A}^{-1}\mathbf{x}$. Thus, the computation of the inverse of the new matrix reduces to the simple operations of calculation of $\tilde{\mathbf{x}}$ and its outer product.

The matrices $\hat{\mathbf{R}}[k]$ involved in the RLS algorithms are conjugate symmetric, and (5.48) is precisely the setting addressed by the matrix inversion lemma, with $\mathbf{A} = \lambda \hat{\mathbf{R}}[k-1]$ and $\mathbf{x} = \mathbf{r}[k]$. It is convenient to define

$$\mathbf{P}[k] = (\hat{\mathbf{R}}[k])^{-1}. \tag{5.50}$$

Applying (5.49) to (5.48), we obtain, upon simplification, the following recursive formula for the required inverse:

$$\mathbf{P}[k] = \lambda^{-1} \left(\mathbf{P}[k-1] - \frac{\tilde{\mathbf{r}}[k] (\tilde{\mathbf{r}}[k])^H}{\lambda + (\mathbf{r}[k])^H \tilde{\mathbf{r}}[k]} \right), \quad \text{where } \tilde{\mathbf{r}}[k] = \mathbf{P}[k-1]\mathbf{r}[k]. \tag{5.51}$$

The vector $\hat{\mathbf{p}}[k]$ in (5.47) is easy to compute recursively, since

$$\mathbf{p}[k] = \lambda \mathbf{p}[k-1] + b^*[k]\mathbf{r}[k]. \tag{5.52}$$

We can now compute the correlator at the kth iteration as

$$\mathbf{c}[k] = \mathbf{P}[k]\mathbf{p}[k]. \tag{5.53}$$

Further algebraic manipulations of (5.53) based on (5.51) and (5.52) yield the following recursion for the correlator sequence $\{\mathbf{c}[k]\}$:

$$\mathbf{c}[k] = \mathbf{c}[k-1] + \frac{e^*[k]\tilde{\mathbf{r}}[k]}{\lambda + (\mathbf{r}[k])^H \tilde{\mathbf{r}}[k]}, \tag{5.54}$$

where

$$e[k] = b[k] - (\mathbf{c}[k-1])^H \mathbf{r}[k] \tag{5.55}$$

is the instantaneous error in tracking the desired sequence $\{b[k]\}$.

Least mean squares algorithm When deriving (5.38), we showed that the gradient of the cost function is given by

$$\nabla_{\mathbf{c}^*} J(\mathbf{c}) = \mathbb{E}[\mathbf{r}[n]((\mathbf{r}[n])^H \mathbf{c} - b^*[n])] = \mathbf{Rc} - \mathbf{p}.$$

One approach to optimizing the cost function $J(\mathbf{c})$, therefore, is to employ gradient descent:

$$\mathbf{c}[k] = \mathbf{c}[k-1] - \mu \, \nabla_{\mathbf{c}^*} J(\mathbf{c}[k-1])$$
$$= \mathbf{c}[k-1] - \mathbb{E}\left[\mathbf{r}[n]\left((\mathbf{r}[n])^H \mathbf{c}[k-1] - b^*[n]\right)\right],$$

where the parameter μ can be adapted as a function of k. The LMS algorithm is a *stochastic gradient* algorithm obtained by dropping the statistical expectation above, using the instantaneous value of the term being averaged: at iteration k, the generic terms $\mathbf{r}[n]$, $b[n]$ are replaced by their current values $\mathbf{r}[k]$, $b[k]$. We can therefore write an iteration of the LMS algorithm as

$$\mathbf{c}[k] = \mathbf{c}[k-1] - \mu \mathbf{r}[k]\left((\mathbf{r}[k])^H \mathbf{c}[k-1] - b^*[k]\right) = \mathbf{c}[k-1] + \mu e^*[k]\mathbf{r}[k], \tag{5.56}$$

where $e[k]$ is the instantaneous error (5.55) and μ is a constant that determines the speed of adaptation. Too high a value of μ leads to instability, while too small a value leads to very slow adaptation (which may be inadequate for tracking channel time variations).

The variant of LMS that is most commonly used in practice is the normalized LMS (NLMS) algorithm. To derive this algorithm, suppose that we scale the received vectors $\{\mathbf{r}[k]\}$ by a factor of A (which means that the power of the received signal scales by A^2). Since $\mathbf{c}^H \mathbf{r}[k]$ must track $b[k]$, this implies that $\mathbf{c}$ must be scaled by a factor of $1/A$, making $\mathbf{c}^H \mathbf{r}[k]$, and hence $e[k]$, scale-invariant. From (5.56), we see that the update to $\mathbf{c}[k-1]$ scales by μA: for this to have the desired $1/A$ scaling, the constant μ must scale as $1/A^2$. That is, the adaptation constant must scale inversely as the received power. The NLMS algorithm implements this as follows:

$$\mathbf{c}[k] = \mathbf{c}[k-1] + \frac{\mu}{P[k]} e^*[k]\mathbf{r}[k], \tag{5.57}$$

where $P[k]$ is adaptively updated to scale with the power of the received signal, while μ is chosen to be a scale-invariant constant (typically $0 < \mu < 1$). A common choice for $P[k]$ is the instantaneous power $P[k] = (\mathbf{r}[k])^H \mathbf{r}[k] + \alpha$, where $\alpha > 0$ is a small constant providing a lower bound for $P[k]$. Another choice is an exponentially weighted average of $(\mathbf{r}[k])^H \mathbf{r}[k]$.

Our goal here was to provide a sketch of the key ideas underlying some common adaptive algorithms. Problem 5.15 contains further exploration of these algorithms. However, there is a huge body of knowledge regarding both the theory and implementation of these algorithms and their variants, that is beyond the scope of this book.

5.6.2 Performance analysis

The output (5.26) of a linear equalizer $\mathbf{c}$ can be rewritten as

$$Z[n] = A_0 b[n] + \sum_{i \neq 0} A_i b[n+i] + W[n],$$

where $A_0 = \langle \mathbf{c}, \mathbf{u}_0 \rangle$ is the amplitude of the desired symbol, $A_i = \langle \mathbf{c}, \mathbf{u}_i \rangle$, $i \neq 0$ are the amplitudes of the terms corresponding to the *residual ISI* at the

5.6 Linear equalization

correlator output, and $W[n]$ is zero mean Gaussian noise with variance $v^2 = \sigma^2 ||\mathbf{c}||^2$ per dimension. If there is no residual ISI (i.e., $A_i \equiv 0$ for $i \neq 0$), as for a ZF equalizer, then error probability computation is straightforward. However, the residual ISI is nonzero for both MMSE equalization and imperfect ZF equalization. We illustrate the methodology for computing the probability of error in such situations for a BPSK ($\{b[k]\}$ i.i.d., ± 1 with equal probability), real baseband system. Generalizations to complex-valued constellations are straightforward. The exact error probability computation involves conditioning on, and then averaging out, the ISI, which is computationally complex if the number of ISI terms is large. A useful *Gaussian approximation*, which is easy to compute, involves approximating the residual ISI as a Gaussian random variable.

BPSK system The bit estimate is given by

$$\hat{b}[n] = \text{sign}(Z[n])$$

and the error probability is given by

$$P_e = P[\hat{b}[n] \neq b[n]].$$

By symmetry, we can condition on $b[n] = +1$, getting

$$P_e = P[Z[n] > 0 | b[n] = +1].$$

Computation of this probability involves averaging over the distribution of both the noise and the ISI. For the exact error probability, we condition further on the ISI bits $\mathbf{b}_I = \{b_{n+i}, i \neq 0\}$.

$$\begin{aligned} P_{e|\mathbf{b}_I} &= P[Z[n] > 0 | b[n] = +1, \mathbf{b}_I] \\ &= P[W[n] > -(A_0 + \sum_{i \neq 0} A_i b[n+i])] \\ &= Q\left(\frac{A_0 + \sum_{i \neq 0} A_i b[n+i]}{v}\right). \end{aligned}$$

We can now average over $\mathbf{b}_I$ to obtain the average error probability:

$$P_e = \mathbb{E}[P_{e|\mathbf{b}_I}].$$

The complexity of computing the exact error probability as above is exponential in the number of ISI bits: if there are K ISI bits, then $\mathbf{b}_I$ takes 2^K different values with equal probability under our model. An alternative approach, which is accurate when there are a moderately large number of residual ISI terms, each of which takes small values, is to apply the central limit theorem to approximate the residual ISI as a Gaussian random variable. The variance of this Gaussian random variable is given by

$$v_I^2 = \text{var}\left(\sum_{i\neq 0} A_i b[n+i]\right) = \sum_{i\neq 0} A_i^2.$$

We therefore get the approximate model

$$Z[n] = A_0 b[n] + N(0, v_I^2 + v^2).$$

The corresponding approximation to the error probability is

$$P_e \approx Q\left(\frac{A_0}{\sqrt{v_I^2 + v^2}}\right) = Q(\sqrt{\text{SIR}}),$$

recognizing that the SIR is given by

$$\text{SIR} = \frac{A_0^2}{v_I^2 + v^2} = \frac{|\langle \mathbf{c}, \mathbf{u}_0 \rangle|^2}{\sum_{i\neq 0}\langle \mathbf{c}, \mathbf{u}_i \rangle + \sigma^2 ||\mathbf{c}||^2}.$$

5.7 Decision feedback equalization

Linear equalizers suppress ISI by projecting the received signal in a direction orthogonal to the interference space: the ZF equalizer does this exactly, the MMSE equalizer does this approximately, taking into account the noise–ISI tradeoff. The resulting noise enhancement can be substantial, if the desired signal vector component orthogonal to the interference subspace is small. The DFE, depicted in Figure 5.10, alleviates this problem by using feedback from prior decisions to *cancel* the interference due to the past symbols, and linearly suppressing only the ISI due to future symbols. Since fewer ISI vectors are being suppressed, the noise enhancement is reduced. The price of this is *error propagation*: an error in a prior decision can cause errors in the current decision via the decision feedback.

The DFE employs a feedforward correlator $\mathbf{c}_{FF}$ to suppress the ISI due to future symbols. This can be computed based on either the ZF or MMSE criteria: the corresponding DFE is called the ZF-DFE or MMSE-DFE, respectively. To compute this correlator, we simply ignore ISI from the past symbols (assuming that they will be canceled perfectly by decision feedback), and

Figure 5.10 A typical architecture for implementing a decision feedback equalizer.

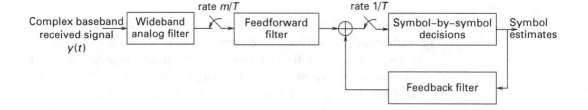

5.7 Decision feedback equalization

work with the following reduced model including only the ISI from future symbols:
$$\mathbf{r}_n^f = b[n]\mathbf{u}_0 + \sum_{j>0} b[n+j]\mathbf{u}_j + \mathbf{w}[n]. \tag{5.58}$$

The corresponding matrix of signal vectors, containing $\{\mathbf{u}_j, j \geq 0\}$ is denoted by $\mathbf{U}_f$. The ZF and MMSE solutions for $\mathbf{c}_{FF}$ can be computed simply by replacing $\mathbf{U}$ by $\mathbf{U}_f$ in (5.30) and (5.42), respectively.

Running example For the model (5.21), (5.22), (5.23) corresponding to our running example, we have

$$\mathbf{U}_f = \begin{pmatrix} 0 & 0 \\ 1 & 0 \\ \frac{1}{2} & 0 \\ -\frac{1}{2} & 1 \\ 0 & \frac{1}{2} \end{pmatrix}. \tag{5.59}$$

Now that $\mathbf{c}_{FF}$ is specified, let us consider its output:
$$\mathbf{c}_{FF}^H \mathbf{r}[n] = b[n]\mathbf{c}_{FF}^H \mathbf{u}_0 + \left\{ \sum_{j>0} b[n+j]\mathbf{c}_{FF}^H \mathbf{u}_j + \mathbf{c}_{FF}^H \mathbf{w}[n] \right\} + \sum_{j>0} b[n-j]\mathbf{c}_{FF}^H \mathbf{u}_{-j}. \tag{5.60}$$

By optimizing $\mathbf{c}_{FF}$ for the reduced model (5.58), we have suppressed the contribution of the term within $\{\}$ above, but the set of terms on the extreme right-hand side, which corresponds to the ISI due to past symbols at the output of the feedforward correlator, can be large. Decision feedback is used to cancel these terms. Setting $\mathbf{c}_{FB}[j] = -\mathbf{c}_{FF}^H \mathbf{u}_{-j}, j > 0$, the DFE decision statistic is given by
$$Z_{\text{DFE}}[n] = \mathbf{c}_{FF}^H \mathbf{r}[n] + \sum_{j>0} \mathbf{c}_{FB}[j]\hat{b}[n-j]. \tag{5.61}$$

Note that
$$Z_{\text{DFE}}[n] = b[n]\mathbf{c}_{FF}^H \mathbf{u}_0 + \left\{ \sum_{j>0} b[n+j]\mathbf{c}_{FF}^H \mathbf{u}_j + \mathbf{c}_{FF}^H \mathbf{w}[n] \right\}$$
$$+ \sum_{j>0} (b[n-j] - \hat{b}[n-j])\mathbf{c}_{FF}^H \mathbf{u}_{-j},$$

so that the contribution of the past symbols is perfectly canceled if the feedback is correct.

Setting $\mathbf{U}_p$ as the matrix with the past ISI vectors $\{\mathbf{u}_{-1}, \mathbf{u}_{-2}, \ldots\}$ as columns, we can write the feedback filter taps in more concise fashion as
$$\mathbf{c}_{FB} = -\mathbf{c}_{FF}^H \mathbf{U}_p, \tag{5.62}$$

where we define $\mathbf{c}_{FB} = (\mathbf{c}_{FB}[K_p], \ldots, \mathbf{c}_{FB}[1])^T$, where K_p are the number of past symbols being fed back.

> **Running example** We compute the ZF-DFE, so as to avoid dependence on the noise variance. The feedforward filter is given by
> $$\mathbf{c}_{\text{FF}} = \mathbf{U}_f \left(\mathbf{U}_f^H \mathbf{U}_f\right)^{-1} \mathbf{e}.$$
> Using (5.59), we obtain
> $$\mathbf{c}_{\text{FF}} = \frac{1}{13}(0, 10, 5, -1, 2)^T.$$
> Since there is only one past ISI vector, we obtain a single feedback tap
> $$\mathbf{c}_{\text{FB}} = -\mathbf{c}_{\text{FF}}^H \mathbf{U}_p = \frac{5}{13},$$
> since
> $$\mathbf{U}_p = (\frac{1}{2}, -\frac{1}{2}, 0, 0, 0)^T.$$

Unified notation for feedforward and feedback taps We can write (5.61) in vector form by setting $\hat{\mathbf{b}}_n = (\hat{b}[n - K_p], \ldots, \hat{b}[n - 1])^T$ as the vector of decisions on past symbols, and $\mathbf{c}_{\text{FB}} = (\mathbf{c}_{\text{FB}}[K_p], \ldots, \mathbf{c}_{\text{FB}}[1])^T$, to obtain

$$Z_{\text{DFE}}[n] = \mathbf{c}_{\text{FF}}^H \mathbf{r}[n] + \mathbf{c}_{\text{FB}}^H \hat{\mathbf{b}}_n = \mathbf{c}_{\text{DFE}}^H \tilde{\mathbf{r}}[n], \tag{5.63}$$

where the extreme right-hand side corresponds to an interpretation of the DFE output as the output of a single correlator

$$\mathbf{c}_{\text{DFE}} = \begin{pmatrix} \mathbf{c}_{\text{FF}} \\ \mathbf{c}_{\text{FB}} \end{pmatrix},$$

whose input is the concatenation of the received vector and the vector of past decisions, given by

$$\tilde{\mathbf{r}}[n] = \begin{pmatrix} \mathbf{r}[n] \\ \hat{\mathbf{b}}[n] \end{pmatrix}.$$

This interpretation is useful for adaptive implementation of the DFE; for example, by replacing $\mathbf{r}[n]$ by $\tilde{\mathbf{r}}[n]$ in (5.44) to obtain an LS implementation of the MMSE-DFE.

5.7.1 Performance analysis

Computing the exact error probability for the DFE is difficult because of the error propagation it incurs. However, we can get a quick idea of its performance based on the following observations about its behavior for typical channels. When all the feedback symbols are correct, then the probability of error equals that of the linear equalizer $\mathbf{c}_{\text{FF}}$ for the reduced model (5.58), since the past ISI is perfectly canceled out. This error probability, $P_{e,\text{FF}}$, can be exactly computed or estimated using the techniques of Section 5.6.2.

Starting from correct feedback, if an error does occur, then it initiates an *error propagation event*. The error propagation event terminates when the feedback again becomes correct (i.e., when there are L_{FB} consecutive correct decisions, where L_{FB} is the number of feedback taps). The number of symbols for which an error propagation event lasts T_e, and the number of symbol errors N_e incurred during an error propagation event, are random variables whose distributions are difficult to characterize. However, the number of symbols between two successive error propagation events is much easier to characterize. When the feedback is correct, if we model the effect of residual ISI and noise for the reduced model (5.58) as independent from symbol to symbol (an excellent approximation in most cases), then symbol errors occur independently. That is, the time T_c between error propagation events is well modeled as a geometric random variable with parameter $P_{e,FF}$:

$$P[T_c = k] = P_{e,FF}(1 - P_{e,FF})^{k-1},$$

with mean $\mathbb{E}[T_c] = 1/P_{e,FF}$. We can now estimate the error probability of the DFE as the average number of errors in an error propagation event, divided by the average length of the error-free and error propagation periods:

$$P_{e,DFE} = \frac{\mathbb{E}[N_e]}{\mathbb{E}[T_e] + \mathbb{E}[T_c]} \approx \mathbb{E}[N_e]P_{e,FF}, \quad (5.64)$$

noting that the average length of an error propagation event, $\mathbb{E}[T_e]$ is typically much shorter than the average length of an error-free period, $\mathbb{E}[T_c] \approx 1/P_{e,FF}$. The average number of errors $\mathbb{E}[N_e]$ for an error propagation event can be estimated by simulations in which we inject an error and let it propagate (which is more efficient than directly simulating DFE performance, especially for moderately high SNR).

The estimate (5.64) allows us to draw important qualitative conclusions about DFE performance relative to the performance of a linear equalizer. Since $\mathbb{E}[N_e]$ is typically quite small, the decay of error probability with SNR is governed by the term $P_{e,FF}$. Thus, the gain in performance of a DFE over a linear equalizer can be quickly estimated by simply comparing the error probability, for linear equalization, of the reduced system (5.58) with that for the original system. In particular, comparing the ZF-DFE and the ZF linear equalizer, the difference in noise enhancement for the reduced and original systems is the dominant factor determining performance.

5.8 Performance analysis of MLSE

We now discuss performance analysis of MLSE. This is important not only for understanding the impact of ISI, but the ideas presented here also apply to analysis of the Viterbi algorithm in other settings, such as ML decoding of convolutional codes.

For concreteness, we consider the continuous-time system model

$$y(t) = \sum_n b[n]p(t - nT) + n(t), \tag{5.65}$$

where n is WGN. We also restrict attention to real-valued signals and BPSK modulation, so that $b[n] \in \{-1, +1\}$.

Notation change To avoid carrying around complicated subscripts, we write the noiseless received signal corresponding to the sequence $\mathbf{b}$ as $s(\mathbf{b})$, dropping the time index. This is the same signal denoted earlier by $s_\mathbf{b}$:

$$s(\mathbf{b}) = s_\mathbf{b} = \sum_n b[n]p(t - nT), \tag{5.66}$$

so that the received signal, conditioned on $\mathbf{b}$ being sent, is given by

$$y = s(\mathbf{b}) + n.$$

Note that this model also applies to the whitened discrete-time model (5.17) in Section 5.4.1, with $s(\mathbf{b}) = \{\sum_{n=0}^{L} f[n]b[k-n] : k \text{ integer}\}$. The analysis is based on the basic results for M-ary signaling in AWGN developed in Chapter 3, which applies to both continuous-time and discrete-time systems.

Let $\Lambda(\mathbf{b})$ denote the log likelihood function being optimized by the Viterbi algorithm, and let L denote the channel memory. As before, the state at time n is denoted by $s[n] = (b[n-L], \ldots, b[n-1])$. Let $\hat{\mathbf{b}}_{\text{ML}}$ denote the MLSE output. We want to estimate

$$P_e(k) = P[\hat{b}_{\text{ML}}[k] \neq b[k]],$$

the probability of error in the kth bit.

5.8.1 Union bound

We first need the notion of an error sequence.

Definition 5.8.1 (Error sequence) *The error sequence corresponding to an estimate $\hat{\mathbf{b}}$ and transmitted sequence $\mathbf{b}$ is defined as*

$$\mathbf{e} = \frac{\mathbf{b} - \hat{\mathbf{b}}}{2}, \tag{5.67}$$

so that

$$\hat{\mathbf{b}} = \mathbf{b} + 2\mathbf{e}. \tag{5.68}$$

For BPSK, the elements of $\mathbf{e} = \{e[n]\}$ take values in $\{0, -1, +1\}$. It is also easy to verify the following consistency condition.

Consistency condition If $e[n] \neq 0$ (i.e., $\hat{b}_n \neq b[n]$), then $e[n] = b[n]$.

5.8 Performance analysis of MLSE

Definition 5.8.2 (Valid error sequence) *An error sequence* **e** *is valid for a transmitted sequence* **b** *if the consistency condition is satisfied for all elements of the error sequence.*

The probability that a given error sequence **e** is valid for a randomly selected sequence **b** is

$$P[\mathbf{e} \text{ is valid for } \mathbf{b}] = 2^{-w(\mathbf{e})}, \tag{5.69}$$

where $w(\mathbf{e})$ denotes the *weight* of **e** (i.e., the number of nonzero elements in **e**). This is because, for any nonzero element of **e**, say $e[n] \neq 0$, we have $P[b[n] = e[n]] = 1/2$.

We can now derive a union bound for $P_e(k)$ by summing over all error sequences that could cause an error in bit $b[k]$. The set of such sequences is denoted by $E_k = \{\mathbf{e} : e[k] \neq 0\}$. Since there are too many such sequences, we tighten it using an "intelligent" union bound which sums over an appropriate subset of E_k.

The *exact* error probability is given by summing over E_k as follows:

$$P_e(k) = \sum_{\mathbf{e} \in E_k} P\left[\mathbf{b} + 2\mathbf{e} = \hat{\mathbf{b}}_{\text{ML}} | \mathbf{e} \text{ valid for } \mathbf{b}\right] P[\mathbf{e} \text{ valid for } \mathbf{b}]$$

$$= \sum_{\mathbf{e} \in E_k} P\left[\Lambda(\mathbf{b}+2\mathbf{e}) = \arg\max_{\mathbf{a}} \Lambda(\mathbf{a}) | \mathbf{e} \text{ valid for } \mathbf{b}\right] 2^{-w(\mathbf{e})}.$$

We can now bound this as we did for M-ary signaling by noting that

$$P[\Lambda(\mathbf{b}+2\mathbf{e}) = \arg\max_{\mathbf{a}} \Lambda(\mathbf{a}) | \mathbf{e} \text{ valid for } \mathbf{b}]$$
$$\leq P[\Lambda(\mathbf{b}+2\mathbf{e}) \geq \Lambda(\mathbf{b}) | \mathbf{e} \text{ valid for } \mathbf{b}]. \tag{5.70}$$

The probability on the right-hand side above is simply the pairwise error probability for binary hypothesis testing between $\mathbf{y} = s(\mathbf{b}+2\mathbf{e}) + \mathbf{n}$ versus $\mathbf{y} = s(\mathbf{b}) + \mathbf{n}$, which we know to be

$$Q\left(\frac{\|s(\mathbf{b}+2\mathbf{e}) - s(\mathbf{b})\|}{2\sigma}\right).$$

It is easy to see, from (5.66), that

$$s(\mathbf{b}+2\mathbf{e}) - s(\mathbf{b}) = 2s(\mathbf{e}),$$

so that the pairwise error probability becomes

$$P[\Lambda(\mathbf{b}+2\mathbf{e}) \geq \Lambda(\mathbf{b}) | \mathbf{e} \text{ valid for } \mathbf{b}] = Q\left(\frac{\|s(\mathbf{e})\|}{\sigma}\right). \tag{5.71}$$

Combining (5.70) and (5.71), we obtain the union bound

$$P_e(k) \leq \sum_{\mathbf{e} \in E_k} Q\left(\frac{\|s(\mathbf{e})\|}{\sigma}\right) 2^{-w(\mathbf{e})}. \tag{5.72}$$

We now want to prune the terms in (5.72) to obtain an "intelligent union bound." To do this, consider Figure 5.11, which shows a simplified schematic

Channel equalization

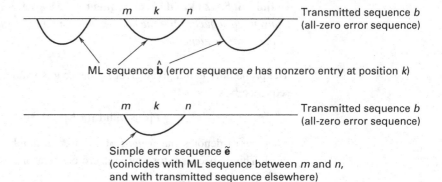

Figure 5.11 Correct path and MLSE output as paths on the error sequence trellis. The correct path corresponds to the all-zero error sequence. In the scenario depicted, the MLSE output makes an error in bit $b[k]$. I also show the *simple* error sequence, which coincides with the MLSE output where it diverges from the correct path around bit $b[k]$, and coincides with the correct path elsewhere.

of the ML sequence $\hat{\mathbf{b}}$ and the true sequence $\mathbf{b}$ as paths through a trellis. Instead of considering a trellis corresponding to the symbol sequence (as in the development of the Viterbi algorithm), it is now convenient to consider a trellis in which a symbol sequence is represented by its error sequence relative to the transmitted sequence. This trellis has 3^L states at each time, and the transmitted sequence corresponds to the all-zero path. Two paths in the trellis merge when L successive symbols for the path are the same. Thus, a path in our error sequence trellis merges with the all-zero path corresponding to the transmitted sequence if there are L consecutive zeros in the error sequence. In the figure, the ML sequence is in E_k, and is shown to diverge and remerge with the transmitted sequence in several segments. Consider now the error sequence $\tilde{\mathbf{e}}$, which coincides with the segment of the ML sequence which diverges from the true sequence around the bit of interest, $b[k]$, and coincides with the true sequence otherwise. Such a sequence has the property that, once it remerges with the all-zero path, it never diverges again. We call such sequences *simple* error sequences, and characterize them as follows.

Definition 5.8.3 (Simple error sequence) *An error sequence $\mathbf{e}$ is simple if there are no more than $L-1$ zeros between any two successive nonzero entries. The set of simple error sequences with $e[k] \neq 0$ is denoted by S_k.*

We now state and prove that the union bound (5.72) can be pruned to include only simple error sequences.

Proposition 5.8.1 (Intelligent union bound using simple error sequences)
The probability of bit error is bounded as

$$P_e(k) \leq \sum_{\mathbf{e} \in S_k} Q\left(\frac{\|s(\mathbf{e})\|}{\sigma}\right) 2^{-w(\mathbf{e})}. \tag{5.73}$$

5.8 Performance analysis of MLSE

Proof Consider the scenario depicted in Figure 5.11. Since the ML sequence and the true sequence have the same state at times m and n, by the principle of optimality, the sum of the branch metrics between times m and n must be strictly greater for the ML path. That is, denoting the sum of branch metrics from m to n as $\Lambda_{m,n}$, we have

$$\Lambda_{m,n}(\hat{\mathbf{b}}) > \Lambda_{m,n}(\mathbf{b}). \tag{5.74}$$

The sequence $\tilde{\mathbf{b}}$ corresponding to the simpler error sequence satisfies

$$\Lambda_{m,n}(\tilde{\mathbf{b}}) = \Lambda_{m,n}(\hat{\mathbf{b}}) \tag{5.75}$$

by construction, since it coincides with the ML sequence $\hat{\mathbf{b}}$ from m to n. Further, since $\tilde{\mathbf{b}}$ coincides with the true sequence $\mathbf{b}$ prior to m and after n, we have, from (5.74) and (5.75)

$$\Lambda(\tilde{\mathbf{b}}) - \Lambda(\mathbf{b}) = \Lambda_{m,n}(\tilde{\mathbf{b}}) - \Lambda_{m,n}(\mathbf{b}) > 0.$$

This shows that, for any $\mathbf{e} \in E_k$, if $\hat{\mathbf{b}} = \mathbf{b} + 2\mathbf{e}$ is the ML estimate, then there exists $\tilde{\mathbf{e}} \in S_k$ such that

$$\Lambda(\mathbf{b} + 2\tilde{\mathbf{e}}) > \Lambda(\mathbf{b}).$$

This implies that

$$P_e(k) = \sum_{\mathbf{e} \in E_k} P\left[\Lambda(\mathbf{b}+2\mathbf{e}) = \arg \max_{\mathbf{a}} \Lambda(\mathbf{a}) \Big| \mathbf{e} \text{ valid for } \mathbf{b}\right] 2^{-w(\mathbf{e})}$$
$$\leq \sum_{\mathbf{e} \in S_k} P[\Lambda(\mathbf{b}+2\mathbf{e}) \geq \Lambda(\mathbf{b}) | \mathbf{e} \text{ valid for } \mathbf{b}] 2^{-w(\mathbf{e})},$$

which proves the desired result upon using (5.71). $\square$

We now consider methods for computing (5.73). To this end, we first recognize that there is nothing special about the bit k whose error probability we are computing. For any times k and l, an error sequence $\mathbf{e}$ in S_k has a one-to-one correspondence with a unique error sequence $\mathbf{e}'$ in S_l obtained by time-shifting $\mathbf{e}$ by $l - k$. To enumerate the error sequences in S_k efficiently, therefore, we introduce the notion of error event.

Definition 5.8.4 (Error event) *An error event is a simple error sequence whose first nonzero entry is at a fixed time, say at 0. The set of error events is denoted by $\mathcal{E}$.*

For $L = 2$, two examples of error events are

$$\mathbf{e}_1 = (\pm 1, 0, 0, 0, \ldots), \quad \mathbf{e}_2 = (\pm 1, 0, \pm 1, 0, \pm 1, 0, 0, \ldots).$$

On the other hand, $\mathbf{e}_3 = (\pm 1, 0, 0, \pm 1, 0, 0, \ldots)$ is not an error event, since it is not a simple error sequence for $L = 2$.

Note that $\mathbf{e}_1$ can be time-shifted so as to line up its nonzero entry with bit $b[k]$, thus creating a simple error sequence in S_k. On the other hand, $\mathbf{e}_2$ can be time-shifted in three different ways, corresponding to its three nonzero entries, to line up with bit $b[k]$; it can therefore generate three distinct members of S_k. In general, an error event of weight $w(\mathbf{e})$ can generate $w(\mathbf{e})$ distinct elements in S_k. Clearly, all members of S_k can be generated in this fashion using error events. We can therefore express the bound (5.73) in terms of error events as follows:

$$P_e(k) \leq \sum_{\mathbf{e} \in \mathcal{E}} Q\left(\frac{||s(\mathbf{e})||}{\sigma}\right) w(\mathbf{e}) 2^{-w(\mathbf{e})}, \qquad (5.76)$$

where the contribution of a given error event $\mathbf{e}$ is scaled by its weight, $w(\mathbf{e})$ corresponding to the number of simple error sequences in S_k it represents.

High SNR asymptotics The high SNR asymptotics of the error probability are determined by the term in the union bound that decays most slowly as the SNR gets large. This corresponds to the smallest Q-function argument, which is determined by

$$\epsilon_{\min}^2 = \min_{\mathbf{e} \in \mathcal{E}} ||s(\mathbf{e})||^2. \qquad (5.77)$$

Proceeding as in the development of the Viterbi algorithm, and specializing to real signals, we have

$$||s(\mathbf{e})||^2 = \sum_n [h[0]e^2[n] + 2e[n] \sum_{m=n-L}^{n-1} h[n-m]e[m]] = \sum_n \lambda(s[n] \to s[n+1]), \qquad (5.78)$$

where $s[n] = (e_{n-L}, \ldots, e[n-1])$ is the state in the error sequence trellis, and where the branch metric λ is implicitly defined above. We can now use the Viterbi algorithm on the error sequence trellis to compute $\epsilon_{\min}^2$. We therefore have the high SNR asymptotics

$$P_e \sim \exp\left(-\frac{\epsilon_{\min}^2}{2\sigma^2}\right), \quad \sigma^2 \to 0.$$

Compare this with the performance without ISI. This corresponds to the error sequence $\mathbf{e}_1 = (\pm 1, 0, 0, \ldots)$, which gives

$$||s(\mathbf{e}_1)||^2 = ||p||^2 = h[0]. \qquad (5.79)$$

Asymptotic efficiency The asymptotic efficiency of MLSE, relative to a system with no ISI, can be defined as the ratio of the error exponents of the error probability in the two cases, given by

$$\eta = \lim_{\sigma^2 \to 0} \frac{-\log P_e(\text{MLSE})}{-\log P_e(\text{no ISI})} = \frac{\min_{\mathbf{e} \in \mathcal{E}} ||s(\mathbf{e})||^2}{||s(\mathbf{e}_1)||^2} = \frac{\epsilon_{\min}^2}{h[0]}. \qquad (5.80)$$

5.8 Performance analysis of MLSE

Even for systems operating at low to moderate SNR, the preceding high SNR asymptotics of MLSE shed some light on the structure of the memory imposed by the channel, analogous to the concept of minimum distance in understanding the structure of a signaling set.

> **Example 5.8.1 (Channels with unit memory)** For $L=1$, error events must only have consecutive nonzero entries (no more than $L-1$ zeros between successive nonzero entries). For an error event of weight w, show that
>
> $$||s(\mathbf{e})||^2_{\min} = wh[0] - 2|h[1]|(w-1) = h[0] + (w-1)(h[0]-2|h[1]|). \quad (5.81)$$
>
> We infer from this, letting w get large, that
>
> $$2|h[1]| \leq h[0]. \quad (5.82)$$
>
> Note that this is a stronger result than that which can be obtained by the Cauchy–Schwartz inequality, which only implies that $|h[1]| \leq h[0]$. We also infer from (5.82) that the minimum in (5.81) is achieved for $w=1$. That is, $\epsilon^2_{\min} = h[0]$, so that, from (5.80), we see that the asymptotic efficiency $\eta = 1$. Thus, we have shown that, for $L=1$, there is no asymptotic penalty due to ISI as long as optimal detection is employed.

Computation of union bound Usually, the bound (5.76) is truncated after a certain number of terms, exploiting the rapid decay of the Q function. The error sequence trellis can be used to compute the energies $||s(\mathbf{e})||^2$ using (5.78). Next, we discuss an alternative approach, which leads to the *transfer function bound*.

5.8.2 Transfer function bound

The transfer function bound includes *all* terms of the intelligent union bound, rather than truncating it at a finite number of terms. There are two steps to computing this bound: first, represent each error event as a path in a state diagram, beginning and ending at the all-zero state; second, replace the Q function by an upper bound which can be evaluated as a product of branch gains as we traverse the state diagram. Specifically, we employ the upper bound

$$Q\left(\frac{||s(\mathbf{e})||}{\sigma}\right) \leq \frac{1}{2}\exp\left(-\frac{||s(\mathbf{e})||^2}{2\sigma^2}\right),$$

which yields

$$P_e \le \frac{1}{2} \sum_{\mathbf{e} \in \mathcal{E}} \exp\left(-\frac{||s(\mathbf{e})||^2}{2\sigma^2}\right) w(\mathbf{e}) 2^{-w(\mathbf{e})}. \tag{5.83}$$

From (5.78), we see that $||s(\mathbf{e})||^2$ can be computed as the sum of additive metrics as we go from state to state in an error sequence trellis. Instead of a trellis, we can consider a state diagram that starts from the all-zero state, contains $3^L - 1$ nonzero states, and then ends at the all-zero state: an error event is a specific path from the all-zero start state to the all-zero end state. The idea now is to associate a branch gain with each state transition, and to compute the net transfer function from the all-zero start state to the all-zero end state, thus summing over all possible error events. By an appropriate choice of the branch gains, we show that the bound (5.83) can be computed as a function of such a transfer function. We illustrate this for $L = 1$ below.

Example 5.8.2 (Transfer function bound for $L=1$) For $L=1$, (5.78) specializes to

$$||s(\mathbf{e})||^2 = \sum_n [h[0]e^2[n] + 2h[1]e[n]e[n-1]].$$

We can therefore rewrite (5.83) as

$$P_e \le \frac{1}{2} \sum_{\mathbf{e} \in \mathcal{E}} w(\mathbf{e}) \prod_n 2^{-|e[n]|} \exp\left(-\frac{h[0]e^2[n] + 2h[1]e[n]e[n-1]}{2\sigma^2}\right).$$

If it were not for the term $w(\mathbf{e})$ inside the summation, the preceding function could be written as the sum of products of branch gains in the state transition diagram. To handle the offending term, we introduce a dummy variable, and consider the following transfer function, which can be computed as a sum of products of branch gains using a state diagram:

$$T(X) = \sum_{\mathbf{e} \in \mathcal{E}} X^{w(\mathbf{e})} \prod_n 2^{-|e[n]|} \exp\left(-\frac{h[0]e^2[n] + 2h[1]e[n]e[n-1]}{2\sigma^2}\right)$$

$$= \sum_{\mathbf{e} \in \mathcal{E}} \prod_n \left(\frac{X}{2}\right)^{|e[n]|} \exp\left(-\frac{h[0]e^2[n] + 2h[1]e[n]e[n-1]}{2\sigma^2}\right). \tag{5.84}$$

Differentiating (5.84) with respect to X, we see that (5.83) can be rewritten as

$$P_e \le \frac{1}{2} \frac{d}{dX} T(X)\big|_{X=1}. \tag{5.85}$$

5.8 Performance analysis of MLSE

We can now label the state diagram for $L = 1$ with branch gains specified as in (5.84): the result is shown in Figure 5.12, with

$$a_0 = \exp\left(-\frac{h[0]}{2\sigma^2}\right),$$
$$a_1 = \exp\left(-\frac{h[0]+2h[1]}{2\sigma^2}\right),$$
$$a_2 = \exp\left(-\frac{h[0]-2h[1]}{2\sigma^2}\right).$$

A systematic way to compute the transfer function from the all-zero start state A to the all-zero end state D is to solve simultaneous equations that relate the transfer functions from the start state to all other states. For example, any path from A to D is a path from A to B, plus the branch BD, or a path from A to C, plus the branch CD. This gives

$$T_{AD}(X) = T_{AB}(X)b_{BD} + T_{AC}(X)b_{CD},$$

where $b_{BD} = 1$ and $b_{CD} = 1$ are the branch gains from B to D and C to D, respectively. Similarly, we obtain

$$T_{AB}(X) = T_{AA}(X)b_{AB} + T_{AB}(X)b_{BB} + T_{AC}(X)b_{CB},$$
$$T_{AC}(X) = T_{AA}(X)b_{AC} + T_{AB}(X)b_{BC} + T_{AC}(X)b_{CC}.$$

Plugging in the branch gains from Figure 5.12, and the initial condition $T_{AA}(X) = 1$, we obtain the simultaneous equations

$$T_{AD}(X) = T_{AB}(X) + T_{AC}(X),$$
$$T_{AB}(X) = a_0\tfrac{X}{2} + a_1\tfrac{X}{2}T_{AB}(X) + a_2\tfrac{X}{2}T_{AC}(X), \quad (5.86)$$
$$T_{AC}(X) = a_0\tfrac{X}{2} + a_2\tfrac{X}{2}T_{AB}(X) + a_1\tfrac{X}{2}T_{AC}(X),$$

which can be solved to obtain that

$$T(X) = T_{AD}(X) = \frac{a_0 X}{1 - \tfrac{1}{2}(a_1+a_2)X}. \quad (5.87)$$

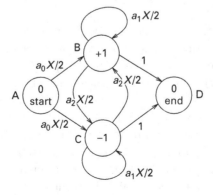

Figure 5.12 State transition diagram for $L = 1$.

Substituting into (5.85), we obtain

$$P_e \leq \frac{\frac{1}{2}a_0}{[1-\frac{1}{2}(a_1+a_2)]^2} = \frac{\frac{1}{2}\exp\left(-\frac{h[0]}{2\sigma^2}\right)}{\left[1-\frac{1}{2}\left(\exp\left(-\frac{h[0]+2h[1]}{2\sigma^2}\right)+\exp\left(-\frac{h[0]-2h[1]}{2\sigma^2}\right)\right)\right]^2}.$$

We can infer from (5.82) that the denominator is bounded away from zero as $\sigma^2 \to 0$, so that the high SNR asymptotics of the transfer function bound are given by $\exp(-(h[0])/(2\sigma^2))$. This is the same conclusion that we arrived at earlier using the dominant term of the union bound.

The computation of the transfer function bound for $L > 1$ is entirely similar to that for the preceding example, with the transfer function defined as

$$T(X) = \sum_{\mathbf{e} \in \mathcal{E}} \prod_n \left(\frac{X}{2}\right)^{|e[n]|} \beta(s[n] \to s[n+1]),$$

where

$$\beta(s[n] \to s[n+1]) = \exp\left(-\frac{h[0]e^2[n]+2e[n]\sum_{n-L}^{n-1}h[n-m]e[m]}{2\sigma^2}\right).$$

The bound (5.85) applies in this general case as well, and the simultaneous equations relating the transfer function from the all-zero start state to all other states can be written down and solved as before. However, solving for $T(X)$ as a function of X can be difficult for large L. An alternative strategy is to approximate (5.85) numerically as

$$P_e \leq \frac{1}{2}\frac{T(1+\delta)-T(1)}{\delta},$$

where $\delta > 0$ is small. Simultaneous equations such as (5.86) can now be solved numerically for $X = 1+\delta$ and $X = 1$, which is simpler than solving algebraically for the function $T(X)$.

5.9 Numerical comparison of equalization techniques

To illustrate the performance of the equalization schemes discussed here, let us consider a numerical example for a somewhat more elaborate channel model than in our running example. Consider a rectangular transmit pulse $g_{TX}(t) = I_{[0,1]}(t)$ and a channel impulse response given by $g_C(t) = 2\delta(t-0.5) - 3\delta(t-2)/4 + j\delta(t-2.25)$. The impulse response of the cascade of the transmit pulse and channel filter is denoted by $p(t)$ and is displayed in Figure 5.13. Over this channel, we transmit Gray coded QPSK symbols taking values $b[n] \in \{1+j, 1-j, -1-j, -1+j\}$ at a rate of 1 symbol per unit time. At the receiver front end, we use the optimal matched filter, $g_{RX}(t) = p^*(-t)$.

It can be checked that the channel memory $L = 2$, so that MLSE requires $4^2 = 16$ states. For a linear equalizer, suppose that we use an observation

5.9 Numerical comparison of equalization techniques

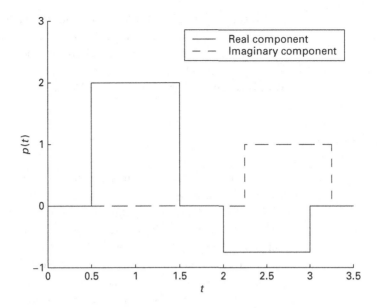

Figure 5.13 The received pulse $p(t)$ formed by the cascade of the transmit and channel filters.

interval that exactly spans the impulse response $\{h[n]\}$ for the desired symbol: this is of length $2L+1 = 5$. It can be seen that the ZF equalizer does not exist, and that the LMMSE equalizer will have an error floor due to unsuppressed ISI. The ZF-DFE and MMSE-DFE can be computed as described in the text: the DFE has four feedback taps, corresponding to the four "past" ISI vectors. A comparison of the performance of all of the equalizers, obtained by averaging over multiple 500 symbol packets, is shown in Figure 5.14. Note that MLSE performance is almost indistinguishable from ISI-free performance. The MMSE-DFE is the best suboptimal equalizer, about 2 dB away from

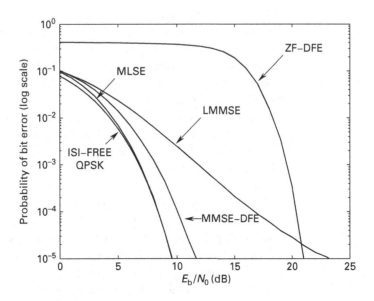

Figure 5.14 Numerical comparison of the performance of various equalizers.

MLSE performance. The LMMSE performance exhibits an error floor, since it does not have enough dimensions to suppress all of the ISI as the SNR gets large. The ZF performance is particularly poor here: the linear ZF equalizer does not exist, and the ZF-DFE performs more poorly than even the linear MMSE equalizer over a wide range of SNR.

5.10 Further reading

The treatment of MLSE in this chapter is based on some classic papers that are still recommended reading. The MLSE formulation followed here is that of Ungerboeck [27], while the alternative whitening-based approach was proposed earlier by Forney [28]. Forney was also responsible for naming and popularizing the Viterbi algorithm in his paper [29]. The sharpest known performance bounds for MLSE (sharper than the ones developed here) are in the paper by Verdu [30]. The geometric approach to finite-complexity equalization, in which the ISI is expressed as interference vectors, is adapted from the author's own work on multiuser detection [31, 32], based on the analogy between intersymbol interference and multiuser interference. For example, the formulation of the LMMSE equalizer is exactly analogous to the MMSE interference suppression receiver described in [31]. It is worth noting that a geometric approach was first suggested for infinite-length equalizers in a classic two-part paper by Messerschmitt [33], which is still recommended reading. A number of papers have addressed the problem of analyzing DFE performance, the key difficulty in which lies in characterizing the phenomenon of error propagation; see [34] and the references therein.

Discussions on the benefits of fractionally spaced equalization can be found in [35]. Detailed discussion of adaptive algorithms for equalization is found in the books by Haykin [36] and Honig and Messerschmitt [37].

While we discuss three broad classes of equalizers, linear, DFE, and MLSE, many variations have been explored in the literature, and we mention a few below. Hybrid equalizers employing MLSE with decision feedback can be used to reduce complexity, as pointed out in [38]. The performance of the DFE can be enhanced by running it in both directions and then arbitrating the results [39]. For long, sparse, channels, the number of equalizer taps can be constrained, but their location optimized [40]. A method for alleviating error propagation in a DFE by using parallelism, and a high-rate error correction code, is proposed in [41].

While the material in this chapter, and in the preceding references, discusses broad principles of channel equalization, creative modifications are required in order to apply these ideas to specific contexts such as wireless channels (e.g., handling time variations due to mobility), magnetic recording channels (e.g., handling runlength constraints), and optical communication channels (e.g., handling nonlinearities). We do not attempt to give specific citations from the vast literature on these topics.

5.11 Problems

5.11.1 MLSE

Problem 5.1 Consider a digitally modulated system using QPSK signaling at *bit rate* $2/T$, and with transmit filter, channel, and receive filter specified as follows:

$$g_{TX}(t) = I_{[0, \frac{T}{2}]} - I_{[\frac{T}{2}, T]}; \quad g_C(t) = \delta(t) - \frac{1}{2}\delta(t - \frac{T}{2}); \quad g_{RX}(t) = I_{[0, \frac{T}{2}]}.$$

Let $z[k]$ denote the receive filter output sample at time $kT_s + \tau$, where T_s is a sampling interval to be chosen.

(a) Show that ML sequence detection using the samples $\{z[k]\}$ is possible, given an appropriate choice of T_s and τ. Specify the corresponding choice of T_s and τ.

(b) How many states are needed in the trellis for implementing ML sequence detection using the Viterbi algorithm?

Problem 5.2 Consider the transmit pulse $g_{TX}(t) = \text{sinc}(\frac{t}{T})\text{sinc}(\frac{t}{2T})$, which is Nyquist at symbol rate $1/T$.

(a) If $g_{TX}(t)$ is used for Nyquist signaling using 8-PSK at 6 Mbit/s, what is the minimum required channel bandwidth?

(b) For the setting in (a), suppose that the complex baseband channel has impulse response $g_C(t) = \delta(t - 0.5T) - \frac{1}{2}\delta(t - 1.5T) + \frac{1}{4}\delta(t - 2.5T)$. What is the minimum number of states in the trellis for MLSE using the Viterbi algorithm?

Problem 5.3 (MLSE performance analysis) For BPSK ± 1 signaling in the standard MLSE setting, suppose that the channel memory $L = 1$, with $h_0 = 1$, $h_1 = -0.3$.

(a) What is the maximum pairwise error probability, as a function of the received E_b/N_0, for two bit sequences that differ only in the first two bits? Express your answer in terms of the Q function.

(b) Plot the transfer function bound (log scale) as a function of E_b/N_0 (dB). Also plot the error probability of BPSK without ISI for comparison.

Problem 5.4 Consider a received signal of the form $y(t) = \sum_l b[l]p(t - lT) + n(t)$, where $b[l] \in \{-1, 1\}$, $n(t)$ is AWGN, and $p(t)$ has Fourier transform given by

$$P(f) = \begin{cases} \cos \pi fT & |f| \le \frac{1}{2T}, \\ 0 & \text{else}. \end{cases} \quad (5.88)$$

(a) Is p a Nyquist pulse for signaling at rate $1/T$?
(b) Suppose that the receive filter is an ideal lowpass filter with transfer function
$$G_{\text{RX}}(f) = \begin{cases} 1 & |f| \leq \frac{1}{2T}, \\ 0 & \text{else}. \end{cases} \quad (5.89)$$
Note that G_{RX} is not the matched filter for P. Let $r(t) = (y * g_{\text{RX}})(t)$ denote the output of the receive filter, and define the samples $r[l] = r(lT_s - \tau)$. Show that it is possible to implement MLSE based on the original continuous-time signal $y(t)$ using only the samples $\{r[l]\}$, and specify a choice of T_s and τ that makes this possible.
(c) Draw a trellis for implementing MLSE, and find an appropriate branch metric assuming that the Viterbi algorithm searches for a *minimum* weight path through the trellis.
(d) What is the asymptotic efficiency (relative to the ISI-free case) of MLSE?
(e) What is the asymptotic efficiency of one-shot detection (which ignores the presence of ISI)?
(f) For E_b/N_0 of 10 dB, evaluate the exact error probability of one-shot detection (condition on the ISI bits, and then remove the conditioning) and the transfer function bound on the error probability of MLSE, and compare with the ISI-free error probability benchmark.

Problem 5.5 (Noise samples at the output of a filter) Consider complex WGN $n(t)$ with PSD σ^2 per dimension, passed through a filter $g(t)$ and sampled at rate $1/(T_s)$. The samples are given by
$$N[k] = (n * g)(kT_s).$$

(a) Show that $\{N[k]\}$ is a stationary proper complex Gaussian random process with zero mean and autocorrelation function
$$R_N[l] = \mathbb{E}\left[N[k]N^*[k-l]\right] = \sigma^2 r_g[l],$$
where
$$r_g[l] = \int g(t)g^*(t - lT_s)\mathrm{d}t$$
is the sampled autocorrelation function of $g(t)$.
(b) Show that $\{r_g[l]\}$ and $\{R_N[l]\}$ are conjugate symmetric.
(c) Define the PSD of N as the z-transform
$$S_N(z) = \sum_{k=-\infty}^{\infty} R_N[l] z^{-l}$$
(setting $z = e^{j2\pi f}$ yields the discrete-time Fourier transform). Show that $S_N(z) = S_N^*(z^{*-1})$.
(d) Conclude that, if z_0 is a root of $S_N(z)$, then so is $1/z_0^*$.

5.11 Problems

(e) Assuming a finite number of roots $\{a_k\}$ inside the unit circle, show that

$$S_N(z) = A \prod_k (1 - a_k z^{-1})(1 - a_k^* z),$$

where A is a constant. Note that the factors $(1 - a_k z^{-1})$ are causal and causally invertible for $|a_k| < 1$.

(f) Show that $\{N[k]\}$ can be generated by passing discrete-time WGN through a causal filter (this is useful for simulating colored noise).

(g) Show that $\{N[k]\}$ can be whitened by passing it through an anticausal filter (this is useful for algorithms predicated on white noise).

Problem 5.6 (MLSE simulation) We would like to develop a model for simulating the symbol rate sampled matched filter outputs for linear modulation through a dispersive channel. That is, we wish to generate the samples $z[k] = (y * p_{\mathrm{MF}})(kT)$ for $y(t) = \sum_n b[n] p(t - nT) + n(t)$, where n is complex WGN.

(a) Show that the signal contribution to $z[k]$ can be written as

$$z_s[k] = \sum_n b[n] h[k-n] = (b * h)[k],$$

where $h[l] = (p * p_{\mathrm{MF}})(lT)$ as before.

(b) Show that the noise contribution to $z[k]$ is a WSS, proper complex Gaussian random process $\{z_n[k]\}$ with zero mean and covariance function

$$C_{z_n}[k] = E[z_n[l] z_n^*[l-k]] = 2\sigma^2 h[k].$$

For real-valued symbols, signals and noise, $h[k]$ are real, and

$$C_{z_n}[k] = E[z_n[l] z_n[l-k]] = \sigma^2 h[k].$$

(c) Now, specialize to the running example in Figure 5.1, with BPSK signaling ($b[n] \in \{-1, +1\}$). We can now restrict y, p and n to be real-valued. Show that the results of (a) specialize to

$$z_s[k] = \frac{3}{2} b[k] - \frac{1}{2}(b[k-1] + b[k+1]).$$

Show that the results of (b) specialize to

$$S_{z_n}(z) = \sigma^2 \left(\frac{3}{2} - \frac{1}{2}(z + z^{-1}) \right),$$

where the PSD $S_{z_n}(z) = \sum_k C_{z_n}[k] z^{-k}$ is the z-transform of $C_{z_n}[k]$.

(d) Suppose that $\{w[k]\}$ are i.i.d. $N(0, 1)$ random variables. Show that this discrete-time WGN sequence can be filtered to generate $z_n[k]$ as follows:
$$z_n[k] = g[0]w[k] + g[1]w[k-1].$$
Find the coefficients $g[0]$ and $g[1]$ such that $z_n[k]$ has statistics as specified in (c).

Hint Factorize $S_{z_n}(z) = (a+bz)(a^* + b^*z^{-1})$ by finding the roots, and use one of the factors to specify the filter.

(e) Use these results to simulate the performance of MLSE for the running example. Compare the resulting BER with that obtained using the transfer function bound.

Problem 5.7 (Alternative MLSE formulation for running example) In Problem 5.6, it is shown that the MF output for the running example satisfies:
$$z[k] = z_s[k] + z_n[k],$$
where $z_s[k] = \frac{3}{2}b[k] - \frac{1}{2}(b[k-1] + b[k+1])$, and $\{z_n[k]\}$ is zero mean colored Gaussian noise with PSD $S_{z_n}(z) = \left(\frac{3}{2} - \frac{1}{2}(z + z^{-1})\right)$ (set $\sigma^2 = 1$ for convenience, absorbing the effect of SNR into the energies of the symbol stream $\{b[k]\}$). In Problem 5.6, we factorized this PSD in order to be able to simulate colored noise by putting white noise through a filter. Now, we use the same factorization to whiten the noise to get an alternative MLSE formulation.

(a) Show that $S_{z_n}(z) = A(1 + az^{-1})(1 + a^*z)$, where $|a| < 1$ and $A > 0$. Note that the first factor is causal (and causally invertible), and the second is anticausal (and anticausally invertible).
(b) Define a whitening filter $Q(z) = 1/(\sqrt{A}(1 + a^*z))$. Observe that the corresponding impulse response is anticausal. Show that $\{z_n[k]\}$ passed through the filter Q yields discrete-time WGN.
(c) Show that $\{z_s[k]\}$ passed through the filter Q yields the symbol sequence $\{b[k]\}$ convolved with $\sqrt{A}(1 + az^{-1})$.
(d) Conclude that passing the matched filter output $\{z[k]\}$ through the whitening filter Q yields a new sequence $\{y[k]\}$ obeying the following model
$$y[k] = \sqrt{A}(b[k] + ab[k-1]) + w[k],$$
where $w[k] \sim N(0, 1)$ are i.i.d. WGN samples. This is an example of the alternative whitened model (5.17).
(e) Is there any information loss due to the whitening transformation?

Problem 5.8 For our running example, how does the model (5.21) change if the sampling times at the output of the receive filter are shifted by 1/2? (Assume that we still use a block of five samples for each symbol decision.) Find a ZF solution and compute its noise enhancement in dB. How sensitive is the performance to the offset in sampling times?

Remark The purpose of this problem is to show the relative insensitivity of the performance of fractionally spaced equalization to sampling time offsets.

Problem 5.9 (Properties of linear MMSE reception) Prove each of the following results regarding the linear MMSE correlator specified by (5.37)–(5.38). For simplicity, restrict attention to real-valued signals and noise and ± 1 BPSK symbols in your proofs.

(a) The MMSE is given by
$$\text{MMSE} = 1 - \mathbf{p}^T \mathbf{c}_{\text{mmse}} = 1 - \mathbf{p}^T \mathbf{R}^{-1} \mathbf{p}.$$

(b) For the model (5.25), the MMSE receiver maximizes the SIR as defined in (5.43).

Hint Consider the problem of maximizing SIR subject to $\langle \mathbf{c}, \mathbf{u}_0 \rangle = \alpha$. Show that the achieved maximum SIR is independent of α. Now choose $\alpha = \langle \mathbf{c}_{\text{mmse}}, \mathbf{u}_0 \rangle$.

(c) Show that the SIR attained by the MMSE correlator is given by
$$\text{SIR}_{\text{max}} = \frac{1}{\text{MMSE}} - 1,$$
where the MMSE is given by (a).

(d) Suppose that the noise covariance is given by $\mathbf{C}_w = \sigma^2 \mathbf{I}$, and that the desired vector $\mathbf{u}_0$ is linearly independent of the interference vectors $\{\mathbf{u}_j, j \neq 0\}$. Prove that the MMSE solution tends to the zero-forcing solution as $\sigma^2 \to 0$. That is, show that
$$\langle \mathbf{c}, \mathbf{u}_0 \rangle \to 1,$$
$$\langle \mathbf{c}, \mathbf{u}_j \rangle \to 0, \quad j = 2, \ldots, K.$$

Hint Show that a correlator satisfying the preceding limits asymptotically (as $\sigma^2 \to 0$) satisfies the necessary and sufficient condition characterizing the MMSE solution.

(e) For the model (5.24), show that a linear correlator $\mathbf{c}$ maximizing $\mathbf{c}^T \mathbf{u}_0$, subject to $\mathbf{c}^T \mathbf{R} \mathbf{c} = 1$, is proportional to the LMMSE correlator.

Hint Write down the Lagrangian for the given constrained optimization problem, and use the fact that $\mathbf{p}$ in (5.37)–(5.38) is proportional to $\mathbf{u}_0$ for the model (5.24).

Remark The correlator in (e) is termed the constrained minimum output energy (CMOE) detector, and has been studied in detail in the context of linear multiuser detection.

Problem 5.10 The discrete-time end-to-end impulse response for a linearly modulated system sampled at three times the symbol rate is $\ldots, 0, -\frac{1+j}{2}, \frac{1-j}{4}, 1+2j, \frac{1}{2}, 0, -\frac{j}{4}, \frac{1}{4}, \frac{1+2j}{4}, \frac{3-j}{2}, 0, \ldots$. Assume that the noise at the output of the sampler is discrete-time AWGN.

(a) Find a length 9 ZF equalizer where the desired signal vector is exactly aligned with the observation interval. What is the noise enhancement?

(b) Express the channel as three parallel symbol rate channels $\{H_i(z), i = 1, 2, 3\}$. Show that the equalizer you found satisfies a relation of the form $\sum_{i=1}^{3} H_i(z) G_i(z) = z^{-d}$, specifying $\{G_i(z), i = 1, 2, 3\}$ and d.

(c) If you were using a rectangular 16-QAM alphabet over this channel, estimate the symbol error rate and the BER (with Gray coding) at E_b/N_0 of 15 dB.

(d) Plot the noise enhancement in dB as you vary the equalizer length between 9 and 18, keeping the desired signal vector in the "middle" of the observation interval (this does not uniquely specify the equalizers in all cases). As a receiver designer, which length would you choose?

Problem 5.11 Consider the setting of Problem 5.10. Answer the following questions for a linear MMSE equalizer of length 9, where the desired signal vector is exactly aligned with the observation interval. Assume that the modulation format is rectangular 16-QAM. Fix E_b/N_0 at 15 dB.

(a) Find the coefficients of the MMSE equalizer, assuming that the desired symbol sequence being tracked is normalized to unit average energy ($\sigma_b^2 = 1$).

(b) Generate and plot a histogram of the I and Q components of the residual ISI at the equalizer output. Does the histogram look zero mean Gaussian?

(c) Use a Gaussian approximation for the residual ISI to estimate the symbol error rate and the BER (with Gray coding) at the output of the equalizer. Compare the performance with that of the ZF equalizer in Problem 5.10(c).

(d) Compute the normalized inner product between the MMSE correlator, and the corresponding ZF equalizer in Problem 5.10. Repeat at E_b/N_0 of 5 dB and at 25 dB, and comment on the results.

Problem 5.12 Consider again the setting of Problem 5.10. Answer the following questions for a DFE in which the feedforward filter is of length 9, with the desired signal vector exactly aligned with the observation interval. Assume that the modulation format is rectangular 16-QAM. Fix E_b/N_0 at 15 dB.

(a) How many feedback taps are needed to cancel out the effect of all "past" symbols falling into the observation interval?

(b) For a number of feedback taps as in (a), find the coefficients of the feedforward and feedback filters for a ZF-DFE.

(c) Repeat (b) for an MMSE-DFE.

(d) Estimate the expected performance improvement in dB for the DFE, relative to the linear equalizers in Problems 5.10 and 5.11. Assume moderately high SNR, and ignore error propagation.

Problem 5.13 Consider the channel of Problem 5.10, interpreted as three parallel symbol-spaced subchannels, with received samples $\{r_i[n]\}$ for the ith subchannel, $i = 1, 2, 3$. We wish to perform MLSE for a QPSK alphabet.

(a) What is the minimum number of states required in the trellis?
(b) Specify the form of the additive metric to be used.

Problem 5.14 (Computer simulations of equalizer performance) For the channel model in Problem 5.10, suppose that we use a QPSK alphabet with Gray coding. Assume that we send 500 byte packets (i.e., 4000 bits per packet). Estimate the BER incurred by averaging within and across packets for the linear MMSE and MMSE-DFE, for a range of error probabilities 10^{-1}–10^{-4}.

(a) Plot the BER (log scale) versus E_b/N_0 (dB). Provide for comparison the BER curve without ISI.
(b) From a comparison of the curves, estimate the approximate degradation in dB due to ISI at BER of 10^{-2}. Can this be predicted by computing the noise enhancement for the corresponding ZF and ZF-DFE equalizers (e.g., using the results from Problems 5.10 and 5.12)?

Problem 5.15 (Computer simulations of adaptive equalization) Consider the packetized system of Problem 5.14. Suppose that the first 100 symbols of every packet are a randomly generated, but known, training sequence.

(a) Implement the normalized LMS algorithm (5.57) with $\mu = 0.5$, and plot the MSE as a function of the number of iterations. (Continue running the equalizer in decision-directed mode after the training sequence is over.)
(b) Simulate over multiple packets to estimate the BER as a function of E_b/N_0 (dB). Compare with the results in Problem 5.14 and comment on the degradation due to the adaptive implementation.
(c) Implement a block least squares equalizer based on the training sequence alone. Estimate the BER and compare with the results in Problem 5.14. Does it work better or worse than NLMS?
(d) Implement the RLS algorithm, using both training and decision-directed modes. Plot the MSE as a function of the number of iterations.
(e) Plot the BER as a function of E_b/N_0 of the RLS implementation, and compare it with the other results.

Problem 5.16 (BER for linear equalizers) The decision statistic at the output of a linear equalizer is given by

$$y[n] = b[n] + 0.1b[n-1] - 0.05b[n-2] - 0.1b[n+1] - 0.05b[n+2] + w[n],$$

where $\{b[k]\}$ are independent and identically distributed symbols taking values ± 1 with equal probability, and $\{w[k]\}$ is real WGN with zero mean and variance σ^2. The decision rule employed is

$$\hat{b}[n] = \text{sign}(y[n]).$$

(a) Find a numerical value for the following limit:

$$\lim_{\sigma^2 \to 0} \sigma^2 \log P(\hat{b}_n \neq b_n).$$

(b) Find the approximate error probability for $\sigma^2 = 0.16$, modeling the sum of the ISI and the noise contributions to $y[n]$ as a Gaussian random variable.

(c) Find the exact error probability for $\sigma^2 = 0.16$.

Problem 5.17 (Software project) This project is intended to give hands-on experience of complexity and performance tradeoffs in channel equalization by working through the example in Section 5.9. Expressing time in units of the symbol time, we take the symbol rate to be 1 symbol per unit time. We consider Gray coded QPSK with symbols $b[n]$ taking values in $\{\pm 1 \pm j\}$. The transmit filter has impulse response

$$g_T(t) = I_{[0,1]}(t).$$

The channel impulse response is given by

$$g_C(t) = 2\delta(t - 0.5) - \frac{3}{4}\delta(t - 2) + j\delta(t - 2.25)$$

(this can be varied to see the effect of the channel on equalizer performance). The receive filter is matched to the cascade of the transmit and channel filters, and is sampled at the symbol rate so as to generate sufficient statistics for symbol demodulation.

You are to evaluate the performance of MLSE as well as of suboptimal equalization schemes, as laid out in the steps below. The results should be formatted in a report that supplies all the relevant information and formulas required for reproducing your results, and a copy of the simulation software should be attached.

The range of error probabilities of interest is 10^{-3} or higher, and the range of E_b/N_0 of interest is 0–30 dB. In plotting your results, choose your range of E_b/N_0 based on the preceding two factors. For all error probability computations, average over multiple 500 symbol packets, with enough additional symbols at the beginning and end to ensure that MLSE starts and ends with a state consisting of $1 + j$ symbols. In all your plots, include the error probability curve for QPSK over the AWGN channel without ISI for reference. In (c) and (d), nominal values for the number of equalizer taps are suggested, but you are encouraged to experiment with other values if they work better.

(a) Set up a discrete-time simulation, in which, given a sequence of symbols and a value of received E_b/N_0, you can generate the corresponding sampled matched filter outputs $\{z[n]\}$. To generate the signal contribution to the output, first find the discrete-time impulse response seen by a single symbol at the output of the sampler. To generate the colored noise at the output, pass discrete-time WGN through a suitable discrete-time filter.

5.11 Problems

Specify clearly how you generate the signal and noise contributions in your report.

(b) For symbol rate sampling, and for odd values of L ranging from 5 to 21, compute the MMSE as a function of the number of taps for an L-tap LMMSE receiver with decision delay chosen such that the symbol being demodulated falls in the middle of the observation interval. What choice of L would you recommend?

Note In finding the MMSE solution, make sure you account for the fact that the noise at the matched filter output is colored.

(c) For $L = 11$, find by computer simulations the bit error rate (BER) of the LMMSE equalizer. Plot the error probability (on log scale) against E_b/N_0 in dB, simulating over enough symbols to get a smooth curve.

(d) Compute the coefficients of an MMSE-DFE with five symbol-spaced feedforward taps, with the desired symbol falling in the middle of the observation interval used by the feedforward filter. Choose the number of feedback taps equal to the number of past symbols falling within the observation interval. Simulate the performance for QPSK as before, and compare the BER with the results of (a).

(e) Find the BER of MLSE by simulation, again considering QPSK with Gray coding. Compare with the results from (b) and (c), and with the performance with no ISI. What is the dB penalty due to ISI at high SNR? Can you predict this based on analysis of MLSE?

Problem 5.18 (Proof of matrix inversion lemma) If we know the inverse of a matrix $\mathbf{A}$, then the matrix inversion lemma (5.49) provides a simple way of updating the inverse to compute $\mathbf{B} = (\mathbf{A} + \mathbf{x}\mathbf{x}^H)^{-1}$. Derive this result as follows. For an arbitrary vector $\mathbf{y}$, consider the equation

$$(\mathbf{A} + \mathbf{x}\mathbf{x}^H)\mathbf{z} = \mathbf{y}. \tag{5.90}$$

Finding $\mathbf{B}$ is equivalent to finding a formula for $\mathbf{z}$ of the form $\mathbf{z} = \mathbf{B}\mathbf{y}$.

(a) Premultiply both sides of (5.90) by $\mathbf{A}^{-1}$ and obtain

$$\mathbf{z} = \mathbf{A}^{-1}\mathbf{y} - \mathbf{A}^{-1}\mathbf{x}\mathbf{x}^H\mathbf{z}. \tag{5.91}$$

(b) Premultiply both sides of (5.91) by $\mathbf{x}^H$ and then solve for $\mathbf{x}^H\mathbf{z}$ in terms of $\mathbf{x}$, $\mathbf{A}$, and $\mathbf{y}$.

(c) Substitute into (5.91) and manipulate to bring into the desired form $\mathbf{z} = \mathbf{B}\mathbf{y}$. Read off the expression for $\mathbf{B}$ to complete the proof.

CHAPTER 6
Information-theoretic limits and their computation

Information theory (often termed Shannon theory in honor of its founder, Claude Shannon) provides fundamental benchmarks against which a communication system design can be compared. Given a channel model and transmission constraints (e.g., on power), information theory enables us to compute, at least in principle, the highest rate at which reliable communication over the channel is possible. This rate is called the *channel capacity*.

Once channel capacity is computed for a particular set of system parameters, it is the task of the communication link designer to devise coding and modulation strategies that approach this capacity. After 50 years of effort since Shannon's seminal work, it is now safe to say that this goal has been accomplished for some of the most common channel models. The proofs of the fundamental theorems of information theory indicate that Shannon limits can be achieved by random code constructions using very large block lengths. While this appeared to be computationally infeasible in terms of both encoding and decoding, the invention of turbo codes by Berrou *et al.* in 1993 provided implementable mechanisms for achieving just this. Turbo codes are random-looking codes obtained from easy-to-encode convolutional codes, which can be decoded efficiently using iterative decoding techniques instead of ML decoding (which is computationally infeasible for such constructions). Since then, a host of "turbo-like" coded modulation strategies have been proposed, including rediscovery of the low density parity check (LDPC) codes invented by Gallager in the 1960s. These developments encourage us to postulate that it should be possible (with the application of sufficient ingenuity) to devise capacity-achieving turbo-like coded modulation strategies for a very large class of channels. Thus, it is more important than ever to characterize information-theoretic limits when setting out to design a communication system, both in terms of setting design goals and in terms of gaining intuition on design parameters (e.g., size of constellation to use). The goal of this chapter, therefore, is to provide enough exposure to Shannon theory to enable computation of capacity benchmarks, with the focus on the AWGN channel and some variants. There is no attempt to give a complete, or completely rigorous, exposition. For this purpose, the reader is referred to information theory textbooks mentioned in Section 6.5.

The techniques discussed in this chapter are employed in Chapter 8 in order to obtain information-theoretic insights into wireless systems. Constructive coding strategies, including turbo-like codes, are discussed in Chapter 7.

We note that the *law of large numbers (LLN)* is a key ingredient of information theory: if $X_1, \ldots, X_n$ are i.i.d. random variables, then their empirical average $(X_1 + \cdots + X_n)/n$ tends to the statistical mean $\mathbb{E}[X_1]$ (with probability one) as $n \to \infty$ under rather general conditions. Moreover, associated with the LLN are *large deviations* results that say that the probability of $O(1)$ deviation of the empirical average from the mean decays exponentially with n. These can be proved using the Chernoff bound (see Appendix B). In this chapter, when we invoke the LLN to replace an empirical average or sum by its statistical counterpart, we implicitly rely on such large deviations results as an underlying mathematical justification, although we do not provide the technical details behind such justification.

Map of this chapter In Section 6.1, we compute the capacity of the continuous and discrete-time AWGN channels using geometric arguments, and discuss the associated power–bandwidth tradeoffs. In Section 6.2, we take a more systematic view, discussing some basic quantities and results of Shannon theory, including the discrete memoryless channel model and the channel coding theorem. This provides a framework for the capacity computations in Section 6.3, where we discuss how to compute capacity under input constraints (specifically focusing on computing AWGN capacity with standard constellations such as PAM, QAM, and PSK). We also characterize the capacity for parallel Gaussian channels, and apply it for modeling dispersive channels. Finally, Section 6.4 provides a glimpse of optimization techniques for computing capacity in more general settings.

6.1 Capacity of AWGN channel: modeling and geometry

In this section, we discuss fundamental benchmarks for communication over a bandlimited AWGN channel.

Theorem 6.1.1 *For an AWGN channel of bandwidth W and received power P, the channel capacity is given by the formula*

$$C = W \log_2 \left(1 + \frac{P}{N_0 W}\right) \quad \text{bit/s.} \qquad (6.1)$$

Let us first discuss some implications of this formula, and then provide some insight into why the formula holds, and how one would go about achieving the rate promised by (6.1).

Consider a communication system that provides an information rate of R bit/s. Denoting by E_b the energy per information bit, the transmitted power is $P = E_b R$. For reliable transmission, we must have $R < C$, so that we have from (6.1):

$$R < W \log_2\left(1 + \frac{E_b R}{N_0 W}\right).$$

Defining $r = R/W$ as the *spectral efficiency*, or information rate per unit of bandwidth, of the system, we obtain the condition

$$r < \log_2\left(1 + \frac{E_b r}{N_0}\right).$$

This implies that, for reliable communication, the signal-to-noise ratio must exceed a threshold that depends on the operating spectral efficiency:

$$\frac{E_b}{N_0} > \frac{2^r - 1}{r}. \tag{6.2}$$

"Reliable communication" in an information-theoretic context means that the error probability tends to zero as codeword lengths get large, while a practical system is deemed reliable if it operates at some desired, nonzero but small, error probability level. Thus, we might say that a communication system is operating 3 dB away from Shannon capacity at a bit error probability of 10^{-6}, meaning that the operating E_b/N_0 for a BER of 10^{-6} is 3 dB higher than the minimum required based on (6.2).

Equation (6.2) brings out a fundamental tradeoff between power and bandwidth. The required E_b/N_0, and hence the required power (assuming that the information rate R and noise PSD N_0 are fixed) increase as we increase the spectral efficiency r, while the bandwidth required to support a given information rate decreases if we increase r. Taking the log of both sides of (6.2), we see that the spectral efficiency and the required E_b/N_0 in dB have an approximately linear relationship. This can be seen from Figure 6.1, which plots achievable spectral efficiency versus E_b/N_0 (dB). Reliable communication is not possible above the curve. In comparing a specific coded modulation scheme with the Shannon limit, we compare the E_b/N_0 required to attain a certain reference BER (e.g., 10^{-5}) with the minimum possible E_b/N_0, given by (6.2) at that spectral efficiency (excess bandwidth used in the modulating pulse is not considered, since that is a heavily implementation-dependent parameter). With this terminology, uncoded QPSK achieves a BER of 10^{-5} at an E_b/N_0 of about 9.5 dB. For the corresponding spectral efficiency $r = 2$, the Shannon limit given by (6.2) is 1.76 dB, so that uncoded QPSK is about 7.8 dB away from the Shannon limit at a BER of 10^{-5}. A similar gap also exists for uncoded 16-QAM. As we shall see in the next chapter, the gap to Shannon capacity can be narrowed considerably by the use of channel coding. For example, suppose that we use a rate 1/2 binary code (1 information bit/2 coded bits), with the coded bits mapped to a QPSK constellation (2 coded bits/channel use). Then the spectral efficiency

6.1 Capacity of AWGN channel: modeling and geometry

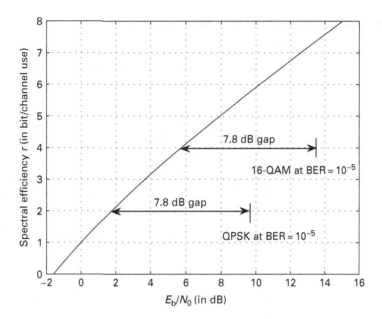

Figure 6.1 Spectral efficiency as a function of E_b/N_0 (dB). The large gap to capacity for uncoded constellations (at a reference BER of 10^{-5}) shows the significant potential benefits of channel coding, which I discuss in Chapter 7.

is $r = 1/2 \times 2 = 1$, and the corresponding Shannon limit is 0 dB. We now know how to design turbo-like codes that get within a fraction of a dB of this limit.

The preceding discussion focuses on spectral efficiency, which is important when there are bandwidth constraints. What if we have access to unlimited bandwidth (for a fixed information rate)? As discussed below, even in this scenario, we cannot transmit at arbitrarily low powers: there is a fundamental limit on the smallest possible value of E_b/N_0 required for reliable communication.

Power-limited communication As we let the spectral efficiency $r \to 0$, we enter a power-limited regime. Evaluating the limit (6.2) tells us that, for reliable communication, we must have

$$\frac{E_b}{N_0} > \ln 2 \quad (-1.6 \text{ dB}) \quad \textbf{Minimum required for reliable communication}.$$
(6.3)

That is, even if we let bandwidth tend to infinity for a fixed information rate, we cannot reduce E_b/N_0 below its minimum value of -1.6 dB. As we have seen in Chapters 3 and 4, M-ary orthogonal signaling is asymptotically optimum in this power-limited regime, both for coherent and noncoherent communication.

Let us now sketch an intuitive proof of the capacity formula (6.1). While the formula refers to a continuous-time channel, both the proof of the capacity formula, and the kinds of constructions we typically employ to try to achieve capacity, are based on discrete-time constructions.

6.1.1 From continuous to discrete time

Consider an ideal complex WGN channel bandlimited to $[-W/2, W/2]$. If the transmitted signal is $s(t)$, then the received signal

$$y(t) = (s*h)(t) + n(t),$$

where h is the impulse response of an ideal bandlimited channel, and $n(t)$ is complex WGN. We wish to design the set of possible signals that we would send over the channel so as to maximize the rate of reliable communication, subject to a constraint that the signal $s(t)$ has average power at most P.

To start with, note that it does not make sense for $s(t)$ to have any component outside of the band $[-W/2, W/2]$, since any such component would be annihilated once we pass it through the ideal bandlimited filter h. Hence, without loss of generality, $s(t)$ must be bandlimited to $[-W/2, W/2]$ for an optimal signal set design. We now recall the discussion on modulation degrees of freedom from Chapter 2 in order to obtain a discrete-time model.

By the sampling theorem, a signal bandlimited to $[-W/2, W/2]$ is completely specified by its samples at rate W, $\{s(i/W)\}$. Thus, signal design consists of specifying these samples, and modulation for transmission over the ideal bandlimited channel consists of invoking the interpolation formula. Thus, once we have designed the samples, the complex baseband waveform that we send is given by

$$s(t) = \sum_{i=\infty}^{\infty} s(i/W) p\left(t - \frac{i}{W}\right), \tag{6.4}$$

where $p(t) = \text{sinc}(Wt)$ is the impulse response of an ideal bandlimited pulse with transfer function $P(f) = \frac{1}{W} I_{[-\frac{W}{2}, \frac{W}{2}]}$. As noted in Chapter 2, this is linear modulation at symbol rate W with symbol sequence $\{s(i/W)\}$ and transmit pulse $p(t) = \text{sinc}(Wt)$, which is the minimum bandwidth Nyquist pulse at rate W. The translates $\{p(t - i/W)\}$ form an orthogonal basis for the space of ideally bandlimited functions, so that (6.4) specifies a basis expansion of $s(t)$.

For signaling under a power constraint P over a (large) interval T_o, the transmitted signal energy should satisfy

$$\int_0^{T_o} |s(t)|^2 dt \approx PT_o.$$

Let $P_s = \mathbb{E}[|s(1/W)|^2]$ denote the average power per sample. Since energy is preserved under the basis expansion (6.4), and we have about $T_o W$ samples in this interval, we also have

$$T_o W P_s ||p||^2 \approx PT_o.$$

For $p(t) = \text{sinc}(Wt)$, we have $||p||^2 = 1/W$, so that $P_s = P$. That is, for the scaling adopted in (6.4), the samples obey the same power constraint as the continuous-time signal.

When the bandlimited signal s passes through the ideally bandlimited complex AWGN channel, we get

$$y(t) = s(t) + n(t), \tag{6.5}$$

where n is complex WGN. Since s is linearly modulated at symbol rate W using modulating pulse p, we know that the optimal receiver front end is to pass the received signal through a filter matched to $p(t)$, and to sample at the symbol rate W. For notational convenience, we use a receive filter transfer function $G_R(f) = I_{[-\frac{W}{2},\frac{W}{2}]}$ which is a scalar multiple of the matched filter $P^*(f) = P(f) = \frac{1}{W}I_{[-\frac{W}{2},\frac{W}{2}]}$. This ideal bandlimited filter lets the signal $s(t)$ through unchanged, so that the signal contributions to the output of the receive filter, sampled at rate W, are $\{s(i/W)\}$. The noise at the output of the receive filter is bandlimited complex WGN with PSD $N_0 I_{[-\frac{W}{2},\frac{W}{2}]}$, from which it follows that the noise samples at rate W are independent complex Gaussian random variables with covariance $N_0 W$. To summarize, the noisy samples at the receive filter output can be written as

$$y[i] = s(i/W) + N[i], \tag{6.6}$$

where the signal samples are subject to an average power constraint $\mathbb{E}[|s(i/W)|^2] \leq P$, and $\{N[i]\}$ are i.i.d., zero mean, proper complex Gaussian noise samples with $\mathbb{E}[|N[i]|^2] = N_0 W$.

Thus, we have reduced the continuous-time bandlimited passband AWGN channel model to the discrete-time complex WGN channel model (6.6) that we get to use W times per second if we employ bandwidth W. We can now characterize the capacity of the discrete-time channel, and then infer that of the continuous-time bandlimited channel.

6.1.2 Capacity of the discrete-time AWGN channel

Since the real and imaginary part of the discrete-time complex AWGN model (6.6) can be interpreted as two uses of a real-valued AWGN channel, we consider the latter first.

Consider a discrete-time real AWGN channel in which the output at any given time

$$Y = X + Z, \tag{6.7}$$

where X is a real-valued input satisfying $\mathbb{E}[X^2] \leq S$, and $Z \sim N(0, N)$ is real-valued AWGN. The noise samples over different channel uses are i.i.d. This is an example of a discrete memoryless channel, where $p(Y|X)$ is specified for a single channel use, and the channel outputs for multiple channel uses are conditionally independent given the inputs. A signal, or codeword, over such a channel is a vector $\mathbf{X} = (X_1, \ldots, X_n)^T$, where X_i is the input for the ith channel use. A code of rate R bits per channel use can be constructed by designing a set of 2^{nR} such signals $\{\mathbf{X}^k, k = 1, \ldots, 2^{nR}\}$, with each signal having an equal probability of being chosen for transmission over the channel. Thus, nR bits are conveyed over n channel uses. Capacity is defined as the largest rate R for which the error probability tends to zero as $n \to \infty$.

Shannon has provided a general framework for computing capacity for a discrete memoryless channel, which we discuss in Section 6.3. However, we provide here a heuristic derivation of capacity for the AWGN channel (6.7), that specifically utilizes the geometry induced by AWGN.

Sphere packing based derivation of capacity formula For a transmitted signal $\mathbf{X}^j$, the n-dimensional output vector $\mathbf{Y} = (Y_1, \ldots, Y_n)^T$ is given by

$$\mathbf{Y} = \mathbf{X}^j + \mathbf{Z}, \quad \mathbf{X}^j \text{ sent},$$

where $\mathbf{Z}$ is a vector of i.i.d. $N(0, N)$ noise samples. For equal priors, the MPE and ML rules are equivalent. The ML rule for the AWGN channel is the minimum distance rule

$$\delta_{\text{ML}}(\mathbf{Y}) = \arg\min_{1 \leq k \leq 2^{nR}} ||\mathbf{Y} - \mathbf{X}^k||^2.$$

Now, the noise vector $\mathbf{Z}$ that perturbs the transmitted signal has energy

$$||\mathbf{Z}||^2 = \sum_{i=1}^{N} Z_i^2 \approx n\mathbb{E}[Z_1^2] = nN,$$

where we have invoked the LLN. This implies that, if we draw a sphere of radius $\sqrt{nN}$ around a signal $\mathbf{X}^j$, then, with high probability, the received vector $\mathbf{Y}$ lies within the sphere when $\mathbf{X}^j$ is sent. Calling such a sphere the "decoding sphere" for $\mathbf{X}^j$, the minimum distance rule would lead to very small error probability if the decoding spheres for different signals were disjoint. We now wish to estimate how many such decoding spheres we can come up with; this gives the value of 2^{nR} for which reliable communication is possible.

Since X is independent of Z (the transmitter does not know the noise realization) in the model (6.7), the input power constraint implies an output power constraint

$$\mathbb{E}[Y^2] = \mathbb{E}[(X+Z)^2] = \mathbb{E}[X^2] + \mathbb{E}[Z^2] + 2\mathbb{E}[X]\mathbb{E}[Z] = \mathbb{E}[X^2] + \mathbb{E}[Z^2] \leq S + N. \tag{6.8}$$

Invoking the law of large numbers again, the received signal energy satisfies

$$\mathbb{E}[||\mathbf{Y}||^2] \approx n(S+N),$$

so that, with high probability, the received signal vector lies within an n-dimensional sphere with radius $R_n = \sqrt{n(S+N)}$. The problem of signal design for reliable communication now boils down to packing disjoint decoding spheres of radius $r_n = \sqrt{nN}$ within a sphere of radius R_n, as shown in Figure 6.2. The volume of an n-dimensional sphere of radius r equals $K_n r^n$,

6.1 Capacity of AWGN channel: modeling and geometry

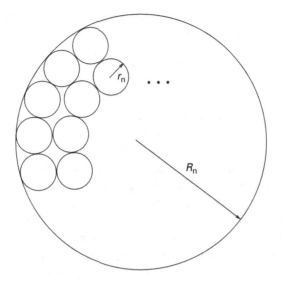

Figure 6.2 Decoding spheres of radius $r_n = \sqrt{nN}$ are packed inside a sphere of radius $R_n = \sqrt{n(S+N)}$.

and the number of decoding spheres we can pack is roughly the following ratio of volumes:

$$\frac{K_n R_n^n}{K_n r_n^n} = \frac{K_n(\sqrt{n(S+N)})^n}{K_n(\sqrt{nN})^n} \approx 2^{nR}.$$

Solving, we obtain that the rate $R = 1/2 \log_2(1 + S/N)$. We show in Section 6.3 that this rate exactly equals the capacity of the discrete-time real AWGN channel. (It is also possible to make the sphere packing argument rigorous, but we do not attempt that here.) We now state the capacity formula formally.

Theorem 6.1.2 (Capacity of discrete-time real AWGN channel) *The capacity of the discrete-time real AWGN channel (6.7) is*

$$C_{\text{AWGN}} = \frac{1}{2} \log_2(1 + \text{SNR}) \text{ bit/channel use}, \qquad (6.9)$$

where $\text{SNR} = S/N$ *is the signal-to-noise ratio.*

Thus, capacity grows approximately logarithmically with SNR, or approximately linearly with SNR in dB.

6.1.3 From discrete to continuous time

For the continuous-time bandlimited complex baseband channel that we considered earlier, we have $2W$ uses per second of the discrete-time real AWGN channel (6.7). With the normalization we employed in (6.4), we have that, per real-valued sample, the average signal energy $S = P/2$ and the noise energy

$N = N_0 W/2$, where P is the power constraint on the continuous-time signal. Plugging in, we get

$$C_{\text{cont-time}} = 2W \, \frac{1}{2} \log_2(1 + \frac{P}{N_0 W}) \text{ bit/s},$$

which gives (6.1).

As the invocation of the LLN in the sphere packing based derivation shows, capacity for the discrete-time channel is achieved by using codewords that span a large number of symbols. Suppose, now, that we have designed a capacity-achieving strategy for the discrete-time channel; that is, we have specified a good code, or signal set. A codeword from this set is a discrete-time sequence $\{s[i]\}$. We can now translate this design to continuous time by using the modulation formula (6.4) to send the symbols $\{s[i] = s(i/W)\}$. Of course, as we discussed in Section 2, the sinc pulse used in this formula cannot be used in practice, and should be replaced by a modulating pulse whose bandwidth is larger than the symbol rate employed. A good choice would be a square root Nyquist modulating pulse at the transmitter, and its matched filter at the receiver, which again yields the ISI-free discrete-time model (6.6) with uncorrelated noise samples.

In summary, good codes for the discrete-time AWGN channel (6.6) can be translated into good signal designs for the continuous-time bandlimited AWGN channel using practical linear modulation techniques; this corresponds to using translates of a square root Nyquist pulse as an orthonormal basis for the signal space. It is also possible to use an entirely different basis: for example, orthogonal frequency division multiplexing, which we discuss in Chapter 8, employs complex sinusoids as basis functions. In general, the use of appropriate signal space arguments allows us to restrict attention to discrete-time models, both for code design and for deriving information-theoretic benchmarks.

Real baseband channel The preceding observations also hold for a physical (i.e., real-valued) baseband channel. That is, both the AWGN capacity formula (6.1) and its corollary (6.2) hold, where W for a physical baseband channel refers to the bandwidth occupancy for positive frequencies. Thus, a real baseband signal $s(t)$ occupying a bandwidth W actually spans the interval $[-W, W]$, with the constraint that $S(f) = S^*(-f)$. Using the sampling theorem, such a signal can be represented by $2W$ real-valued samples per second. This is the same result as for a passband signal of bandwidth W, so that the arguments made so far, relating the continuous-time model to the discrete-time real AWGN channel, apply as before. For example, suppose that we wish to find out how far uncoded binary antipodal signaling at BER of 10^{-5} is from Shannon capacity. Since we transmit at 1 bit per sample, the information rate is $2W$ bits per second, corresponding to a spectral efficiency of $r = R/W = 2$. This corresponds to a Shannon limit of 1.8 dB E_b/N_0, using (6.2). Setting the BER of $Q\left(\sqrt{(2E_b/N_0)}\right)$ for binary antipodal signaling to

10^{-5}, we find that the required E_b/N_0 is 9.5 dB, which is 7.7 dB away from the Shannon limit. There is good reason for this computation looking familiar: we obtained exactly the same result earlier for uncoded QPSK on a passband channel. This is because QPSK can be interpreted as binary antipodal modulation along the I and Q channels, and is therefore exactly equivalent to binary antipodal modulation for a real baseband channel.

At this point, it is worth mentioning the potential for confusion when dealing with Shannon limits in the literature. Even though PSK is a passband technique, the term BPSK is often used when referring to binary antipodal signaling on a real baseband channel. Thus, when we compare the performance of BPSK with rate 1/2 coding to the Shannon limit, we should actually be keeping in mind a real baseband channel, so that $r = 1$, corresponding to a Shannon limit of 0 dB E_b/N_0. (On the other hand, if we had literally interpreted BPSK as using only the I channel in a passband system, we would have gotten $r = 1/2$.) That is, whenever we consider real-valued alphabets, we restrict ourselves to the real baseband channel for the purpose of computing spectral efficiency and comparing Shannon limits. For a passband channel, we can use the same real-valued alphabet over the I and Q channels (corresponding to a rectangular complex-valued alphabet) to get exactly the same dependence of spectral efficiency on E_b/N_0.

6.1.4 Summarizing the discrete-time AWGN model

In previous chapters, we have used constellations over the AWGN channel with a finite number of signal points. One of the goals of this chapter is to be able to compute Shannon theoretic limits for performance when we constrain ourselves to using such constellations. In Chapters 3 to 5, when sampling signals corrupted by AWGN, we model the discrete-time AWGN samples as having variance $\sigma^2 = N_0/2$ per dimension. On the other hand, the noise variance in the discrete-time model in Section 6.1.3 depends on the system bandwidth W. We would now like to reconcile these two models, and use a notation that is consistent with that in the prior chapters.

Real discrete-time AWGN channel Consider the following model for a real-valued discrete-time channel:

$$Y = X + Z, \quad Z \sim N(0, \sigma^2) \qquad (6.10)$$

where X is a power-constrained input, $\mathbb{E}[X^2] \leq E_s$, as well as possibly constrained to take values in a given alphabet (e.g., BPSK or 4-PAM). This notation is consistent with that in Chapter 3, where we use E_s to denote the average energy per symbol. Suppose that we compute the capacity of this discrete-time model as C_d bits per channel use, where C_d is a function of $\text{SNR} = E_s/\sigma^2$. If E_b is the energy per information bit, we must have $E_s = E_b C_d$ joules per channel use. Now, if this discrete-time channel arose from a real baseband channel of bandwidth W, we would have $2W$ channel uses per second, so that the capacity of the continuous-time channel is $C_c = 2WC_d$ bits per second. This means that the spectral efficiency is given by

$$r = \frac{C_c}{W} = 2C_d \quad \text{Real discrete-time channel}. \tag{6.11}$$

Thus, the SNR for this system is given by

$$\text{SNR} = \frac{E_s}{\sigma^2} = 2C_d \frac{E_b}{N_0} = r\frac{E_b}{N_0} \quad \text{Real discrete-time channel}. \tag{6.12}$$

Thus, we can restrict attention to the real discrete-time model (6.10), which is consistent with our notation in prior chapters. To apply the results to a bandlimited system as in Sections 6.1.1 and 6.1.3, all we need is the relationship (6.11) which specifies the spectral efficiency (bits per Hz) in terms of the capacity of the discrete-time channel (bits per channel use).

Complex discrete-time AWGN model The real-valued model (6.10) can be used to calculate the capacity for rectangular complex-valued constellations such as rectangular 16-QAM, which can be viewed as a product of two real-valued 4-PAM constellations. However, for constellations such as 8-PSK, it is necessary to work directly with a two-dimensional observation. We can think of this as a complex-valued symbol, plus proper complex AWGN (discussed in Chapter 4). The discrete-time model we employ for this purpose is

$$Y = X + Z, \quad Z \sim CN(0, 2\sigma^2), \tag{6.13}$$

where $\mathbb{E}[|X|^2] \leq E_s$ as before. However, we can also express this model in terms of a two-dimensional real-valued observation (in which case, we do not need to invoke the concepts of proper complex Gaussianity covered in Chapter 4):

$$Y_c = X_c + Z_c, \quad Y_s = X_s + Z_s, \tag{6.14}$$

with Z_c, Z_s i.i.d. $N(0, \sigma^2)$, and $\mathbb{E}[X_c^2 + X_s^2] \leq E_s$. The capacity C_d bits per channel use for this system is a function of the SNR, which is given by $E_s/2\sigma^2$, as well as any other constraints (e.g., that X is drawn from an 8-PSK constellation). If this discrete-time channel arises from a passband channel of bandwidth W, we have W channel uses per second for the corresponding complex baseband channel, so that the capacity of the continuous-time channel is $C_c = WC_d$ bits per second, so that the spectral efficiency is given by

$$r = \frac{C_c}{W} = C_d \quad \text{Complex discrete-time channel}. \tag{6.15}$$

The SNR is given by

$$\text{SNR} = \frac{E_s}{2\sigma^2} = C_d \frac{E_b}{N_0} = r\frac{E_b}{N_0} \quad \text{Complex discrete-time channel}. \tag{6.16}$$

Comparing (6.12) with (6.16), we note that the relation of SNR with E_b/N_0 and spectral efficiency is the same for both systems. The relations (6.11) and

(6.15) are also consistent: if we get a given capacity for a real-valued model, we should be able to double that in a consistent complex-valued model by using the real-valued model twice.

6.2 Shannon theory basics

From the preceding sphere packing arguments, we take away the intuition that we need to design codewords so as to achieve a good packing of decoding spheres in n dimensions. A direct approach to trying to realize this intuition is not easy (although much progress has been made in recent years in the encoding and decoding of lattice codes that attempt to implement the sphere packing prescription directly). We are interested in determining whether standard constellations (e.g., PSK, QAM), in conjunction with appropriately chosen error-correcting codes, can achieve the same objectives. In this section, we discuss just enough of the basics of Shannon theory to enable us to develop elementary capacity computation techniques. We introduce the general discrete memoryless channel model, for which the model (6.7) is a special case. Key information-theoretic quantities such as entropy, mutual information, and divergence are discussed. We end this section with a statement and partial proof of the channel coding theorem.

While developing this framework, we emphasize the role played by the LLN as the fundamental basis for establishing information-theoretic benchmarks: roughly speaking, the randomness that is inherent in one channel use is averaged out by employing signal designs spanning multiple independent channel uses, thus leading to reliable communication. We have already seen this approach at work in the sphere packing arguments in Section 6.1.2.

Definition 6.2.1 (Discrete memoryless channel) *A discrete memoryless channel is specified by a transition density or probability mass function $p(y|x)$ specifying the conditional distribution of the output y given the input x. For multiple channel uses, the outputs are conditionally independent given the inputs. That is, if $x_1, \ldots, x_n$ are the inputs, and $y_1, \ldots, y_n$ denote the corresponding outputs, for n channel uses, then*

$$p(y_1, \ldots, y_n | x_1, \ldots, x_n) = p(y_1|x_1) \cdots p(y_n|x_n).$$

Real AWGN channel For the real Gaussian channel (6.10), the channel transition density is given by

$$p(y|x) = \frac{e^{-\frac{(y-x)^2}{2\sigma^2}}}{\sqrt{2\pi\sigma^2}}, \quad y \text{ real}. \tag{6.17}$$

Here both the input and the output are real numbers, but we typically constrain the input to average symbol energy E_s. In addition, we can constrain the input

x to be drawn from a finite constellation: for example, for BPSK, the input would take values $x = \pm\sqrt{E_s}$.

Complex AWGN channel For the complex Gaussian channel (6.13) or (6.14), the channel transition density is given by

$$p(y|x) = \frac{e^{-\frac{|y-x|^2}{2\sigma^2}}}{2\pi\sigma^2} = \frac{e^{-\frac{(y_c-x_c)^2}{2\sigma^2}}}{\sqrt{2\pi\sigma^2}} \frac{e^{-\frac{(y_s-x_s)^2}{2\sigma^2}}}{\sqrt{2\pi\sigma^2}}, \tag{6.18}$$

where the output $y = y_c + jy_s$ and input $x = x_c + jx_s$ can be viewed as complex numbers or two-dimensional real vectors. We typically constrain the input to average symbol energy E_s, and may also constrain it to be drawn from a finite constellation: for example, for M-ary PSK, the input $x \in \{\sqrt{E_s}e^{j2\pi i/M}, \ i = 0, 1, \ldots, M-1\}$. Equation (6.18) makes it transparent that the complex AWGN model is equivalent to two uses of the real model (6.17), where the I component x_c and the Q component x_s of the input may be correlated due to constraints on the input alphabet.

Binary symmetric channel (BSC) In this case, x and y both take values in $\{0, 1\}$, and the transition "density" is now a probability mass function:

$$p(y|x) = \begin{cases} 1-p, & y = x, \\ p, & y = 1-x. \end{cases} \tag{6.19}$$

That is, the BSC is specified by a "crossover" probability p, as shown in Figure 6.3.

Consider BPSK transmission over an AWGN channel. When we make symbol-by-symbol ML decisions, we create a BSC with crossover probability $p = Q(\sqrt{2E_b/N_0})$. Of course, we know that such symbol-by-symbol *hard decisions* are not optimal; for example, ML decoding using the Viterbi algorithm for a convolutional code involves real-valued observations, or *soft decisions*. In Problem 6.10, we quantify the fundamental penalty for hard decisions by comparing the capacity of the BSC induced by hard decisions to the maximum achievable rate on the AWGN channel with BPSK input.

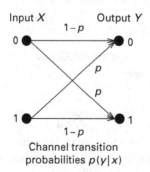

Figure 6.3 Binary symmetric channel with crossover probability p.

6.2.1 Entropy, mutual information, and divergence

We now provide a brief discussion of relevant information-theoretic quantities and discuss their role in the law of large numbers arguments invoked in information theory.

Definition 6.2.2 (Entropy) *For a discrete random variable (or vector) X with probability mass function p(x), the entropy H(X) is defined as*

$$H(X) = -\mathbb{E}[\log_2 p(X)] = -\sum_i p(x_i) \log_2 p(x_i) \quad \textbf{Entropy}, \quad (6.20)$$

where $\{x_i\}$ is the set of values taken by X.

Entropy is a measure of the information gained from knowing the value of the random variable X. The more uncertain we are regarding the random variable from just knowing its distribution, the more information we gain when its value is revealed, and the larger its entropy. The information is measured in bits, corresponding to the base 2 used in the logarithms in (6.20).

> **Example 6.2.1 (Binary entropy)** We set aside the special notation $H_B(p)$ for the entropy of a Bernoulli random variable X with $P[X=1] = p = 1 - P[X=0]$. From (6.20), we can compute this entropy as
>
> $$H_B(p) = -p \log_2 p - (1-p) \log_2 (1-p) \quad \textbf{Binary entropy function}. \quad (6.21)$$
>
> Note that $H_B(p) = H_B(1-p)$: as expected, the information content of X does not change if we switch the labels 0 and 1. The binary entropy function is plotted in Figure 6.4. The end points $p=0$ and $p=1$ correspond to certainty regarding the value of the random variable, so that no information is gained by revealing its value. On the other hand, $H_B(p)$ attains its maximum value of 1 bit at $p = 1/2$, which corresponds to maximal uncertainty regarding the value of the random variable (which maximizes the information gained by revealing its value).

Law of large numbers interpretation of entropy Let $X_1, \ldots, X_n$ be i.i.d. random variables, each with pmf $p(x)$. Then their joint pmf satisfies

$$\frac{1}{n} \log_2 p(X_1, \ldots, X_n) = \frac{1}{n} \sum_{i=1}^{n} \log_2 p(X_i) \to \mathbb{E}[\log_2 p(X_1)] = -H(X), \quad n \to \infty. \quad (6.22)$$

We can therefore infer that, with high probability, we see the "typical" behavior

$$p(X_1, \ldots, X_n) \approx 2^{-nH(X)} \quad typical \text{ behavior.} \quad (6.23)$$

Figure 6.4 The binary entropy function.

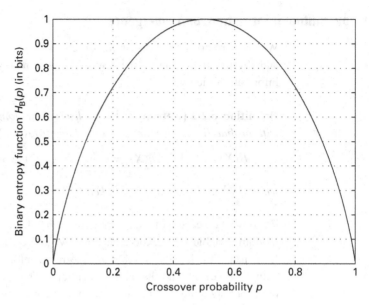

A sequence that satisfies this behavior is called a *typical* sequence. The set of such sequences is called the *typical set*. The LLN implies that

$$P[(X_1, \ldots, X_n) \text{ is typical}] \to 1, \quad n \to \infty. \tag{6.24}$$

That is, any sequence of length n that is not typical is extremely unlikely to occur. Using (6.23) and (6.24), we infer that there must be approximately $2^{nH(X)}$ sequences in the typical set. We have thus inferred a very important principle, called the *asymptotic equipartition property (AEP)*, stated informally as follows.

Asymptotic equipartition property (Discrete random variables) For a length n sequence of i.i.d. discrete random variables $X_1, \ldots, X_n$, where n is large, the typical set consists of about $2^{nH(X)}$ sequences, each occurring with probability approximately $2^{-nH(X)}$. Sequences outside the typical set occur with negligible probability for large n.

Since $nH(X)$ bits are required to specify the $2^{nH(X)}$ typical sequences, the AEP tells us that describing n i.i.d. copies of the random variable X requires about $nH(X)$ bits, so that the average number of bits per copy of the random variable is $H(X)$. This gives a concrete interpretation for what we mean by entropy measuring information content. The implications for data compression (not considered in detail here) are immediate: by arranging i.i.d. copies of the source in long blocks, we can describe it at rates approaching $H(X)$ per source symbol, by only assigning bits to represent the typical sequences.

We have defined entropy for discrete random variables. We also need an analogous notion for continuous random variables, termed *differential entropy*, defined as follows.

6.2 Shannon theory basics

Definition 6.2.3 (Differential entropy) *For a continuous random variable (or vector) X with probability density function p(x), the differential entropy h(X) is defined as*

$$h(X) = -\mathbb{E}[\log_2 p(X)] = -\int p(x) \log_2 p(x)\, dx \quad \textbf{Differential entropy.}$$

Example 6.2.2 (Differential entropy for a Gaussian random variable)
For $X \sim N(m, v^2)$,

$$-\log_2 p(x) = \frac{(x-m)^2}{2v^2}(\log_2 e) + \frac{1}{2}\log_2(2\pi v^2).$$

Thus, we obtain

$$h(X) = -\mathbb{E}[\log_2 p(X)] = \mathbb{E}\left[\frac{(X-m)^2}{2v^2}(\log_2 e) + \frac{1}{2}\log_2(2\pi v^2)\right]$$

$$= \frac{1}{2}\log_2 e + \frac{1}{2}\log_2(2\pi v^2).$$

We summarize as follows:

$$h(X) = \frac{1}{2}\log_2(2\pi e v^2) \quad \textbf{Differential entropy for Gaussian } N(m, v^2) \textbf{ random variable.} \quad (6.25)$$

Note that the differential entropy does not depend on the mean, since that is a deterministic parameter that can be subtracted out from X without any loss of information.

Cautionary note There are key differences between entropy and differential entropy. While entropy must be nonnegative, this is not true of differential entropy (e.g., set $v^2 < 1/2\pi e$ in Example 6.2.2). While entropy is scale-invariant, differential entropy is not, even though scaling a random variable by a known constant should not change its information content. These differences can be traced to the differences between probability mass functions and probability density functions. Scaling changes the location of the mass points for a discrete random variable, but does not change their probabilities. On the other hand, scaling changes both the location and size of the infinitesimal intervals used to define a probability density function for a continuous random variable. However, such differences between entropy and differential entropy are irrelevant for our main purpose of computing channel capacities, which, as we shall see, requires computing *differences* between unconditional and conditional entropies or differential entropies. The effect of scale factors "cancels out" when we compute such differences.

Law of large numbers interpretation of differential entropy Let $X_1, \ldots, X_n$ be i.i.d. random variables, each with density $f(x)$, then their joint density satisfies

$$\frac{1}{n} \log_2 f(X_1, \ldots, X_n) = \frac{1}{n} \sum_{i=1}^n \log_2 f(X_i) \to \mathbb{E}[\log_2 f(X_1)] = -h(X), \quad n \to \infty. \tag{6.26}$$

We now define "typical" behavior in terms of the value of the joint density

$$f(X_1, \ldots, X_n) \approx 2^{-nh(X)} \quad \text{typical behavior}, \tag{6.27}$$

and invoke the LLN to infer that (6.24) holds. Since the "typical" value of the joint density is a constant, $2^{-nh(X)}$, we infer that the typical set must have volume approximately $2^{nh(X)}$, in order for the joint density to integrate to one. This leads to the AEP for continuous random variables stated below.

Asymptotic equipartition property (Continuous random variables) For a length n sequence of i.i.d. continuous random variables $X_1, \ldots, X_n$, where n is large, the joint density takes value approximately $2^{-nh(X)}$ over a typical set of volume $2^{nh(X)}$. The probability mass outside the typical set is negligible for large n.

Joint entropy and mutual information The entropy $H(X, Y)$ of a pair of random variables (X, Y) (e.g., the input and output of a channel) is called the joint entropy of X and Y, and is given by

$$H(X, Y) = -\mathbb{E}[\log_2 p(X, Y)], \tag{6.28}$$

where $p(x, y) = p(x)p(y|x)$ is the joint pmf. The mutual information between X and Y is defined as

$$I(X; Y) = H(X) + H(Y) - H(X, Y). \tag{6.29}$$

Conditional entropy The conditional entropy $H(Y|X)$ is defined as

$$H(Y|X) = -\mathbb{E}[\log_2 p(Y|X)] = -\sum_x \sum_y p(x, y) \log_2 p(y|x). \tag{6.30}$$

Since $p(y|x) = p(x, y)/p(x)$, we have

$$\log_2 p(Y|X) = \log_2 p(X, Y) - \log_2 p(X).$$

Taking expectations and changing sign, we get

$$H(Y|X) = H(X, Y) - H(X).$$

Substituting into (6.29), we get an alternative formula for the mutual information (6.29): $I(X; Y) = H(Y) - H(Y|X)$. By symmetry, we also have

6.2 Shannon theory basics

$I(X; Y) = H(X) - H(X|Y)$. For convenience, we state all of these formulas for mutual information together:

$$\begin{aligned} I(X; Y) &= H(Y) - H(Y|X) \\ &= H(X) - H(X|Y) \\ &= H(X) + H(Y) - H(X, Y). \end{aligned} \quad (6.31)$$

It is also useful to define the entropy of Y conditioned on a particular value of $X = x$, as follows:

$$H(Y|X = x) = -\mathbb{E}[\log_2 p(Y|X)|X = x] = -\sum_y p(y|x) \log_2 p(y|x),$$

and note that

$$H(Y|X) = \sum_x p(x) H(Y|X = x). \quad (6.32)$$

The preceding definitions and formulas hold for continuous random variables as well, with entropy replaced by differential entropy.

One final concept that is closely related to entropies is information-theoretic *divergence*, also termed the *Kullback–Leibler (KL) distance*.

Divergence The divergence $D(P||Q)$ between two distributions P and Q (with corresponding densities $p(x)$ and $q(x)$) is defined as

$$D(P||Q) = \mathbb{E}_P\left[\log\left(\frac{p(X)}{q(X)}\right)\right] = \sum_x p(x) \log\left(\frac{p(x)}{q(x)}\right),$$

where $\mathbb{E}_P$ denotes expectation computed using the distribution P (i.e., X is a random variable with distribution P).

Divergence is nonnegative The divergence $D(P||Q) \geq 0$, with equality if and only if $P \equiv Q$.
The proof is as follows:

$$\begin{aligned} -D(P||Q) &= \mathbb{E}_P\left[\log\left(\frac{q(X)}{p(X)}\right)\right] = \sum_{x:p(x)>0} p(x) \log\left(\frac{q(x)}{p(x)}\right) \\ &\leq \sum_{x:p(x)>0} p(x) \left(\frac{q(x)}{p(x)} - 1\right) = \left(\sum_{x:p(x)>0} q(x)\right) - 1 \leq 0, \end{aligned}$$

where the first inequality is because $\log x \leq x - 1$. Since equality in the latter inequality occurs if and only if $x = 1$, the first inequality is an equality if and only if $q(x)/p(x) = 1$ wherever $p(x) > 0$. The second inequality follows from the fact that q is a pmf, and is an equality if and only if $q(x) = 0$ wherever $p(x) = 0$. Thus, we find that $D(P||Q) = 0$ if and only if $p(x) = q(x)$ for all x (for continuous random variables, the equalities would only need to hold "almost everywhere").

Mutual information as a divergence The mutual information between two random variables can be expressed as a divergence between their joint distribution, and a distribution corresponding to independent realizations of these random variables, as follows:

$$I(X;Y) = D(P_{X,Y} \| P_X P_Y). \tag{6.33}$$

This follows by noting that

$$\begin{aligned}
I(X;Y) &= H(X) + H(Y) - H(X,Y) \\
&= -\mathbb{E}[\log p(X)] - \mathbb{E}[\log p(Y)] + \mathbb{E}[\log p(X,Y)] \\
&= \mathbb{E}\left[\log\left(\frac{p(X,Y)}{p(X)p(Y)}\right)\right],
\end{aligned}$$

where the expectation is computed using the joint distribution of X and Y.

6.2.2 The channel coding theorem

We first introduce joint typicality, which is the central component of a random coding argument for characterizing the maximum achievable rate on a DMC.

Joint typicality Let X and Y have joint density $p(x,y)$. Then the law of large numbers can be applied to n channel uses with i.i.d. inputs $X_1, \ldots, X_n$, leading to outputs $Y_1, \ldots, Y_n$, respectively. Note that the pairs (X_i, Y_i) are i.i.d., as are the outputs $\{Y_i\}$. We thus get three LLN-based results:

$$\frac{1}{n} \log_2 p(X_1, \ldots, X_n) \to -H(X)$$

$$\frac{1}{n} \log_2 p(Y_1, \ldots, Y_n) \to -H(Y) \tag{6.34}$$

$$\frac{1}{n} \log_2 p(X_1, Y_1, \ldots, X_n, Y_n) \to -H(X,Y).$$

For an input sequence $\mathbf{x} = (x_1, \ldots, x_n)^T$ and an output sequence $\mathbf{y} = (y_1, \ldots, y_n)^T$, the pair $(\mathbf{x}, \mathbf{y})$ is said to be *jointly typical* if its empirical characteristics conform to the statistical averages in (6.34); that is, if

$$\begin{aligned}
p(\mathbf{x}) &\approx 2^{-nH(X)} \\
p(\mathbf{y}) &\approx 2^{-nH(Y)} \\
p(\mathbf{x}, \mathbf{y}) &\approx 2^{-nH(X,Y)}.
\end{aligned} \tag{6.35}$$

We also infer that there are about $2^{nH(X,Y)}$ jointly typical sequences, since

$$\sum_{\mathbf{x},\mathbf{y}\text{ jointly typical}} p(\mathbf{x}, \mathbf{y}) \approx 1.$$

In the following, we apply the concept of joint typicality to a situation in which X is the input to a DMC, and Y its output. In this case, $p(x,y) = p(x)p(y|x)$, where $p(x)$ is the marginal pmf of X, and $p(y|x)$ is the channel transition pmf.

Random coding For communicating at rate R bit/channel use over a DMC $p(y|x)$, we use 2^{nR} codewords, where a codeword of the form $\mathbf{X} = (X_1, \ldots, X_n)^T$ is sent using n channel uses (input X_i sent for ith channel use). The elements $\{X_i\}$ are chosen to be i.i.d., drawn from a pmf $p(x)$. Thus, all elements in all codewords are i.i.d., hence the term random coding (of course, the encoder and decoder both know the set of codewords once the random codebook choice has been made). All codewords are equally likely to be sent.

Joint typicality decoder While ML decoding is optimal for equiprobable transmission, it suffices to consider the following *joint typicality* decoder for our purpose. This decoder checks whether the received vector $\mathbf{Y} = (Y_1, \ldots, Y_n)$ is jointly typical with any codeword $\hat{\mathbf{X}} = (\hat{X}_1, \ldots, \hat{X}_n)^T$. If so, and if there is exactly one such codeword, then the decoder outputs $\hat{\mathbf{X}}$. If not, it declares decoding failure. Decoding error occurs if $\hat{\mathbf{X}} \neq \mathbf{X}$, where $\mathbf{X}$ is the transmitted codeword. Let us now estimate the probability of decoding error or failure.

If $\mathbf{X}$ is the transmitted codeword, and $\hat{\mathbf{X}}$ is any other codeword, then $\hat{\mathbf{X}}$ and the output $\mathbf{Y}$ are independent by our random coding construction, so that $p(\hat{\mathbf{X}}, \mathbf{Y}) = p(\hat{\mathbf{X}})p(\mathbf{Y}) \approx 2^{-n(H(X)+H(Y))}$ if $\hat{\mathbf{X}}$ and $\mathbf{Y}$ are typical. Now, the probability that they are jointly typical is

$$P[\hat{\mathbf{X}}, \mathbf{Y} \text{ jointly typical}] = \sum_{\mathbf{x,y} \text{ jointly typical}} p(\hat{\mathbf{X}})p(\mathbf{Y})$$

$$\approx 2^{nH(X,Y)} 2^{-n(H(X)+H(Y))} = 2^{-nI(X;Y)}.$$

Since there are $2^{nR} - 1$ possible incorrect codewords, the probability of at least one of them being jointly typical with the received vector can be estimated using the union bound

$$(2^{nR} - 1)2^{-nI(X;Y)} \leq 2^{-n(I(X;Y)-R)}, \tag{6.36}$$

which tends to zero as $n \to \infty$, as long as $R < I(X; Y)$.

There are some other possible events that lead to decoding error that we also need to estimate (but that we omit here). However, the estimate (6.36) is the crux of the random coding argument for the "forward" part of the noisy channel coding theorem, which we now state below.

Theorem 6.2.1 (Channel coding theorem: achievability)

(a) For a DMC with channel transition pmf $p(y|x)$, we can use i.i.d. inputs with pmf $p(x)$ to communicate reliably, as long as the code rate satisfies

$$R < I(X; Y).$$

(b) The preceding achievable rate can be maximized over the input density $p(x)$ to obtain the channel capacity

$$C = \max_{p(x)} I(X; Y).$$

We omit detailed discussion and proof of the "converse" part of the channel coding theorem, which states that it is not possible to do better than the achievable rates promised by the preceding theorem.

Note that, while we considered discrete random variables for concreteness, the preceding discussion goes through unchanged for continuous random variables (as well as for mixed settings, such as when X is discrete and Y is continuous), by appropriately replacing entropy by differential entropy.

6.3 Some capacity computations

We are now ready to undertake some example capacity computations. In Section 6.3.1, we discuss capacity computations for guiding the choice of signal constellations and code rates on the AWGN channel. Specifically, for a given constellation, we wish to establish a benchmark on the best rate that it can achieve on the AWGN channel as a function of SNR. Such a result is nonconstructive, saying only that there is *some* error-correcting code which, when used with the constellation, achieves the promised rate (and that no code can achieve reliable communication at a higher rate). However, as mentioned earlier, it is usually possible with a moderate degree of ingenuity to obtain a turbo-like coded modulation scheme that approaches these benchmarks quite closely. Thus, the information-theoretic benchmarks provide valuable guidance on choice of constellation and code rate. We then discuss the parallel Gaussian channel model, and its application to modeling dispersive channels, in Section 6.3.2. The optimal "waterfilling" power allocation for this model is an important technique that appears in many different settings.

6.3.1 Capacity for standard constellations

We now compute mutual information for some examples. We term the maximum mutual information attained under specific input constraints as the channel capacity under those constraints. For example, we compute the capacity of the AWGN channel with BPSK signaling and a power constraint. This is, of course, smaller than the capacity of power-constrained AWGN signaling when there are no constraints on the input alphabet, which is what we typically refer to as *the* capacity of the AWGN channel.

Binary symmetric channel capacity Consider the BSC with crossover probability p as in Figure 6.3. Given the symmetry of the channel, it is plausible that the optimal input distribution is to send 0 and 1 with equal probability (see Section 6.4 for techniques for validating such guesses, as well as for computing optimal input distributions when the answer is not "obvious"). We now calculate $C = I(X;Y) = H(Y) - H(Y|X)$. By symmetry, the resulting output distribution is also uniform over $\{0,1\}$, so that

$$H(Y) = -\frac{1}{2}\log_2\frac{1}{2} - \frac{1}{2}\log_2\frac{1}{2} = 1.$$

Now,

$$H(Y|X=0) = -p(Y=1|X=0)\log_2 p(Y=1|X=0) - p(Y=0|X=0) \times$$
$$\log_2 p(Y=0|X=0)$$
$$= -p\log_2 p - (1-p)\log_2(1-p) = H_B(p),$$

where $H_B(p)$ is the entropy of a Bernoulli random variable with probability p of taking the value one. By symmetry, we also have $H(Y|X=1) = H_B(p)$, so that, from (6.32), we get

$$H(Y|X) = H_B(p).$$

We therefore obtain the capacity of the BSC with crossover probability p as

$$C_{\text{BSC}}(p) = 1 - H_B(p). \tag{6.37}$$

AWGN channel capacity Consider the channel model (6.10), with the observation

$$Y = X + Z$$

with input $\mathbb{E}[X^2] \leq E_s$ and $Z \sim N(0, \sigma^2)$. We wish to compute the capacity

$$C = \max_{p(x):\mathbb{E}[X^2]\leq E_s} I(X;Y).$$

Given $X = x$, $h(Y|X=x) = h(Z)$, so that $h(Y|X) = h(Z)$. We therefore have

$$I(X;Y) = h(Y) - h(Z), \tag{6.38}$$

so that maximizing mutual information is equivalent to maximizing $h(Y)$. Since X and Z are independent (the transmitter does not know the noise realization Z), we have $E[Y^2] = E[X^2] + E[Z^2] \leq E_s + \sigma^2$. Subject to this constraint, it follows from Problem 6.3 that $h(Y)$ is maximized if Y is zero mean Gaussian. This is achieved if the input distribution is $X \sim N(0, E_s)$, independent of the noise Z, which yields $Y \sim N(0, E_s + \sigma^2)$. Substituting the expression (6.25) for the entropy of a Gaussian random variable into (6.38), we obtain the capacity:

$$I(X;Y) = \frac{1}{2}\log_2(2\pi e(E_s + \sigma^2)) - \frac{1}{2}\log_2(2\pi e\sigma^2)$$
$$= \frac{1}{2}\log_2(1 + \frac{E_s}{\sigma^2}) = \frac{1}{2}\log_2(1 + \text{SNR});$$

the same formula that we got from the sphere packing arguments. We have now in addition proved that this capacity is attained by Gaussian input $X \sim N(0, E_s)$.

We now consider the capacity of the AWGN channel when the signal constellation is constrained.

Example 6.3.1 (AWGN capacity with BPSK signaling) Let us first consider BPSK signaling, for which we have the channel model

$$Y = \sqrt{E_s}X + Z, \quad X \in \{-1, +1\}, \quad Z \sim N(0, \sigma^2).$$

It can be shown (e.g., using the techniques to be developed in Section 6.4.1) that the mutual information $I(X; Y)$, subject to the constraint of BPSK signaling, is maximized for equiprobable signaling. Let us now compute the mutual information $I(X; Y)$ as a function of the signal power E_s and the noise power σ^2. We first show that, as with the capacity without an input alphabet constraint, the capacity for BPSK also depends on these parameters only through their ratio, the SNR E_s/σ^2. To show this, replace Y by Y/σ to get the model

$$Y = \sqrt{\text{SNR}}\, X + Z, \quad X \in \{-1, +1\}, \quad Z \sim N(0, 1). \tag{6.39}$$

For notational simplicity, set $A = \sqrt{\text{SNR}}$. We have

$$p(y|+1) = \frac{1}{\sqrt{2\pi}} e^{-(y-A)^2/2},$$

$$p(y|-1) = \frac{1}{\sqrt{2\pi}} e^{-(y+A)^2/2},$$

and

$$p(y) = \frac{1}{2}p(y|+1) + \frac{1}{2}p(y|-1). \tag{6.40}$$

We can now compute

$$I(X; Y) = h(Y) - h(Y|X).$$

As before, we can show that $h(Y|X) = h(Z) = 1/2 \log_2(2\pi e)$. We can now compute

$$h(Y) = -\int \log_2(p_Y(y))\, p_Y(y)\, dy$$

by numerical integration, plugging in (6.40). An alternative approach, which is particularly useful for more complicated constellations and channel models, is to use Monte Carlo integration (i.e., simulation-based empirical averaging) for computing the expectation $h(Y) = -\mathbb{E}[\log_2 p(Y)]$. For this method, we generate i.i.d. samples $Y_1, \ldots, Y_n$ using the model (6.39), and then use the estimate

$$\hat{h}(Y) = -\frac{1}{n}\sum_{i=1}^{n} \log_2 p(Y_i).$$

We can also use the alternative formula

$$I(X; Y) = H(X) - H(X|Y)$$

to compute the capacity. For equiprobable binary input, $H(X) = H_B(\frac{1}{2}) = 1$ bit/symbol. It remains to compute

$$H(X|Y) = \int H(X|Y=y) p_Y(y)\, dy. \qquad (6.41)$$

By Bayes' rule, we have

$$P[X = +1|Y = y] = \frac{P[X = +1]p(y|+1)}{p(y)}$$

$$= \frac{P[X = +1]p(y|+1)}{P[X = +1]p(y|+1) + P[X = -1]p(y|-1)}$$

$$= \frac{e^{Ay}}{e^{Ay} + e^{-Ay}} \quad \text{(equal priors).}$$

We also have

$$P[X = -1|Y = y] = 1 - P[X = +1|Y = y] = \frac{e^{-Ay}}{e^{Ay} + e^{-Ay}}.$$

Such a posteriori probability computations can be thought of as *soft decisions* on the transmitted bits, and are employed extensively when we discuss iterative decoding. We can now use the binary entropy function to compute

$$H(X|Y = y) = H_B(P[X = +1|Y = y]).$$

The average in (6.41) can now be computed by direct numerical integration or by Monte Carlo integration as before. The latter, which generalizes better to more complex models, gives the estimate

$$\hat{H}(X|Y) = \frac{1}{n}\sum_{i=1}^{n} H_B(P[X = +1|Y_i = y_i]).$$

The preceding methodology generalizes in a straightforward manner to PAM constellations. For complex-valued constellations, we need to consider the complex discrete-time AWGN channel model (6.13). For rectangular QAM constellations, one use of the complex channel with QAM input is equivalent to two uses of a real channel using PAM input, so that the same methodology applies again. However, for complex-valued constellations that cannot be decomposed in this fashion (e.g., 8-PSK and other higher order PSK alphabets), we must work directly with the complex AWGN channel model (6.18).

Example 6.3.2 (Capacity with PSK signaling) For PSK signaling over the complex AWGN channel (6.13), we have the model:

$$Y = X + Z,$$

where $X \in A = \{\sqrt{\text{SNR}}\, e^{j2\pi i/M}, i = 0, 1, \ldots, M-1\}$ and $Z \sim CN(0, 1)$ with density

$$p_Z(z) = \frac{e^{-|z|^2}}{\pi}, \quad z \text{ complex-valued,}$$

where we have normalized the noise to obtain a scale-invariant model. As before, for an additive noise model in which the noise is independent of the input, we have $h(Y|X) = h(Z)$. The differential entropy of the proper complex Gaussian random variable Z can be inferred from that for a real Gaussian random variable given by (6.25), or by specializing the formula in Problem 6.4(b). This yields that

$$h(Z) = \log_2 \pi e.$$

Furthermore, assuming that a uniform distribution achieves capacity (this can be proved using the techniques in Section 6.4.2), we have

$$p_Y(y) = \frac{1}{M} \sum_{x \in A} p_Z(y - x) = \frac{1}{M} \sum_{i=0}^{M-1} \frac{1}{\pi} \exp\left(-\left|y - \sqrt{\text{SNR}}\, e^{j2\pi i/M}\right|^2\right).$$

We can now use Monte Carlo integration to compute $h(Y)$, and then compute the mutual information $I(X; Y) = h(Y) - h(Z)$.

Figure 6.5 plots the capacity (in bits per channel use) for QPSK, 16PSK and 16-QAM versus SNR (dB).

Power–bandwidth tradeoffs Now that we know how to compute capacity as a function of SNR for specific constellations using the discrete-time AWGN channel model, we can relate it, as discussed in Section 6.1.4, back to the

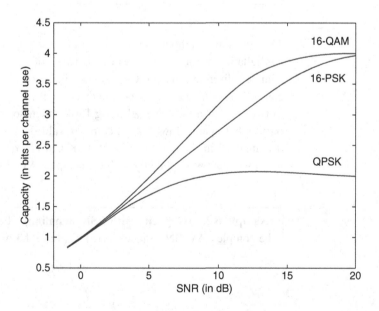

Figure 6.5 The capacity of the AWGN channel with different constellations as a function of SNR.

Figure 6.6 The capacity of the AWGN channel with different constellations as a function of E_b/N_0.

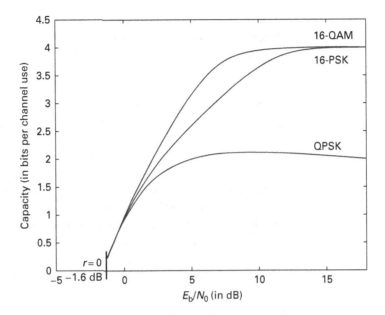

continuous-time AWGN channel to understand the tradeoff between power efficiency and spectral efficiency. As shown in Section 6.1.4, we have, for both the real and complex models,

$$\text{SNR} = r\frac{E_b}{N_0},$$

where r is the spectral efficiency in bit/s per Hz (not accounting for excess bandwidth requirements imposed by implementation considerations).

For a given complex-valued constellation A, suppose that the capacity in discrete time is $C_A(\text{SNR})$ bit/channel use. Then, using (6.15), we find that the feasible region for communication using this constellation is given, as a function of E_b/N_0, by

$$r < C_A\left(r\frac{E_b}{N_0}\right), \quad \text{complex-valued constellation } A. \quad (6.42)$$

Figure 6.6 expresses the plots in Figure 6.5 in terms of E_b/N_0 (dB), obtained by numerically solving for equality in (6.42). (Actually, for each value of SNR, we compute the capacity, and then the corresponding E_b/N_0 value, and then plot the latter two quantities against each other.)

For real-valued constellations such as 4-PAM, we would use (6.11), and obtain

$$r < 2C_A\left(r\frac{E_b}{N_0}\right), \quad \text{real-valued constellation } A. \quad (6.43)$$

6.3.2 Parallel Gaussian channels and waterfilling

A useful generalization of the AWGN channel model is the parallel Gaussian channel model (we can use this to model both dispersive channels and colored noise, as we shall see shortly), stated as follows. We have access to K parallel complex Gaussian channels, with the output of the kth channel, $1 \leq k \leq K$, modeled as

$$Y_k = h_k X_k + Z_k,$$

where h_k is the channel gain, $Z_k \sim CN(0, N_k)$ is the noise on the kth channel, and $E[|X_k|^2] = P_k$, with a constraint on the total power:

$$\sum_{k=1}^{K} P_k \leq P. \qquad (6.44)$$

The noises $\{Z_k\}$ are independent across channels, as well as across time. The channel is characterized by the gains $\{h_k\}$ and the noise variances $\{N_k\}$. The goal is to derive the capacity of this channel model for fixed $\{P_k\}$, which is appropriate when the transmitter does not know the channel characteristics, as well as to optimize the capacity over the power allocation $\{P_k\}$ when the channel characteristics are known at the transmitter.

The mutual information between the input vector $\mathbf{X} = (X_1, \ldots, X_k)$ and the output vector $\mathbf{Y} = (Y_1, \ldots, Y_K)$ is given by

$$I(\mathbf{X}; \mathbf{Y}) = h(\mathbf{Y}) - h(\mathbf{Y}|\mathbf{X}) = h(\mathbf{Y}) - h(\mathbf{Z}).$$

Owing to the independence of the noises, $h(\mathbf{Z}) = \sum_{k=1}^{K} h(Z_k)$. Furthermore, we can bound the joint differential entropy of the output as the sum of the individual differential entropies:

$$h(\mathbf{Y}) = h(Y_1, \ldots, Y_K) \leq \sum_{k=1}^{K} h(Y_k)$$

with equality if $Y_1, \ldots, Y_K$ are independent. Thus, we obtain

$$I(\mathbf{X}; \mathbf{Y}) \leq \sum_{k=1}^{K} (h(Y_k) - h(Z_k)).$$

Each of the K terms on the right-hand side can be maximized as for a standard Gaussian channel, by choosing $X_k \sim CN(0, P_k)$. We therefore find that, for a given power allocation $\mathbf{P} = (P_1, \ldots, P_K)$, the capacity is given by

$$C(\mathbf{P}) = \sum_{k=1}^{K} \log_2 \left(1 + \frac{|h_k|^2 P_k}{N_k}\right) \quad \text{Fixed power allocation} \qquad (6.45)$$

(the received signal power on the kth channel is $S_k = |h_k|^2 P_k$). The preceding development also holds for real-valued parallel Gaussian channels, except that a factor of 1/2 must be inserted in (6.45).

Optimizing the power allocation We can now optimize $C(\mathbf{P})$ subject to the constraint (6.44) by maximizing the Lagrangian

$$J(\mathbf{P}) = C(\mathbf{P}) + \lambda \sum_{k=1}^{K} P_k = \sum_{k=1}^{K} \log_2 \left(1 + \frac{|h_k|^2 P_k}{N_k}\right) - \lambda \sum_{k=1}^{K} P_k. \qquad (6.46)$$

6.3 Some capacity computations

We discuss the theory of such convex optimization problems very briefly in Section 6.4.1, but for now, it suffices to note that we can optimize $J(\mathbf{P})$ by setting the partial derivative with respect to P_k to zero. This yields

$$0 = \frac{\partial J(\mathbf{P})}{\partial P_k} = \frac{|h_k|^2/N_k}{1 + \frac{|h_k|^2 P_k}{N_k}} - \lambda,$$

so that

$$P_k = a - \frac{N_k}{|h_k|^2}$$

for some constant a. However, we must also satisfy the constraint that $P_k \geq 0$. We therefore get the following solution.

Waterfilling power allocation

$$P_k = \begin{cases} a - \dfrac{N_k}{|h_k|^2}, & \dfrac{N_k}{|h_k|^2} \leq a, \\[2ex] 0, & \dfrac{N_k}{|h_k|^2} > a, \end{cases} \quad (6.47)$$

where a is chosen so as to satisfy the power constraint with equality:

$$\sum_{k=1}^{K} P_k = P.$$

This has the waterfilling interpretation depicted in Figure 6.7. The water level a is determined by pouring water until the net amount equals the power budget P. Thus, if the normalized noise level $N_k/|h_k|^2$ is too large for a channel, then it does not get used (the corresponding $P_k = 0$ in the optimal allocation).

Application to dispersive channels The parallel Gaussian model provides a means of characterizing the capacity of a dispersive channel with impulse response $h(t)$ (we are working in complex baseband now), by signaling across parallel frequency bins. Let us assume colored proper complex Gaussian noise with PSD $S_n(f)$. Then a frequency bin of width Δf around f_k follows the model

$$Y_k = H(f_k) X_k + Z_k,$$

where $N_k \sim CN(0, S_n(f_k)\Delta f)$ and $E[|X_k|^2] = S_s(f_k)\Delta f$, where $S_s(f)$ is the PSD of the input (which must be proper complex Gaussian to achieve capacity). Note that the SNR on the channel around f_k is

$$\text{SNR}(f_k) = \frac{|H(f_k)|^2 S_s(f_k)\Delta f}{S_n(f_k)\Delta f} = \frac{|H(f_k)|^2 S_s(f_k)}{S_n(f_k)}.$$

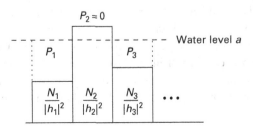

Figure 6.7 Waterfilling power allocation for the parallel Gaussian channel.

The net capacity is now given by

$$\sum_k \Delta f \log_2\left(1 + \mathrm{SNR}(f_k)\right).$$

By letting $\Delta f \to 0$, the sum above tends to an integral, and we obtain the capacity as a function of the input PSD:

$$C(S_s) = \int_{-\frac{W}{2}}^{\frac{W}{2}} \log_2\left(1 + \frac{|H(f)|^2 S_s(f)}{S_n(f)}\right) df, \qquad (6.48)$$

where W is the channel bandwidth, and the input power is given by

$$\int_{-\frac{W}{2}}^{\frac{W}{2}} S_s(f)\, df = P. \qquad (6.49)$$

This reduces to the formula (6.1) for the complex baseband AWGN channel by setting $H(f) \equiv 1$, $S_s(f) \equiv P/W$ and $S_n(f) \equiv N_0$ for $-W/2 \le f \le W/2$.

Waterfilling can now be used to determine the optimum input PSD as follows:

$$S_s(f) = \begin{cases} a - \dfrac{S_n(f)}{|H(f)|^2}, & \dfrac{S_n(f)}{|H(f)|^2} \le a, \\ 0, & \dfrac{S_n(f)}{|H(f)|^2} > a, \end{cases} \qquad (6.50)$$

with a chosen to satisfy the power constraint (6.49).

An important application of the parallel Gaussian model is orthogonal frequency division multiplexing (OFDM), also called discrete multitone, in which data are modulated onto a discrete set of subcarriers in parallel. Orthogonal frequency division multiplexing is treated in detail in Chapter 8, where we focus on its wireless applications. However, OFDM has also been successfully applied to dispersive wireline channels such as digital subscriber loop (DSL). In such settings, the channel can be modeled as time-invariant, and can be learnt by the transmitter using channel sounding and receiver feedback. Waterfilling, appropriately modified to reflect practical constraints such as available constellation choices and the gap to capacity for the error correction scheme used, then plays an important role in optimizing the constellations to be used on the different subcarriers, with larger constellations being used on subcarriers seeing a better channel gain.

6.4 Optimizing the input distribution

We have shown how to compute mutual information between the input and output of a discrete memoryless channel (DMC) for a given input distribution. For finite constellations over the AWGN channel, we have, for example, considered input distributions that are uniform over the alphabet. This is an intuitively pleasing and practical choice, and indeed, it is optimal in certain situations, as we shall show. However, the optimal input distribution is by no means obvious in all cases, hence it is important to develop a set of tools for characterizing and computing it in general. The key ideas behind developing such tools are as follows:

(a) Mutual information is a concave function of the input distribution, hence a unique maximizing input distribution exists.
(b) There are necessary and sufficient conditions for optimality that can easily be used to check guesses regarding the optimal input distribution. However, directly solving for the optimal input distribution based on these conditions is difficult.
(c) An iterative algorithm to find the optimal input distribution can be obtained by writing the maximum mutual information as the solution to a two-stage maximization problem, such that it is easy to solve each stage. Convergence to the optimal input distribution is obtained by alternating between the two stages. This algorithm is referred to as the Blahut–Arimoto algorithm.

We begin with a brief discussion of concave functions and their maximization. We apply this to obtain necessary and sufficient conditions that must be satisfied by the optimal input distribution. We end with a discussion of the Blahut–Arimoto algorithm.

6.4.1 Convex optimization

A set $\mathbf{C}$ is *convex* if, given $\mathbf{x}_1, \mathbf{x}_2 \in \mathbf{C}$, $\lambda \mathbf{x}_1 + (1-\lambda)\mathbf{x}_2 \in \mathbf{C}$ for any $\lambda \in [0, 1]$. We are interested in optimizing mutual information over a set of probability distributions, which is a convex set. Thus, we consider functions whose arguments lie in a convex set.

A function $f(\mathbf{x})$ (whose argument may be a real or complex vector $\mathbf{x}$ in a convex set $\mathbf{C}$) is *convex* (also termed *convex up*) if

$$f(\lambda \mathbf{x}_1 + (1-\lambda)\mathbf{x}_2) \leq \lambda f(\mathbf{x}_1) + (1-\lambda)f(\mathbf{x}_2) \qquad (6.51)$$

for any $\mathbf{x}_1$, $\mathbf{x}_2$, and any $\lambda \in [0, 1]$. That is, the line joining any two points on the graph of the function lies above the function.

Similarly, $f(\mathbf{x})$ is concave (also termed *convex down*) if

$$f(\lambda \mathbf{x}_1 + (1-\lambda)\mathbf{x}_2) \geq \lambda f(\mathbf{x}_1) + (1-\lambda)f(\mathbf{x}_2). \qquad (6.52)$$

From the preceding definitions, it is easy to show that linear combinations of convex (concave) functions are convex (concave). Also, the negative of a convex function is concave, and vice versa. Affine functions (i.e., linear functions plus constants) are both convex and concave, since they satisfy (6.51) and (6.52) with equality.

Example 6.4.1 A twice differentiable function $f(x)$ with a one-dimensional argument x is convex if $f''(x) \geq 0$, and concave if $f''(x) \leq 0$. Thus, $f(x) = x^2$ is convex, $f(x) = \log x$ is concave, and a line has second derivative zero, and is therefore both convex and concave.

Entropy is a concave function of the probability density/mass function
The function $f(x) = -x \log x$ is concave (verify by differentiating twice). Use this to show that

$$-\sum_x p(x) \log p(x)$$

is concave in the probability mass function $\{p(x)\}$, where the latter is viewed as a real-valued vector.

Mutual information between input and output of a DMC is a concave function of the input probability distribution The mutual information is given by $I(X; Y) = H(Y) - H(Y|X)$. The output entropy $H(Y)$ is a concave function of p_Y, and p_Y is a linear function of p_X. It is easy to show, proceeding from the definition (6.52) that $H(Y)$ is a concave function of p_X. The conditional entropy $H(Y|X)$ is easily seen to be a linear function of p_X.

Kuhn–Tucker conditions for constrained maximization of a concave function We state without proof necessary and sufficient conditions for optimality for a special case of constrained optimization, which are specializations of the so-called Kuhn–Tucker conditions for constrained convex optimization. Suppose that $f(\mathbf{x})$ is a concave function to be maximized over $\mathbf{x} = (x_1, \ldots, x_m)^T$, subject to the constraints $x_k \geq 0$, $1 \leq k \leq m$, and $\sum_{k=1}^m x_k = c$, where c is a constant. Then the following conditions are necessary and sufficient for optimality: for $1 \leq k \leq m$, we have

$$\begin{aligned} \frac{\partial f}{\partial x_k} &= \lambda, \quad x_k > 0, \\ \frac{\partial f}{\partial x_k} &\leq \lambda, \quad x_k = 0, \end{aligned} \quad (6.53)$$

for a value of λ such that $\sum_k x_k = c$.

We can interpret the Kuhn–Tucker conditions in terms of the Lagrangian for the constrained optimization problem at hand:

$$J(\mathbf{x}) = f(\mathbf{x}) - \lambda \sum_k x_k.$$

For $x_k > 0$, we set $\partial/\partial x_k J(\mathbf{x}) = 0$. For a point on the boundary with $x_k = 0$, the performance must get worse when we move in from the boundary by increasing x_k, so that $\partial/\partial x_k J(\mathbf{x}) \leq 0$.

We apply these results in the next section to characterize optimal input distributions for a DMC.

6.4.2 Characterizing optimal input distributions

A capacity-achieving input distribution must satisfy the following conditions.

Necessary and sufficient conditions for optimal input distribution For a DMC with transition probabilities $p(y|x)$, an input distribution $p(x)$ is optimal, achieving a capacity C, if and only if, for each input x_k,

6.4 Optimizing the input distribution

$$D(P_{Y|X=x_k}||P_Y) = C, \quad p(x_k) > 0,$$
$$D(P_{Y|X=x_k}||P_Y) \leq C, \quad p(x_k) = 0. \tag{6.54}$$

Interpretation of optimality condition We show that the mutual information is the average of the terms $D(P_{Y|X=x}||P_Y)$ as follows.

$$\begin{aligned} I(X;Y) &= D(P_{X,Y}||P_X P_Y) = \sum_{x,y} p(x,y) \log\left(\frac{p(x,y)}{p(x)p(y)}\right) \\ &= \sum_x p(x) \sum_y p(y|x) \log\left(\frac{p(y|x)}{p(y)}\right) \\ &= \sum_x p(x) D(P_{Y|X=x}||P_Y). \end{aligned} \tag{6.55}$$

The optimality conditions state that each term making a nontrivial contribution to the average mutual information must be equal. That is, each term equals the average, which for the optimal input distribution equals the capacity C. Terms corresponding to $p(x) = 0$ do not contribute to the average, and are smaller (otherwise we could get a bigger average by allocating probability mass to them). We now provide a proof of these conditions.

Proof The Kuhn–Tucker conditions for capacity maximization are as follows:

$$\frac{\partial}{\partial p(x)} I(X;Y) - \lambda = 0, \quad p(x) > 0,$$
$$\frac{\partial}{\partial p(x)} I(X;Y) - \lambda \leq 0, \quad p(x) = 0. \tag{6.56}$$

Now to evaluate the partial derivatives of $I(X;Y) = H(Y) - H(Y|X)$. Since

$$H(Y) = -\sum_y p(y) \log p(y),$$

we have, using the chain rule,

$$\begin{aligned} \frac{\partial}{\partial p(x_k)} H(Y) &= \sum_y \left[\frac{\partial H(Y)}{\partial p(y)} \frac{\partial p(y)}{\partial p(x_k)}\right] \\ &= \sum_y [-1 - \log p(y)] \, p(y|x_k) \\ &= -1 - \sum_y p(y|x_k) \log p(y). \end{aligned} \tag{6.57}$$

Also,

$$H(Y|X) = -\sum_{x,y} p(x) p(y|x) \log p(y|x),$$

so that

$$\frac{\partial}{\partial p(x_k)} H(Y|X) = -\sum_y p(y|x_k) \log p(y|x_k) = H(Y|X = x_k). \tag{6.58}$$

Using (6.57) and (6.58), we obtain

$$\frac{\partial}{\partial p(x_k)} I(X;Y) = -1 + \sum_y p(y|x_k) \log \frac{p(y|x_k)}{p(y)} = -1 + D(P_{Y|X=x_k} \| P_Y).$$
(6.59)

Plugging into (6.56), we get $D(P_{Y|X=x_k} \| P_Y) \leq \lambda + 1$, with equality for $p(x_k) > 0$. Averaging over the input distribution, we realize from (6.55) that we must have $\lambda + 1 = C$, completing the proof. □

Remark While we prove all results for discrete random variables, their natural extensions to continuous random variables hold, with probability mass functions replaced by probability density functions, and summations replaced by integrals. In what follows, we use the term *density* to refer to either probability mass function or probability density function.

Symmetric channels The optimality conditions (6.54) impose a symmetry in the input–output relation. When the channel transition probabilities exhibit a natural symmetry, it often suffices to pick an input distribution that is uniform over the alphabet to achieve capacity. Rather than formally characterizing the class of symmetric channels for which this holds, we leave it to the reader to check, for example, that uniform inputs work for the BSC, and for PSK constellations over the AWGN channel.

While the conditions (6.54) are useful for checking guesses as to the optimal input distribution, they do not provide an efficient computational procedure for obtaining the optimal input distribution. For channels for which guessing the optimal distribution is difficult, a general procedure for computing it is provided by the Blahut–Arimoto algorithm, which we describe in the next section.

6.4.3 Computing optimal input distributions

A key step in the Blahut–Arimoto algorithm is the following lemma, which expresses mutual information as the solution to a maximization problem with an explicit solution. This enables us to write the maximum mutual information, or capacity, as the solution to a double maximization that can be obtained by an alternating maximization algorithm.

Lemma 6.4.1 The mutual information between X and Y can be written as

$$I(X;Y) = \max_{q(x|y)} \sum_{x,y} p(x) p(y|x) \log \frac{q(x|y)}{p(x)},$$

where $\{q(x|y)\}$ is a set of conditional densities for X (that is, $\sum_x q(x|y) = 1$ for each y). The maximum is achieved by the conditional distribution $p(x|y)$ that is consistent with $p(x)$ and $p(y|x)$. That is, the optimizing q is given by

$$q^*(x|y) = \frac{p(x) p(y|x)}{\sum_{x'} p(x') p(y|x')}.$$

6.4 Optimizing the input distribution

Proof We show that the difference in values attained by q^* and any other q is nonnegative as follows:

$$\sum_{x,y} p(x)p(y|x) \log \frac{q^*(x|y)}{p(x)} - \sum_{x,y} p(x)p(y|x) \log \frac{q(x|y)}{p(x)}$$

$$= \sum_{x,y} p(x)p(y|x) \log \frac{q^*(x|y)}{q(x|y)}$$

$$= \sum_{y} p(y) \sum_{x} q^*(x|y) \log \frac{q^*(x|y)}{q(x|y)}$$

$$= \sum_{y} p(y) D(Q^*(\cdot|y) || Q(\cdot|y)) \geq 0,$$

where we have used $p(x,y) = p(x)p(y|x) = p(y)p(x|y) = p(y)q^*(x|y)$, and where $Q^*(\cdot|y)$, $Q(\cdot|y)$ denote the conditional distributions corresponding to the conditional densities $q^*(x|y)$ and $q(x|y)$, respectively. $\square$

The capacity of a DMC characterized by transition densities $\{p(y|x)\}$ can now be written as

$$C = \max_{p(x)} I(X; Y) = \max_{p(x)} \max_{q(x|y)} \sum_{x,y} p(x)p(y|x) \log \frac{q(x|y)}{p(x)}.$$

We state without proof that an alternating maximization algorithm, which maximizes over $q(x|y)$ keeping $p(x)$ fixed, and then maximizes over $p(x)$ keeping $q(x|y)$ fixed, converges to the global optimum. The utility of this procedure is that each maximization can be carried out explicitly. The lemma provides an explicit form for the optimal $q(x|y)$ for fixed $p(x)$. It remains to provide an explicit form for the optimal $p(x)$ for fixed $q(x|y)$. To this end, consider the Lagrangian

$$J(p) = \sum_{x,y} p(x)p(y|x) \log \frac{q(x|y)}{p(x)} - \lambda \sum_{x} p(x)$$

$$= \sum_{x,y} [p(x)p(y|x) \log q(x|y) - p(x)p(y|x) \log p(x)] - \lambda \sum_{x} p(x), \quad (6.60)$$

corresponding to the usual sum constraint $\sum_{x} p(x) = 1$. Setting partial derivatives to zero, we obtain

$$\frac{\partial}{\partial p(x_k)} J(p) = \sum_{y} [p(y|x_k) \log q(x_k|y) - p(y|x_k) - p(y|x_k) \log p(x_k)] - \lambda = 0.$$

Noting that $\sum_{y} p(y|x_k) = 1$, we get

$$\log p(x_k) = -\lambda - 1 + \sum_{y} p(y|x_k) \log q(x_k|y),$$

from which we conclude that

$$p^*(x_k) = K \exp(\sum_{y} p(y|x_k) \log q(x_k|y)) = K \Pi_y (q(x_k|y))^{p(y|x_k)},$$

where the constant K is chosen so that $\sum_{x} p^*(x) = 1$.

We can now state the Blahut–Arimoto algorithm for computing optimal input distributions.

Blahut–Arimoto algorithm

Step 0 Choose an initial guess $p(x)$ for input distribution, ensuring that there is nonzero probability mass everywhere that the optimal input distribution is expected to have nonzero probability mass (e.g., for finite alphabets, a uniform distribution is a safe choice).

Step 1 For the current $p(x)$, compute the optimal $q(x|y)$ using
$$q^*(x|y) = \frac{p(x)p(y|x)}{\sum_{x'} p(x')p(y|x')}.$$
Set this to be the current $q(x|y)$.

Step 2 For the current $q(x|y)$, compute the optimal $p(x)$ using
$$p^*(x) = \frac{\Pi_y (q(x|y))^{p(y|x)}}{\sum_{x'} \Pi_y (q(x'|y))^{p(y|x')}}.$$
Set this to be the current $p(x)$. Go back to Step 1.

Alternate Steps 1 and 2 until convergence (using any sensible stopping criterion to determine when the changes in $p(x)$ are sufficiently small).

Example 6.4.2 (Blahut–Arimoto algorithm applied to BSC) For a BSC with crossover probability α, we know that the optimal input distribution is uniform. However, let us apply the Blahut–Arimoto algorithm, starting with an arbitrary input distribution $P[X=1] = p = 1 - P[X=0]$, where $0 < p < 1$. We can now check that Step 1 yields
$$q(1|0) = \frac{p\alpha}{p\alpha + (1-p)(1-\alpha)} = 1 - q(0|0)$$
$$q(0|1) = \frac{(1-p)\alpha}{(1-p)\alpha + p(1-\alpha)} = 1 - q(1|1)$$
and Step 2 yields
$$p = p(1) = \frac{q(1|0)^\alpha q(1|1)^{1-\alpha}}{q(1|0)^\alpha q(1|1)^{1-\alpha} + q(0|1)^\alpha q(0|0)^{1-\alpha}}.$$
Iterating these steps should yield $p \to 1/2$.

Extensions of the basic Blahut–Arimoto algorithm Natural extensions of the Blahut–Arimoto algorithm provide methods for computing optimal input distributions that apply in great generality. As an example of a simple extension, Problem 6.16 considers optimization of the input probabilities for a 4-PAM alphabet $\{\pm d, \pm 3d\}$ over the AWGN channel with a power constraint. The Blahut–Arimoto iterations must now account for the fact that the signal power depends both on d and the input probabilities.

6.5 Further reading

The information theory textbook by Cover and Thomas [14] provides a lucid exposition of the fundamental concepts of information theory, and is perhaps the best starting point for delving further into this field. The classic text by Gallager [42] is an important reference for many topics. The text by Csiszar and Korner [43] is the definitive work on the use of combinatorial techniques and the method of types in proving fundamental theorems of information theory. Other notable texts providing in-depth treatments of information theory include Blahut [44], McEliece [45], Viterbi and Omura [12], and Wolfowitz [46]. Shannon's original work [47, 48] is a highly recommended read, because of its beautiful blend of intuition and rigor in establishing the foundations of the field.

For most applications, information-theoretic quantities such as capacity must be computed numerically as solutions to optimization problems. The Blahut–Arimoto algorithm discussed here [49, 50] is the classical technique for optimizing input distributions. More recently, however, methods based on convex optimization and duality [51, 52] and on linear programming [53] have been developed for deeper insight into, and efficient solution of, optimization problems related to the computation of information-theoretic quantities. Much attention has been focused in recent years on information-theoretic limits for the wireless channel, as discussed in Chapter 8.

Good sources for recent results in information theory are the *Proceedings of the International Symposium on Information Theory (ISIT)*, and the journal *IEEE Transactions on Information Theory*. The October 1998 issue of the latter commemorates the fiftieth anniversary of Shannon's seminal work, and provides a perspective on the state of the field at that time.

6.6 Problems

Problem 6.1 (Estimating the capacity of a physical channel) Consider a line of sight radio link with free space propagation. Assume transmit and receive antenna gains of 10 dB each, a receiver noise figure of 6 dB, and a range of 1 km. Using the Shannon capacity formula for AWGN channels, what is the transmit power required to attain a link speed of 1 gigabit/s using a bandwidth of 1.5 GHz (assuming 50% excess bandwidth)?

Problem 6.2 (Entropy for an M-ary random variable) Suppose that X is a random variable taking one of M possible values (e.g., X may be the index of the transmitted signal in an M-ary signaling scheme).

(a) What is the entropy of X, assuming all M values are equally likely?
(b) Denoting the pmf for the uniform distribution in (a) by $q(x)$, suppose now that X is distributed according to pmf $p(x)$. Denote the entropy of X under pmf p by $H_p(X)$. Show that the divergence between p and q equals

$$D(p||q) = \mathbb{E}_p[\log_2 \frac{p(X)}{q(X)}] = \log_2 M - H_p(X).$$

(c) Infer from (b) that the maximum possible entropy for X is $\log_2 M$, which is achieved by the uniform distribution.

Problem 6.3 (Differential entropy is maximum for Gaussian random variables) Consider a zero mean random variable X with density $p(x)$ and variance v^2. Let $q(x)$ denote the density of an $N(0, v^2)$ random variable with the same mean and variance.

(a) Compute the divergence $D(p||q)$ in terms of $h(X)$ and v^2.
(b) Use the nonnegativity of divergence to show that

$$h(X) \le \frac{1}{2}\log_2(2\pi e v^2) = h\left(N(0, v^2)\right).$$

That is, the Gaussian density maximizes the differential entropy over all densities with the same variance.

Remark This result, and the technique used for proving it, generalizes to random vectors, with Gaussian random vectors maximizing differential entropy over all densities with the same covariance.

Problem 6.4 (Differential entropy for Gaussian random vectors) Derive the following results.

(a) If $\mathbf{X} \sim N(\mathbf{m}, \mathbf{C})$ is an n-dimensional Gaussian random vector with mean vector $\mathbf{m}$ and covariance matrix $\mathbf{C}$, then its differential entropy is given by

$$h(\mathbf{X}) = \frac{1}{2}\log_2\left((2\pi e)^n |\mathbf{C}|\right) \quad \text{Differential entropy for real Gaussian}.$$

(b) If $\mathbf{X} \sim CN(\mathbf{m}, \mathbf{C})$ is an n-dimensional proper complex Gaussian random vector with mean vector $\mathbf{m}$ and covariance matrix $\mathbf{C}$, then its differential entropy is given by

$$h(\mathbf{X}) = \log_2\left((\pi e)^n |\mathbf{C}|\right) \quad \text{Differential entropy for proper complex Gaussian}.$$

Problem 6.5 (Entropy under simple transformations) Let X denote a random variable, and a, b denote arbitrary constants.

(a) If X is discrete, how are the entropies $H(aX)$ and $H(X+b)$ related to $H(X)$?
(b) If X is continuous, how are the differential entropies $h(aX)$ and $h(X+b)$ related to $h(X)$?

Figure 6.8 The binary (symmetric) erasures channel.

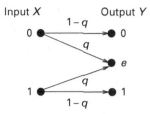

Channel transition probabilities $p(y|x)$

Figure 6.9 The binary (symmetric) errors and erasures channel.

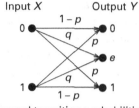

Channel transition probabilities $p(y|x)$

Problem 6.6 (Binary erasures channel) Show that the channel capacity of the binary erasures channel with erasure probability q, as shown in Figure 6.8, is given by $1-q$.

Problem 6.7 (Binary errors and erasures channel) Find the channel capacity of the binary errors and erasures channel with error probability p and erasures probability q, as shown in Figure 6.9.

Problem 6.8 (AWGN capacity plots for complex constellations) Write computer programs for reproducing the capacity plots in Figures 6.5 and 6.6.

Problem 6.9 (Shannon theory for due diligence) A binary noncoherent FSK system is operating at an E_b/N_0 of 5 dB, and passes hard decisions (i.e., decides whether 0 or 1 was sent) up to the decoder. The designer claims that her system achieves a BER of 10^{-5} using a powerful rate 1/2 code. Do you believe her claim?

Problem 6.10 (BPSK with errors and erasures) Consider BPSK signaling with the following scale-invariant model for the received samples:
$$Y = A(-1)^X + Z,$$
where $X \in \{0, 1\}$ is equiprobable, and $Z \sim N(0, 1)$, with $A^2 = \text{SNR}$.

(a) Find the capacity in bits per channel use as outlined in the text, and plot it as a function of SNR (dB).

(b) Specify the BSC induced by hard decisions. Find the capacity in bits per channel use and plot it as a function of SNR (dB).
(c) What is the degradation in dB due to hard decisions at a rate of 1/4 bits per channel use?
(d) What is the E_b/N_0 (dB) corresponding to (c), for both soft and hard decisions?
(e) Now suppose that the receiver supplements hard decisions with erasures. That is, the receiver declares an erasure when $|Y| < \alpha$, with $\alpha \geq 0$. Find the error and erasure probabilities as a function of α and SNR.
(f) Apply the result of Problem 6.7 to compute the capacity as a function of α and SNR. Set SNR at 3 dB, and plot capacity as a function of α. Compare with the capacity for hard decisions.
(g) Find the best value of α for SNR of 0 dB, 3 dB and 6 dB. Is there a value of α that works well over the range 0–6 dB?

Problem 6.11 (Gray coded two-dimensional modulation with hard decisions) A communication system employs Gray coded 16-QAM, with the demodulator feeding hard decisions to an outer binary code.

(a) What is a good channel model for determining information-theoretic limits on the rate of the binary code as a function of E_b/N_0?
(b) We would like to use the system to communicate at an information rate of 100 Mbps using a bandwidth of 150 MHz, where the modulating pulse uses an excess bandwidth of 50%. Use the model in (a) to determine the minimum required value of E_b/N_0 for reliable communication.
(c) Now suppose that we use QPSK instead of 16-QAM in the setting of (b). What is the minimum required value of E_b/N_0 for reliable communication?

Problem 6.12 (Parallel Gaussian channels) Consider two parallel complex Gaussian channels with channel gains $h_1 = 1+j$, $h_2 = -3j$ and noise covariances $N_1 = 1$, $N_2 = 2$. Assume that the transmitter knows the channel characteristics.

(a) At "low" SNR, which of the two channels would you use?
(b) For what values of net input power P would you start using both channels?
(c) Plot the capacity as a function of net input power P using the waterfilling power allocation.

Also plot for comparison the capacity attained if the transmitter does not know the channel characteristics, and splits power evenly across the two channels.

Problem 6.13 (Waterfilling for a dispersive channel) A real baseband dispersive channel with colored Gaussian noise is modeled as in Figure 6.10.

Figure 6.10 Channel characteristics for Problem 6.13.

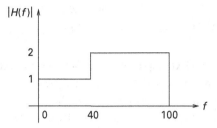

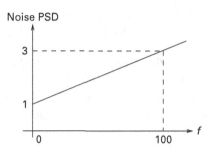

We plan to use the channel over the band $[0, 100]$. Let $N = \int_0^{100} S_n(f)\,df$ denote the net noise power over the band. If the net signal power is P, then we define SNR as P/N.

(a) Assuming an SNR of 10 dB, find the optimal signal PSD using waterfilling. Find the corresponding capacity.
(b) Repeat (a) for an SNR of 0 dB.
(c) Repeat (a) and (b) assuming that signal power is allocated uniformly over the band $[0, 100]$.

Problem 6.14 (Multipath channel) Consider a complex baseband multipath channel with impulse response

$$h(t) = 2\delta(t-1) - \frac{j}{2}\delta(t-2) + (1+j)\delta(t-3.5).$$

The channel is used over the band $[-W/2, W/2]$. Let $C_W(\text{SNR})$ denote the capacity as a function of bandwidth W and SNR, assuming that the input power is spread evenly over the bandwidth used and that the noise is AWGN.

(a) Plot $C_W(\text{SNR})/W$ versus W over the range $1 < W < 20$, fixing the SNR at 10 dB. Do you notice any trends?
(b) For $W = 10$, find the improvement in capacity due to waterfilling at an SNR of 10 dB.

Problem 6.15 (Blahut–Arimoto iterations for BSC) Consider the binary symmetric channel with crossover probability 0.1. Starting from an initial input distribution with $P[X = 1] = p = 0.3$, specify the values of p obtained in the first five iterations of the Blahut–Arimoto algorithm. Comment on whether the iterations are converging to the result you would expect.

Problem 6.16 (Extension of Blahut–Arimoto algorithm for constellation optimization) Consider a 4-PAM alphabet $\{\pm d, \pm 3d\}$ to be used on the real, discrete-time AWGN channel. Without loss of generality, normalize the noise variance to one. Assuming that the input distribution satisfies a natural symmetry condition:

$$P[X = \pm d] = p, \quad P[X = \pm 3d] = \frac{1}{2} - p.$$

(a) What is the relation between p and d at SNR of 3 dB?
(b) Starting from an initial guess of $p = 1/4$, iterate the Blahut–Arimoto algorithm to find the optimal input distribution at SNR of 3 dB, modifying as necessary to satisfy the SNR constraint.
(c) Comment on how the optimal value of p varies with SNR by running the Blahut–Arimoto algorithm for a few other values.

CHAPTER

7 Channel coding

In this chapter, we provide an introduction to some commonly used channel coding techniques. The key idea of channel coding is to introduce redundancy in the transmitted signal so as to enable recovery from channel impairments such as errors and erasures. We know from the previous chapter that, for any given set of channel conditions, there exists a Shannon capacity, or maximum rate of reliable transmission. Such Shannon-theoretic limits provide the ultimate benchmark for channel code design. A large number of error control techniques are available to the modern communication system designer, and in this chapter, we provide a glimpse of a small subset of these. Our emphasis is on convolutional codes, which have been a workhorse of communication link design for many decades, and turbo-like codes, which have revolutionized communication systems by enabling implementable designs that approach Shannon capacity for a variety of channel models.

Map of this chapter We begin in Section 7.1 with binary convolutional codes. We introduce the trellis representation and the Viterbi algorithm for ML decoding, and develop performance analysis techniques. The structure of the memory introduced by a convolutional code is similar to that introduced by a dispersive channel. Thus, the techniques are similar to (but simpler than) those developed for MLSE for channel equalization in Chapter 5. Concatenation of convolutional codes leads to turbo codes, which are iteratively decoded by exchanging soft information between the component convolutional decoders. We discuss turbo codes in Section 7.2. While the Viterbi algorithm gives the ML sequence, we need soft information regarding individual bits for iterative decoding. This is provided by MAP decoding using the BCJR algorithm, discussed in Section 7.2.1. The logarithmic version of the BCJR algorithm, which is actually more useful both practically and conceptually, is discussed in Section 7.2.2. Once this is done, we can specify both parallel and serial concatenated turbo codes quite easily, and this is done in Section 7.2.3. The performance of turbo codes is discussed in Sections 7.2.4, 7.2.5 and 7.2.6. An especially intuitive way of visualizing the progress of iterative decoding, as well as to predict the SNR threshold at which the BER starts decreasing steeply, is the method of EXIT charts introduced by ten Brink, which is discussed in Section 7.2.5. Another important class of "turbo-like" codes, namely, low density parity check (LDPC) codes, is discussed in Section 7.3. Section 7.4 discusses channel code design for two-dimensional modulation. A broadly applicable approach is the use of bit interleaved coded modulation (BICM), which allows us to employ powerful binary codes in conjunction with higher order modulation formats: the output of a binary encoder is scrambled and then mapped to the signaling constellation, typically with a Gray-like encoding that minimizes the number of bits changing across nearest neighbors. We also discuss another approach that couples coding and modulation more tightly: trellis coded modulation (TCM). Finally, in Section 7.5, we provide a quick exposure to the role played in communication system design by codes such as Reed–Solomon codes, which are constructed using finite-field algebra. We attempt to provide an operational understanding of what we can do with such codes, without getting into the details of the code construction, since the required background in finite fields is beyond the scope of this book.

7.1 Binary convolutional codes

Binary convolutional codes are important not only because they are deployed in many practical systems, but also because they form a building block for other important classes of codes, such as trellis coded modulation and a variety of "turbo-like" codes. Such codes can be interpreted as convolving a binary information sequence through a filter, or "code generator," with binary coefficients (with addition and multiplication over the binary field). They therefore have a structure very similar to the dispersive channels discussed earlier, and are therefore amenable to similar techniques for decoding (using the Viterbi algorithm) and performance analysis (union bounds using error events, and transfer function bounds). We discuss these techniques in the following, focusing on examples rather than on the most general development.

Consider a binary information sequence $u[k] \in \{0, 1\}$, which we want to send reliably using BPSK over an AWGN channel. Instead of directly sending the information bits (e.g., sending the BPSK symbols $\{(-1)^{u[k]}\}$), we first use $\mathbf{u} = \{u[k]\}$ to generate a coded binary sequence, termed a codeword, which includes redundancy. This operation is referred to as *encoding*. We then send this new *coded* bit sequence using BPSK, over an AWGN channel. The *decoder* at the receiver exploits the redundancy to recover the information bits from the noisy received signal. The *code* is the set of all possible codewords that can be obtained in this fashion. The encoder, or the mapping between information bits and coded bits, is not unique for a given code, and bit error rate attained by the code, as well as its role as a building block for more complex codes, can depend on the mapping. The encoder mapping is termed *nonrecursive*, or feedforward, if the codeword is obtained by passing the information sequence through a finite impulse response feedforward filter. It is termed *recursive* if codeword generation involves the use of feedback. The encoder is *systematic* if the information sequence appears directly as one component of the codeword, and it is termed *nonsystematic* otherwise. In addition to the narrow definition of the term "code" as the set of all possible codewords, we often also employ the term more broadly, to refer to both the set of codewords and the encoder.

For the purpose of this introductory development, it suffices to restrict attention to two classes of convolutional codes, based on how the encoding is done: nonrecursive, nonsystematic codes and recursive, systematic codes.

7.1.1 Nonrecursive nonsystematic encoding

Consider the following nonrecursive nonsystematic convolutional code: for an input sequence $\{u[k]\}$, the encoded sequence $c[k] = (y_1[k], y_2[k])$, where

$$y_1[k] = u[k] + u[k-1] + u[k-2],$$
$$y_2[k] = u[k] + u[k-2],$$ (7.1)

where the addition is modulo 2. A shift register implementation of the encoder is depicted in Figure 7.1.

The output sequences $\mathbf{y}_1 = \{y_1[k]\}$, $\mathbf{y}_2 = \{y_2[k]\}$ are generated by convolving the input sequence $\mathbf{u}$ with two "channels" using binary arithmetic. The output $y[k] = (y_1[k], y_2[k])$ at time k depends on the input $u[k]$ at time k, and the encoder state $s[k] = (u[k-1], u[k-2])$. A codeword is any sequence $\mathbf{y} = (\mathbf{y}_1, \mathbf{y}_2)$ that is a valid output of such a system. The rate R of the code equals the ratio of the number of information bits to the number of coded bits. In our example, $R = 1/2$, since two coded bits $y_1[k], y_2[k]$ are generated per information bit $u[k]$ coming in.

Nomenclature Some common terminology used to describe convolutional encoding is summarized below. It is common to employ the *D*-transform in the literature on convolutional codes; this is the same as the *z*-transform commonly used in signal processing, except that the delay operator $D = z^{-1}$.

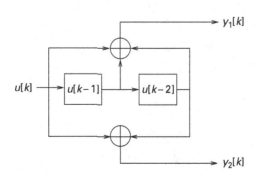

Figure 7.1 Shift register implementation of convolutional encoder for the running example. The outputs $(y_1[k], y_2[k])$ at time k are a function of the input $u[k]$ and the shift register state $s[k] = (u[k-1], u[k-2])$.

Channel coding

For a discrete-time sequence $\{x[k]\}$, let $x(D) = \sum_k x[k]D^k$ denote the D-transform. The encoding operation (7.1) can now be expressed as

$$y_1(D) = u(D)(1+D+D^2),$$
$$y_2(D) = u(D)(1+D^2).$$

Thus, we can specify the convolutional encoder by the set of two *generator polynomials*

$$G(D) = \left[g_1(D) = 1+D+D^2, \quad g_2(D) = 1+D^2 \right]. \tag{7.2}$$

The input polynomial $u(D)$ is multiplied by the generator polynomials to obtain the codeword polynomials. The generator polynomials are often specified in terms of their coefficients. Thus, the generator vectors corresponding to the polynomials are

$$[\mathbf{g}_1 = (1\ 1\ 1), \quad \mathbf{g}_2 = (1\ 0\ 1)]. \tag{7.3}$$

Often, we specify the encoder even more compactly by representing the preceding coefficients in octal format. Thus, in our example, the generators are specified as $[7, 5]$.

Trellis representation As for a dispersive channel, we can introduce a trellis that represents the code. The trellis has four states at each time k, corresponding to the four possible values of $s[k]$. The transition from $s[k]$ to $s[k+1]$ is determined by the value of the input $u[k]$. In Figure 7.2, we show a section of the trellis between time k and $k+1$, with each branch labeled with the input and outputs associated with it: $u[k]/y_1[k]y_2[k]$. Each path through the trellis corresponds to a different information sequence u and a corresponding *codeword* y.

We use this nonrecursive, nonsystematic code as a running example for our discussions of ML decoding and its performance analysis.

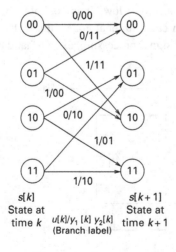

Figure 7.2 A section of the trellis representation of the code, showing state transitions between time k and $k+1$. Each trellis branch is labeled with the input and outputs associated with it, $u[k]/y_1[k]y_2[k]$.

7.1.2 Recursive systematic encoding

The same set of codewords as in Section 7.1.1 can be obtained using a recursive systematic encoder, simply by dividing the generator polynomials in (7.2) by $g_1(D)$. That is, we use the set of generators

$$G(D) = \left[1, \frac{g_2(D)}{g_1(D)} = \frac{1+D^2}{1+D+D^2}\right]. \tag{7.4}$$

Thus, the encoder outputs two sequences, the information sequence $\{u[k]\}$, and a parity sequence $v[k]$ whose D-transform satisfies

$$v(D) = u(D)\frac{1+D^2}{1+D+D^2}.$$

The code can still be specified in octal notation as [7, 5], where we understand that, for a recursive systematic code, the parity generating polynomial is obtained by dividing the second polynomial by the first one.

We would now like to specify a shift register implementation for generating the parity sequence $\{v[k]\}$. The required transfer function we wish to implement is $(1+D^2)/(1+D+D^2)$. Let us do this in two stages, first by implementing the transfer function $1/(1+D+D^2)$, which requires feedback, and then the feedforward transfer function $1+D^2$. To this end, define

$$y(D) = \frac{u(D)}{1+D+D^2}$$

as the output of the first stage. We see that

$$y(D) + Dy(D) + D^2 y(D) = u(D),$$

so that

$$y[k] + y[k-1] + y[k-2] = u[k].$$

Thus, in binary arithmetic, we have

$$y[k] = u[k] + y[k-1] + y[k-2].$$

We now have to pass $\{y[k]\}$ through the feedforward transfer function $1+D^2$ to get $\{v[k]\}$. That is,

$$v[k] = y[k] + y[k-2].$$

The resulting encoder implementation is depicted in Figure 7.3.

We employ this recursive, systematic code as a running example in our later discussions of maximum a posteriori probability (MAP) decoding and turbo codes.

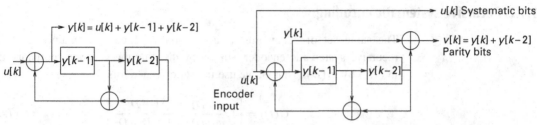

(a) Shift register realization of feedback transfer function $1/(1+D+D^2)$

(b) Shift register realization of encoder for recursive systematic code (cascade feedforward transfer function with feedback function in (a) to generate parity bits)

Figure 7.3 Shift register implementation of a [7,5] recursive systematic code. The state of the shift register at time k is $s[k] = (y[k-1], y[k-2])$, and the outputs at time k depend on the input $u[k]$ and the state $s[k]$.

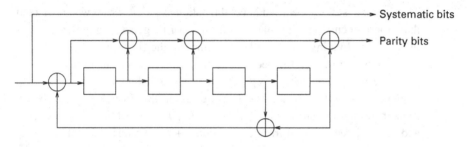

Figure 7.4 Recursive systematic encoder for a [23,35] code.

Another example code While our running example is a 4-state code, in practice, we often use more complex codes; for example, a 16-state code is shown in Figure 7.4. This code is historically important because it was a component code for the turbo code invented by Berrou et al. in 1993. This example also gives us the opportunity to clarify our notation for the code generators. Note that $g_1 = 10011$ (specifies the feedback taps) and $g_2 = 11101$ (specifies the feedforward taps), reading the shift register tap settings from left to right. The convention for the octal notation for specifying the generators is to group the bits specifying the taps in groups of three, from right to left. This yields $g_1 = 23$ and $g_2 = 35$. Thus, the code in Figure 7.4 is a [23,35] recursive systematic code.

7.1.3 Maximum likelihood decoding

We use the rate 1/2 [7,5] code as a running example in our discussion. For both the encoders shown in Figures 7.1 and 7.4, an incoming input bit $u[k]$ at time k results in two coded bits, say $c_1[k]$ and $c_2[k]$, that depend on $u[k]$ and the state $s[k]$ at time k. Also, there is a unique mapping between

7.1 Binary convolutional codes

$(u[k], s[k])$ and $(s[k], s[k+1])$, since, given $s[k]$, there is a one-to-one mapping between the input bit and the next state $s[k+1]$. Thus, $c_1[k]$ and $c_2[k]$ are completely specified given either $(u[k], s[k])$ or $(s[k], s[k+1])$. This observation is important in the development of efficient algorithms for ML decoding.

Let us now consider the example of BPSK transmission for sending these coded bits over an AWGN channel.

BPSK transmission Letting E_b denote the received energy per information bit, the energy per code symbol is $E_s = E_b R$, where R is the code rate. For BPSK transmission over a discrete-time real WGN channel, therefore, the noisy received sequence $\mathbf{z}[k] = (z_1[k], z_2[k])$ is given by

$$z_1[k] = \sqrt{E_s}(-1)^{c_1[k]} + n_1[k], \qquad (7.5)$$
$$z_2[k] = \sqrt{E_s}(-1)^{c_2[k]} + n_2[k],$$

where $\{n_1[k]\}, \{n_2[k]\}$ are i.i.d. $N(0, \sigma^2)$ random variables ($\sigma^2 = N_0/2$). *Hard* decision decoding corresponds to only the signs of the received sequence $\{\mathbf{z}[k]\}$ being passed up to the decoder. *Soft* decisions correspond to the real values $\{z_i[k]\}$, or multilevel quantization (number of levels greater than two) of some function of these values, being passed to the decoder. We now discuss maximum likelihood decoding when the real values $\{z_i[k]\}$ are available to the decoder.

Maximum likelihood decoding with soft decisions An ML decoder for the AWGN channel must minimize the minimum distance between the noisy received signal and the set of possible transmitted signals. For any given information sequence $\mathbf{u} = \{u[k]\}$ (with corresponding codeword $\mathbf{c} = \{(c_1[k], c_2[k])\}$), this distance can be written as

$$D(\mathbf{u}) = \sum_k \left\{ \left(z_1[k] - \sqrt{E_s}(-1)^{c_1[k]}\right)^2 + \left(z_2[k] - \sqrt{E_s}(-1)^{c_2[k]}\right)^2 \right\}.$$

Recalling that $(c_1[k], c_2[k])$ are determined completely by $s[k]$ and $s[k+1]$, we can denote the kth term in the above sum by

$$\lambda_k(s[k], s[k+1]) = \left(z_1[k] - \sqrt{E_s}(-1)^{c_1[k]}\right)^2 + \left(z_2[k] - \sqrt{E_s}(-1)^{c_2[k]}\right)^2.$$

The ML decoder must therefore minimize an additive *distance squared* metric to obtain the sequence

$$\hat{\mathbf{u}}_{ML} = \arg \min_{\mathbf{u}} D(\mathbf{u}) = \arg \min_{\mathbf{u}} \sum_k \lambda_k(s[k], s[k+1]).$$

An alternative form of the metric is obtained by noting that

$$\left(z_i[k] - \sqrt{E_s}(-1)^{c_i[k]}\right)^2 = z_i^2[k] + E_s - 2\sqrt{E_s} z_i[k](-1)^{c_i[k]},$$

with the first two terms on the right-hand side independent of $\mathbf{u}$. Dropping these terms, and scaling and changing the sign of the third term, we can therefore define an alternative *correlator* branch metric:

$$\gamma_k(s[k], s[k+1]) = z_i[k](-1)^{c_i[k]},$$

where the objective is now to *maximize* the sum of the branch metrics.

For an information sequence of length K, a direct approach to ML decoding would require comparing the metrics for 2^K possible sequences; this exponential complexity in K makes the direct approach infeasible even for moderately large values of K. Fortunately, ML decoding can be accomplished much more efficiently, with complexity linear in K, using the Viterbi algorithm described below.

The basis for the Viterbi algorithm is the *principle of optimality* for additive metrics, which allows us to prune drastically the set of candidates when searching for the ML sequence. Let

$$\Lambda(m:n, \mathbf{u}) = \sum_{k=m}^{n} \lambda_k(s_\mathbf{u}[k], s_\mathbf{u}[k+1])$$

denote the running sum of the branch metrics between times m and n, where $\{s_\mathbf{u}[k]\}$ denotes the sequence of trellis states corresponding to $\mathbf{u}$.

Principle of optimality Suppose that two sequences $\mathbf{u}_1$ and $\mathbf{u}_2$ have the same state at times m and n (i.e., $s_{\mathbf{u}_1}[m] = s_{\mathbf{u}_2}[m]$ and $s_{\mathbf{u}_1}[n] = s_{\mathbf{u}_2}[n]$), as shown in Figure 7.5. Then the sequence that has a worse running sum between m and n cannot be the ML sequence.

Proof For concreteness, suppose that we seek to maximize the sum metric, and that $\Lambda(m:n, \mathbf{u}_1) > \Lambda(m:n, \mathbf{u}_2)$. Then we claim that $\mathbf{u}_2$ cannot be the ML sequence. To see this, note that the additive nature of the metric implies that

$$\Lambda(\mathbf{u}_2) = \Lambda(1:m-1, \mathbf{u}_2) + \Lambda(m:n, \mathbf{u}_2) + \Lambda(n+1:K, \mathbf{u}_2). \tag{7.6}$$

Since $\mathbf{u}_2$ and $\mathbf{u}_1$ have the same states at times m and n, and the branch metrics depend only on the states at either end of the branch, we can replace the segment of $\mathbf{u}_2$ between m and n by the corresponding segment from $\mathbf{u}_1$

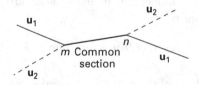

Figure 7.5 Two paths through a trellis with a common section between times m and n. The principle of optimality states that the path with the worse metric in the common section cannot be the ML path.

7.1 Binary convolutional codes

without changing the first and third terms in (7.6). We get a new sequence $\mathbf{u}_3$ with metric

$$\Lambda(\mathbf{u}_3) = \Lambda(1:m-1,\mathbf{u}_2) + \Lambda(m:n,\mathbf{u}_1) + \Lambda(n+1:K,\mathbf{u}_2) > \Lambda(\mathbf{u}_2).$$

Since $\mathbf{u}_3$ has a better metric than $\mathbf{u}_2$, we have shown that $\mathbf{u}_2$ cannot be the ML sequence. □

We can now state the Viterbi algorithm.

Viterbi algorithm Assume that the starting state of the encoder $s[0]$ is known. Now, all sequences through the trellis meeting at state $s[k]$ can be directly compared, using the principle of optimality between times 0 and k, and all sequences except the one with the best running sum can be discarded. If the trellis has S states at any given time (the algorithm also applies to time-varying trellises where the number of states can depend on time), we have exactly S surviving sequences, or *survivors*, at any given time. We need to keep track of only these S sequences (i.e., the sequence of states through the trellis, or equivalently, the input sequence, that they correspond to) up to the current time. We apply this principle successively at times $k = 1, 2, 3, \ldots$. Consider the S survivors at time k. Let $F(s')$ denote the set of possible values of the next state $s[k+1]$, given that the current state is $s[k] = s'$. For example, for a convolutional code with one input bit per unit time, for each possible value of $s[k] = s'$, there are two possible values of $s[k+1]$; that is, $F(s')$ contains two states. Denote the running sum of metrics up to time k for the survivor at $s[k] = s'$ by $\Lambda^*(1:k, s')$. We now extend the survivors by one more time step as follows:

Add step   For each state s', extend the survivor at s' in all admissible ways, and add the corresponding branch metric to the current running sum to get

$$\Lambda_0(1:k+1, s' \to s) = \Lambda^*(1:k, s') + \lambda_{k+1}(s', s), \quad s \in F(s').$$

Compare step   After the "add" step, each possible state $s[k+1] = s$ has a number of candidate sequences coming into it, corresponding to different possible values of the prior state. We compare the metrics for these candidates and choose the best as the survivor at $s[k+1] = s$. Denote by $P(s)$ the set of possible values of $s[k] = s'$, given that $s[k+1] = s$. For example, for a convolutional code with one input bit per unit time, $P(s)$ has two elements. We can now update the metric of the survivor at $s[k+1] = s$ as follows (assuming for concreteness that we wish to maximize the running sum)

$$\Lambda^*(1:k+1, s) = \max_{s' \in P(s)} \Lambda_0(1:k+1, s' \to s)$$

and store the maximizing s' for each $s[k+1] = s$. (When we wish to minimize the metric, the maximization above is replaced by minimization.)

At the end of the add and compare steps, we have extended the set of S survivors by one more time step. If the information sequence is chosen such that the terminating state is fixed, then we simply pick the survivor with the best metric at the terminal state as the ML sequence. The complexity of this algorithm is $O(S)$ per time step; that is, it is exponential in the encoder complexity but linear in the (typically much larger) number of transmitted symbols. Contrast this with brute force ML estimation, which is exponential in the number of transmitted symbols.

The Viterbi algorithm is often simplified further in practical implementations. For true ML decoding, we must wait until the terminal state to make bit decisions, which can be cumbersome in terms of both decoding delay and memory (we need to keep track of S surviving information sequences) for long information sequences. However, we can take advantage of the fact that the survivors at time k typically have merged at some point in the past, and make hard decisions on the bits corresponding to this common section with the confidence that this section must be part of the ML solution. In practice, we may impose a hard constraint on the decoding delay d and say that, if the Viterbi algorithm is at time step k, then we must make hard decisions on all information bits prior to time step $k - d$. If the survivors at time k have not merged by time step $k - d$, therefore, we must employ heuristic rules for making bit decisions: for example, we may make decisions prior to $k - d$ corresponding to the survivor with the best metric at time k. Alternatively, some form of majority logic, or weighted majority logic, may be used to combine the information contained in all survivors at time k.

General applicability of the Viterbi algorithm The Viterbi algorithm applies whenever there is an additive metric that depends only on the current time and the state transition, and is an example of *dynamic programming*. In the case of BPSK transmission over the AWGN channel, it is easy to see, for example, how the Viterbi algorithm applies if we quantize the channel outputs. Referring back to (7.5), suppose that we pass back to the decoder the quantized observation $\mathbf{r}[k] = \mathcal{Q}(\mathbf{z}[k])$, where $\mathcal{Q}$ is a memoryless transformation. An example of this is hard decisions on the code bits $c_i[k]$; that is $\mathbf{r}[k] = (\hat{c}_1[k], \hat{c}_2[k])$, where

$$\hat{c}_i[k] = 1_{z_i[k]<0}, \quad i = 1, 2.$$

Regardless of the choice of $\mathcal{Q}$, we can characterize the equivalent discrete memoryless channel $p(\mathbf{r}[k]|\mathbf{c}[k])$ that it induces. By the independence of the noise at different time units, we can write the ML decoding metric in terms of maximizing the log likelihood ratio,

$$\sum_k \log p(\mathbf{r}[k]|\mathbf{c}[k]), \tag{7.7}$$

over all possible codewords $\mathbf{c}$ (or equivalently, information sequences $\mathbf{u}$).

As discussed in Problem 7.7, ML decoding for the BSC induced by hard decisions for BPSK over an AWGN channel takes a particularly simple and intuitive form, minimizing the Hamming distance between the hard decision estimate and the codewords.

7.1.4 Performance analysis of ML decoding

We develop a general framework for performance analysis, rather than considering a code with a specific structure. Let $\mathbf{c} = \{c[i]\}$ denote a binary codeword in $\mathbf{C}$, the set of all possible codewords. Let $\mathbf{s} = \{s_i\}$ denote the BPSK signal corresponding to it. Thus,

$$s[i] = \sqrt{E_s}(-1)^{c[i]}.$$

We first analyze the performance of ML decoding with channel outputs directly available to the decoder (i.e., with unquantized soft decisions). We then note that the same methods apply when the decoder only has access to quantized channel outputs.

The first step is to determine the pairwise error probability, or the probability that, given that $\mathbf{c}_1$ is the transmitted codeword, the ML decoder outputs a different codeword $\mathbf{c}_2$. We know that, over an AWGN channel, this probability is determined by the Euclidean distance between the BPSK signals $\mathbf{s}_1$ and $\mathbf{s}_2$ corresponding to these two codewords. Now, $s_2[i] - s_1[i] = \pm 2\sqrt{E_s}$ if $c_2[i] \neq c_1[i]$. Let $d_H(\mathbf{c}_1, \mathbf{c}_2)$ denote the Hamming distance between $\mathbf{c}_1$ and $\mathbf{c}_2$, which is defined as the number of bits in which they differ. We can also write the Hamming distance as $w(\mathbf{c}_2 - \mathbf{c}_1)$, the weight of the difference between the two codewords (computed, of course, using binary arithmetic, so that $\mathbf{c}_2 - \mathbf{c}_1 = \mathbf{c}_2 + \mathbf{c}_1$). Thus,

$$\|\mathbf{s}_2 - \mathbf{s}_1\|^2 = \sum_i |s_2[i] - s_1[i]|^2 = 4E_s w(\mathbf{c}_2 - \mathbf{c}_1) = 4E_b R\, w(\mathbf{c}_2 - \mathbf{c}_1),$$

where R is the code rate. We can now write the pairwise error probability of decoding to $\mathbf{c}_2$ when $\mathbf{c}_1$ is sent as

$$P[\mathbf{c}_1 \to \mathbf{c}_2] = Q\left(\frac{\|\mathbf{s}_2 - \mathbf{s}_1\|}{2\sigma}\right) = Q\left(\sqrt{\frac{2E_b R\, w(\mathbf{c}_2 - \mathbf{c}_1)}{N_0}}\right). \quad (7.8)$$

Condition on sending the all-zero codeword Now, note that a convolutional code is a linear code, in that, if $\mathbf{c}_1$ and $\mathbf{c}_2$ are codewords, then so is $\mathbf{c}_2 - \mathbf{c}_1 = \mathbf{c}_2 + \mathbf{c}_1$. We can therefore subtract out $\mathbf{c}_1$ from the code, leaving it (and hence the relative geometry of the corresponding BPSK modulated signal set) unchanged. This means that we can condition, without loss of generality, on the all-zero codeword being sent. Let us therefore set $\mathbf{c}_1$ equal to the all-zero codeword, and set $\mathbf{c}_2 = \mathbf{c}$.

We wish to estimate $P_e(k)$, the probability that the kth information bit, $u[k]$ is decoded incorrectly. Since we have sent the all-zero codeword, the correct information sequence is the all-zero sequence. Thus, to make an error in the kth information bit, the ML decoder must output a codeword such that $u[k] = 1$. Let $\mathbf{C}[k]$ denote the set of all such codewords:

$$\mathbf{C}[k] = \{\mathbf{c} \in \mathbf{C} : u[k] = 1\}.$$

Clearly,

$$P_e(k) = P[\text{ML decoder chooses some } \mathbf{c} \in \mathbf{C}_k].$$

From (7.8), we see that the pairwise error probability that a codeword $\mathbf{c}$ has a higher likelihood than the all-zero codeword, when the all-zero codeword is sent, is given by $Q(\sqrt{(2E_b R w(\mathbf{c})/N_0)})$. Since this depends only on the weight of $\mathbf{c}$, it is convenient to define

$$q(x) = Q\left(\sqrt{\frac{2E_b R x}{N_0}}\right) \qquad (7.9)$$

as the pairwise error probability for a codeword of weight x relative to the all-zero codeword.

We start with a loose union bound on the bit error probability

$$P_e(k) \leq \sum_{\mathbf{c} \in \mathbf{C}[k]} q(w(\mathbf{c})). \qquad (7.10)$$

We now want to prune the terms in (7.10) to obtain an "intelligent union bound." To do this, consider Figure 7.6, which shows a simplified schematic of the ML codeword $\mathbf{c}$ and the transmitted all-zero codeword as paths through a trellis. Any nonzero codeword must diverge from the all-zero codeword on the code trellis at some point. Such a codeword may or may not remerge with the all-zero codeword at some later point, and in general, may diverge and

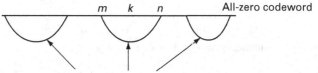

ML codeword having an error in information bit k must diverge from all-zero codeword around k
ML codeword may diverge and remerge from all-zero codeword several times

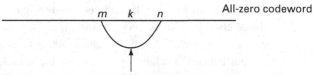

Simple codeword coinciding with ML codeword between m and n
Has better metric than all-zero codeword, and has an error in information bit k

Figure 7.6 Transmitted all-zero codeword and ML codeword as paths on the trellis. In the scenario depicted, the input bit $u[k]$ corresponding to the ML codeword is incorrect. I also show the *simple* codeword which coincides with the ML codeword where it diverges from the all-zero path around bit k, and coincides with the all-zero codeword elsewhere.

7.1 Binary convolutional codes

remerge several times. We define a *simple* codeword as a nonzero codeword that, if it remerges with the all-zero codeword, never diverges again. Let $\mathbf{C}_s[k]$ denote all simple codewords that belong to $\mathbf{C}[k]$; that is,

$$\mathbf{C}_s[k] = \{\mathbf{c} \in \mathbf{C} : u[k] = 1, \mathbf{c} \text{ simple}\}.$$

We now state and prove that the union bound (7.10) can be pruned to include only simple codewords.

Proposition 7.1.1 (Intelligent union bound using simple codewords) *The probability of bit error is bounded as*

$$P_e(k) \leq \sum_{\mathbf{c} \in \mathbf{C}_s[k]} q(w(\mathbf{c})). \tag{7.11}$$

Proof Consider the scenario depicted in Figure 7.6. Since the ML codeword and the all-zero codeword have the same state at times m and n, by the principle of optimality, the sum of the branch metrics between times m and n must be strictly greater for the ML path. That is,

$$\Lambda(m:n, \mathbf{u}_{\text{ML}}) > \Lambda(m:n, \mathbf{0}),$$

where $\mathbf{u}_{\text{ML}}$ denotes the information sequence corresponding to the ML codeword. Thus, the accumulated metric for the simple codeword which coincides with the ML codeword between m and n, and with the all-zero path elsewhere, must be bigger than that of the all-zero path, since the difference in their metrics is precisely the difference accumulated between m and n, given by

$$\Lambda(m:n, \mathbf{u}) - \Lambda(m:n, \mathbf{0}) > 0.$$

This shows that the ML codeword $\mathbf{c} \in \mathbf{C}[k]$ if and only if there is *some* simple codeword $\tilde{\mathbf{c}} \in \mathbf{C}_s[k]$ which has a better metric than the all-zero codeword. A union bound on the latter event is given by

$$P_e(k) \leq \sum_{\mathbf{c} \in \mathbf{C}_s[k]} P[\mathbf{c} \text{ has better metric than } \mathbf{0}|\mathbf{0} \text{ sent}] = \sum_{\mathbf{c} \in \mathbf{C}_s[k]} q(w(\mathbf{c})),$$

which proves the desired result. □

We now want to count simple codewords efficiently for computing the above bound. To this end, we use the concept of *error event*, defined via the trellis representation of the code.

Definition 7.1.1 (Error event) *An error event $\mathbf{c}$ is a simple codeword which diverges on the trellis from the all-zero codeword for the first time at time zero.*

Figure 7.7 An error event for our running example of a nonrecursive, nonsystematic rate 1/2 code with generator [7, 5].

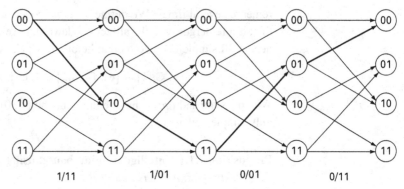

(Inputs and outputs along the path shown in bold)
Input weight $i = 2$
Output weight $w = 6$

For the rate 1/2 nonrecursive, nonsystematic code [7, 5] in Section 7.1.1 that serves as my running example, Figure 7.7 shows an error event marked as a path in bold through the trellis. Note that the error event is a nonzero codeword that diverges from the all-zero path at time zero, and remerges four time units later (never to diverge again).

Let **E** denote the set of error events. Suppose that a given codeword $\mathbf{c} \in \mathbf{E}$ has output weight x and input weight i. That is, the input sequence that generates **c** has i nonzero elements, and the codeword **c** has x nonzero elements. Then we can translate **c** to create i simple error events in $\mathbf{C}_s[k]$, by lining up each of the nonzero input bits in turn with $u[k]$. The corresponding pairwise error probability $q(x)$ depends only on the output weight w. Now, suppose there are $A(i, x)$ error events with input weight i and output weight x. We can now rewrite the bound (7.11) as follows:

Union bound using error event weight enumeration

$$P_e(k) \leq \sum_{i=1}^{\infty} \sum_{x=1}^{\infty} i A(i, x) q(x). \qquad (7.12)$$

If $\{A(i, x)\}$, the *weight enumerator* function of the code, is known, then the preceding bound can be directly computed, truncating the infinite summations in i and x at moderate values, exploiting the rapid decay of the Q function with its argument.

We can also use a "nearest neighbor" approximation, in which we only consider the minimum weight codewords in the preceding sum. The minimum possible weight for a nonzero codeword is called the *free distance* of the code, d_{free}. That is,

$$d_{\text{free}} = \min\{x > 0 : A(i, x) > 0 \text{ for some } i\}.$$

7.1 Binary convolutional codes

Then the nearest neighbor approximation is given by

$$P_e(k) \approx Q\left(\sqrt{\frac{2E_b R d_{\text{free}}}{N_0}}\right) \sum_i i A(i, d_{\text{free}}). \tag{7.13}$$

This provides information on the high SNR asymptotics of the error probability. The exponent of decay of error probability with E_b/N_0 relative to uncoded BPSK is better by a factor of Rd_{free}, which is termed the coding gain (typically expressed in dB). Of course, this provides only coarse insight; convolutional codes are typically used at low enough SNR that it is necessary to go beyond the nearest neighbors approximation to estimate the error probability accurately.

We now show how $A(i, x)$ can be computed using the transfer function method. We also slightly loosen the bound (7.12) to get a more explicit form that can be computed using the transfer function method without truncation of the summations in i and x.

Transfer function Define the transfer function

$$T(I, X) = \sum_{i=1}^{\infty} \sum_{x=1}^{\infty} A(i, x) I^i X^x. \tag{7.14}$$

This transfer function can be computed using a state diagram representation for the convolutional code. We illustrate this procedure using our running example, the nonrecursive, nonsystematic encoder depicted in Figure 7.1 in Section 7.1.1. The state diagram is depicted in Figure 7.8. We start from the all-zero state START and end at the all-zero state END, but the states in between are all nonzero. Thus, a path from START to END is an error event, or a codeword that diverges from the all-zero codeword for the first time at time zero, and does not diverge again once it remerges with the all-zero codeword. By considering all possible paths from START to END, we can enumerate all possible error events. If a state transition corresponds to a nonzero input bits and b nonzero output bits, then the branch gain for that transition is $I^a X^b$. For an error event of input weight i and output weight x, the product of all branch gains along the path equals $I^i X^x$. Thus, summing over all possible paths gives us the transfer function $T(I, X)$ between START and END.

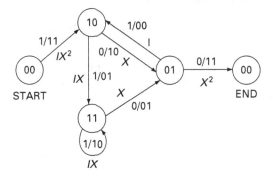

Figure 7.8 State diagram for running example. Each transition is labeled with the input and output bits, as well as a branch gain $I^a X^b$, where a is the input weight, and b the output weight.

The transfer function for our running example equals

$$T(I, X) = \frac{IX^5}{1 - 2IX}. \tag{7.15}$$

A formal expansion yields

$$T(I, X) = IX^5 \sum_{k=0}^{\infty} 2^k I^k X^k. \tag{7.16}$$

Comparing the coefficients of terms of the form $I^i X^x$, we can now read off $A(i, x)$, and then compute (7.12). We can also see that the free distance $d_{\text{free}} = 5$, corresponding to the smallest power of X that appears in (7.16). Thus, the coding gain Rd_{free} relative to uncoded BPSK is $10\log_{10}(5/2) \approx 4$ dB. Note that the running example is meant to illustrate basic concepts, and that better coding gains can be obtained at the same rate by increasing the code memory (with a corresponding penalty in terms of decoding complexity, which is proportional to the number of trellis states).

Transfer function bound We now develop a transfer function based bound that can be computed without truncating the sum over paths from START to END. Using the bound $Q(x) \leq \frac{1}{2} e^{-x^2/2}$ in (7.9), we have

$$q(x) \leq ab^x, \tag{7.17}$$

where $a = 1/2$ and $b = e^{-\frac{E_b R}{N_0}}$. Plugging into (7.12), we get the slightly weaker bound

$$P_e(k) \leq a \sum_{i=1}^{\infty} \sum_{x=1}^{\infty} iA(i, x)b^x. \tag{7.18}$$

From (7.14), we see that

$$\frac{\partial}{\partial I} T(I, X) = \sum_{i=1}^{\infty} \sum_{x=1}^{\infty} A(i, w) i I^{i-1} X^x.$$

We can now rewrite (7.18) as follows:

Transfer function bound

$$P_e(k) \leq a \frac{\partial}{\partial I} T(I, X)|_{I=1, X=b} \tag{7.19}$$

($a = 1/2$, $b = e^{-\frac{E_b R}{N_0}}$ for soft decisions).

For our running example, we can evaluate the transfer function bound (7.19) using (7.15) to get

$$P_e \leq \frac{1}{2} \frac{e^{-\frac{5E_b}{2N_0}}}{\left(1 - 2e^{-\frac{E_b}{2N_0}}\right)^2}. \tag{7.20}$$

For moderately high SNR, this is close to the nearest neighbors (7.13), which is given by

$$P_e \approx Q\left(\sqrt{\frac{5E_b}{N_0}}\right) \leq \frac{1}{2}e^{-\frac{5E_b}{2N_0}},$$

where we have used (7.16) to infer that $d_{\text{free}} = 5$, $A(i, d_{\text{free}}) = 1$ for $i = 1$, and $A(i, d_{\text{free}}) = 0$ for $i > 1$.

7.1.5 Performance analysis for quantized observations

We noted in Section 7.1.3 that the Viterbi algorithm applies in great generality, and can be used in particular for ML decoding using quantized observations. We now show that the performance analysis methods we have discussed are also directly applicable in this setting. To see this, consider a single coded bit c sent using BPSK over an AWGN channel. The corresponding real-valued received sample is $z = \sqrt{E_s}(-1)^c + N$, where $N \sim N(0, \sigma^2)$. A quantized version $r = \mathcal{Q}(z)$ is then sent up to the decoder. The equivalent discrete memoryless channel has transition densities $p(r|1)$ and $p(r|0)$. When running the Viterbi algorithm to maximize the log likelihood, the branch metric corresponding to r is $\log p(r|0)$ for a trellis branch with $c = 0$, and $\log p(r|1)$ for a trellis branch with $c = 1$.

The quantized observations inherit the symmetry of the noise and the signal around the origin, as long as the quantizer is symmetric. That is, $p(r|0) = p(-r|1)$ for a symmetric quantizer. Under this condition, it can be shown with a little thought that there is no loss of generality in assuming in our performance analysis that the all-zero codeword is sent.

Next, we discuss computation of pairwise error probabilities. A given nonzero codeword $\mathbf{c}$ is more likely than the all-zero codeword if

$$\sum_i \log p(r_i|c_i) > \sum_i \log p(r_i|0),$$

where c_i denotes the ith code symbol, and r_i the corresponding quantized observation. Canceling the common terms corresponding to $c_i = 0$, we see that $\mathbf{c}$ is more likely than the all-zero codeword if

$$\sum_{i:c_i=1} \log \frac{p(r_i|1)}{p(r_i|0)} > 0.$$

If $\mathbf{c}$ has weight x, then there are x terms in the summation above. These terms are independent and identically distributed, conditioned on the all-zero codeword being sent. A typical term is of the form

$$V = \log \frac{p(r|1)}{p(r|0)}, \qquad (7.21)$$

where, conditioned on the code bit $c = 0$,

$$r = \mathcal{Q}(\sqrt{E_s} + N), \quad N \sim N(0, \sigma^2).$$

It is clear that the pairwise error probability depends only on the codeword weight x, hence we denote it as before by $q(x)$, where

$$q(x) = P_0[V_1 + \cdots + V_x > 0], \qquad (7.22)$$

with P_0 denoting the distribution conditioned on zero code bits being sent. Given the equivalent channel model, we can compute $q(x)$ exactly.

We now note that the intelligent union bound (7.12) applies as before, since its derivation only used the principle of optimality and the fact that the pairwise error probability for a codeword depends only on its weight. Only the value of $q(x)$ depends on the specific form of quantization employed.

The transfer function bound (7.19) is also directly applicable in this more general setting. This is because, for sums of i.i.d. random variables as in (7.22), we can find Chernoff bounds (see Appendix B) of the form

$$q(x) \leq ab^x$$

for constants $a > 0$ and $b \leq 1$. A special case of the Chernoff bound that is useful for random variables which are log likelihood ratios, as in (7.21) is the Bhattacharya bound, introduced in Problem 7.9, and applied in Problems 7.10 and 7.11.

> **Example 7.1.1 (Performance with hard decisions)** Consider a BPSK system with hard decisions. The hard decision $r = \hat{c} = I_{\{z<0\}}$, where $z = \sqrt{E_s}(-1)^c + N$ is the noisy observation corresponding to the transmitted code symbol $c \in \{0, 1\}$, where $N \sim N(0, \sigma^2)$. Clearly, we can model r as the output when c is passed through an equivalent BSC with crossover probability $p = Q(\sqrt{2E_b R/N_0})$. In Problems 7.8 and 7.10, we derive the following upper bound on the pairwise error probability for ML decoding over a BSC:
>
> $$q(x) \leq (2\sqrt{p(1-p)})^x.$$

This can now be plugged into the transfer function bound to estimate the BER with hard decisions as follows:

$$P_e \leq \frac{\partial}{\partial I} T(I, X)\Big|_{I=1, X=b}, \qquad (7.23)$$

where $b = 2\sqrt{p(1-p)}$. For large SNR, $p \doteq e^{-E_b R/N_0}$, so that $b = 2\sqrt{p(1-p)} \approx 2\sqrt{p} \doteq e^{-E_b R/2N_0}$. Comparing with (7.19), where $b \approx e^{-E_b R/N_0}$, we see that hard decisions incur a 3 dB degradation in performance relative to soft decisions, asymptotically at high SNR (at low SNR, the degradation is smaller – about 2 dB). However, most of this deficit can be made up by using observations quantized using relatively few levels. Problem 7.11 explores this comment in further detail.

7.2 Turbo codes and iterative decoding

We now describe turbo codes, which employ convolutional codes as building blocks for constructing random-looking codes that perform very close to Shannon-theoretic limits. Since the randomization leads to dependencies among code bits that are very far apart in time, a finite state representation of the code amenable to ML decoding using the Viterbi algorithm is infeasible. Thus, a crucial component of this breakthrough in error correction coding is a mechanism for suboptimal iterative decoding, in which simple decoders for the component convolutional codes exchange information over several iterations. The component decoders are based on the BCJR algorithm (named in honor of its inventors, Bahl, Cocke, Jelinek, and Raviv), which provides an estimate of the a posteriori probability of each bit in a codeword, based on the received signal, and taking into account the constraints imposed by the code structure. Therefore, we first describe the BCJR algorithm in a great deal of detail. This makes our subsequent description of turbo codes and iterative decoding quite straightforward. We consider both parallel concatenated codes (the original turbo codes) and serial concatenated codes (often found to yield superior performance).

7.2.1 The BCJR algorithm: soft-in, soft-out decoding

Maximum likelihood decoding using the Viterbi algorithm chooses the most likely codeword. If all codewords are equally likely to be sent, this also minimizes the probability of choosing the wrong codeword. In contrast, the BCJR algorithm provides estimates of the posterior distribution of each bit in the codeword. The method applies to both the information bits and coded bits for all classes of convolutional codes that I have discussed so far. The major part of the computation is in running two Viterbi-like algorithms, one forward and one backward, through the trellis. The complexity of the BCJR algorithm is therefore somewhat higher than that of the Viterbi algorithm. For any given bit $b \in \{0, 1\}$, the output of the BCJR algorithm can be summarized by the log likelihood ratio (LLR) $L_{\text{out}}(b) = \log(P[b=0|\mathbf{y}])/(P[b=1|\mathbf{y}])$, where $\mathbf{y}$ denotes the observations fed to the BCJR algorithm. Note that computation of the posterior distribution of b requires knowledge of the prior distribution of b, which can also be summarized in terms of an LLR $L_{\text{in}}(b) = \log(P[b=0])/(P[b=1])$. These LLRs provide *soft* information regarding b, with our confidence on our knowledge of b increasing with their magnitude (the LLR takes value $+\infty$ if $b=0$ and value $-\infty$ if $b=1$).

We begin with an exposition of the original BCJR algorithm, followed by a detailed discussion of its logarithmic version, which is preferred in practice because of its numerical stability. The logarithmic implementation also provides more insight into the nature of the different kinds of information being used by the BCJR algorithm. This is important in our later discussion of how to use the BCJR algorithm as a building block for iterative decoding.

Notation: BPSK version of a bit Since we focus on BPSK modulation, it is convenient to define for any bit b taking values in $\{0, 1\}$ its BPSK counterpart,

$$\tilde{b} = (-1)^b, \tag{7.24}$$

taking values in $\{+1, -1\}$, with 0 mapped to $+1$ and 1 to -1. Thus,

$$L(b) = \log \frac{P[b=0]}{P[b=1]} = \log \frac{P[\tilde{b}=+1]}{P[\tilde{b}=-1]}. \tag{7.25}$$

After running the algorithm, we can make a maximum a posteriori probability (MAP) *hard* decision for b, if we choose to, as follows:

$$\hat{b}_{\text{MAP}} = \begin{cases} 0, & L(b) > 0, \\ 1, & L(b) < 0, \end{cases}$$

with ties broken arbitrarily. From the theory of hypothesis testing, we know that such MAP decoding minimizes the probability of error, so that the BCJR algorithm can be used to implement the bitwise minimum probability of error (MPE) rule. However, this in itself does not justify the additional complexity relative to the Viterbi algorithm: for a typical convolutional code, the BER obtained using ML decoding is almost as good as that obtained using MAP decoding. This is why the BCJR algorithm, while invented in 1974, did not have a major impact on the practice of decoding until the invention of turbo codes in 1993. We now know that the true value of the BCJR algorithm, and of a number of its suboptimal, lower-complexity, variants, lies in their ability to accept soft inputs and produce soft outputs. Interchange of soft information between such soft-in, soft-out (SISO) modules is fundamental to iterative decoding.

As a running example in this section, we consider the rate 1/2 RSC code with generator [7, 5] introduced earlier. A trellis section for this code is shown in Figure 7.9.

The fundamental quantity to be computed by the BCJR algorithm is the posterior probability of a given branch of the trellis being traversed by the transmitted codeword, given the received signal and the priors. Given the posterior probabilities of all allowable branches in a trellis section, we can compute posterior probabilities for the bits associated with these branches. For example, we see from Figure 7.9 that the input bit $u_k = 0$ corresponds to exactly four of the eight branches in the trellis section, so the posterior probability that $u_k = 0$ can be written as:

$$P[u_k = 0|\mathbf{y}] = P[s_k = 00, s_{k+1} = 00|\mathbf{y}] + P[s_k = 01, s_{k+1} = 10|\mathbf{y}]$$
$$+ P[s_k = 10, s_{k+1} = 11|\mathbf{y}] + P[s_k = 11, s_{k+1} = 01|\mathbf{y}]. \tag{7.26}$$

7.2 Turbo codes and iterative decoding

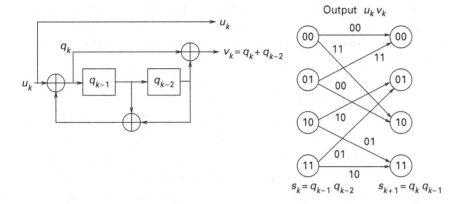

Figure 7.9 Shift register implementation and trellis section for our running example of a rate 1/2 recursive systematic code.

Similarly, the posterior probability that $u_k = 1$ can be obtained by summing up the posterior probability of the other four branches in the trellis section:

$$P[u_k = 1|\mathbf{y}] = P[s_k = 00, s_{k+1} = 10|\mathbf{y}] + P[s_k = 01, s_{k+1} = 00|\mathbf{y}]$$
$$+ P[s_k = 10, s_{k+1} = 01|\mathbf{y}] + P[s_k = 11, s_{k+1} = 11|\mathbf{y}].$$
(7.27)

The posterior probability $P[s_k = A, s_{k+1} = B|\mathbf{y}]$ of a branch $A \to B$ is proportional to the joint probability $P[s_k = A, s_{k+1} = B, \mathbf{y}]$, which is more convenient to compute. Note that we have abused notation in denoting this as a probability: $\mathbf{y}$ is often a random vector with continuous-valued components (e.g., the output of an AWGN channel with BPSK modulation), so that this "joint probability" is really a mixture of a probability mass function and a probability density function. That is, we get the value one when we sum over branches and integrate over $\mathbf{y}$. However, since we are interested in posterior distributions conditioned on $\mathbf{y}$, we never need to integrate out $\mathbf{y}$. We therefore do not need to be careful about this issue. On the other hand, the *posterior* probabilities of all branches in a trellis section do add up to one. Since the posterior probability of a branch is proportional to the joint probability, this gives us the normalization condition that we need. That is, for any state transition $A \to B$,

$$P[s_k = A, s_{k+1} = B|\mathbf{y}] = \xi_k P[s_k = A, s_{k+1} = B, \mathbf{y}], \quad (7.28)$$

where ξ_k is a normalization constant such that the posterior probabilities of all branches in the kth trellis section sum up to one:

$$\xi_k = \frac{1}{\sum_{s_k = s', s_{k+1} = s} P[s_k = s', s_{k+1} = s, \mathbf{y}]},$$

where only the eight branches $s' \to s$ that are feasible under the code constraints appear in the summation above. For example, the transition $10 \to 00$ does not appear, since it is not permitted by the code constraints in the code trellis.

Explicit computation of the normalization constant ξ_k is often not required (e.g., if we are interested in LLRs). For example, if we compute the output LLR of u_k from (7.26) and (7.27), and plug in (7.28), we see that ξ_k cancels out and we get

$$L_{\text{out}}(u_k) = \log \frac{P[u_k = 0|\mathbf{y}]}{P[u_k = 1|\mathbf{y}]}$$

$$= \log \left(\frac{P[s_k = 00, s_{k+1} = 00, \mathbf{y}] + P[s_k = 01, s_{k+1} = 10, \mathbf{y}]}{P[s_k = 00, s_{k+1} = 10, \mathbf{y}] + P[s_k = 01, s_{k+1} = 00, \mathbf{y}]} \right.$$

$$\left. \frac{+P[s_k = 10, s_{k+1} = 11, \mathbf{y}] + P[s_k = 11, s_{k+1} = 01, \mathbf{y}]}{+P[s_k = 10, s_{k+1} = 01, \mathbf{y}] + P[s_k = 11, s_{k+1} = 11, \mathbf{y}]} \right).$$

Let us now express this in more compact notation. Denote by U_0 and U_1 the branches in the trellis section corresponding to $u_k = 0$ and $u_k = 1$, respectively, given by

$$U_0 = \{(s', s) : s_k = s', s_{k+1} = s, u_k = 0\},$$
$$U_1 = \{(s', s) : s_k = s', s_{k+1} = s, u_k = 1\}. \tag{7.29}$$

In our example, we have

$$U_0 = \{(00, 00), (01, 10), (10, 11), (11, 01)\},$$
$$U_1 = \{(00, 10), (01, 00), (10, 01), (11, 11)\}.$$

(Since we consider a time-invariant trellis, the sets U_0 and U_1 do not depend on k. However, the method of computing bit LLRs from branch posteriors applies just as well to time-varying trellises.)

Writing

$$P_k(s', s, \mathbf{y}) = P[s_k = s', s_{k+1} = s, \mathbf{y}],$$

we can now provide bit LLRs as the output of the BCJR algorithm, as follows.

Log likelihood ratio computation

$$L_{\text{out}}(u_k) = \log \left(\frac{\sum_{(s',s) \in U_0} P_k(s', s, \mathbf{y})}{\sum_{(s',s) \in U_1} P_k(s', s, \mathbf{y})} \right). \tag{7.30}$$

This method applies to any bit associated with a given trellis section. For example, the LLR for the parity bit v_k output by the BCJR algorithm is computed by partitioning the branches according to the value of v_k:

$$L_{\text{out}}(v_k) = \log \left(\frac{\sum_{(s',s) \in V_0} P_k(s', s, \mathbf{y})}{\sum_{(s',s) \in V_1} P_k(s', s, \mathbf{y})} \right), \tag{7.31}$$

where, for our example, we see from Figure 7.9, that

$$V_0 = \{(s', s) : s_k = s', s_{k+1} = s, v_k = 0\}$$
$$= \{(00, 00), (01, 10), (10, 01), (11, 11)\},$$
$$V_1 = \{(s', s) : s_k = s', s_{k+1} = s, v_k = 1\}$$
$$= \{(00, 10), (01, 00), (10, 11), (11, 01)\}.$$

We now discuss how to compute the joint "probabilities" $P_k(s', s, \mathbf{y})$. Before doing this, let us establish some more notation. Let $\mathbf{y}_k$ denote the received signal corresponding to the bits sent in the kth trellis section, and let $\mathbf{y}_a^b$ denote $(\mathbf{y}_a, \mathbf{y}_{a+1}, \ldots, \mathbf{y}_b)$, the received signals corresponding to trellis sections a through b. Suppose that there are K trellis sections in all, numbered from 1 through K. Considering our rate 1/2 running example, suppose that we use BPSK modulation over an AWGN channel, and that we feed the unquantized channel outputs directly to the decoder. Then the received signal $\mathbf{y}_k = (y_k(1), y_k(2))$ in the kth trellis section is given by the two real-valued samples:

$$y_k(1) = A(-1)^{u_k} + N_{1,k} = A\tilde{u}_k + N_{1,k}, \\ y_k(2) = A(-1)^{v_k} + N_{2,k} = A\tilde{v}_k + N_{2,k}, \quad (7.32)$$

where $N_{1,k}, N_{2,k}$ are i.i.d. $N(0, \sigma^2)$ noise samples, and $A = \sqrt{E_s} = \sqrt{E_b/2}$ is the modulating amplitude.

Applying the chain rule for joint probabilities, we can now write

$$\begin{aligned} P_k(s', s, \mathbf{y}) &= P[s_k = s', s_{k+1} = s, \mathbf{y}_1^{k-1}, \mathbf{y}_k, \mathbf{y}_{k+1}^K] \\ &= P[\mathbf{y}_{k+1}^K | s_k = s', s_{k+1} = s, \mathbf{y}_1^{k-1}, \mathbf{y}_k] \\ &\quad \times P[s_{k+1} = s, \mathbf{y}_k | s_k = s', \mathbf{y}_1^{k-1}] P[s_k = s', \mathbf{y}_1^{k-1}]. \end{aligned} \quad (7.33)$$

We can now simplify the preceding expression as follows. Given $s_{k+1} = s$, the channel outputs $\mathbf{y}_{k+1}^K$ are independent of the values of the prior channel outputs $\mathbf{y}_1^k$ and the prior state s_k, because the channel is memoryless, and because future outputs of a convolutional encoder are determined completely by current state and future inputs. Thus, we have $P[\mathbf{y}_{k+1}^K | s_k = s', s_{k+1} = s, \mathbf{y}_1^{k-1}, \mathbf{y}_k] = P[\mathbf{y}_{k+1}^K | s_{k+1} = s]$. Following the notation in the original exposition of the BCJR algorithm, we define this quantity as

$$\beta_k(s) = P[\mathbf{y}_{k+1}^K | s_{k+1} = s]. \quad (7.34)$$

The memorylessness of the channel also implies that $P[s_{k+1} = s, \mathbf{y}_k | s_k = s', \mathbf{y}_1^{k-1}] = P[s_{k+1} = s, \mathbf{y}_k | s_k = s']$, since, given the state at time k, the past channel outputs do not tell us anything about the present and future states and channel outputs. We define this quantity as

$$\gamma_k(s', s) = P[s_{k+1} = s, \mathbf{y}_k | s_k = s']. \quad (7.35)$$

Finally, let us define the quantity

$$\alpha_{k-1}(s') = P[s_k = s', \mathbf{y}_1^{k-1}]. \quad (7.36)$$

We can now rewrite (7.33) as follows.

Branch probability computation

$$P_k(s', s, \mathbf{y}) = \beta_k(s) \gamma_k(s', s) \alpha_{k-1}(s'). \quad (7.37)$$

Note that α, β, γ do not have the interpretation of probability mass functions over a discrete space, since all of them involve the probability density of a possibly continuous-valued observation. Indeed, these functions can be scaled arbitrarily (possibly differently for each k), as long as the scaling is independent of the states. By virtue of (7.37), these scale factors get absorbed into the joint probability $P_k(s', s, \mathbf{y})$, which also does not have the interpretation of probability mass function over a discrete space. However, the *posterior* branch probabilities $P_k(s', s|\mathbf{y})$ must indeed sum to one over (s', s), so that the arbitrary scale factors are automatically resolved using the normalization (7.28). By the same reasoning, arbitrary (state-independent) scale factors in α, β, γ leave posterior bit probabilities and LLRs unchanged.

We now develop a forward recursion for α_k in terms of α_{k-1}, and a backward recursion for β_{k-1} in terms of β_k. Let us assume that the trellis sections are numbered from $k = 0, \ldots, K-1$, with initial all-zero state $s[0] = \mathbf{0}$. The final state s_k is also terminated at $\mathbf{0}$ (although we will have occasion to revisit this condition in the context of turbo codes).

We can rewrite $\alpha_k(s)$ using the law of total probability as follows:

$$\alpha_k(s) = P[s_{k+1} = s, \mathbf{y}_1^k] = \sum_{s'} P[s_{k+1} = s, \mathbf{y}_1^k, s_k = s'], \tag{7.38}$$

considering all possible prior states s' (the set of states s' that need to be considered is restricted by code constraints, as we illustrate in an example shortly). A typical term in the preceding summation can be rewritten as

$$P[s_{k+1} = s, \mathbf{y}_1^k, s_k = s'] = P[s_{k+1} = s, \mathbf{y}_k, \mathbf{y}_1^{k-1}, s_k = s']$$
$$= P[s_{k+1} = s, \mathbf{y}_k | \mathbf{y}_1^{k-1}, s_k = s'] P[\mathbf{y}_1^{k-1}, s_k = s'].$$

Now, given the present state $s_k = s'$, the future states and observations are independent of the past observations $\mathbf{y}_1^{k-1}$, so that

$$P[s_{k+1} = s, \mathbf{y}_k | \mathbf{y}_1^{k-1}, s_k = s'] = P[s_{k+1} = s, \mathbf{y}_k | s_k = s'] \gamma_k(s', s).$$

We now see that

$$P[s_{k+1} = s, \mathbf{y}_1^k, s_k = s'] = \gamma_k(s', s) \alpha_{k-1}(s').$$

Substituting into (7.38), we get the forward recursion in compact form.

Forward recursion

$$\alpha_k(s) = \sum_{s'} \gamma_k(s', s) \alpha_{k-1}(s'), \tag{7.39}$$

which is the desired forward recursion for α. If the initial state is known to be, say, the all-zero state $\mathbf{0}$, then we would initialize the recursion with

$$\alpha_0(s) = \begin{cases} 0, & s \neq \mathbf{0}, \\ c, & s = \mathbf{0}, \end{cases} \tag{7.40}$$

where the constant $c > 0$ can be chosen arbitrarily, since we only need to know $\alpha_k(s)$ for any given k up to a scale factor. We often set $c = 1$ for convenience.

7.2 Turbo codes and iterative decoding

Forward recursion for running example Consider now the forward recursion for $\alpha_k(s)$, $s = 01$. Referring to Figure 7.9, we see that the two possible values of prior state s' permitted by the code constraints are $s' = 10$ and $s' = 11$. We therefore get

$$\alpha_k(01) = \gamma_k(10, 01)\alpha_{k-1}(10) + \gamma_k(11, 01)\alpha_{k-1}(11). \tag{7.41}$$

Similarly, we can rewrite $\beta_{k-1}(s')$ using the law of total probability, considering all possible future states s, as follows:

$$\beta_{k-1}(s') = P[\mathbf{y}_k^K | s_k = s'] = \sum_s P[\mathbf{y}_k^K, s_{k+1} = s | s_k = s']. \tag{7.42}$$

A typical term in the summation above can be written as

$$P[\mathbf{y}_k^K, s_{k+1} = s | s_k = s'] = P[\mathbf{y}_{k+1}^K, \mathbf{y}_k, s_{k+1} = s | s_k = s']$$
$$= P[\mathbf{y}_{k+1}^K | s_{k+1} = s, s_k = s', \mathbf{y}_k] P[s_{k+1} = s, \mathbf{y}_k | s_k = s'].$$

Given the state s_{k+1}, the future observations $\mathbf{y}_{k+1}^K$ are independent of past states and observations, so that

$$P[\mathbf{y}_{k+1}^K | s_{k+1} = s, s_k = s', \mathbf{y}_k] = P[\mathbf{y}_{k+1}^K | s_{k+1} = s] = \beta_k(s).$$

This shows that

$$P[\mathbf{y}_k^K, s_{k+1} = s | s_k = s'] = \beta_k(s) \gamma_k(s', s).$$

Substituting into (7.42), we obtain the backward recursion.

Backward recursion

$$\beta_{k-1}(s') = \sum_s \beta_k(s) \gamma_k(s', s). \tag{7.43}$$

Often, we set the terminal state of the encoder to be the all-zero state, $\mathbf{0}$, in which case the initial condition for the backward recursion is given by

$$\beta_K(s) = \begin{cases} 0, & s \neq \mathbf{0}, \\ c > 0, & s = \mathbf{0}, \end{cases} \tag{7.44}$$

where we often set $c = 1$.

Backward recursion for running example Consider $\beta_{k-1}(s')$ for $s' = 11$. The two possible values of next state s are 01 and 11, so that

$$\beta_{k-1}(11) = \gamma_k(11, 01)\beta_k(01) + \gamma_k(11, 11)\beta_k(11).$$

While termination in the all-zero code for a nonrecursive encoder is typically a matter of sending several zero information bits at the end of the information payload, for a recursive code, the terminating sequence of bits may depend on the payload, as illustrated next for our running example.

Table 7.1 Terminating information bits required to obtain $s_k = 00$ are a function of the state $s[K-2]$, as shown for the RSC running example.

$s[K-2]$	$u[K-2]$	$u[K-1]$
00	0	0
01	1	0
10	1	1
11	0	1

> **Trellis termination for running example** To get an all-zero terminal state $s(K) = 00$, we see from Figure 7.9 that, for different values of the state $s[K-2]$, we need different choices of information bits $u[K-2]$ and $u[K-1]$, as listed in Table 7.1.

It remains to specify the computation of $\gamma_k(s', s)$, which we can rewrite as

$$\gamma_k(s', s) = P[\mathbf{y}_k, s_{k+1} = s | s_k = s']$$
$$= P[\mathbf{y}_k | s_{k+1} = s, s_k = s'] P[s_{k+1} = s | s_k = s']. \quad (7.45)$$

Given the states s_{k+1} and s_k, the code output corresponding to the trellis section k is completely specified as $\mathbf{c}_k(s', s)$. The probability

$$P[\mathbf{y}_k | s_{k+1} = s, s_k = s'] = P[\mathbf{y}_k | \mathbf{c}_k(s', s)]$$

is a function of the modulation and demodulation employed, and the channel model (i.e., how code bits are mapped to channel symbols, how the channel output and input are statistically related, and how the received signal is processed before sending the information to the decoder). The probability $P[s_{k+1} = s | s_k = s']$ is the prior probability that the input to the decoder is such that, starting from state s', we transition to state s. Letting $u_k(s', s)$ denote the value of the input corresponding to this transition, we have

$$\gamma_k(s', s) = P[\mathbf{y}_k, s_{k+1} = s | s_k = s'] = P[\mathbf{y}_k | \mathbf{c}_k(s', s)] P[u_k(s', s)]. \quad (7.46)$$

Thus, γ_k incorporates information from the priors and the channel outputs. Note that, if prior information about parity bits is available, then it should also be incorporated into γ_k. For the moment, we ignore this issue, but we return to it when we consider iterative decoding of serially concatenated convolutional codes.

> **Computation of $\gamma_k(s', s)$ for running example** Assume BPSK modulation of the bits u_k and v_k corresponding to the kth trellis section as in (7.32). There is a unique mapping between the states s_k, s_{k+1} and the output bits

u_k, v_k. Thus, the first term in the extreme right-hand side of (7.46) can be written as

$$P[\mathbf{y}_k|u_k(s',s), v_k(s',s)] = P[y_k(1)|u_k(s',s)]P[y_k(2)|v_k(s',s)],$$

using the independence of the channel noise samples. Since the noise is Gaussian, we get

$$P[y_k(1)|u_k] = \frac{1}{\sqrt{2\pi\sigma^2}} \exp\left(-[y_k(1) - A\tilde{u}_k]^2\right),$$

$$P[y_k(2)|v_k] = \frac{1}{\sqrt{2\pi\sigma^2}} \exp\left(-[y_k(2) - A\tilde{v}_k]^2\right).$$

Note that we can scale these quantities arbitrarily, as long as the scale factor is independent of the states. Thus, we can discard the factor $1/\sqrt{(2\pi\sigma^2)}$. Further, expanding the exponent in the expression for $P[y_k(1)|u_k]$, we have

$$[y_k(1) - A(-1)^{u_k}]^2 = y_k^2(1) + A^2 - 2Ay_k(1)\tilde{u}_k.$$

Only the third term, which is a correlation between the received signal and the hypothesized transmitted signal, is state-dependent. The other two terms contribute state-independent multiplicative factors that can be discarded. The same reasoning applies to the expression for $P[y_k(2)|v_k]$. We can therefore write

$$P[y_k(1)|u_k] = \mu_k \exp\left(\frac{Ay_k(1)\tilde{u}_k}{\sigma^2}\right),$$

$$P[y_k(2)|v_k] = \nu_k \exp\left(\frac{Ay_k(2)\tilde{v}_k}{\sigma^2}\right),$$
(7.47)

where μ_k, ν_k are constants that are implicitly evaluated or cancelled when we compute posterior probabilities.

For computing the second term on the extreme right-hand side of (7.46), we note that, given $s_k = s'$, the information bit u_k uniquely defines the next state $s_{k+1} = s$. Thus, we have

$$P[s_{k+1} = s|s_k = s'] = P[u_k(s',s)] = \begin{cases} P(u_k = 0), & (s',s) \in U_0, \\ P(u_k = 1), & (s',s) \in U_1. \end{cases}$$
(7.48)

Using (7.47) and (7.48), we can now write down an expression for $\gamma_k(s',s)$ as follows:

$$\gamma_k(s',s) = \zeta_k \exp\left(\frac{A}{\sigma^2}[y_k(1)\tilde{u}_k + y_k(2)\tilde{v}_k]\right) P[u_k],$$
(7.49)

where the dependence of the bits on (s',s) has been suppressed from the notation.

We can now summarize the BCJR algorithm as follows.

Summary of BCJR algorithm

Step 1 Using the received signal and the priors, compute $\gamma_k(s', s)$ for all k, and for all (s', s) allowed by the code constraints.

Step 2 Run the forward recursion (7.39) and the backward recursion (7.43), scaling the outputs at any given time in state-independent fashion as necessary to avoid overflow or underflow.

Step 3 Compute the LLRs of the bits of interest. Substituting (7.37) into (7.30), we get

$$L_{\text{out}}(u_k) = \log\left(\frac{\sum_{(s',s)\in U_0} \alpha_{k-1}(s')\gamma_k(s',s)\beta_k(s)}{\sum_{(s',s)\in U_1} \alpha_{k-1}(s')\gamma_k(s',s)\beta_k(s)}\right). \quad (7.50)$$

(A similar equation holds for v_k, with U_i replaced by V_i, $i = 0, 1$.)

Hard decisions, if needed, are made based on the sign of the LLRs: for a generic bit b, we make the hard decision

$$\hat{b} = \begin{cases} 0, & L(b) > 0, \\ 1, & L(b) < 0. \end{cases} \quad (7.51)$$

We now discuss the logarithmic implementation of the BCJR algorithm, which is not only computationally more stable, but also reveals more clearly the role of the various sources of soft information.

7.2.2 Logarithmic BCJR algorithm

We propagate the log of the intermediate variables α, β and γ, defined as

$$a_k(s) = \log \alpha_k(s),$$
$$b_k(s) = \log \beta_k(s),$$
$$g_k(s) = \log \gamma_k(s).$$

We can now rewrite a typical forward recursion (7.41) for our running example as follows:

$$a_k(00) = \log(e^{g_k(10,01)+a_{k-1}(10)} + e^{g_k(11,01)+a_{k-1}(11)}). \quad (7.52)$$

To obtain a more compact notation, as well as to better understand the nature of the preceding computation, it is convenient to define a new function, max*, as follows.

The max* operation For real numbers $x_1, \ldots, x_n$, we define

$$\max{}^*(x_1, x_2, \ldots, x_n) = \log(e^{x_1} + e^{x_2} + \cdots + e^{x_n}). \quad (7.53)$$

For two arguments, the max* operation can be rewritten as

$$\max{}^*(x, y) = \max(x, y) + \log(1 + e^{-|x-y|}). \quad (7.54)$$

7.2 Turbo codes and iterative decoding

This relation is easy to see by considering $x > y$ and $y \geq x$ separately. For $x > y$,

$$\begin{aligned}\max{}^*(x, y) &= \log\left(e^x(1+e^{-(x-y)})\right) \\ &= \log e^x + \log\left((1+e^{-(x-y)})\right) \\ &= x + \log\left((1+e^{-(x-y)})\right),\end{aligned}$$

while for $y \geq x$, we similarly obtain

$$\max{}^*(x, y) = y + \log\left((1+e^{-(y-x)})\right).$$

The second term in (7.54) has a small range, from 0 to $\log 2$. It can therefore be computed efficiently using a look-up table. Thus, the max* operation can be viewed as a maximization operation together with a correction term.

Properties of max* We list below two useful properties of the max* operation.

Associativity The max* operation can be easily shown to be associative, so that its efficient computation for two arguments can be applied successively to evaluate it for multiple arguments:

$$\max{}^*(x, y, z) = \max{}^*\left(\max{}^*(x, y), z\right). \tag{7.55}$$

Translation of arguments It is also easy to check that common additive constants in the arguments of max* can be pulled out. That is, for any real number c,

$$\max{}^*(x_1 + c, x_2 + c, \ldots, x_n + c) = c + \max{}^*(x_1, x_2, \ldots, x_n). \tag{7.56}$$

We can now rewrite the computation (7.52) as

$$a_k(00) = \max{}^*\{g_k(10, 01) + a_{k-1}(10), g_k(11, 01) + a_{k-1}(11)\}.$$

Thus, the forward recursion is analogous to the Viterbi algorithm, in that we add a branch metric g_k to the accumulated metric a_{k-1} for the different branches entering the state 00. However, instead of then picking the maximum from among the various branches, we employ the max* operation. Similarly, the backward recursion is a Viterbi algorithm running backward through the trellis, with maximum replaced by max*. If we drop the correction term in (7.54) and approximate max* by max, the recursions reduce to the standard Viterbi algorithm.

We now specify computation of the logarithmic version of γ_k. From (7.46), we can write

$$g_k(s', s) = \log P[\mathbf{y}_k | \mathbf{c}_k(s', s)] + \log P[u_k(s', s)].$$

For our running example, we can write a more explicit expression, based on (7.49) as follows:

$$g_k(s', s) = \frac{A}{\sigma^2}[y_k(1)\tilde{u}_k + y_k(2)\tilde{v}_k] + \log P[u_k(s', s)]. \tag{7.57}$$

To express the role of priors in a convenient fashion, we now derive a convenient relation between LLR and log probabilities. For a generic bit b taking values in $\{0, 1\}$, the LLR L is given by

$$L = \log \frac{P[b=0]}{P[b=1]},$$

from which we can infer that

$$P[b=0] = \frac{e^L}{e^L+1} = \frac{e^{L/2}}{e^{L/2}+e^{-L/2}}.$$

Similarly,

$$P[b=1] = \frac{1}{e^L+1} = \frac{e^{-L/2}}{e^{L/2}+e^{-L/2}}.$$

Taking logarithms, we have

$$\log P(b) = \begin{cases} L/2 + \log(e^{L/2}+e^{-L/2}), & b=0, \\ -L/2 + \log(e^{L/2}+e^{-L/2}), & b=1. \end{cases}$$

This can be summarized as

$$\log P(b) = \tilde{b} L/2 + \log(e^{L/2}+e^{-L/2}), \tag{7.58}$$

where $\tilde{b}$ is the BPSK version of b. The second term on the right-hand side is the same for both $b=0$ and $b=1$ in the expressions above. Thus, it can be discarded as a state-independent constant when it appears in quantities such as $g_k(s',s)$.

The channel information can also be conveniently expressed in terms of an LLR. Suppose that y is a channel observation corresponding to a transmitted bit b. Assuming uniform priors for b, the LLR for b that we can compute from the observation is as follows:

$$L_{\text{channel}}(b) = \log \frac{p[y|b=0]}{p[y|b=1]}.$$

For BPSK signaling over an AWGN channel, we have

$$y = A\tilde{b} + N,$$

where $N \sim N(0, \sigma^2)$, from which it is easy to show that

$$L_{\text{channel}}(b) = \frac{2A}{\sigma^2} y. \tag{7.59}$$

For our running example, we have

$$L_{\text{channel}}(u_k) = \frac{2A}{\sigma^2} y_k(1), \qquad L_{\text{channel}}(v_k) = \frac{2A}{\sigma^2} y_k(2). \tag{7.60}$$

Returning to (7.57), suppose that the prior information for u_k is specified in the form of an *input LLR* $L_{\text{in}}(u_k)$. We can replace the prior term $\log P(u_k)$ in (7.57) by $\tilde{u}_k L_{\text{in}}(u_k)/2$, using the first term in (7.58). Further, we use the channel LLRs (7.60) to express the information obtained from the channel. We can now rewrite (7.57) as

$$g_k(s',s) = \tilde{u}_k [L_{\text{in}}(u_k) + L_{\text{channel}}(u_k)]/2 + \tilde{v}_k L_{\text{channel}}(v_k)/2. \tag{7.61}$$

7.2 Turbo codes and iterative decoding

If prior information is available about the parity bit v_k (e.g., from another component decoder in an iteratively decoded turbo code), then we incorporate it into (7.61) as follows:

$$g_k(s',s) = \tilde{u}_k[L_{\text{in}}(u_k) + L_{\text{channel}}(u_k)]/2 + \tilde{v}_k[L_{\text{in}}(v_k) + L_{\text{channel}}(v_k)]/2. \quad (7.62)$$

We now turn to the computation of the *output* LLR for a given information bit u_k. We transform (7.50) to logarithmic form to get

$$L_{\text{out}}(u_k) = \max_{(s',s) \in U_0}{}^* \{a_{k-1}(s') + g_k(s's) + b_k(s)\} \\ - \max_{(s',s) \in U_1}{}^* \{a_{k-1}(s') + g_k(s's) + b_k(s)\}. \quad (7.63)$$

We now write this in a form that makes transparent the roles played by different sources of information about u_k. Specializing to our running example for concreteness, we rewrite (7.62) in more detail:

$$g_k(s',s) = \begin{cases} [L_{\text{in}}(u_k) + L_{\text{channel}}(u_k)]/2 + \tilde{v}_k[L_{\text{in}}(v_k) \\ \quad + L_{\text{channel}}(v_k)]/2, & (s's) \in U_0, \\ -[L_{\text{in}}(u_k) + L_{\text{channel}}(u_k)]/2 + \tilde{v}_k[L_{\text{in}}(v_k) \\ \quad + L_{\text{channel}}(v_k)]/2, & (s's) \in U_1, \end{cases}$$

since $\tilde{u}_k = +1$ ($u_k = 0$) for $(s's) \in U_0$, and $\tilde{u}_k = -1$ ($u_k = 1$) for $(s's) \in U_1$. The common contribution due to the input and channel LLRs for u_k can therefore be pulled out of the max* operations in (7.63), and we get

$$L_{\text{out}}(u_k) = L_{\text{in}}(u_k) + L_{\text{channel}}(u_k) + L_{\text{code}}(u_k), \quad (7.64)$$

where we define the code LLR $L_{\text{code}}(u_k)$ as

$$L_{\text{code}}(u_k) = \max_{(s',s) \in U_0}{}^* \{a_{k-1}(s') + \tilde{v}_k[L_{\text{in}}(v_k) + L_{\text{channel}}(v_k)]/2 + b_k(s)\} \\ - \max_{(s',s) \in U_1}{}^* \{a_{k-1}(s') + \tilde{v}_k[L_{\text{in}}(v_k) + L_{\text{channel}}(v_k)]/2 + b_k(s)\}.$$

This is the information obtained about u_k from the prior and channel information about *other* bits, invoking the code constraints relating u_k to these bits.

Equation (7.64) shows that the output LLR is a sum of three LLRs: the input (or prior) LLR, the channel LLR, and the code LLR. We emphasize that the code LLR for u_k does not depend on the input and channel LLRs for u_k. The quantity $a_{k-1}(s')$ summarizes information from bits associated with trellis sections before time k, and the quantity $b_k(s)$ summarizes information from bits associated with trellis sections after time k. The remaining information in $L_{\text{code}}(u_k)$ comes from the prior and channel information regarding other bits in the kth trellis section (in our example, this corresponds to $L_{\text{in}}(v_k)$ and $L_{\text{channel}}(v_k)$).

We are now ready to summarize the logarithmic BCJR algorithm.

Step 0 (Input LLRs) Express prior information, if any, in the form of input LLRs $L_{\text{in}}(b)$. If no prior information is available for bit b, then set $L_{\text{in}}(b) = 0$.

Step 1 (Channel LLRs) Use the received signal to compute $L_{\text{channel}}(b)$ for all bits sent over the channel. For BPSK modulation over the AWGN channel with received signal $y = A\tilde{b} + N$, $N \sim N(0, \sigma^2)$, we have $L_{\text{channel}}(b) = 2Ay/\sigma^2$.

Step 2 (Branch gains) Compute the branch gains $g_k(s', s)$ using the prior and channel information for all bits associated with that branch, adding terms of the form $\tilde{b}L(b)/2$. For our running example,

$$g_k(s', s) = \tilde{u}_k[L_{\text{in}}(u_k) + L_{\text{channel}}(u_k)]/2 + \tilde{v}_k[L_{\text{in}}(v_k) + L_{\text{channel}}(v_k)]/2.$$

Step 3 (Forward and backward recursions) Run Viterbi-like algorithms forward and backward, using max* instead of maximization.

$$a_k(s) = \max_{s'}{}^*\{a_{k-1}(s') + g_k(s', s)\}.$$

(Initial condition: $a_0(s) = -C$, $s \neq \mathbf{0}$ and $a_0(\mathbf{0}) = 0$, where $C > 0$ is a large positive number.)

$$b_{k-1}(s') = \max_{s}{}^*\{b_k(s) + g_k(s', s)\}.$$

(Initial condition: $b_K(s) = -C$, $s \neq \mathbf{0}$ and $b_K(\mathbf{0}) = 0$, where $C > 0$ is a large positive number.)

Step 4 (Output LLRs and hard decisions) Compute output LLRs for each bit of interest as

$$L_{\text{out}}(b) = L_{\text{in}}(b) + L_{\text{channel}}(b) + L_{\text{code}}(b),$$

where $L_{\text{code}}(b)$ is a summary of prior and channel information for bits other than b, using the code constraints. For my running example, I have

$$L_{\text{code}}(u_k) = \max_{(s',s) \in U_0}{}^*\{a_{k-1}(s') + \tilde{v}_k[L_{\text{in}}(v_k) + L_{\text{channel}}(v_k)]/2 + b_k(s)\}$$
$$- \max_{(s',s) \in U_1}{}^*\{a_{k-1}(s') + \tilde{v}_k[L_{\text{in}}(v_k) + L_{\text{channel}}(v_k)]/2 + b_k(s)\},$$

$$L_{\text{code}}(v_k) = \max_{(s',s) \in V_0}{}^*\{a_{k-1}(s') + \tilde{u}_k[L_{\text{in}}(u_k) + L_{\text{channel}}(u_k)]/2 + b_k(s)\}$$
$$- \max_{(s',s) \in V_1}{}^*\{a_{k-1}(s') + \tilde{u}_k[L_{\text{in}}(u_k) + L_{\text{channel}}(u_k)]/2 + b_k(s)\}.$$

Once output LLRs have been computed, hard decisions are obtained using $\hat{b} = 1_{L_{\text{out}}(b) < 0}$.

Note on Step 3 In case the trellis is not terminated in the all-zero state (or some other fixed state), then the initial condition stated earlier for the backward recursion need not be satisfied. In this case, one practical approach is to use the result of the forward recursion as an initial condition for the backward recursion (e.g., $\beta_K(s) \equiv \alpha_K(s)$).

7.2.3 Turbo constructions from convolutional codes

We know from Shannon theory that random coding can be used to attain capacity. However, optimum (e.g., ML or MPE) decoding of random codes with no structure is computationally infeasible, unlike, for example, Viterbi or BCJR decoding for convolutional codes that can be described by a trellis with a manageable number of states. Thus, the basic contradiction in coding theory prior to the invention of turbo codes was that, while random codes are known to be good, all known codes were highly structured and "not good" (i.e., far from Shannon-theoretic limits). Turbo codes avoid this dilemma by using long interleavers to obtain a random-looking code based on structured component codes, and then exploiting the structure of the component codes to obtain a very effective suboptimal iterative decoding algorithm. For concatenated convolutional codes, iterative decoding may involve information exchange between two or more decoders running the BCJR algorithm, or approximations thereof. A given decoder sends another decoder *extrinsic* information that is approximately independent of the information available to the second decoder, and that serves as a prior for the second decoder. For example, if Decoder 1 sends Decoder 2 the LLR for a given bit b, this becomes $L_{\text{in}}(b)$ for Decoder 2. Decoder 2 then applies the BCJR algorithm to compute $L_{\text{out}}(b) = L_{\text{in}}(b) + L_{\text{channel}}(b) + L_{\text{code}}(b)$. However, it does not send $L_{\text{out}}(b)$ back to Decoder 1, since it includes $L_{\text{in}}(b)$, the information that came from Decoder 1. The extrinsic information that Decoder 2 sends back to Decoder 1 will either be $L_{\text{channel}}(b) + L_{\text{code}}(b)$ (if Decoder 1 does not have direct access to the channel observation regarding bit b), or $L_{\text{code}}(b)$ (if Decoder 1 does have access to the same channel information regarding bit b that Decoder 2 does). We clarify these concepts in the context of two turbo constructions built from simple convolutional component codes: parallel concatenation (the original turbo codes) and serial concatenation.

Parallel concatenated codes Parallel concatenation of convolutional codes is depicted in Figure 7.10. An information sequence u is fed into a convolutional encoder, Encoder 1. To get good performance, it turns out that the encoder should be chosen to be recursive. We employ our recursive systematic rate 1/2 code in Figure 7.9 as a running example. The information sequence is then permuted and fed to another convolutional encoder, Encoder 2. Encoders 1 and 2 can be, and often are, chosen to be identical. The information sequence and the two parity sequences are then modulated and sent

Figure 7.10 Encoder and decoder for a parallel concatenated turbo code.

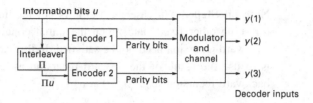

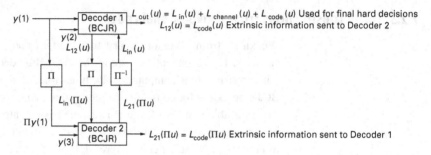

through the channel (we use BPSK over an AWGN channel as a running example). Thus, if the constituent encoders are rate 1/2, then the overall turbo code thus obtained is of rate 1/3. A higher rate can be achieved using the same construction simply by not transmitting some of the bits generated by the encoders. This procedure is referred to as *puncturing*. Note that a punctured convolutional code can be decoded in the same manner as one without puncturing, by interpreting the bits not sent as erasures (set $L_{\text{channel}}(b) = 0$ in the logarithmic BCJR algorithm). For simplicity of notation, however, we do not consider puncturing in our discussion here.

In the iterative decoding depicted in Figure 7.10. both decoders see the channel output $\mathbf{y}(1)$ for the information sequence. Decoder 1 sees the channel output $\mathbf{y}(2)$ for the parity sequence for Encoder 1, while Decoder 2 sees the channel output $\mathbf{y}(3)$ for the parity sequence from Encoder 2. The decoders exchange information about the information sequence $\mathbf{u}$.

For a typical information bit u, the two decoders function as follows.

Channel LLRs These are computed as in the standard BCJR algorithm for both information and parity bits. For example, for a bit b sent using BPSK over AWGN, we have $L_{\text{channel}}(b) = 2Ay/\sigma^2$, where $y = Ab + N$, $N \sim N(0, \sigma^2)$.

Decoder 1 Operates using channel outputs $\mathbf{y}(1)$ and $\mathbf{y}(2)$, and extrinsic information from Decoder 2.

Step 1 Receive extrinsic information from Decoder 2 regarding the LLRs of information bits. Use these as $L_{\text{in}}(u)$ for BCJR algorithm (set $L_{\text{in}}(u) = 0$ for first iteration). Set $L_{\text{in}}(v) = 0$ for parity bits.

7.2 Turbo codes and iterative decoding

Step 2 Run forward and backward recursions.

Step 3 Compute $L_{\text{code}}(u)$ for each information bit u. Feed $L_{\text{code}}(u)$ back as extrinsic information to Decoder 2 (note that $L_{\text{code}}(u)$ does not depend on $L_{\text{in}}(u)$ or $L_{\text{channel}}(u)$, and hence is not directly available to Decoder 2).

Decoder 2 Operation is identical to that of Decoder 1, except that it works with the permuted information sequence, and with the received signals $\mathbf{y}(1)$ (permuted) and $\mathbf{y}(3)$.

Iteration and termination Decoders 1 and 2 interchange information in this fashion until some termination condition is satisfied (e.g., a maximum number of iterations is reached, or the LLRs have large enough magnitude, or a CRC check is satisfied). Then they make hard decisions based on $L_{\text{out}}(u) = L_{\text{in}}(u) + L_{\text{channel}}(u) + L_{\text{code}}(u)$ for each information bit based on the output of Decoder 1.

Serial concatenated codes Serial concatenation of convolutional codes is shown in Figure 7.11. The output from the first convolutional encoder is interleaved and then fed as input to a second convolutional encoder. The output from the second encoder is then modulated and transmitted over the channel.

In the iterative decoding method depicted in Figure 7.11, only Decoder 2 sees the channel outputs. Thus, extrinsic information sent by Decoder 2 is given by $L_{\text{code}} + L_{\text{channel}}$, since Decoder 1 does not have access to the channel. Decoder 1 employs this extrinsic information to compute L_{code}, and sends it back to Decoder 2 as extrinsic information. The final decisions are based on the output LLRs L_{out} from Decoder 1, since the information sequence is the

Figure 7.11 Encoder and decoder for a serial concatenated turbo code.

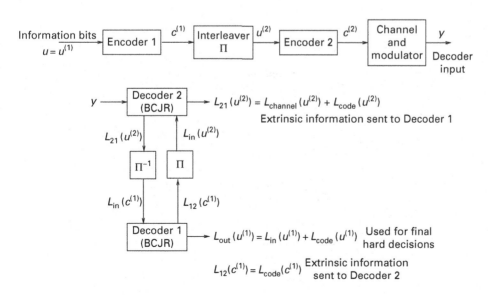

input to Encoder 1. Note that $L_{\text{in}}(u^{(1)}) = 0$ if the outer code is nonsystematic, since the extrinsic information available to the outer decoder is regarding the output bits of the outer code. Thus, for a nonsystematic outer code, the only information extracted regarding the input bits $u^{(1)}$ is $L_{\text{code}}(u^{(1)})$.

7.2.4 The BER performance of turbo codes

Figure 7.12 shows the BER for a rate 1/3 parallel concatenated turbo code. The component convolutional code is our familiar rate 1/2 convolutional code with generator [7, 5]. Despite the simplicity of this code (each component decoder employs only four states), parallel concatenation gets to within 1 dB of the Shannon limit at a BER of 10^{-4}, as shown in Figure 7.12. The steep decrease in BER is termed the "waterfall" region. As the BER gets smaller (not shown in the figure), we eventually hit an "error floor" region (not evident from the figure) where the decrease in BER becomes less steep. Turbo code design requires an understanding of how code construction impacts the SNR threshold for the waterfall region and the BER floor. As shown in Section 7.2.5, the SNR threshold can be understood in terms of averaged trajectories for iterative decoding which are termed extrinsic information transfer (EXIT) charts. The slope of the error floor region, on the other hand, is governed by the "most likely" error events, the characterizing of which requires investigation of the code weight distribution, as discussed in Section 7.2.6.

While Figure 7.12 shows the BER after a relatively large number of iterations, the dependence of BER on the number of iterations is shown in

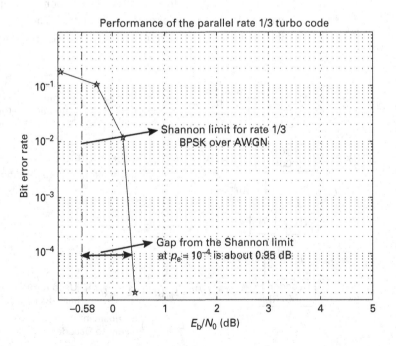

Figure 7.12 Bit error rate for a rate 1/3 parallel concatenated turbo code obtained by concatenating two identical rate 1/2 [7, 5] convolutional codes. The block length is $2^{14} \approx 16\,000$.

Figure 7.13 Bit error rate as a function of number of iterations for the turbo code in Figure 7.12.

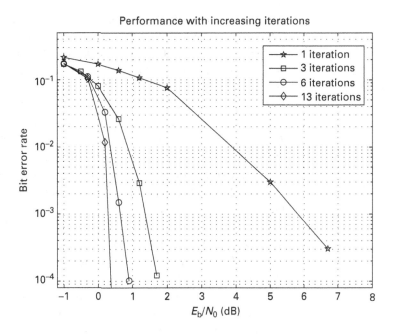

Figure 7.13. There is a large improvement in performance in the first few iterations, after which the progress is slower. Typically, the improvement in performance is insignificant after about 10–15 iterations.

Similar observations also hold for serially concatenated turbo codes, hence we omit BER plots for these.

7.2.5 Extrinsic information transfer charts

Bit error rate curves for turbo codes have a distinct waterfall region: once E_b/N_0 crosses a threshold, the BER decays sharply. Extrinsic information transfer (EXIT) charts provide a computationally efficient tool for visualizing the progress of iterative decoding, showing what happens before and after the waterfall region without requiring exhaustive simulations for estimating the BER. These EXIT charts are therefore a very useful design tool for optimizing the component codes in parallel or serial concatenated turbo codes.

Let us illustrate the concept by considering the parallel concatenated turbo code shown in Figure 7.10. Decoders 1 and 2 exchange extrinsic information about the information bits. Let X denote a particular information bit. Let E_1 denote the extrinsic information (expressed as an LLR) regarding X at the output of Decoder 1, and let E_2 denote the LLR corresponding to the extrinsic information regarding X at the output of Decoder 2. Similarly, let A_1 denote the a priori information regarding X at the input of Decoder 1, and A_2 the a priori information regarding X at the input of Decoder 2. For iterative decoding, $A_1[n] = E_2[n-1]$ and $A_2[n] = E_1[n-1]$, where n denotes the nth iteration. The mutual information $I(X; E_1)$ is a measure of the quality of the information regarding X at the output of Decoder 1, and equals the mutual information $I(X; A_2)$ at the input of Decoder 2 at the next iteration. Similarly, $I(X; E_2)$ is a measure of the quality of the output of Decoder 2, and equals $I(X; A_1)$ at the input of Decoder 1 at the next iteration.

The information transfer function for Decoder j ($j = 1, 2$) plots the mutual information at its output, $I(X; E_j)$, versus the mutual information at its input, $I(X; A_j)$. Let us call this curve $T_j(i)$, where i is the mutual information at the input, and $T_j(i)$ the mutual information at the output. The output extrinsic information for Decoder 1 is the input extrinsic information for Decoder 2, and vice versa. Thus, to plot both decoder characteristics using the same set of axes, we flip the roles of input and output for one of them. Specifically, let us plot $I(X; E_1)$ and $I(X; A_2)$ on the y-axis, and $I(X; A_1)$ and $I(X; E_2)$ on the x-axis. That is, we plot $T_1(i)$ versus i, and $T_2^{-1}(i)$ versus i. Figure 7.14 shows an example of such a plot. We can now visualize the progress of iterative decoding as shown. We have $I(X; A_1) = 0$ at the beginning (no input from Decoder 2). In this case, Decoder 1 generates extrinsic information E_1 of nonzero quality $i = I(X; E_1)$ using its code constraints and the information available from the channel. This provides extrinsic information A_2 of quality $i = I(X; A_2)$ at the input to Decoder 2. This is then used by Decoder 2, in conjunction with the channel information, to generate extrinsic information E_2 of quality $T_2(i) = I(X; E_2)$, which is now fed as input A_1 for Decoder 1. Figure 7.14 indicates that iterative decoding should be successful, since the quality of the extrinsic information keeps increasing, approaching the $(1, 1)$ point on the EXIT chart. This is

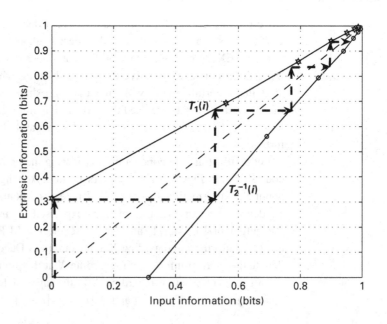

Figure 7.14 An EXIT chart showing decoding progress for the running example of a parallel concatenated code when the SNR is high enough ($E_b/N_0 = 0.8$ dB) that the iterations converge.

7.2 Turbo codes and iterative decoding

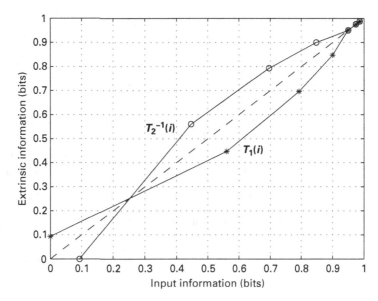

Figure 7.15 An EXIT chart for the running example of a parallel concatenated code, showing that iterative decoding gets stuck when the SNR is too low ($E_b/N_0 = -0.7$ dB).

consistent with the BER curve in Figure 7.12: the E_b/N_0 value of 0.8 dB used in the EXIT chart is beyond the onset of the waterfall region. In contrast, Figure 7.15 shows a scenario where iterative decoding gets stuck at the intersection of the curves for Decoders 1 and 2, since the "tunnel" required to progress towards the $(1, 1)$ point is not available. Referring back to the BER curve in Figure 7.12, we see that the E_b/N_0 value of -0.7 dB used in the EXIT chart is prior to the onset of the waterfall region.

The shape of the information transfer functions depends, of course, on the quality of the information obtained from the channel. Thus, the tunnel between the curves for Decoders 1 and 2 should be open at large enough E_b/N_0. The value of E_b/N_0 at which the tunnel is barely open is when we expect iterative decoding to begin to work, and provides an excellent estimate of the beginning of the waterfall portion of the BER curve for a turbo code with long enough block length. The beauty of this approach is that we do not need exhaustive simulations of the turbo code; we only need to simulate the performance of the individual decoders to generate their input–output characteristics. Let us now discuss how to do this.

Estimation of mutual information We first discuss computation of mutual information between an information bit and an LLR. Specifically, if we run a decoder with a given set of priors and generate a set of extrinsic LLRs $\{L_n\}$ corresponding to information bits $\{X_n\}$, how do we estimate the mutual information $I(X; L)$ from the pairs (X_n, L_n)? The mutual information between a bit X taking values 0 and 1 with equal probability, and a random variable L is given by

Channel coding

$$I(X; L) = \sum_{x=0}^{1} p(x) \int_{-\infty}^{\infty} p(l|x) \log_2 \frac{p(l|x)}{p(l)} \, dl$$

$$= \frac{1}{2} \sum_{x=0}^{1} \int_{-\infty}^{\infty} p(l|x) \log_2 \frac{2p(l|x)}{p(l|0) + p(l|1)} \, dl. \quad (7.65)$$

Suppose now that the random variable L is an LLR for X obtained using an observation Y whose distribution conditioned on X satisfies the following symmetry condition: $p(y|0) = p(-y|1)$. Note that $L(y) = \log_2(P[X=0|Y=y])/(P[X=1|Y=y])$. As shown in Problem 7.12, the conditional distribution of L given X obeys the following consistency condition:

$$p(l|0) = p(-l|1) = e^l p(-l|0) \quad \textbf{Consistency condition for LLRs.} \quad (7.66)$$

Plugging (7.66) into (7.65), we obtain

$$I(X; L) = \frac{1}{2} \int_{-\infty}^{\infty} p(l|0) \log_2 \frac{2p(l|0)}{p(l|0) + p(l|1)} \, dl$$

$$+ \frac{1}{2} \int_{-\infty}^{\infty} p(l|1) \log_2 \frac{2p(l|1)}{p(l|0) + p(l|1)} \, dl$$

$$= \frac{1}{2} \int_{-\infty}^{\infty} p(l|0) \log_2 \frac{2p(l|0)}{p(l|0) + p(-l|0)} \, dl$$

$$+ \frac{1}{2} \int_{-\infty}^{\infty} p(-l|0) \log_2 \frac{2p(-l|0)}{p(l|0) + p(-l|0)} \, dl$$

$$= \int_{-\infty}^{\infty} p(l|0) \log_2 \frac{2p(l|0)}{p(l|0) + p(-l|0)} \, dl$$

$$= \int_{-\infty}^{\infty} p(l|0) \log_2 \frac{2}{1 + e^{-l}} \, dl$$

$$= 1 - \int_{-\infty}^{\infty} p(l|0) \log_2(1 + e^{-l}) \, dl.$$

In summary, under the consistency condition (7.66), the mutual information between a bit X and its LLR L is given by

$$I(X; L) = 1 - E[\log_2(1 + e^{-L})|X = 0] = 1 - \int_{-\infty}^{\infty} p(l|0) \log_2(1 + e^{-l}) \, dl. \quad (7.67)$$

Under the consistency condition, it also follows that

$$E[\log_2(1 + e^{-L})|X = 0] = E[\log_2(1 + e^{L})|X = 1],$$

so that

$$I(X; L) = 1 - E[\log_2(1 + e^{-(-1)^X L})|X = x]. \quad (7.68)$$

To measure the mutual information empirically using independent pairs (X_n, L_n), $n = 1, \ldots, N$, we can replace (7.68) by the following empirical average:

$$\hat{I}(X; L) = 1 - \frac{1}{N} \sum_{n=1}^{N} \log_2\left(1 + e^{-(-1)^{X_n} L_n}\right). \quad (7.69)$$

7.2 Turbo codes and iterative decoding

For symmetric channels, it suffices to consider the all-zero codeword in order to measure the mutual information, in which case, we can use the empirical average corresponding to the formula (7.67):

$$\hat{I}(X; L) = 1 - \frac{1}{N} \sum_{n=1}^{N} \log_2 \left(1 + e^{-L_n}\right) \quad \text{\textbf{All-zero codeword sent.}} \quad (7.70)$$

The Gaussian approximation for computing information transfer functions We now know how to measure the mutual information between the information bits and the extrinsic information at the output of a decoder. To run the decoder, however, we need to generate the LLRs at the input to the decoder. Assuming large enough code block lengths and good enough interleaving, the LLRs corresponding to different information bits can be modeled as conditionally independent, conditioned on the value of the information bit. Thus, we can generate these LLRs if we know their conditional marginal distributions. For parallel concatenation, the LLR for a given bit X at the input to the decoder is given by $L_{\text{in}} = L_{\text{channel}} + L_{\text{prior}}$. In turn, L_{channel} and L_{prior} can be modeled as conditionally independent, given X. The conditional distribution of L_{channel} can be computed based on the channel statistics. Let us consider the example of BPSK transmission of bit $X \in \{0, 1\}$ over an AWGN channel to obtain observation Y:

$$Y = A(-1)^X + N(0, \sigma^2).$$

It can be shown that the channel LLR is given by

$$L(y) = \log_2 \frac{P[X=0|Y=y]}{P[X=1|Y=y]} = \frac{2Ay}{\sigma^2}.$$

Since $L(Y)$ depends linearly on Y, it inherits the conditional Gaussianity of Y. We have $L \sim N\left(2A^2/\sigma^2, 4A^2/\sigma^2\right)$ conditioned on $X = 0$, and $L \sim N\left(-2A^2/\sigma^2, 4A^2/\sigma^2\right)$ conditioned on 1. That is, the conditional mean and variance of the LLR depend on a scale-invariant SNR parameter, and are linearly related. In particular, we can take $\sigma_L^2 = 4A^2/\sigma^2$ as a measure of the quality of the LLR, and with the means under the two hypotheses given by $\pm \sigma_L^2/2$.

Thus, all that remains is to specify the conditional distribution of L_{prior}, which is actually the extrinsic information from some other decoder. A key simplification results from modeling the extrinsic conditional distribution as Gaussian; simulations show that this is an excellent model for the AWGN channel. A complete explanation for this observed Gaussianity is not yet available. The Gaussianity of the underlying channel may be a factor. Another factor, perhaps, is that a particular extrinsic LLR involves contributions from many LLRs, so that some form of central limit theorem might be at work (although the operations in the log BCJR algorithm cannot be interpreted simply as an arithmetic sum, which is the standard setting for application of central limit theorem approximations). Indeed, an excellent model of the extrinsic LLR L is to assume that it comes from BPSK modulation over a virtual AWGN channel. This yields the following model for the conditional distribution of the extrinsic LLR.

Gaussian model for LLR

$$L \sim N\left(\tfrac{\sigma_L^2}{2}, \sigma_L^2\right) \quad \text{if } X = 0, \quad L \sim N\left(-\tfrac{\sigma_L^2}{2}, \sigma_L^2\right) \quad \text{if } X = 1. \qquad (7.71)$$

As shown in Problem 7.12, if we assume that the conditional LLRs are Gaussian, then the consistency condition (7.66) can be used to infer that the conditional means and variances must be related as in (7.71).

Using (7.71), we can generate the decoder characteristic very simply. The input LLR for bit X is given by

$$L_{\text{in}} = L_{\text{channel}} + L_{\text{prior}},$$

where $L_{\text{channel}} \sim N\left((-1)^X \sigma_c^2/2, \sigma_c^2\right)$ and $L_{\text{prior}} \sim N\left((-1)^X \sigma_p^2/2, \sigma_p^2\right)$ are conditionally independent. The parameter $\sigma_c^2 = 4A^2/\sigma^2$ for BPSK signaling, whereas σ_p^2 is a parameter that we vary from 0 to ∞ to vary the quality of the input extrinsic information (which would be obtained from the other decoder during iterative decoding). The quality of the input extrinsic information is given by the mutual information (7.67), where

$$p(l|0) = \frac{1}{\sqrt{2\pi\sigma_p^2}} \exp\left(-\left(l - \frac{\sigma_p^2}{2}\right)^2 / 2\sigma_p^2\right).$$

Let us term this mutual information $I_{\text{in}}(\sigma_p^2)$. We can now simulate BCJR decoding and generate the output extrinsic LLRs $\{L_n\}$ for the bits $\{X_n\}$. We could model these as conditionally Gaussian, but we do not need to. We can empirically estimate the mutual information using the formula (7.69). We term this $I_{\text{out}}(\sigma_p^2)$. The information transfer function is simply the plot of $T(i) = I_{\text{out}}(\sigma_p^2)$ versus $i = I_{\text{in}}(\sigma_p^2)$ as the parameter σ_p^2 varies from 0 to ∞. The function $T(i)$ depends on E_b/N_0, but we have suppressed this dependence from the notation.

The results shown in Figures 7.14 and 7.15 were obtained using the preceding procedure to compute the information transfer functions for the component decoders. Of course, since we consider identical component codes, we only need to find a single information transfer function. Thus, the curve for decoder 1 is $T(i)$ versus i, whereas the curve for decoder 2 is $T^{-1}(i)$ versus i.

The EXIT charts apply to serially concatenated codes as well. Consider the serial concatenated code shown in Figure 7.11. Even though the bits of ultimate interest are the information bits $u^{(1)}$ at the input to the outer encoder, for the EXIT chart, we must consider the mutual information of the bits $c^{(1)}$ at the output of the outer encoder, which are also the inputs $u^{(2)}$ to the inner encoder (after interleaving), since the decoders exchange extrinsic information regarding these bits. This leads to information transfer functions

7.2 Turbo codes and iterative decoding

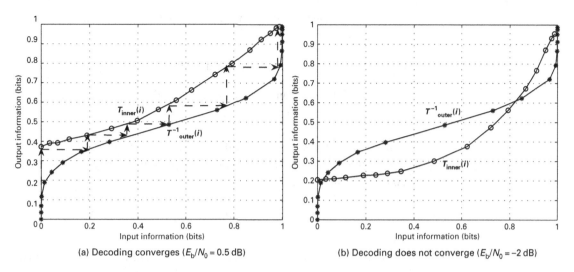

Figure 7.16 The EXIT charts for a rate 1/4 serial concatenated turbo code. Inner and outer codes are identical recursive systematic rate 1/2 [7, 5] convolutional codes.

$T_{\text{outer}}(i)$ and $T_{\text{inner}}(i)$ for the outer and inner decoders, respectively. The inner decoder starts decoding first, since it has access to the channel measurements. As before, the extrinsic inputs can be modeled as Gaussian when generating the information transfer curves. The EXIT chart plots $T_{\text{inner}}(i)$ and $T_{\text{outer}}^{-1}(i)$ versus i. In this case, the $T_{\text{inner}}(i)$ must lie above $T_{\text{outer}}^{-1}(i)$, to provide the tunnel needed for iterative decoding to converge. Figure 7.16 shows an EXIT chart for a serially concatenated code constructed from our rate 1/2 running example convolutional code. Since the outer decoder does not have access to the channel measurements, its curve does not change with E_b/N_0. On the other hand, $T_{\text{inner}}(i)$ moves upward as we increase E_b/N_0. Thus, we can again estimate the threshold corresponding to the waterfall region as the E_b/N_0 at which the inner decoder's curve $T_{\text{inner}}(i)$ is barely above the curve $T_{\text{outer}}^{-1}(i)$ for the outer decoder.

Area properties When the extrinsic information can be modeled as coming from an erasures channel, it is possible to relate the area under the information transfer function of a code to quantities such as channel capacity and code rate. Moreover, empirical results indicate that such area properties hold more generally (e.g., over AWGN channels in which the extrinsic information is well approximated as Gaussian). We state without proof two such results:

Code without channel access For a binary code of rate R that does not see the channel observations (e.g., the outer code in a serially concatenated turbo code), we have

$$\int_0^1 T(i)\,di = 1 - R.$$

Since the information transfer function is plotted in a box of area one, this also implies that

$$\int_0^1 T^{-1}(i)\,\mathrm{d}i = R. \tag{7.72}$$

From Figure 7.16, we can see that the area under $T_{\text{out}}^{-1}(i)$ does appear to satisfy the preceding result ($R = 1/2$).

Unit rate code For a unit rate code in which the input follows the optimal distribution (e.g., equally likely binary input for a channel which obeys symmetry conditions), we have

$$\int_0^1 T(i)\,\mathrm{d}i = C, \tag{7.73}$$

where C is the channel capacity.

Consider, for example, serial concatenation of a convolutional outer code of rate R_{out} with an inner unit rate binary DPSK encoder, to obtain an overall code of rate $R = R_{\text{out}}$. From (7.72), the area under the outer code's EXIT chart is given by

$$\int_0^1 T_{\text{outer}}^{-1}(i)\,\mathrm{d}i = R_{\text{out}} = R.$$

From (7.73), the area under the inner code's EXIT chart equals C. For convergence of iterative decoding, the inner code's curve must lie above the outer code's curve, which implies that the area under the inner code's curve must be larger than that under the outer code's curve. Thus, a necessary condition for convergence of iterative decoding is the intuitively pleasing condition that $R < C$. More interestingly, this example illustrates how EXIT charts can help in code design. The closer we can fit the outer code's curve to the inner code's curve, the closer we can operate to capacity. That is, suitable application of the area property means that code design (for a fixed code rate) becomes a matter of designing the shape of information transfer curves, subject to a constraint on the area under the curve. Details of such design are beyond the scope of this book, but references for further reading are provided in Section 7.6.

7.2.6 Turbo weight enumeration

The EXIT charts predict the SNR threshold at which we expect iterative decoding to converge, at which point the BER starts dropping steeply. At SNRs significantly above this threshold (i.e., when the decoding tunnel in the EXIT chart is wide open), turbo codes exhibit a BER floor that is governed by the weight distribution of the turbo code (which depends strongly on the code length as well as the structure of the component codes). The EXIT chart analysis does not shed light on this behavior: at low BER, the mutual information attained by a decoder is very close to one. To understand the behavior at high SNR, therefore, we must revert to signal space concepts and look for the "most likely" error events that determine the BER floor. We now seek to develop an understanding of the role of the choice of the component codes and the codeword length on these error events. The approach is to employ a union bound on ML decoding using an appropriately chosen weight enumerator function for the turbo code. Of course, ML decoding is computationally infeasible for turbo codes because of their random-like structure. However, comparison of simulation results for suboptimal iterative decoding to approximate union bounds for ML decoding indicates that iterative decoding performs almost as well as ML decoding. Thus, insights based on analysis of ML decoding provide valuable guidelines for the design of codes to be decoded using iterative decoding.

The key reason why turbo codes work is that the number of low-weight codewords for a turbo code is much smaller than for the constituent convolutional codes. Thus, turbo codes reduce the *multiplicity* of low-weight codewords, rather than increasing the minimum possible codeword weight as in conventional design. The following discussion is based on a series of influential papers by Benedetto *et al.* We focus on parallel concatenated codes, but also summarize design rules for serial concatenated codes.

Consider parallel concatenation of two identical rate $1/2$ convolutional codes to get a rate $1/3$ turbo code. Neglecting edge effects due to trellis termination, we have K information bits as input, and $3K$ bits sent over the channel (we assume BPSK over an AWGN channel): the K information bits, K parity bits from the first encoder, and K parity bits from the second encoder (which sees an interleaved version of the input bits at its input). The resulting turbo code is a block code with codeword length $3K$ bits. Let $A_{\text{turbo}}(w, p)$ be the number of codewords of this block code which have input weight w and parity weight p. We now consider a union bound on the bit error probability. As in our earlier analysis of ML decoding of convolutional codes, we assume, without loss of generality, that the all-zero codeword is sent. The pairwise error probability that a codeword of input weight w and parity weight p is more likely than the all-zero word is

$$q(w, p) = Q\left(\sqrt{\frac{2E_b R(w+p)}{N_0}}\right). \tag{7.74}$$

Such a codeword error causes error in w out of K information bits. We can now write the following union bound for the probability of bit error:

$$P_e = P(\text{information bit error}) \leq \sum_{w,p} \frac{w}{K} A_{\text{turbo}}(w, p) \, Q\left(\sqrt{\frac{2E_b R(w+p)}{N_0}}\right). \tag{7.75}$$

Let us now define the conditional parity weight enumeration function (where we condition on the weight w of the information sequence) as

$$A_{\text{turbo}}(P|w) = \sum_p A_{\text{turbo}}(w, p) P^p, \tag{7.76}$$

where P is a dummy variable. We now consider a looser version of the bound in (7.75) using the inequality

$$Q(x) \leq e^{-\frac{x^2}{2}} \qquad (7.77)$$

(we can tighten this further by a factor of $1/2$, but we choose not to carry this around in what follows). Using (7.77) in (7.75), and substituting the definition (7.76), we get

$$P_e \leq \sum_w \frac{w}{K} W^w A_{\text{turbo}}(P|w) \Big|_{W=P=e^{-\frac{E_b R}{N_0}}}. \qquad (7.78)$$

The exponential decay of the term W^w with w means that we can concentrate on the terms with small w to get insight into performance. To do this, we need to characterize $A_{\text{turbo}}(P|w)$.

Consider the conditional parity weight enumerating function $A(P|w)$ for the constituent convolutional RSC (i.e., this enumerates the parity weight on one of the two branches of the parallel concatenated structure). A nonzero codeword for this code can be decomposed into a number of error events. Let $A(P|w, n)$ denote the parity weight enumerating function for an input sequence of weight w, considering codewords that consist of exactly n error events, as shown in Figure 7.17. Any nonzero codeword that is not an error event has a very large weight, and hence has a very small pairwise error probability contribution to the union bounds (7.75) or (7.78).

If we know $A(P|w, n)$, then we can write down the following approximate expression for $A(P|w)$:

$$A(P|w) \approx \sum_{n=1}^{n_{\max}} A(P|w, n) \binom{K}{n},$$

by counting the number of ways in which the input bits starting each of the n error events can be located among the K input bits, and neglecting the lengths of the error events compared to K. In the above equation, $n_{\max} = n_{\max}(w)$ is the largest number of error events that can be generated by an input sequence of weight w.

We now wish to find $A_{\text{turbo}}(P|w)$. Finding the weights resulting for a specific interleaver is complicated, but it is relatively easy to find the weight enumerator averaging over all possible interleavers. Thus, for any parity sequence generated by Encoder 1 for an input sequence of weight w, we can get any other parity sequence from Encoder 2 that can be generated by

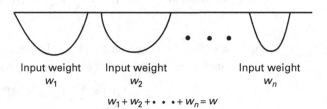

Figure 7.17 Nonzero convolutional codeword consisting of n error events, with total input weight w.

an input sequence of weight w with probability $1/\binom{K}{w}$. In terms of weight enumerator functions, we therefore obtain

$$A_{\text{turbo}}(P|w) = \frac{(A(P|w))^2}{\binom{K}{w}}$$

$$\approx \sum_{n_1=1}^{n_{\max}} \sum_{n_2=1}^{n_{\max}} \frac{\binom{K}{n_1}\binom{K}{n_2}}{\binom{K}{w}} A(P|w, n_1) A(P|w, n_2). \quad (7.79)$$

The union bound depends linearly on $A(P|w)$. Thus, the union bound for the averaged weight enumerator function above is an upper bound on the BER for the best interleaver. Therefore, there exists some interleaver for which the BER is at least as good as the union bound computed using the averaged weight enumerator above.

We can now approximate the combinatorial coefficients in (7.79) using the following result:

$$\binom{K}{l} \approx \frac{K^l}{l!} \quad \text{for large } K.$$

Substituting into (7.79), we have

$$A_{\text{turbo}}(P|w) \approx \sum_{n_1=1}^{n_{\max}} \sum_{n_2=1}^{n_{\max}} \frac{w!}{n_1! \, n_2!} K^{n_1+n_2-w} A(P|w, n_1) A(P|w, n_2).$$

We approximate the preceding summation by its dominant term, $n_1 = n_2 = n_{\max}$, which corresponds to the highest power of K. We therefore get

$$A_{\text{turbo}}(P|w) \approx \frac{w!}{(n_{\max}!)^2} K^{2n_{\max}-w} (A(P|w, n_{\max}))^2. \quad (7.80)$$

Substituting into (7.78), we get

$$P_e \precsim \sum_{w=w_{\min}}^{K} \frac{w}{K} W^w \frac{w!}{(n_{\max}!)^2} K^{2n_{\max}-w} (A(P|w, n_{\max}))^2 \Big|_{W=P=e^{-\frac{E_b R}{N_0}}}, \quad (7.81)$$

where $w_{\min}$ is the lowest weight input that can result in an error event.

Nonrecursive component codes do not work We can now immediately show that choosing a nonrecursive component code is a bad idea. For such a code, a finite weight input leads to a finite weight output, i.e., to an error event. Thus, $w_{\min} = 1$, and the maximum number of error events resulting from an input sequence of weight w is w. That is, $n_{\max}(w) = w$, and

$$A(P|w, n_{\max}) = (A(P|1, 1))^w,$$

since we consider w error events, each with input weight 1. Substituting in (7.81), we have

$$P_e \precsim \sum_{w=1}^{K} \frac{K^{w-1}}{(w-1)!} W^w (A(P|1,1))^{2w} \Big|_{W=P=e^{-\frac{E_b R}{N_0}}} \quad \textbf{Nonrecursive component code.} \quad (7.82)$$

The dominant term above is $w=1$, which is independent of K. Thus, there is no performance gain from using longer input sequences (and hence interleavers) for parallel concatenation of nonrecursive codes. As we show next, however, the use of recursive component codes does lead to performance improvement as a function of interleaver length K.

Interleaving gain for recursive component codes Now, consider a systematic recursive component code with generator

$$G(D) = \left[1 \ \frac{n(D)}{d(D)}\right].$$

For such a code, an input sequence of weight 1 leads to an infinite impulse response, so that the minimum input weight for an error event, $w_{\min}$ must be larger than one. In fact, we can show that $w_{\min} = 2$. This is because there exists i such that $d(D)$ divides $1+D^i$; in fact, the smallest such i is the period of a linear shift register with connection polynomial $d(D)$. Thus, a weight 2 input $u(D) = 1+D^i$ leads to a finite length output $u(D)G(D)$.

Since $w_{\min} = 2$, the maximum number of error events that can be generated by a weight w input is

$$n_{\max}(w) = \left\lfloor \frac{w}{2} \right\rfloor.$$

We now consider w even and odd separately.

Case 1 $w = 2k$ even. Then $n_{\max} = k$, and these k error events each correspond to a weight 2 subsequence of the input sequence. Since there are k such subsequences, we have $A(P|w, n_{\max}) = A(P|2k, k) = (A(P|2, 1))^k$. The corresponding term in the union bound (7.81) is

$$(2k)\frac{(2k)!}{(k!)^2} K^{2k-2k-1} W^{2k} \left(A(P|2, 1)\right)^{2k} \quad \textbf{Even weight term in union bound}.$$
(7.83)

The rate of decay with the interleaver length K is seen to be K^{-1}. This is referred to as the *interleaver gain*.

Case 2 $w = 2k+1$ odd. Then $n_{\max} = k$. The corresponding term in the union bound (7.81) is given by

$$(2k+1)\frac{(2k+1)!}{(k!)^2} K^{2k-(2k+1)-1} W^{2k+1} \left(A(P|2k+1, k)\right)^2$$

Odd weight term in union bound. (7.84)

7.2 Turbo codes and iterative decoding

The dependence on the interleaver length K is therefore K^{-2}, which is faster than the decay for w even. We can therefore neglect terms corresponding to odd w in the union bound for large K.

We can now concentrate on terms corresponding to even w. Let us first evaluate $A(P|2, 1)$ for our running example RSC code with generator $G(D) = [1, (1+D^2)/(1+D+D^2)]$. From the trellis or state diagram for this code, we see that the minimum parity weight for an error event resulting from a weight 2 input is $p_{\min} = 4$. This corresponds to the states 00, 10, 11, 01, 00 being traversed in succession, corresponding to the input sequence 1001. More generally, noting that the state traversal 10, 11, 01 forms a cycle corresponding to zero input weight and parity weight $p_{\min} - 2 = 2$, we can get an error event of parity weight $2 + l(p_{\min} - 2)$, $l \geq 1$ from an input sequence of weight 2 by traversing this zero input weight cycle l times. We therefore obtain the following expression for $A(P|2, 1)$:

$$A(P|2, 1) = \sum_{l=1}^{\infty} P^{2+l(p_{\min}-2)} = \frac{P^{p_{\min}}}{1 - P^{p_{\min}-2}}. \qquad (7.85)$$

It turns out that this formula, which we have derived for our running example, holds for any choice of parity generating polynomials, although the value of $p_{\min}$ depends on the specific choice of polynomials.

Substituting (7.85) and (7.83) into (7.81), we get

$$P_e \lesssim K^{-1} \sum_{k=1}^{\lfloor \frac{K}{2} \rfloor} 2k \binom{2k}{k} \frac{(S^{2+2p_{\min}})^k}{(1 - S^{p_{\min}-2})^{2k}} \bigg|_{S=e^{-\frac{E_b R}{N_0}}}, \qquad (7.86)$$

where we have set $W = P = S = e^{-\frac{E_b R}{N_0}}$, and we have neglected terms corresponding to odd w. For $k = 1$, the decay with SNR has exponent $2 + 2p_{\min}$.

Interleaving gain and effective free distance The approximate union bound (7.86) clearly shows the interleaving gain K^{-1} from increasing the length of the turbo code. In addition, from the $k = 1$ term in (7.86), we see that the "high SNR" asymptotics are governed by an effective free distance $d_{\text{eff}} = 2 + 2p_{\min}$. For our running example, we have $p_{\min} = 4$ and hence $d_{\text{eff}} = 10$.

Design rules for parallel concatenation We now summarize the design rules that we have derived.

(a) The component codes should be recursive.
(b) The interleaver length K should be as large as possible (subject to memory and delay constraints), to maximize interleaving gain.
(c) The component code should be chosen so as to maximize $p_{\min}$, and hence the effective free distance $d_{\text{eff}} = 2 + 2p_{\min}$. This is typically accomplished by choosing as large a memory for the constituent code as possible, subject to complexity constraints (this is similar to the conventional design prescriptions for convolutional codes).

An EXIT analysis yields further insights and design rules: while we have considered parallel concatenation of identical component codes, not choosing the component codes to be identical can have advantages in trading off convergence thresholds and error floors. We do not discuss this in detail here.

Design rules for serial concatenation A similar analysis based on weight enumerator functions can also be carried out for serial concatenation of convolutional codes. We omit this development, but the results are summarized below:

(a) The inner code must be recursive.
(b) If N is the number of bits at the output of the outer code (and hence the length of the interleaver), then the interleaving gain is given by $N^{-\lfloor \frac{d_{\text{free}}(\text{outer})+1}{2} \rfloor}$, where $d_{\text{free}}(\text{outer})$ is the free distance of the outer convolutional code.
(c) The outer code is preferably nonrecursive, since this leads to fewer input errors for error events at the free distance (which dominate the overall code performance). This means, for example, that we could improve on the code considered in Figure 7.16 by making the outer code the nonrecursive version of our running example.

To summarize, both parallel and serial concatenation of convolutional codes provide interleaving gains by decreasing the *multiplicity* of low-weight codewords in inverse proportion to the interleaver length. This is contrast to classical design, where the effort is to increase the minimum distance. While turbo codes do display an error floor, this can be pushed down by increasing the code length, with residual errors handled by an outer error detection code (e.g., a CRC code) or a high rate error correction code (e.g., a Reed–Solomon code).

7.3 Low density parity check codes

Low density parity check (LDPC) codes were introduced by Gallager in the 1960s, but were essentially forgotten for three decades after that. After turbo codes were introduced in 1993, it was realized by MacKay that LDPC codes form a compelling alternative for approaching Shannon limits. Since then, there has been a flurry of effort on design and analysis of LDPC codes for various channels. We describe here the code structure, iterative decoding, and some approaches to performance analysis for LDPC codes. These codes have the distinction of being one of the few codes for which theorems giving guarantees on asymptotic performance for large block lengths are available, with an analytical framework for evaluating how far away the performance is from Shannon capacity. As a practical consequence, it is possible to design LDPC codes that, for large block lengths, outperform turbo codes formed from serial or parallel concatenation of convolutional codes. Encoding for LDPC codes is somewhat more complex than for turbo codes, but decoding has the advantage of being more amenable to parallelized implementation.

7.3.1 Some terminology from coding theory

An (n, k) binary block code is a mapping of k information bits onto n coded bits, forming a codeword $\mathbf{x} = (x_1, \ldots, x_n)$. Let us denote the set of 2^k codewords by $\mathcal{C}$.

A binary *linear* code satisfies the property that, if $\mathbf{x}_1, \mathbf{x}_2 \in \mathcal{C}$, then $\mathbf{x}_1 + \mathbf{x}_2 \in \mathcal{C}$, where the addition is over the binary field, and corresponds to the XOR operator. Thus, for $a, b, c \in \{0, 1\}$, we have properties such as the following:

$$a + a = 0, \text{ so that } a = -a;$$

$$\text{if } a + b + c = 0, \text{ then } a = b + c.$$

Generator matrix We are familiar with vector spaces and subspaces over the real and complex fields. A binary linear code $\mathcal{C}$ forms a subspace of dimension k within an n-dimensional space over the binary field. Just as with real and complex fields, we can find k linearly independent basis vectors $\mathbf{v}_1, \ldots, \mathbf{v}_k$ for $\mathcal{C}$, such that any codeword $\mathbf{x} \in \mathcal{C}$ can be written as

$$\mathbf{x} = u_1 \mathbf{v}_1 + \cdots + u_k \mathbf{v}_k, \tag{7.87}$$

where $u_1, \ldots, u_k \in \{0, 1\}$ are coefficients of the basis expansion of $\mathbf{x}$ with respect to the chosen basis. Note that we are considering row vectors here, respecting standard convention in coding theory. We can rewrite (7.87) as

$$\mathbf{x} = \mathbf{u}\mathbf{G}, \tag{7.88}$$

where $\mathbf{u} = (u_1, \ldots, u_k)$ is a $1 \times k$ vector of basis expansion coefficients, and $\mathbf{G}$ is a $k \times n$ matrix with the basis vectors $\mathbf{v}_1, \ldots, \mathbf{v}_k$ as the rows. Since $u_1, \ldots, u_k$ can be freely assigned any values from $\{0, 1\}$, there are 2^k possible choices for $\mathbf{u}$, each corresponding to a unique $\mathbf{x}$, by virtue of the linear independence of the basis vectors. Thus, (7.88) can be interpreted as an *encoding map* from information vectors $\mathbf{u}$ of length k to codewords $\mathbf{x}$ of length n. The matrix $\mathbf{G}$ is a generator matrix for $\mathcal{C}$. Since the basis for a linear subspace is not unique, the generator matrix $\mathbf{G}$ and the corresponding encoding map are not unique either, even though the vector space, or the set of codewords $\mathcal{C}$, is fixed.

Dual code and parity check matrix Appealing again to our knowledge of vector spaces over the real and complex fields, we know that, for any

k-dimensional subspace of an n-dimensional space, there is an $(n-k)$-dimensional subspace that is its orthogonal complement. Similarly, we can define an $(n-k)$-dimensional subspace $\mathcal{C}^\perp$ over the binary field for an (n, k) binary linear code $\mathcal{C}$. Clearly, $\mathcal{C}^\perp$ can also be thought of as an $(n, n-k)$ binary linear code, which we term the *dual code* or dual space for $\mathcal{C}$. For $\mathbf{x} \in \mathcal{C}$ and $\mathbf{y} \in \mathcal{C}^\perp$, the inner product

$$\langle \mathbf{x}, \mathbf{y} \rangle = \sum_i x_i y_i = 0 \quad (\text{modulo } 2).$$

Let $\mathbf{H}$ denote a matrix whose rows form a basis for $\mathcal{C}^\perp$. If the basis is chosen to be linearly independent, then the matrix H has dimension $(n-k) \times n$, and can serve as a generator matrix for $\mathcal{C}^\perp$. Now, any codeword $\mathbf{x} \in \mathbf{C}$ must be orthogonal to every element of $\mathcal{C}^\perp$, which is true if and only if it is orthogonal to every row of $\mathbf{H}$. The latter condition can be written as

$$\mathbf{H}\mathbf{x}^T = \mathbf{0}. \tag{7.89}$$

Thus, each row of $\mathbf{H}$ provides a *parity check* equation that any codeword $\mathbf{x}$ must satisfy. Thus, $\mathbf{H}$ is termed a *parity check matrix* for $\mathcal{C}$. The parity check matrix is not unique, since the choice of basis for the dual code $\mathcal{C}^\perp$ is not unique.

Example 7.3.1 (Hamming code) A generator matrix for the (7, 4) Hamming code is given by

$$\mathbf{G} = \begin{pmatrix} 1 & 0 & 0 & 0 & 0 & 1 & 1 \\ 0 & 1 & 0 & 0 & 1 & 0 & 1 \\ 0 & 0 & 1 & 0 & 1 & 1 & 0 \\ 0 & 0 & 0 & 1 & 1 & 1 & 1 \end{pmatrix}. \tag{7.90}$$

This generator matrix is in *systematic* form, i.e., four out of the seven code bits are the information bits, without any modification. The remaining three bits are the parity check bits, formed by taking linear combinations of the information bits. A parity check matrix for the (7, 4) Hamming code is given by

$$\mathbf{H} = \begin{pmatrix} 0 & 1 & 1 & 1 & 1 & 0 & 0 \\ 1 & 0 & 1 & 1 & 0 & 1 & 0 \\ 1 & 1 & 0 & 1 & 0 & 0 & 1 \end{pmatrix}. \tag{7.91}$$

It can be checked that the inner product (using binary arithmetic) of any row of the generator matrix with any row of the parity check matrix is zero.

For any vector space of dimension n, a k-dimensional subspace can be specified either directly, or by specifying its orthogonal complement of dimension $n-k$ within the vector space. Thus, the code $\mathcal{C}$ can be specified compactly by either specifying a generator matrix $\mathbf{G}$ or a parity check matrix $\mathbf{H}$.

Figure 7.18 Tanner graph for (7, 4) Hamming code.

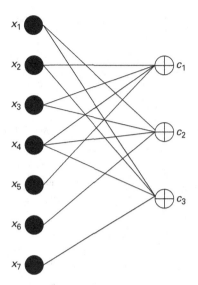

Tanner graph representation of a binary linear code Any binary linear graph can be represented by a Tanner graph using a parity check matrix **H**. A Tanner graph is a bipartite graph with variable nodes on the left corresponding to the different code symbols, and check nodes on the right, one for each check equation, or row of the parity check matrix. A variable node is connected to a check node if it participates in the corresponding parity check equation.

The Tanner graph corresponding to the parity check matrix (7.91) for the (7, 4) Hamming code is depicted in Figure 7.18. For example, the first check node c_1 corresponds to the first row of the parity check matrix:

$$x_2 + x_3 + x_4 + x_5 = 0,$$

so that the variable nodes x_2, x_3, x_4 and x_5 are connected to check node c_1.

Low density parity check codes are described in terms of a parity check matrix **H**, with the term "low density" referring to the sparseness of ones in **H**. Recall that parity check matrices are not unique, so it is also possible to find nonsparse parity check matrices for the same code. However, the sparseness of the parity check matrix **H** that we do use is crucial for both code construction and iterative decoding of LDPC codes.

7.3.2 Regular LDPC codes

A regular (d_v, d_c) LDPC code is one which is specified by a parity check matrix that has d_v ones in each column, and d_c ones in each row. That is, each parity check equation has d_c participating code variables, and each code variable appears in d_v parity check equations. As the code length n grows, keeping the code rate R constant, the dimensions of **H** scale up. However, **H** is sparse, in that the row and column weights remain constant at d_c and

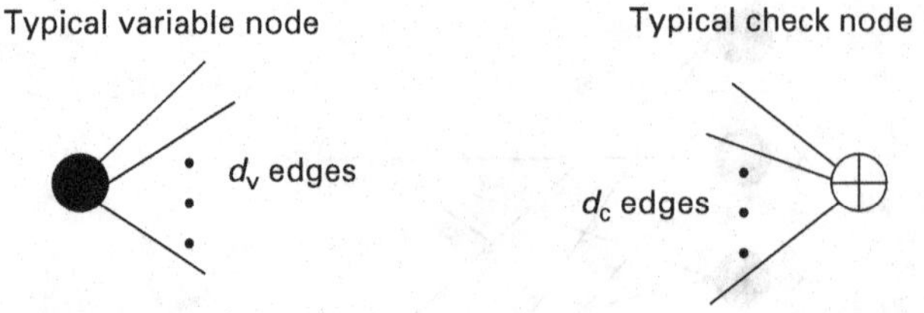

Figure 7.19 Typical variable and check node in a (d_v, d_c) regular LDPC code.

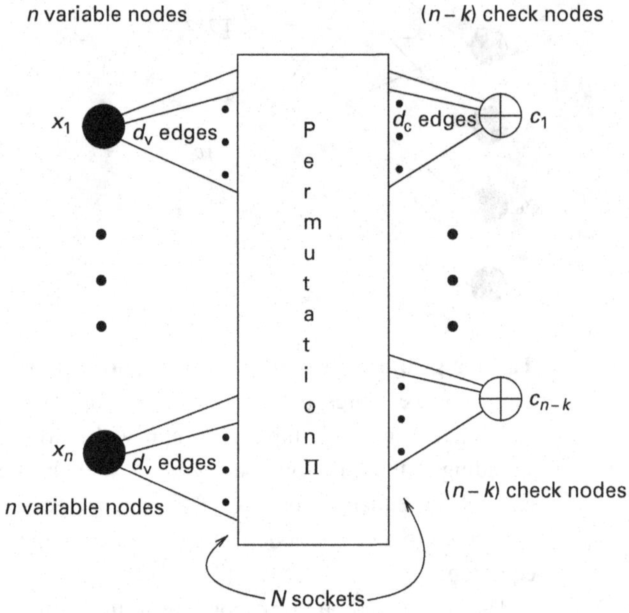

Figure 7.20 Random generation of a (d_v, d_c) regular LDPC code.

d_v, respectively. Consider a Tanner graph for the code, with n variable nodes on the left, and $n - k$ parity check nodes on the right. Typical variable and check nodes in such a graph are shown in Figure 7.19. There are d_v edges emanating from each variable node, so that there are $N = nd_v$ edges in all. Number this in any way (e.g., sequentially, starting from the first edge for the first variable). Similarly, there are d_c edges emanating from each check node, so that there are $N = (n-k)d_c$ edges in all. Since the number of edges N satisfies

$$N = nd_v = (n-k)d_c,$$

we obtain the code rate;

$$R = \frac{k}{n} = 1 - \frac{d_v}{d_c}.$$

Let us now consider how such a code can be generated randomly. On the left side of the Tanner graph, provide d_v "sockets" where edges can go in for each variable node. On the right side, provide d_c "sockets" for each check node. On each side, number the sockets from 1 to N in any order. Then connect socket i on the left with socket $\Pi(i)$ on the right to form the ith edge, $i = 1, \ldots, N$, where Π is a permutation of $\{1, \ldots, N\}$. This specifies the Tanner graph, and a corresponding parity check matrix, for the code. Thus, codes can be generated randomly by random choices of Π. The work of Richardson and Urbanke shows that most such codes are good, with performance clustering around the ensemble-averaged performance as the block length gets large. Thus, one way to get a good code is to generate random instances as above, and use some simple criteria to eliminate obviously bad choices. Of course, we must finally check by computer simulation that the candidate code indeed works well. That is, the chosen code (at some large but finite block length) should have a performance close to the average behavior, which can be analytically characterized in the limit of infinite code block lengths.

7.3.3 Irregular LDPC codes

For irregular LDPC codes, the variable nodes can have a range of degrees, as can the check nodes, where the degrees are chosen in a random manner according to specified *degree distributions*. Rather than specifying the fraction of nodes with a given degree, however, it is more convenient to specify these distributions in terms of what a randomly chosen edge in the Tanner graph sees. Specifically, let

$\lambda_i = P[\text{random edge is incident on variable node of degree } i]$,
$\rho_i = P[\text{random edge is incident on check node of degree } i]$.

It is convenient to specify the degree distributions in the form of polynomials:

$$\lambda(x) = \sum_i \lambda_i x^{i-1},$$
$$\rho(x) = \sum_i \rho_i x^{i-1}.$$

For example, for a $(3, 6)$ regular code,

$$\lambda(x) = x^2, \quad \rho(x) = x^5.$$

In an irregular code, a check node must have a degree of at least two. For a nontrivial check equation, there must be at least two variables participating (otherwise we get an equation of the form $x_l = 0$, which means that the code variable x_l conveys no information). Typically, the variable node degrees are also chosen to be at least two, unless we are considering relatively high rate codes.

Suppose, now, that there are N edges in the Tanner graph. The number of edges connected to degree i variable nodes is therefore equal to $N\lambda_i$. Thus, the number of variable nodes with degree i equals $N\lambda_i/i$, and the total number of variables equals

$$n = N \sum_i \frac{\lambda_i}{i} = N \int_0^1 \lambda(x)\, dx.$$

Similarly, the number of edges connected to check nodes of degree j is $N\rho_j$, and the number of check nodes of degree j equals $N\rho_j/j$. The total number of check nodes is therefore given by

$$n - k = N \sum_j \frac{\rho_j}{j} = N \int_0^1 \rho(x)\,dx.$$

We can therefore infer that the code rate R depends on the degree distributions as follows:

$$R = \frac{k}{n} = 1 - \frac{\int_0^1 \rho(x)\,dx}{\int_0^1 \lambda(x)\,dx}. \qquad (7.92)$$

We can also see that the fraction of variable nodes of degree i is given by

$$\tilde{\lambda}_i = \frac{\frac{\lambda_i}{i}}{\sum_l \frac{\lambda_l}{l}}$$

and the fraction of check nodes of degree j is given by

$$\tilde{\rho}_j = \frac{\frac{\rho_j}{j}}{\sum_l \frac{\rho_l}{l}}.$$

Example 7.3.2 (Irregular LDPC code) Consider an irregular LDPC code with $\lambda(x) = 0.2x + 0.8x^2$ and $\rho(x) = 0.5x^4 + 0.5x^5$. We can compute, using (7.92), the code rate as $R = 1/2$. The fraction of variable nodes of degree 2 is

$$\tilde{\lambda}_2 = \frac{\frac{0.2}{2}}{\frac{0.2}{2} + \frac{0.8}{3}} = \frac{3}{11}$$

and the fraction of check nodes of degree 5 is

$$\tilde{\rho}_5 = \frac{\frac{0.5}{5}}{\frac{0.5}{5} + \frac{0.5}{6}} = \frac{6}{11}.$$

As with regular codes, irregular LDPC codes can be generated by randomly choosing edges on a Tanner graph, as follows. For polynomials $\lambda(x)$ and $\rho(x)$ of a given degree, compute the rate R and the node degree distributions $\{\tilde{\lambda}_i\}$ and $\{\tilde{\rho}_j\}$. For n variable nodes and $n(1-R) = n-k$ check nodes, assign i sockets to $n\tilde{\lambda}_i$ of the variable nodes, and assign j sockets to $(n-k)\tilde{\rho}_j$ of the check nodes. We number the sockets on each side from $1,\ldots,N$, where

$$N = \frac{n}{\int_0^1 \lambda(x)dx} = \frac{n-k}{\int_0^1 \rho(x)dx}$$

is the number of edges to be specified. As before, for a random permutation Π on $\{1,\ldots,N\}$, connect socket i to socket $\Pi(i)$ to form an edge. Throw out or modify obviously bad choices, and verify that the code obtained performs well by simulation.

Maximum likelihood decoding is too complex for LDPC codes of even moderate block lengths, but iterative decoding based on message passing on the Tanner graph provides excellent performance. We now describe some message algorithms and techniques for understanding their performance.

7.3.4 Message passing and density evolution

Consider a regular (d_v, d_c) LDPC code over a BSC with crossover probability ϵ. A variable node x_l is connected to the channel, and receives a bit y_l which is in error with probability ϵ. We now consider a simple iterative bit flipping algorithm for decoding, termed Gallager's algorithm A, where the messages being passed back and forth between the variable and check nodes are binary, with the interpretation of simply being estimates of the code bits $\{x_l\}$. The key principle, which applies also to more complex iterative decoding algorithms, is that, for any node, the message sent out on an edge e is *extrinsic information*, depending only on messages coming in on the other edges incident on the node, and not on the message coming in on e.

Gallager's algorithm A

Initialization (variable node) The only information available at this point to a variable node is the channel bit. Thus, the variable node x_l sends the message y_l to each of the d_v check nodes connected to it.

Iteration 1 (check node) A check node corresponding to a check equation of the form

$$x_{i_1} + \cdots + x_{i_{d_c}} = 0$$

estimates each of the variables involved based on the estimates for the *other* variables, and passes the message back to the corresponding variable nodes. For example, the message passed back to variable node x_{i_1} is

$$\hat{x}_{i_1} = y_{i_2} + \cdots + y_{i_{d_c}}.$$

Iteration 1 (variable node) A variable node x_l receives d_v bit estimates from the check nodes it is connected to. It now uses these, and the channel bit, to compute messages to be sent back to the check nodes it is connected to. The message to be sent back to a given check node c depends only on the $d_v - 1$ messages coming in from the *other* check nodes, and the message from the channel. If all $d_v - 1$ messages from the other check nodes agree, and their estimate differs from the bit received from the channel, then the channel bit is flipped to obtain the message sent to c.

Iteration 2 (check node) As before, estimates to be sent back to the variable nodes are computed by taking the XOR of messages coming in from the other variable nodes.

These iterations continue until convergence, or until a maximum number of iterations is reached. At that point, one can check whether the bit estimates obtained form a valid codeword, and declare decoding failure if not.

Let us now provide a more formal description of the algorithm by describing the operations performed at a typical variable and check node at a given iteration.

Message from variable node Let u_0 denote the message received from the channel, and let $u_1, \ldots, u_{d_v}$ denote the incoming messages (from the check nodes) on the d_v edges incident on the variable node. The message v_i sent out along the ith edge, $i = 1, \ldots, d_v$, is then given by

$$v_i = \begin{cases} \bar{u}_0, & u_1 = \cdots = u_{i-1} = u_{i+1} = \cdots = u_{d_v} = \bar{u}_0, \\ u_0, & \text{else.} \end{cases}$$

The message v_i going out on the ith edge satisfies our concept of extrinsic information, since it does not depend on the message u_i coming in on that edge.

Message from check node Let $v_1, \ldots, v_{d_c}$ denote the incoming messages from the variable nodes on the d_c edges incident on the check node. The message u_j sent out along the jth edge, $j = 1, \ldots, d_c$, is given by

$$u_j = \sum_{l=1, l \neq j}^{d_c} v_l,$$

where the addition is over the binary field. Again, note that the outgoing message u_j is extrinsic information that does not depend on the incoming message v_j on that edge.

Initialization The variable nodes initiate the iterations using the channel outputs, setting $v_i \equiv u_0$ for $i = 1, \ldots, d_v$, since there are no messages yet from the check nodes.

While we consider a regular code as an example, the same algorithm is equally applicable to irregular codes (d_v and d_c above can vary).

Intuitively, we expect this simple bit flipping algorithm to correct all channel errors successively if the initial number of channel errors is small enough. We would now like to understand the performance in more detail. To this end, condition, without loss of generality, on the all-zero codeword being sent. Assume that all messages arriving at a node are independent and identically distributed. Since the initial channel outputs (and hence the messages sent by the variable nodes in iteration 1) are independent, this assumption holds if there are no cycles in the Tanner graph, so that the influence of a message sent out by a node in some iteration is not included in an incoming message to that node at a later iteration. Actually, this tree-based analysis can be shown to yield the right answer for computing ensemble-averaged performance for

7.3 Low density parity check codes

randomly generated Tanner graphs for "long enough" block lengths, for which the cycle lengths in the Tanner graph are, with high probability, long enough to be "tree-like." Since the messages are binary, they are modeled as i.i.d. Bernoulli random variables, whose distribution can be described by a single parameter (e.g., the probability of taking the value 1).

Define

$$p(l) = P[\text{message sent by variable node in iteration } l \text{ is } 1],$$

$$q(l) = P[\text{message sent by check node in iteration } l \text{ is } 1].$$

Note that $p(0) = \epsilon$. Recall that we are conditioning on the all-zero codeword being sent, so that p and q are estimates of the error probability of messages being sent along the edges, as a function of iteration number.

The message sent by a check node on an edge is 1 if and only if an odd number of ones come in on the other edges. At iteration l, the probability of an incoming 1 is $p(l)$. Thus,

$$q(l) = \sum_{j=1, j \text{ odd}}^{d_c - 1} \binom{d_c - 1}{j} (p(l))^j (1 - p(l))^{d_c - 1 - j}.$$

This can be simplified to

$$q(l) = \frac{1 - (1 - 2p(l))^{d_c - 1}}{2}. \tag{7.93}$$

The message sent by a variable node to a check node c is 1 if and only if one of two mutually exclusive events occur: (a) the channel output is 1, and not all incoming messages are 0, or (b) the channel output is 0, and all incoming messages are 1, where the incoming messages are those coming in from all check nodes other than c. These are modeled as i.i.d. Bernoulli, with probability $q(l)$ of taking the value 1. We therefore obtain

$$p(l) = p(0)[1 - (1 - q(l))^{d_v - 1}] + (1 - p(0))(q(l))^{d_v - 1}. \tag{7.94}$$

We can combine (7.93) and (7.94) into a single recursion for p, given by:

$$p(l) = p(0) - p(0) \left\{ \frac{1}{2}[1 + (1 - 2p(l-1))^{d_c - 1}] \right\}^{d_v - 1} \\ + (1 - p(0)) \left\{ \frac{1}{2}[1 - (1 - 2p(l-1))^{d_c - 1}] \right\}^{d_v - 1}, \tag{7.95}$$

with $p(0) = \epsilon$.

The preceding procedure is an example of *density evolution*, where we characterize the message probability density as a function of the iteration number, assuming independent messages coming in. In this case, the density is specified by a single number, which makes the evolution particularly simple.

The bit flipping algorithm converges to the correct solution if $p(l) \to 0$ as $l \to \infty$. By implementing the recursion for various values of ϵ, we can find a threshold value ϵ_t such that the error probability converges to zero for $\epsilon < \epsilon_t$.

We would typically find that this threshold is quite far from the ϵ for which the BSC channel capacity equals the rate of the code being used. However, significant improvement in the performance of bit flipping can be obtained at a relatively small increase in complexity. For example, we can introduce erasures when we cannot decide whether to believe the messages from the check nodes or from the channel, rather than insisting on binary messages. Simple message formats and node operations can also be handcrafted for more complex channel models.

The most powerful form of message passing is when the messages are log likelihood ratios (LLRs). Such a message passing algorithm is referred to as *belief propagation*, where the terminology comes from Pearl's seminal work on Bayesian networks. We discuss this next.

7.3.5 Belief propagation

As before, consider the Tanner graph for a (d_v, d_c) regular LDPC code as an example, with the understanding that the message passing algorithm extends immediately to irregular LDPC codes.

The messages being passed are LLRs for the code bits. The initial message from the channel regarding a bit is denoted by u_0. If x is the bit of interest and y is the corresponding channel output, then

$$u_0 = \log \frac{p[y|x=0]}{p[y|x=1]}.$$

For a BSC with crossover probability ϵ, we have

$$u_0(BSC) = \begin{cases} \log \frac{1-\epsilon}{\epsilon}, & y = 0, \\ -\log \frac{1-\epsilon}{\epsilon}, & y = 1. \end{cases}$$

For BPSK over an AWGN channel, where $y = A(-1)^x + N(0, \sigma^2)$, we have

$$u_0(AWGN) = \frac{2Ay}{\sigma^2},$$

as derived earlier.

Once the messages from the channel are determined, the remainder of the algorithm proceeds without any further need to invoke the channel model (we assume a memoryless channel). As with the bit flipping algorithm, we can specify the belief propagation algorithm by describing the operations performed at a typical variable and check node at a given iteration. We derive these next, and then summarize the operation of iterative decoding with belief propagation.

A message coming in on an edge incident on variable node x conveys an LLR for x. Assuming that the LLRs coming in on different edges are independent, the extrinsic LLR to be sent out on an edge is simply the sum of the LLRs coming in on the other edges, since all these LLRs refer to the same variable x.

7.3 Low density parity check codes

For a check node c, the computation required for message passing is more complicated. Consider, for example, the check equation $x_1 + x_2 + x_3 = 0$. Suppose that we know the LLRs for x_1 and x_2 (these come in as messages m_1 and m_2 along the edges corresponding to x_1 and x_2, respectively). Assuming that these are independent, we wish to compute the LLR for x_3, and then send it out as message m_3 on the edge connected to x_3. Since $x_3 = x_1 + x_2$, we have

$$P[x_3 = 0] = P[x_1 = 0, x_2 = 0] + P[x_1 = 1, x_2 = 1] \quad (7.96)$$
$$= P[x_1 = 0]P[x_2 = 0] + P[x_1 = 1]P[x_2 = 1],$$
$$P[x_3 = 1] = P[x_1 = 0, x_2 = 1] + P[x_1 = 1, x_2 = 0]$$
$$= P[x_1 = 0]P[x_2 = 1] + P[x_1 = 1]P[x_2 = 0].$$

Now, since $m_i = \log \frac{P[x_i=0]}{P[x_i=1]}$, we have

$$P[x_i = 0] = \frac{e^{m_i}}{e^{m_i} + 1}, \quad P[x_i = 1] = \frac{1}{e^{m_i} + 1}.$$

Substituting into (7.96) and simplifying, we obtain

$$e^{m_3} = \frac{P[x_3 = 0]}{P[x_3 = 1]} = \frac{e^{m_1+m_2} + 1}{e^{m_1} + e^{m_2}}, \quad (7.97)$$

from which we can compute m_3 in terms of m_1 and m_2. However, this formula does not generalize to check nodes with larger degrees. To remedy this, instead of considering the LLR, let us consider instead the quantity $P[x = 0] - P[x = 1]$ for a bit x whose LLR is m. We obtain that

$$P[x = 0] - P[x = 1] = \frac{e^m - 1}{e^m + 1} = \frac{e^{m/2} - e^{-m/2}}{e^{m/2} + e^{-m/2}} = \tanh\left(\frac{m}{2}\right). \quad (7.98)$$

We can now use (7.97) to infer that

$$\tanh(m_3/2) = P[x_3 = 0] - P[x_3 = 1] = \frac{e^{m_1+m_2} + 1 - e^{m_1} - e^{m_2}}{e^{m_1+m_2} + 1 + e^{m_1} + e^{m_2}} \quad (7.99)$$
$$= \frac{(e^{m_1} - 1)(e^{m_2} - 1)}{(e^{m_1} + 1)(e^{m_2} + 1)} = \tanh\left(\frac{m_1}{1}\right)\tanh\left(\frac{m_2}{2}\right).$$

This is a formula that does generalize to check nodes with larger degrees, as we can see by induction. While the choice of using $P[x = 0] - P[x = 1]$ appears to have been fortuitous, the fact that this is indeed the right quantity to consider to obtain a generalizable formula is discussed in Problem 7.19.

We can now summarize message passing for belief propagation as follows.

Message from variable node Let u_0 denote the message received from the channel, and let $u_1, \ldots, u_{d_v}$ denote the incoming messages (from the check nodes) on the d_v edges incident on the variable node. Then the message v_p sent out along the pth edge, $p = 1, \ldots, d_v$, is given by

$$v_p = u_0 + \sum_{q=1, q \neq p}^{d_v} u_q. \quad (7.100)$$

Note that v_p does not depend on u_p.

Message from check node Let $v_1, \ldots, v_{d_c}$ denote the incoming messages from the variable nodes on the d_c edges incident on the check node. The message u_p sent out along the pth edge, $p = 1, \ldots, d_c$, is given by the implicit equation

$$\tanh\left(\frac{u_p}{2}\right) = \Pi_{q=1, q \neq p}^{d_c} \tanh\left(\frac{v_q}{2}\right). \tag{7.101}$$

Alternatively, the outgoing message u_p can be represented as a hard decision $\hat{b}_p(\text{out}) = I_{\{u_p < 0\}}$, together with a reliability metric $\Lambda_p(\text{out}) = |\log\tanh(|u_p|/2)|$. By representing the incoming messages in the same fashion, with $\hat{b}_q(\text{in}) = I_{\{v_q < 0\}}$ and $\Lambda_q(\text{in}) = |\log\tanh(|v_q|/2)|$, it is easy to show, from (7.101), that

$$\hat{b}_p(\text{out}) = \sum_{q=1, q\neq p}^{d_c} \hat{b}_q(\text{in}) \quad \text{(binary addition)}, \tag{7.102}$$

$$\Lambda_p(\text{out}) = \sum_{q=1, q\neq p}^{d_c} \Lambda_q(\text{in}) \quad \text{(real addition)}.$$

We can get the LLR L for a bit from its $(\hat{b}, \Lambda)$ representation as follows:

$$L = (-1)^{\hat{b}} \, 2\tanh^{-1}\left(e^{\Lambda}\right).$$

Initialization The variable nodes initiate the iterations using the channel outputs, setting $v_i \equiv u_0$ for $i = 1, \ldots, d_v$, since there are no messages yet from the check nodes.

The preceding description specifies the belief propagation algorithm, where for irregular codes, the degrees d_v and d_c would be variable. The algorithm is initialized by each variable node sending its channel message along all of its edges.

The alternative representation (7.102) is useful for a density evolution analysis of belief propagation. However, we employ a less complex Gaussian approximation to develop an understanding of belief propagation in the following, for which (7.101) suffices.

7.3.6 Gaussian approximation

To evaluate the performance of belief propagation, condition on the all-zero codeword being sent. As in our discussion of how to generate EXIT charts for turbo codes, we again invoke the consistency condition for LLRs. That is, from Problem 7.12, we have the following results. For an output symmetric channel satisfying $p(y|0) = p(-y|1)$, the conditional density $p(z|0)$ of the channel LLR, conditioned on 0 being sent, satisfies $p(z|0) = e^z p(-z|0)$. Assuming that enough mixing is going on at the variable and check nodes, we model the LLR messages as Gaussian, satisfying this consistency condition. If the conditional distribution $Z \sim N(m, v^2)$, conditioned on 0 being sent, the consistency condition becomes equivalent to $v^2 = 2m$ (which automatically

7.3 Low density parity check codes

imposes the requirement that $m \geq 0$). We therefore only need to specify how the means evolve as a function of iteration number.

Regular LDPC codes For a (d_v, d_c) regular LDPC code, model the messages going out from a variable node at iteration l as $N(m_v(l), 2 m_v(l))$, and the messages going out from a check node at iteration l as $N(m_u(l), 2m_u(l))$. From (7.100), it follows immediately that the mean of the output from a variable node is given by

$$m_v(l) = m_{u_0} + (d_v - 1)m_u(l-1), \quad (7.103)$$

where the channel LLR is $N(m_{u_0}, 2 m_{u_0})$. This is exactly true for BPSK over an AWGN channel, but may also be a good approximation to bit LLRs for larger constellations with a Gray coded bit-to-symbol map.

For a check node, a typical outgoing message $u(l)$ at iteration l satisfies

$$\tanh\left(\frac{u(l)}{2}\right) = \Pi_{i=1, i \neq j}^{d_c} \tanh\left(\frac{v_i(l)}{2}\right). \quad (7.104)$$

For $m \geq 0$, it is now convenient to define the function

$$\phi(m) = 1 - E\left[\tanh\left(\frac{Z}{2}\right)\right], \quad Z \sim N(m, 2m). \quad (7.105)$$

Since we model $v_i(l)$ as i.i.d. $N(m_v(l), 2 m_v(l))$, and $u(l) \sim N(m_u(l), 2 m_u(l))$, we have, taking expectations on both sides of (7.104):

$$1 - \phi(m_u(l)) = [1 - \phi(m_v(l))]^{d_c-1},$$

so that

$$m_u(l) = \phi^{-1}\left(1 - [1 - \phi(m_v(l))]^{d_c-1}\right). \quad (7.106)$$

Combining (7.103) and (7.106), we get the following recursion for m_u:

$$m_u(l) = \phi^{-1}\left(1 - [1 - \phi\left(m_{u_0} + (d_v - 1)m_u(l-1)\right)]^{d_c-1}\right), \quad (7.107)$$

with $m_u(0) = 0$.

One can compute thresholds on the channel quality by checking for which values of m_{u_0} we have $m_u(l) \to \infty$.

Some bounds and approximations for $\phi(m)$ can be used to simplify the computation of ϕ and ϕ^{-1}, as well as to understand the behavior of the Gaussian approximation in more depth. We state these below without proof.

$$\left(1 - \frac{3}{m}\right)\sqrt{\frac{\pi}{m}} e^{-\frac{m}{4}} < \phi(m) < \left(1 + \frac{1}{7m}\right)\sqrt{\frac{\pi}{m}} e^{-\frac{m}{4}} \phi(m). \quad (7.108)$$

(The bounds are only useful for $m > 0$. For $m = 0$, observe that $\phi(0) = 1$.)

An approximation that works well for ϕ for $m < 10$ is

$$\phi(m) \approx e^{\alpha m^\gamma + \beta}, \quad (7.109)$$

where $\alpha = -0.4527$, $\beta = 0.0218$, $\gamma = 0.86$. This has the advantage of being analytically invertible. For $m > 10$, an average of the upper and lower bounds in (7.108) has been found to work well.

Thresholds calculated using such techniques are found to be within 0.1 dB of far more complex calculations using density evolution, so that the Gaussian approximation is a very valuable tool for understanding LDPC performance.

Next, we discuss irregular LDPC codes, for which the Gaussian approximation has been employed effectively as a tool for optimizing degree sequences.

Irregular LDPC codes The messages from the channel to the variable nodes are modeled as $N(m_{u_0}, 2m_{u_0})$ (this is exact for BPSK over an AWGN channel). Let us model the input messages to variable nodes at the beginning of iteration l as i.i.d., $N(m_u(l-1), 2m_u(l-1))$. Consider a variable node of degree i. From (7.100), we have that the typical output message from such a node takes the form

$$v = u_0 + \sum_{q=1}^{i-1} u_q,$$

where u_q are messages coming in on the $(i-1)$ other edges, and u_0 is the message from the channel. Modeling u_q as i.i.d. $N(m_u(l-1), 2m_u(l-1))$, it follows that the output message can be modeled as $N(m_{v,i}(l), 2m_{v,i}(l))$, where

$$m_{v,i}(l) = m_{u_0} + (i-1)m_u(l-1). \tag{7.110}$$

These messages are sent to the check nodes. A typical edge incident on a check node sees an edge connected to a degree i variable node with probability λ_i. Thus, a typical input message $v(l)$ to a check node at iteration l is a Gaussian mixture, with the distribution $N(m_{v,i}(l), 2m_{v,i}(l))$ selected with probability λ_i. We denote this as

$$v(l) \sim \sum_i \lambda_i N(m_{v,i}(l), 2m_{v,i}(l)). \tag{7.111}$$

We would now like to characterize the mean of the output message from the check node of degree j. We do this indirectly by instead computing the mean of the hyperbolic tangent of the message, using the check update equation (7.101). Then, assuming that the output message is Gaussian $N(m_{u,j}(l), 2m_{u,j}(l))$, we compute $m_{u,j}(l)$ by inverting the ϕ function. Let us now go through the details. From (7.101), the check update at a degree j node takes the form

$$\tanh\left(\frac{u}{2}\right) = \Pi_{q=1}^{j-1} \tanh\left(\frac{v_q}{2}\right). \tag{7.112}$$

Assuming that v_q follow the Gaussian mixture distribution (7.111), we obtain

$$E\left[\tanh\left(\frac{v_q}{2}\right)\right] = \sum_i \lambda_i \left(1 - \phi(m_{v,i}(l))\right)$$

$$= 1 - \sum_i \lambda_i \phi(m_{v,i}(l)).$$

Modeling v_q as i.i.d., we have, taking expectations on both sides of (7.112),

$$1 - \phi(m_{u,j}(l)) = \left[1 - \sum_i \lambda_i \phi(m_{v,i}(l))\right]^{j-1},$$

where we have used the Gaussian model $u \sim N(m_{u,j}(l), 2m_{u,j}(l))$ for the output message. We therefore obtain

$$m_{u,j}(l) = \phi^{-1}\left(1 - \left[1 - \sum_i \lambda_i \phi(m_{v,i}(l))\right]^{j-1}\right). \quad (7.113)$$

Averaging across the check node degrees, we see that the mean of the messages going into the variable nodes after iteration l is

$$m_u(l) = \sum_j \rho_j m_{u,j}(l) \quad (7.114)$$

$$= \sum_j \rho_j \phi^{-1}\left(1 - \left[1 - \sum_i \lambda_i \phi(m_{v,i}(l))\right]^{j-1}\right).$$

We can now substitute (7.110) into (7.114) to get a recursion for $m_u(l)$ as follows:

$$m_u(l) = \sum_j \rho_j \phi^{-1}\left(1 - \left[1 - \sum_i \lambda_i \phi(m_{u_0} + (i-1)m_u(l-1))\right]^{j-1}\right),$$
(7.115)

with initial condition $m_u(0) = 0$.

Convergence to a good solution occurs if $m_u(l) \to \infty$.

7.4 Bandwidth-efficient coded modulation

Thus far, we have focused on BPSK modulation over an AWGN channel, when illustrating the performance of binary codes. Clearly, these results apply to coherent QPSK modulation directly, since QPSK with Gray coding can be viewed as two BPSK streams sent in parallel over the I and Q channels. However, when we wish to increase bandwidth efficiency at the cost of power efficiency, we must learn how to do channel coding for larger constellations, such as 8-PSK or 16-QAM. There are several approaches for constructing such bandwidth-efficient coded modulation techniques, and we now discuss two of these: bit interleaved coded modulation (BICM) and trellis coded modulation (TCM). Bit interleaved coded modulation exploits the powerful binary codes discussed so far, and essentially achieves a clean separation between coding and modulation, while TCM involves a more detailed co-design of coding and modulation. Given the advances in binary turbo-like codes, BICM probably has the edge for applications in which the code block lengths can be large enough for the turbo effect to kick in.

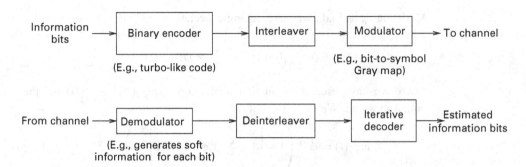

Figure 7.21 Bit interleaved coded modulation with a turbo-like outer binary code, serially concatenated with Gray coded modulation.

7.4.1 Bit interleaved coded modulation

The term BICM has acquired a rather broad usage, but we focus on a specific meaning corresponding to a clean separation of coding and modulation. The information bits are encoded using a binary code. The coded bits are interleaved and then mapped to the modulated signal sent over the channel. A particularly appealing combination, shown in Figure 7.21, is a turbo-like binary code, along with Gray coded modulation. The demodulator then computes soft decisions for each bit corresponding to a symbol, and then feeds them via the deinterleaver to the decoder, which then uses standard iterative decoding techniques. Because of the interleaving, we can think of each coded bit as seeing an equivalent channel which is the cascade of the modulator, the channel, and the demodulator's soft information computation. This effective binary channel is different from the BPSK channel considered so far. However, we do expect turbo-like codes optimized for BPSK over AWGN to work quite well in this context as well. Moreover, once we characterize the effective binary channel, it is also possible to optimize turbo-like codes (e,g., irregular LDPC codes) specifically for it, although this is beyond the scope of this book.

To make the concepts concrete, let us consider the example of BICM using Gray coded 4-PAM, as shown in Figure 7.22. When used over an AWGN channel, the output is given by

$$Y = s(x_1 x_2) + n , \quad n \sim N(0, \sigma^2), \tag{7.116}$$

where the symbol $s(x_1 x_2)$ takes values in $\{\pm A, \pm 3A\}$ as a function of the bits x_1 and x_2 according to a Gray code, as shown in Figure 7.22. Assuming that

Figure 7.22 Gray coded 4-PAM, with symbols labeled $x_1 x_2$, where x_1 and x_2 are bits coming from the interleaver.

10	11	01	00
−3A	−A	+A	+3A

7.4 Bandwidth-efficient coded modulation

x_1, x_2 are i.i.d., taking values in $\{0, 1\}$ with equal probability, we can now compute their LLRs based on Y as follows:

$$L_i(y) = \log \frac{P[x_i = 0|Y = y]}{P[x_i = 1|Y = y]} = \log \frac{p(y|x_i = 0)}{p(y|x_i = 1)}, \quad i = 1, 2.$$

As shown in Problem 7.23, these can be evaluated to be

$$L_1(y) = \frac{2Ay}{\sigma^2} + \log \frac{1 + e^{\frac{2A}{\sigma^2}(y - 2A)}}{1 + e^{-\frac{2A}{\sigma^2}(y + 2A)}}, \tag{7.117}$$

$$L_2(y) = -\frac{4A^2}{\sigma^2} + \log \frac{\cosh \frac{3Ay}{\sigma^2}}{\cosh \frac{Ay}{\sigma^2}}. \tag{7.118}$$

Knowing the values of A and σ^2, the demodulator can compute these LLRs and feed them to the decoder. The energy per coded symbol $E_s = \frac{A^2 + (3A)^2}{2} = 5A^2 = 2E_bR$, where E_b is the energy per *information* bit, and R is the rate of the outer binary code. This allows us to relate the parameters A and σ^2 to E_b/N_0 and R, once we choose a convenient scaling (e.g., $\sigma^2 = N_0/2 = 1$).

Once the LLRs above are fed to the binary decoder, it can perform softinput decoding as usual (e.g., using iterative decoding, if the outer code is a turbo-like code, or using the Viterbi or BCJR algorithm, if it is a convolutional code).

Once the effective binary channels are characterized, it is easy to identify information-theoretic performance limits for BICM. For example, the capacity attained by a 4-PAM-based BICM scheme is the attained by a 4-PAM-based BICM scheme is the sum of the capacities of the bits x_1 and x_2. We can compare this with the capacity of equiprobable 4-PAM over an AWGN channel to quantify the degradation in performance due to restricting the design to BICM.

The capacity for standard 4-PAM with equiprobable signaling can be computed as discussed in Chapter 6. For the effective binary channels for bits x_i, $i = 1, 2$, for a BICM system, the capacity can be computed using the equation

$$C_i = H(X_i) - H(X_i|Y) = 1 - H(X_i|Y),$$

where we use capital letters to denote random variables, and where the quantity $H(X_i|Y)$ can be estimated either by numerical integration or simulation, as discussed in Chapter 6. Specifically, consider noisy observations $y_1, \ldots, y_K$ generated by K i.i.d. uses of the 4-PAM channel (7.116) with the bits X_1, X_2 used to select the symbols chosen i.i.d., equiprobable over $\{0, 1\}$. We obtain the estimates

$$\hat{H}(X_i|Y) = \frac{1}{K} \sum_{j=1}^{K} H_B(P[X_i = 1|Y_j = y_j])$$

$$= \frac{1}{K} \sum_{j=1}^{K} H_B\left(\frac{1}{e^{L_i(y_j)} + 1}\right), \tag{7.119}$$

where $H_B(p) = -p\log_2 p - (1-p)\log_2(1-p)$ is the binary entropy function. The capacity for the BICM scheme is therefore $C_{\text{BICM}} = C_1 + C_2$.

Problem 7.23 compares the capacity of standard 4-PAM with the BICM-based 4-PAM capacity. The capacity is a function of $\text{SNR} = E_s/\sigma^2 = 5A^2/\sigma^2$. This can then be converted into a plot of spectral efficiency versus E_b/N_0 as described in Chapter 6.

7.4.2 Trellis coded modulation

Trellis coded modulation (TCM), pioneered by Ungerboeck in the early 1980s, combines coding and modulation using convolutional code based trellis structures, but with the design criterion being Euclidean distance between the sequences of real-valued or complex-valued symbols, rather than the Hamming distance between binary sequences. We illustrate the basic concepts underlying TCM through a simple example here, and provide pointers for delving deeper in Section 7.6. Suppose that we wish to use a passband channel to signal at a spectral efficiency of 2 bits per complex-valued sample. For uncoded communication, we could achieve this by using uncoded QPSK, which is about 8 dB away from Shannon capacity at a BER of 10^{-5}. To close this gap using coding, clearly we must use a constellation larger than QPSK in order to "make room" for inserting some redundancy. If we expand the constellation by a factor of two, to 8-PSK, then we obtain one extra bit to work with. We now present one of Ungerboeck's original code constructions. The basic idea is as follows: partition the 8-PSK constellation into subsets, as shown in Figure 7.23. The partition is hierarchical, with the bit c_1 partitioning 8-PSK into two QPSK subsets $S(0)$ and $S(1)$, the bit c_2 partitioning the QPSK subsets further into BPSK subsets $S(00)$, $S(01)$, $S(10)$, $S(11)$, and the bit c_3 indexing the elements of the BPSK subsets. This introduces a 3 bit labeling $c_1 c_2 c_3$ for the 8-PSK constellation. Note that this set-partitioning-based labeling is quite different from a Gray map.

The symbol error probability for uncoded 8-PSK is dominated by the minimum distance d_0 for the constellation. An error to a nearest neighbor corresponds to an error in c_1 (i.e., we decode into a QPSK subset which is different from the one the transmitted symbol belongs to). However, if the bit c_1 could be sent error-free (in which case we know which QPSK subset the transmitted symbol belongs to), then the error probability is dominated by the minimum distance d_1 for the QPSK subconstellations. Furthermore, if both c_1 and c_2 could be transmitted error-free, then the error probability would be dominated by the minimum distance d_2 for the BPSK subset.

Suppose, now, that we wish to support the same rate as uncoded QPSK using a TCM scheme based on 8-PSK. If we set the energy per information bit $E_b = 1/2$, the symbol energy becomes normalized to one (i.e., the radius of the constellation is one). In this case, the minimum distance for uncoded QPSK

7.4 Bandwidth-efficient coded modulation

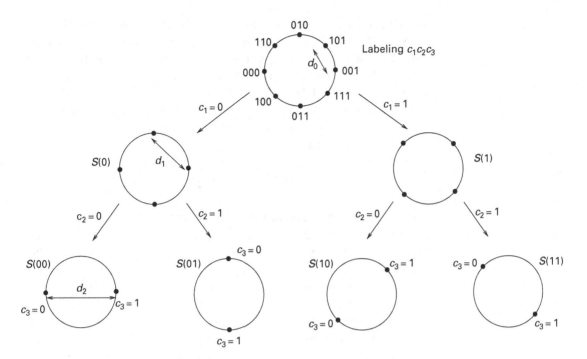

Figure 7.23 Ungerboeck set partitioning and labeling for 8-PSK.

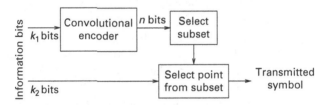

Figure 7.24 A TCM encoder uses convolutionally coded bits to select a subset of points, and uncoded bits to select points within the subset. In the figure, the constellation has 2^{n+k_2} points, divided into 2^n subsets, each with 2^{k_2} points. The convolutional code has rate k_1/n.

satisfies $d_{\min}^2(\text{QPSK}) = d_1^2 = 2$. We would like to devise a TCM scheme for which the minimum distance between codewords is better than this.

Figure 7.24 depicts the generic structure of a TCM encoder. In the following, we consider an 8-PSK-based scheme in which $k_1 = k_2 = 1$ and $n = 2$; thus, the convolutional code has rate $1/2$. The bits c_1 and c_2 in the 8-PSK set partition of Figure 7.23 are provided by the output of the convolutional encoder, while the bit c_3 is left uncoded. Thus, if c_1 and c_2 could be sent error-free, then the error probability is dominated by $d_2^2 = 4$, the minimum distance of the BPSK subconstellation $S(c_1 c_2)$. This is 3 dB better than the squared minimum distance of $d_1^2 = 2$ obtained for uncoded QPSK. Of course, c_1 and c_2 cannot be sent without error. However, we soon show that, by an appropriate choice of the rate 1/2 convolutional code, the minimum distance that dominates the error probability for c_1 and c_2 is larger than d_2. Thus, the error probability for the overall TCM scheme is determined by $d_{\min}(\text{TCM}) = d_2$, which gives a 3 dB gain over uncoded QPSK.

Let us now see how to ensure that the minimum distance corresponding to c_1 and c_2 is larger than d_2. Let us first define the distance between two subconstellations A and B as

$$d(A, B) = \min_{s_1 \in A, s_2 \in B} \|s_1 - s_2\|.$$

Thus, the distance between the QPSK subsets $S(0)$ and $S(1)$ is d_0. The distance between the BPSK subsets $S(00)$ and $S(01)$ is $d_1 > d_0$, while the distance between the BPSK subsets $S(00)$ and $S(10)$ is d_0 again. That is, two BPSK subsets have a larger distance only if their labels agree in the first bit c_1. Let us now consider the minimum Euclidean distance between the subsets chosen as the convolutional codeword dictating how the sequence of $c_1 c_2$ diverges from the all-zero path. Suppose that we restrict attention to a four-state encoder of the form shown in Figure 7.25. The output corresponding to a one at the input to the encoder is dictated by the generators $g_1(D) = g_{10} + g_{11}D + g_{12}D^2$ (for c_1) and $g_2(D) = g_{20} + g_{21}D + g_{22}D^2$ (for c_2). When we diverge from the all-zero path, we would like to maximize the contribution to the Euclidean distance, and hence we wish to keep $c_1 = 0$. Similarly, when we merge back with the all-zero path, we would again like to have $c_1 = 0$. This means that $g_{10} = g_{12} = 0$. Thus, for nontrivial g_1, we must have $g_{11} = 1$, which gives $g_1(D) = D^2$. To ensure that the outputs when we diverge and merge are different from those for the all-zero path, we must have $g_{20} = g_{22} = 1$ (since $g_{10} = g_{12} = 0$). Setting $g_{21} = 0$ (for no particular reason), we get $g_2(D) = 1 + D^2$. The TCM scheme we obtain thus is depicted in Figure 7.26.

A trellis for this code is shown in Figure 7.27. Assuming that the input bits are equally likely, all BPSK subsets $S(c_1 c_2)$ are used equally often. Moreover, both branches leaving or entering a state have the same value of c_1, which maximizes the Euclidean distance between the corresponding subsets. The

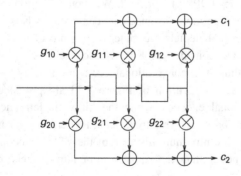

Figure 7.25 A generic rate 1/2, four-state, convolutional encoder.

7.4 Bandwidth-efficient coded modulation

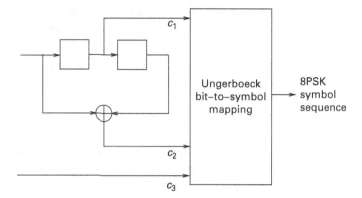

Figure 7.26 Four-state TCM scheme with rate 2 bit/channel use, using an 8-PSK alphabet.

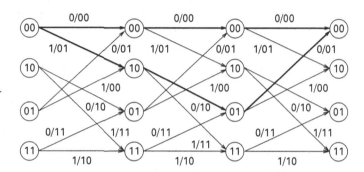

Figure 7.27 Trellis section for the convolutional code in Figure 7.26. The branches are labeled with the input bit and the two output bits corresponding to the transition.

all-zero sequence is highlighted in the trellis. It corresponds to repeatedly using the BPSK subset $S(00)$. Also highlighted in the trellis is a binary codeword whose subset sequence has minimum Euclidean distance to the subset sequence corresponding to the all-zero binary codeword. This binary codeword corresponds to the $(c_1 c_2)$ sequence $(01), (10), (01)$. The squared Euclidean distance between the corresponding subset sequences is

$$d^2_{\text{subset}} = d^2(S(00), S(01)) + d^2(S(00), S(10)) + d^2(S(00), S(01))$$
$$= d_1^2 + d_0^2 + d_1^2 = 4.586,$$

where we have substituted $d_0 = 2\sin\pi/8$ and $d_1 = \sqrt{2}$. Since d_{subset} is bigger than the Euclidean distance $d_2 = 2$ governing the error in the uncoded bit c_3, the performance of the TCM scheme is dominated by d_2. This gives us the promised 3 dB gain over uncoded QPSK.

Figure 7.27 shows only the evolution of the BPSK subset sequences chosen using the bits $c_1 c_2$. The effect of the uncoded bit c_3, which chooses one of the two points in the BPSK subset selected by $c_1 c_2$, is not shown. If we were to include the effect of c_3 in the trellis, then each branch would have to be replaced by a pair of parallel branches, one for $c_3 = 0$ and the other for $c_3 = 1$. These are usually termed *parallel transitions* in the literature. We have designed our TCM scheme so that the performance is dominated by the minimum Euclidean distance d_2 between parallel transitions, by making sure that the Euclidean distance d_{subset} between the subset sequences is larger. But this means that our coding gain is limited by the parallel transitions. In particular, for the TCM scheme considered, even if we try to improve the rate 1/2 convolutional code, we cannot get more than 3 dB gain over uncoded QPSK. Thus, if we wish to increase the coding gain (for the same information rate and the same constellation), we must eliminate parallel transitions. This can be achieved, for example, by using a rate 2/3 convolutional code whose output determines the triple $c_1 c_2 c_3$. Indeed, such TCM codes do exist, and provide gains of up to about 6 dB beyond uncoded QPSK. However, such codes involve a large number of states (e.g., a 256-state code was found by Ungerboeck for 5.75 dB gain).

Our distance computations implicitly assumed that performance can be evaluated by conditioning on the all-zero codeword. How do we know this is valid? This technique works for evaluating the performance for binary linear codes because the set of neighbors of the all-zero codeword is isomorphic to the set of neighbors of any other codeword **c**. That is, the geometric relationship between a codeword $\mathbf{c}_k$ and the all-zero codeword is exactly the same as that between the codewords $\mathbf{c}_k + \mathbf{c}$ (the addition is over the binary field) and **c**. This geometric relationship is defined by the places at which these codewords differ, and is summarized by the Hamming distances between the codewords. When these zeros and ones are sent over a channel in the case of TCM, however, we are interested in the *Euclidean* distance between sequences of subsets chosen based on the output of the convolutional code. There is a significant body of work (see Section 7.6) on characterizing when an analysis of the weight patterns of the component convolutional code is sufficient to determine the performance of the corresponding TCM scheme. We do not discuss such criteria in detail, but summarize the intuition as follows. First, the constellation must be partitioned into subsets that are geometrically congruent: for example, we partitioned 8-PSK into two QPSK subsets that are rotations of each other, and then we similarly partitioned each QPSK subset into BPSK subsets. Second, the labeling we employ for these subsets must obey some natural symmetry conditions. The result of such "geometric uniformity" is that the geometric relationship in Euclidean space of a codeword with its neighbors is the same for all codewords. This means that we can condition on the complex-valued TCM codeword corresponding to the all-zero binary convolutional codeword.

Study of TCM in further detail is beyond the scope of this book, but references for further reading are provided in Section 7.6.

7.5 Algebraic codes

So far, We have considered convolutional codes, turbo codes, and LDPC codes. Classical treatments of coding theory, however, start with algebraic constructions of codes with relatively short block lengths. The term "algebraic" refers to the fact that such code designs often rely on the mathematical properties of finite fields, the study of which is often termed abstract algebra. Since the study of finite fields falls outside the scope of this book, we restrict ourselves here to an *operational* description of some such codes. That is, our goal is to illustrate how such codes fit within the communication system designer's toolbox, rather than to explain the details of code construction.

Finite fields A finite field is a set with a finite number of elements on which one can do arithmetic manipulations similar to those we are accustomed to for real and complex numbers: these include addition, subtraction, multiplication, and division (except by zero). We saw examples of operations over the binary field when we discussed encoders for convolutional codes. In general, we can define finite fields with 2^m elements (m a positive integer), termed *Galois fields* in honor of Evariste Galois, and denoted by $GF(2^m)$. Clearly, m bits can be mapped to an element of a Galois field. An (n, k) block code over $GF(2^m)$ has k information symbols encoded into n code symbols. Thus, a codeword in such a code carries km bits of information.

Minimum distance The minimum Hamming distance between two codewords is denoted by $d_{\min}$. For an (n, k) code, the minimum distance cannot exceed the following bound, termed the *Singleton* bound:

$$d_{\min} \leq n - k + 1, \qquad \textbf{Singleton bound}. \qquad (7.120)$$

Channel model and decoding Traditionally, the decoder for an algebraic code would have access only to hard decisions or erasures. Efficient algebraic decoding algorithms are available for *bounded distance decoding* under such a model: given a received word, decode to a codeword which is within Hamming distance d of it. If no such codeword exists, then declare a *decoding failure*. A *decoding error* occurs when we decode to the wrong codeword. Correct decoding occurs if the received word is within distance d of the transmitted codeword. We can now infer the following facts about bounded distance decoding.

Number of correctable erasures For an erasures-only channel, bounded distance decoding can correct up to $d_{\min} - 1$ erasures.

Number of correctable errors For an errors-only channel, bounded distance decoding can correct up to $\lfloor d_{\min} - 1/2 \rfloor$ errors.

Number of correctable errors and erasures For an errors and erasures channel, bounded distance decoding can correct t_{error} errors and t_{erasure} erasures, as long as the following criterion is satisfied:

$$2t_{\text{error}} + t_{\text{erasure}} \leq d_{\min} - 1. \qquad (7.121)$$

For most algebraic codes, if correct decoding does not occur, then decoding failure is significantly more likely than decoding error (the gaps between the decoding spheres for various codewords occupy a greater "volume" than the decoding spheres themselves). This is fortunate, because knowing that we have failed to decode is usually preferable to possibly incurring a large number of errors. Denoting the probability of correct decoding by P_C, that of decoding failure by P_{DF}, and that of decoding error by P_{DE}, we have

$$P_C + P_{DF} + P_{DE} = 1.$$

For quick analytical calculations, we often approximate the probability of decoding failure by the bound $P_{DF} \leq 1 - P_C$, and ignore the probability of decoding error, since computation of the latter involves a detailed understanding of the code structure.

Some common examples of block codes over the binary field $GF(2)$ include parity check codes, Hamming codes, Reed–Muller codes, Golay codes, and Bose–Choudhary–Hocquenhem (BCH) codes.

Example 7.5.1 (Estimating the probability of decoding failure) A $(15,7)$ BCH code has minimum distance $d_{\min} = 5$. It is used over a binary symmetric channel with crossover probability $p = 0.01$. Find an upper bound on the probability of decoding failure.

The number of errors that can be corrected by bounded distance decoding is $t = \lfloor d_{\min} - 1/2 \rfloor = 2$. The probability of decoding failure is bounded by the probability that we make $t+1$ errors or more:

$$P_{DF} \leq 1 - P_C = \sum_{l=t+1}^{n} \binom{n}{l} p^l (1-p)^{n-l},$$

which yields $P_{DF} \leq 4.16 \times 10^{-4}$.

Reed–Solomon codes Reed–Solomon (RS) codes are codes defined over $GF(2^m)$ that satisfy the following properties:

$$n \leq 2^m - 1, \quad \text{Block length limited by alphabet size,} \quad (7.122)$$

$$d_{\min} = n - k + 1, \quad \text{Singleton bound satisfied with equality.} \quad (7.123)$$

These codes also have the desirable property that the probability of decoding error is extremely small, relative to the probability of decoding failure. Thus, RS codes are a very good option if we wish to employ bounded distance decoding, since they have the best possible minimum distance for a given block length and rate. Unfortunately, the restriction (7.122) on block length means that, to get the averaging effects of increasing block length, we would also be forced to increase the alphabet size, thus increasing the complexity of encoding and decoding. Increasing alphabet size can also adversely impact performance, depending on the channel model. For example, if $GF(2^m)$ symbols are sent over a binary channel, with each symbol encoded into m bits, then a symbol error occurs if any of these m bits are incorrect. Thus, the probability of symbol error increases with m.

For completeness, note that the range of block lengths in (7.122) can be increased by one for an extended RS code to obtain $n = 2^m$, while still satisfying (7.123). However, this does not change the drawbacks associated with linking block length to alphabet size.

For low enough symbol error probabilities, RS codes provide extremely small probabilities of decoding failure with relatively small redundancy. Thus, if we require a very high level of reliability over a low SNR channel, we could serially concatenate an inner channel code designed to give low BER (e.g., a binary convolutional code or a turbo-like code), with an outer RS code. The RS code can then "clean up" the (relatively infrequent) errors at the output of the inner decoder. Reed–Shannon codes are also useful for cleaning up errors in systems without an inner code, if they are operating at high SNR (e.g., on optical links). Finally, they are effective on erasures channels.

While bounded distance decoding with hard decisions (and erasures) is the classical technique for decoding algebraic codes, techniques for ML decoding and for the utilization of soft decisions have also been developed over the years. These techniques increase the range of SNRs over which such codes might be useful. However, for long block lengths, the utility of algebraic codes in communication system design is increasingly threatened by advances in turbo-like codes. For example, while RS codes are very effective at handling erasures, irregular binary LDPC codes with degree sequences optimized for the erasures channel can give almost as good a performance, while allowing for simple decoding and arbitrarily long block lengths. However, algebraic codes should still have a significant role in system design for short code block lengths, for which turbo-like codes are less effective.

7.6 Further reading

Good sources for "classical" coding theory include textbooks by Blahut [54] and Lin and Costello [55]. We also recommend another text by Blahut [13], which provides an excellent perspective on digital communication, including a discussion of the key concepts behind pre-turbo error control coding. For a detailed treatment of performance analysis for coded systems, the classic text by Viterbi and Omura [12] remains an excellent resource. A recent book by Biglieri [56] covers coding with a special emphasis on wireless channels.

Turbo-like codes are discussed in the books by Biglieri [56] and Lin and Costello [55]. Other books on turbo coding include [57] and [58]. However, for a detailed understanding of the rapidly evolving field of turbo-like codes, the research literature is perhaps the best source. In particular, we refer the reader to the papers in the literature from which the material presented here has been drawn, as well as a small sampling of some additional papers of interest. The original papers by Berrou et al., introducing turbo codes, are still highly recommended reading [59, 60]. The connection between iterative decoding and belief propagation (a well established tool in artificial intelligence pioneered by Pearl [61]) was established in [62]. The material on EXIT charts and area properties is drawn from the work of ten Brink and his collaborators [63, 64]. Reference [65] contains related work on the Gaussian approximation for estimating turbo code thresholds and code design. The material on weight distributions for turbo codes is based on the papers of Benedetto et al. [66–68], which we recommend for more detailed exploration of the turbo error floor. Interleaver design and trellis termination are important implementation aspects of turbo codes, for which we refer to [69] and the references therein.

Gallager's work in the 1960s on LDPC codes can be found in [70]. The rediscovery of Gallager's LDPC codes by MacKay subsequent to the invention of turbo codes is documented in [71]. The description of LDPC codes provided here is based on a series of influential papers [72–74] by Richardson, Urbanke, and their co-authors. Efficient encoding for LDPC codes, which we do not consider here, is addressed by Richardson and Urbanke in [75]. There are many other authors who have made significant contributions to the state of the art for LDPC codes, and the February 2001 special issue of *IEEE Transactions on Information Theory* provides a good snapshot of the state of the art at that time. Two companies that have pioneered the use of irregular LDPC codes in practice are Digital Fountain, which focuses on packet loss recovery on the Internet, and Flarion Networks (now part of Qualcomm, Inc.), which employs them in their proposed fourth generation wireless communication system. Tornado codes [76] were an early example of erasure-optimized codes constructed by Digital Fountain. These were followed by the invention of "rateless" codes, which are irregular LDPC codes that do not require prior knowledge of the channel's erasure probability. That is, if k information packets are to be sent, then a rateless code decodes successfully, with high probability, when the receiver gets slightly more than k packets. Examples of rateless codes include LT codes [77] and raptor codes [78].

While iterative decoding relies on information exchange between decoders, the "turbo principle" can be applied more broadly for iterative information exchange between receiver modules. For example, turbo equalization [79, 80] is based on information exchange between a decoder and an equalizer; turbo multiuser detection [81] is based on information exchange between a decoder and a multiuser detector (see Chapter 8 for more on multiuser detection); and turbo noncoherent communication [82] involves information exchange between a decoder and a block noncoherent demodulator (the latter is described in Chapter 4). These are just a few examples of the large, and rapidly expanding, literature on exploring applications of the turbo principle for joint estimation (broadly applied to receiver functions including synchronization, channel estimation, and demodulation) and decoding.

We refer to [83] as the starting point for more detailed investigation of BICM. For further reading on TCM, a good source is the survey paper by Forney and Ungerboeck [84] and the references therein. A description of the successful application of TCM to voiceband modems is provided in [85, 86].

7.7 Problems

Problem 7.1 For the running example of the rate 1/2 [7, 5] convolutional code with nonsystematic, nonrecursive encoding as in Figure 7.1, answer the following questions using what you have learnt about transfer functions:

(a) What is the minimum number of differing input bits corresponding to two codewords that have Hamming distance 7?
(b) What is the maximum output weight that can be generated by an input sequence of weight 4?
(c) For an E_b/N_0 of 7 dB, what is the pairwise error probability for two codewords that have Hamming distance 7? Let **E** denote the set of error events (i.e., paths through the trellis that diverge from the all-zero state at time 0, and never diverge again once they remerge). Define the following sum over all error events:

$$T(W, L) = \sum_{\mathbf{e} \in \mathbf{E}} W^{w(\mathbf{e})} L^{l(\mathbf{e})},$$

where $w(\mathbf{e})$ is the weight of the coded bits corresponding to an error event, and $l(\mathbf{e})$ is the length of the error event (i.e., the number of trellis branches traversed before merging with the all-zero state).
(d) Find a closed form expression for $T(W, L)$.
(e) Is there an error event of weight 100 and length 190?

Problem 7.2 Consider the running example [7, 5] rate 1/2 convolutional code with recursive systematic encoding as in Figure 7.3.

(a) Draw a state diagram for the code.
(b) Draw a trellis section for the code.
(c) Find the transfer function $T(I, X)$ defined in (7.14).
(d) What is the minimum Hamming distance between codewords which differ by one input bit? What if they differ by two input bits?

Problem 7.3 Consider a nonsystematic, nonrecursive, rate 1/2 convolutional code with generator [7, 6].

(a) Draw a state diagram for the code.
(b) Draw a trellis section for the code.

(c) Find the transfer function $T(I, X)$ defined in (7.14).
(d) Find the free distance of the code. How does it compare with the [7, 5] code of the running example?

Problem 7.4 Consider a rate 2/3 convolutional code in which two inputs $(u_1[k], u_2[k])$ come in at every time k, and the three outputs $(y_1[k], y_2[k], y_3[k])$ are emitted at every time k. The input–output relation is given by

$$y_1[k] = u_1[k] + u_1[k-1] + u_2[k-1],$$
$$y_2[k] = u_1[k-1] + u_2[k],$$
$$y_3[k] = u_1[k] + u_2[k].$$

(a) Draw a simple shift register implementation of the preceding encoding function.
(b) Draw a trellis section showing all possible transitions between the encoder states. Label each transition by $u_1[k]u_2[k]/y_1[k]y_2[k]y_3[k]$.
(c) Draw a state diagram for enumerating all error events. Label each branch by $I^i X^x$, where i is the input weight, and x the output weight, of the transition.
(d) Find the free distance of the code. Find the transfer function $T(X)$ for enumerating the output weights of error events. How many error events of weight 5 are there?

Problem 7.5 Consider BPSK transmission over the AWGN channel of the code in Problem 7.1. The encoder starts in state 00, and five randomly chosen input bits are sent, followed by two zero bits to ensure that the encoder state goes back to 00. Thus, there are a total of 14 output bits, in seven groups of two, that are sent out, mapping 0 to +1 and 1 to −1 as usual. Suppose that we get noisy received samples, grouped in seven pairs as below:

$$(-0.5, 1.5; -0.5, -0.8; 1.2, -0.2; 0.2, 0.1; 1, 1; -0.5, 1.5; -1.1, -2).$$

(a) Run the Viterbi algorithm to get an ML estimate of the five bit input payload.
(b) Make hard decisions on the samples and then run the Viterbi algorithm. Compare the result with (a).

Problem 7.6 Consider the rate 1/2 nonrecursive, nonsystematic convolutional code with generator [7, 5] which provides the running example.

(a) Plot the transfer function bound on BER on a log scale as a function of E_b/N_0 (dB) for BPSK transmission over the AWGN channel, with soft decision ML decoding. Also plot the BER estimate obtained by taking the first few terms of the union bound (e.g., up to the code's free distance plus three).

(b) On the same plot, show the performance of uncoded BPSK and comment on the coding gain.

(c) At a BER of 10^{-5}, how far is this code from the Shannon limits for unrestricted input and BPSK input on the AWGN channel? (Use the results from Chapter 6 and compare with the results of (a).)

Problem 7.7 Consider transmission of a binary codeword over a BSC.

(a) Use (7.7) to show that ML decoding for transmission over a BSC is equivalent to minimizing the Hamming distance between the received word and the set of codewords, assuming that the crossover probability $p < 1/2$.

(b) For a binary code of rate R transmitted over an AWGN channel using BPSK, show that hard decisions induce a BSC, and specify the crossover probability in terms of R and E_b/N_0.

Problem 7.8 (Performance of Viterbi decoding with hard decisions) Consider Viterbi decoding with hard decisions for BPSK over an AWGN channel.

(a) For two codewords with Hamming distance x, show that the pairwise error probability for ML decoding with hard decisions takes the form

$$q(x) = \sum_{i=\lceil x/2 \rceil}^{x} \binom{x}{i} p^i (1-p)^{x-i}, \qquad (7.124)$$

where p is the probability of hard decision error.

(b) For $p < \frac{1}{2}$ and $i \geq \lceil x/2 \rceil$, show that $p^i(1-p)^{x-i} \leq (\sqrt{p(1-p)})^x$. Use this to infer that

$$q(x) \leq (2\sqrt{p(1-p)})^x.$$

(c) Follow the steps in Example 7.1.1 to derive a transfer function bound for hard decisions for the running example rate 1/2 [7,5], nonrecursive, nonsystematic code. Plot the BER as a function of E_b/N_0 as in Problem 7.6. Compare with the soft decision BER curve in Problem 7.6 to estimate the dB loss in performance due to hard decisions.

(d) Using the first term in the union bound and a high SNR approximation, estimate analytically the dB loss in performance due to hard decisions.

Problem 7.9 (Bhattacharya bound) Consider ML decoding over a discrete memoryless channel, as discussed in Section 7.1.5. We wish to derive a special case of the Chernoff bound, called the Bhattacharya bound. Consider $V = \log(p(r|1))/(p(r|0))$, the log likelihood ratio corresponding to a particular observation.

(a) Show that

$$\mathbb{E}\left[e^{sV}|0\right] = \int (p(r|1))^s (p(r|0))^{1-s}\, dr = b_0,$$

where the integral becomes a sum over all possible values of r when r takes discrete values. Set $s = 1/2$ to obtain the special case

$$b_0 = \int \sqrt{p(r|1)p(r|0)}\, dr \quad \text{Continuous-valued observation,}$$
$$b_0 = \sum_r \sqrt{p(r|1)p(r|0)} \quad \text{Discrete-valued observation.} \quad (7.125)$$

(b) Consider the pairwise error probability $q(x)$ defined by (7.22). Show that

$$q(x) \leq b_0^x, \quad (7.126)$$

where b_0 is as defined in (7.125). This is the Bhattacharya bound on pairwise error probability with ML decoding.

Problem 7.10 (Bhattacharya bound for BSC) Show that the Bhattacharya bound for pairwise error probability with ML decoding for the special case of the BSC with crossover probability p is given by

$$q(x) \leq \left(2\sqrt{p(1-p)}\right)^x;$$

the same result as in Problem 7.8(b). Is this the best possible Chernoff bound for the BSC?

Problem 7.11 (Performance with 2 bit quantization of observations) Consider BPSK transmission over the AWGN channel with observation corresponding to code symbol c given by

$$z = \sqrt{E_s}(-1)^c + N,$$

where $N \sim N(0, \sigma^2)$. Fix SNR $= E_s/\sigma^2$ to 6 dB, and set $\sigma^2 = 1$ for simplicity.

(a) Find the Bhattacharya bound parameter b_0 given by (7.125) for unquantized observations. Now, suppose that we use 2 bit quantized observations as follows:

$$r = \begin{cases} +3, & z > \gamma, \\ +1, & 0 < z < \gamma, \\ -1, & -\gamma < z < 0, \\ -3, & z < -\gamma, \end{cases}$$

where $\gamma > 0$ is a design parameter. (The values for the quantized observation r could be assigned arbitrarily, but are chosen so as to make obvious the symmetry of the mapping.)

(b) Numerically compute, and plot, the Bhattacharya bound parameter b_0 given by (7.125) as a function of γ from 0 to 3. What is the best choice of γ? How does the Bhattacharya parameter b_0 for the optimum γ compare with that for unquantized observations?

(c) Repeat (b) for SNRs of 0 dB and 10 dB. How does the optimal value of γ depend on SNR?

Problem 7.12 (Consistency condition for LLR distributions) Consider binary transmission (0 or 1 sent) over a symmetric channel with a real-valued output satisfying
$$p(y|0) = p(-y|1).$$
Define the LLR $L(y) = \log(p(y|0))/(p(y|1))$. Let $q(l)$ denote the conditional density of $L(y)$, given that 0 is sent. Derive a consistency condition that $q(l)$ must satisfy using the following steps.

(a) Show that $L(y)$ is an antisymmetric function.
(b) How is the distribution of $L(y)$ conditioned on 0 related to the distribution of $L(-y)$ conditioned on 1?
(c) Show that
$$P[L(y) = l|0] = P[L(y) = -l|1].$$
(For convenience of notation, we treat $L(y)$ as a discrete random variable. Otherwise we would replace $L(y) = l$ by $L(y) \in [l, l+dl]$, etc.)
(d) Using $p(y|0) = e^{L(y)}p(y|1)$, show that the following *consistency condition* holds:
$$q(l) = e^l q(-l).$$
(e) For BPSK transmission through an AWGN channel, show that the LLR satisfies the consistency condition.
(f) Suppose that the LLR conditioned on 0 sent is modeled as Gaussian $N(m, v^2)$. Apply the consistency condition to infer that $v^2 = 2|m|$.

Problem 7.13 (Software project: BCJR algorithm) For the running example of the rate 1/2 [7, 5] RSC convolutional code, implement the log domain BCJR algorithm in software. Assume that you are sending 10 000 information bits, with two more bits for terminating the trellis in the all-zero state. The encoded bits are sent using BPSK over an AWGN channel.

(a) For the all-zero codeword, plot the histograms of L_{code}, $L_{channel}$, and $L_{out} = L_{code} + L_{channel}$ over a packet. Also plot histograms over multiple packets (with independent noise samples across packets, of course). Plot the histograms for several values of E_b/N_0, including 0 dB and 7 dB, and comment on the shapes.
(b) Make hard decisions based on the bit LLRs L_{out}, and estimate the BER by simulation over "enough" packets. Plot the BER as a function of E_b/N_0, starting from the Shannon limit for BPSK at the given rate, and going up to 10 dB. How far away is the code from the Shannon limit for BER of 10^{-4}? Also plot the uncoded BER for BPSK for reference, and comment on the coding gain at BER of 10^{-4}.

(c) Plot the simulated BER with the log BCJR algorithm with the transfer function bound for the BER with the Viterbi algorithm. Comment on the plots.

(d) Now, suppose that the all-zero codeword is sent, and you get nonzero prior LLRs as input to the BCJR algorithm. These prior LLRs are modeled as i.i.d. Gaussian random variables $L_{\text{in}}(u) \sim N(m, 2m)$, where $m > 0$ is a parameter to be varied. Assume that there is no input from the channel. Plot the mean and variance of $L_{\text{code}}(u)$ as a function of m, averaging over multiple packets if needed to get a smooth curve. Is there a relation that you can find between the mean and the variance of $L_{\text{code}}(u)$? Also, plot the histogram of $L_{\text{code}}(u)$ and comment on its shape.

Problem 7.14 (Software project: BCJR algorithm) Repeat Problem 7.13 for the nonrecursive nonsystematic version of the running example rate 1/2 [7, 5] code. In part (d), assume that prior LLRs are available for coded bits, rather than the information bits. (This would be the case when the code is an outer code in a serial concatenated system, as in Problem 7.17.)

Problem 7.15 (Software project: parallel concatenated turbo code) This project builds on the software developed for Problem 7.13. Use parallel concatenation of the rate 1/2 [7, 5] RSC code which is the running example to obtain a rate 1/3 turbo code.

(a) Implement the encoder, using a good interleaver obtained from the literature (e.g., Berrou and Glavieux [60]) or a web search. Use about 10 000 input bits. Discuss whether or not the encoder is (or can be) terminated in the all-zero state on both branches. Try encoding a number of input sequences of weight 2, and report on the codeword weights thus obtained.

(b) Implement iterative decoding with information exchange between two logarithmic BCJR decoders. Simulate performance (assume all-zero codeword sent) by plotting the BER (log scale) versus E_b/N_0 (dB) from 10^{-1} down to 10^{-5}, sending enough packets of 10 000 bits each to get an accurate estimate. Look at the curve to estimate the threshold E_b/N_0 at which iterative decoding converges. At a BER of 10^{-4}, how far is the code from the Shannon limit?

(c) Perform an EXIT analysis at E_b/N_0 which is 0.5 dB smaller and 0.5 dB larger than the convergence threshold estimated from your simulation of the turbo code, and comment on the differences in behavior. The EXIT analysis involves a simulation of an individual BCJR decoder in a manner similar to Problem 7.13(d). Can you predict the convergence threshold using the EXIT analysis?

(d) Plot the histograms for various LLRs (e.g., L_{code} and L_{out}) at various stages of the iterative decoding. Do they look Gaussian? Using the Gaussian assumption, how well can you predict the "SNR" (square of mean, divided

7.7 Problems

by variance) of the LLRs by counting the number of bit errors resulting from hard decisions based on these LLRs?

Problem 7.16 (Software project: serial concatenated turbo code) Repeat Problem 7.15 for the rate 1/4 turbo code obtained by serial concatenation of the rate 1/2 [7, 5] RSC code with itself.

Problem 7.17 (Software project: serial concatenated turbo code) Repeat Problem 7.15 for the rate 1/4 turbo code obtained by serial concatenation with the rate 1/2 [7, 5] nonsystematic, nonrecursive code as outer code, and the RSC version of the same code as inner code. For the EXIT analysis, you now need two different transfer functions, leveraging Problems 7.13 and 7.14.

Problem 7.18 (Turbo weight enumeration) Consider the rate 1/3 turbo code obtained by parallel concatenation of the [7, 5] rate 1/2 RSC code which is our running example.

(a) For $K = 10^3, 10^4, 10^5$, plot the Benedetto bound (7.81) on BER (log scale) versus E_b/N_0 (dB), focusing on the BER range 10^{-2} to 10^{-5}. How far away is this code from the Shannon limit at BER of 10^{-4}?
(b) Repeat part (a) for a [5, 7] rate 1/2 component code, i.e., with generator $G(D) = [1, (1+D+D^2)/(1+D^2)]$. Note that the codewords for this convolutional code are the same as that for the one in part (a).
(c) For $K = 10^4$, plot the BER bounds for both turbo codes on the same graph, and comment on the reason for the differences.

Problem 7.19 (The tanh rule) Consider independent binary random variables X_1 and X_2, with $P[X_i = 0] = p_i[0]$ and $P[X_i = 1] = p_i[1]$, $i = 1, 2$. Of course, we have $p_i[0] + p_i[1] = 1$. Now, consider the random variable

$$X = X_1 + X_2,$$

where the addition is over the binary field (i.e., the sum is actually an exclusive or operation).

(a) Show that the probability mass function for X is given by the cyclic convolution

$$p_X[k] = (p_1 * p_2)[k] = \sum_{n_1+n_2=k \,(\mathrm{mod}\ 2)} p_1[n_1]p_2[n_2].$$

(b) Recall that the DFT of an N-dimensional vector is given by

$$P_i[n] = \sum_{k=0}^{1} p_i[k] e^{-j2\pi kn/N}.$$

For the two-dimensional pmfs p_i, set $N = 2$ to show that the DFT is given by

$$P_i[0] = 1,$$

and that

$$P_i[1] = p_i[0] - p_i[1].$$

(c) Using the result that cyclic convolution in the time domain corresponds to multiplication in the discrete Fourier transform domain, show that

$$p_X[0] - p_X[1] = (p_1[0] - p_1[1])(p_2[0] - p_2[1]).$$

(d) Setting m_1 and m_2 as the LLRs for X_1 and X_2, given by $m_i = \log(p_i[0])/(p_i[1])$, and m equal to the LLR for X, show that

$$\tanh\left(\frac{m}{2}\right) = \tanh\left(\frac{m_1}{2}\right)\tanh\left(\frac{m_2}{2}\right) \quad \textbf{Tanh rule.}$$

(e) Use induction to infer that the preceding tanh rule generalizes to the exclusive or of an arbitrary number of independent binary variables. This yields the formula (7.101) for belief propagation at a check node.

Problem 7.20 (LDPC convergence thresholds) The purpose of this problem is to compute convergence thresholds for LDPC codes.

(a) For a $(3, 6)$ LDPC code over a BSC with crossover probability α, find the threshold for Gallager's algorithm A using density evolution.
(b) For the same $(3, 6)$ code over the same BSC channel, find the threshold for belief propagation using density evolution or a Gaussian approximation for the LLRs passed between the variable and check nodes. For a given α, one possible approach is to use the exact message distribution for the first set of messages from variable to check nodes (the message only takes one of two values at this point). Use this to compute $E[\tanh(u/2)]$ for the messages passed from check nodes to variable nodes, and now invoke the Gaussian approximation to find m_u. Thereafter, use the Gaussian approximation.
(c) Use the Gaussian approximation to find the E_b/N_0 threshold for the $(3, 6)$ code sent using BPSK over an AWGN channel, and decoded using belief propagation. How far is this from the Shannon limit?
(d) Use the results of (b) and (c) to estimate the penalty in dB of using hard decisions with the $(3, 6)$ code, for BPSK transmission over AWGN.
(e) Use the Gaussian approximation to compute the E_b/N_0 threshold for the rate 1/2 irregular LDPC with degree sequences specified by $\lambda(x) = 0.2x + 0.8x^2$, $\rho(x) = 0.5x^4 + 0.5x^5$, for BPSK over AWGN, decoded using belief propagation. How does the result compare with (c)?

7.7 Problems

Problem 7.21 (Irregular LDPC codes) Consider an irregular LDPC code with $\lambda(x) = 0.3x^2 + 0.1x^3 + 0.6x^4$ and $\rho(x) = ax^7 + bx^8$.

(a) Find a and b such that the code has rate 1/2. Use these values for the remaining parts of the problem.
(b) What fraction of the variable nodes have degree 4?
(c) What fraction of the check nodes have degree 9?
(d) Find the E_b/N_0 threshold for belief propagation over an AWGN channel using the Gaussian approximation. Compare with the performance for the codes in Problem 7.20(c) and (e).
(e) Now, compute the E_b/N_0 threshold with hard decisions, decoded using Gallager's algorithm A. Compare with (d).

Problem 7.22 (Message passing with errors and erasures) Consider a (3, 6) binary regular LDPC code. Assume that the all-zero codeword is sent. For any bit x, the channel output z is 1 (error) with probability p_0, e (erasure) with probability q_0, and 0 (correct decision) with probability $1 - p_0 - q_0$. The following suboptimal message passing algorithm is employed by the decoder. Messages take values 0, 1, e, and are updated as follows. The message density is characterized by (p, q), where p is the probability of error and q the probability of erasure.

Variable node computations Take a majority vote of incoming messages (from the check nodes and the channel) to determine whether the outgoing message is 0 or 1. Ignore erasures in the voting. If the vote is deadlocked, or if all incoming messages are erasures, then the output message is an erasure.

Check node computations If any incoming message is an erasure, then the outgoing message is an erasure. Otherwise the outgoing message is the XOR sum of the incoming messages.

Initialization The variable nodes send out the channel output to check nodes on all edges. This is followed by multiple iterations, in which first the check nodes, and then the variable nodes, send messages. In the following, you are asked to go through density evolution for the first iteration.

(a) Find expressions for the check node output message distribution (p_u, q_u) on a given iteration, as a function of the message distribution (p_v, q_v) coming from the variable nodes. Give numerical values for (p_u, q_u) for the first iteration when $p_0 = 0.03$ and $q_0 = 0.2$ ($p_v = p_0$ and $q_v = q_0$ for the first iteration).
(b) Give expressions for the variable node output message distribution (p_v, q_v) as a function of the channel message distribution (p_0, q_0) and the input message distribution (p_u, q_u) from the check nodes. Give a numerical answer for (p_v, q_v) at the end of the first iteration, using $p_0 = 0.03$, $q_0 = 0.2$, and (p_u, q_u) taking the numerical values from (a).

Problem 7.23 (BICM) Consider the Gray coded 4-PAM-based BICM system described in Section 7.4.1.

(a) For the channel model (7.116), derive the expressions (7.117) and (7.118) for the LLRs. Plot these LLRs as a function of the observation y.
(b) Normalizing the noise variance $\sigma^2 = 1$, what is the value of A for a rate 1/2 outer binary code, if E_b/N_0 is 6 dB?
(c) Compute and plot the sum of the capacities of the equivalent binary channels seen by the bits x_1 and x_2 as a function of SNR (dB). Plot for comparison the capacity of equiprobable 4-PAM over the AWGN channel. Comment on the performance degradation due to restricting to a BICM system.
(d) Replot the results of (c) as spectral efficiency versus E_b/N_0 (dB).

Problem 7.24 (TCM) Consider the following TCM scheme based on a rectangular 16-QAM constellation.

(a) Use Ungerboeck set partitioning to obtain four subsets which are translated versions of QPSK. Illustrate the partitioning using a figure analogous to Figure 7.23.
(b) Use the rate 1/2 convolutional code in Figure 7.26 for subset selection, and use a Gray code for the two uncoded bits selecting the symbol within the subset. Compute the asymptotic coding gain in dB (if any) relative to an uncoded 8-QAM constellation which is optimized for power efficiency.

CHAPTER

8 Wireless communication

Freedom from wires is an attractive, and often indispensable, feature for many communication applications. Examples of wireless communication include radio and television broadcast, point-to-point microwave links, cellular communications, and wireless local area networks (WLANs). Increasing integration of transceiver functionality using DSP-centric design has driven down implementation costs, and has led to explosive growth in consumer and enterprise applications of wireless, especially cellular telephony and WLANs.

While the focus of this chapter is on wireless *link* design, we comment briefly on some *system* design issues in this introductory section. In terms of system design, a key difference between wireless and wireline communication is that wireless is a broadcast medium. That is, users "close enough" to each other can "hear," and potentially interfere with, each other. Thus, appropriate resource sharing mechanisms must be put in place if multiple users are to co-exist in a particular frequency band. The wireless channel can be shared among multiple users using several different approaches. One possibility is to eliminate potential interference by assigning different frequency channels to different users; this is termed frequency division multiple access (FDMA). Similarly, we can assign different time slots to different users; this is termed time division multiple access (TDMA). If we use orthogonal multiple access such as FDMA or TDMA, then we can focus on single-user wireless link design. However, there are also nonorthogonal forms of multiple access, in which different users can signal at the same time over the same frequency band. In this case, the users would be assigned different waveforms, or "codes," which leads to the name code division multiple access (CDMA) for these techniques.

In addition to time and bandwidth, another resource available in wireless systems is *space*. For example, if one transmitter–receiver pair is far enough away from another, then the mutual interference between them is attenuated enough so as to be negligible. Thus, wireless resources can be utilized more efficiently by employing *spatial reuse*, which forms the basis for cellular communication systems. The area controlled by a single base station in a cellular system is termed a *cell*. The GSM and IS-54 digital cellular standards use TDMA within a cell, and FDMA to avoid interference between neighboring cells. However, cells that are far enough away can use the same frequency band, with the "reuse pattern" depending on factors such as whether directional antennas are used, and how much interference the modulation technique has been designed to tolerate. On the other hand, the IS-95 and related third generation cellular standards use CDMA with "100% spatial reuse," with the same frequency band used across all cells.

For the remainder of this chapter, we restrict attention to the design of an individual wireless link. There are three major themes: understanding the characteristics of the wireless channel, coverage of modulation formats commonly used in wireless systems, and the design of wireless systems with multiple antennas.

Map of this chapter We begin by discussing wireless channel models in Section 8.1, including the phenomenon of multipath fading and channel time variations due to mobility. In Section 8.2, we discuss the potentially disastrous impact of Rayleigh fading on performance, and discuss how diversity can be used to combat fading. The remainder of this chapter provides an exposure to communication techniques commonly used on wireless channels. We begin with orthogonal frequency division multiplexing (OFDM) in Section 8.3, a technique designed to simplify the task of equalization over a multipath channel. This is the basis for emerging fourth generation cellular systems, including those based on the IEEE 802.16 and 802.20 standards, as well as for the IEEE 802.11a, 802.11g, and 802.11n standards. Direct sequence (DS) spread spectrum signaling is discussed in Section 8.4; DS-CDMA forms the basis for the IS-95 second generation cellular standard, as well as third generation cellular standards. Frequency hop (FH) spread spectrum is briefly discussed in Section 8.5; FH-CDMA is used in military packet radio networks (e.g., the SINCGARS system) due to its robustness to multiple-access interference, fading, and jamming. It is also used in the IEEE 802.15.1 WPAN standard commonly known as Bluetooth. Space–time, or multiple antenna, communication is discussed in Section 8.7. Space–time communication is a key feature of the emerging IEEE 802.11n WLAN standard, and can also be used to enhance the performance of cellular systems.

8.1 Channel modeling

Channel models play a critical role in the design of communication systems: a typical design must go through several iterations driven by computer-based performance evaluations before the expense of building and testing a prototype is taken on. A common approach to channel modeling is to abstract statistical models from a large set of measurements. Computer simulations of

8.1 Channel modeling

performance can then be carried out by drawing channel realizations from the statistical model, evaluating the performance for a specific realization, and averaging the performance attained over many such realizations. In addition to measurements, artificial statistical channel models may also be defined, to test whether the communication system design can withstand a variety of challenges. For example, there are a number of representative channel models defined for the GSM cellular system that a good design must be able to function under. In the following, we discuss some basic elements of statistical channel modeling.

As shown in Figure 8.1, in a typical wireless environment, the receiver sees a superposition of multiple attenuated and phase shifted copies of the transmitted signal. Thus, if $u(t)$ is the transmitted complex baseband signal, the received complex baseband signal has the form

$$y(t) = \sum_{k=1}^{M} A_k e^{j\phi_k} u(t - \tau_k) e^{-j2\pi f_c \tau_k},$$

where, for the kth multipath component, A_k is the amplitude, ϕ_k is the phase (modeling the phase changes due to the scattering undergone by this component), τ_k is relative delay, and $2\pi f_c \tau_k$ is the phase lag caused by this delay, where f_c is the carrier frequency.

Since the carrier frequency f_c is large, small changes in delay τ_k cause large changes in the phase $-2\pi f_c \tau_k$. When expressed modulo 2π, such changes may be viewed as a completely random new choice of phase. Thus, setting $\theta_k = \phi_k - 2\pi f_c \tau_k \bmod 2\pi$ as the phase of the kth component, expressed in the interval $[0, 2\pi]$, we obtain the following model. The complex baseband

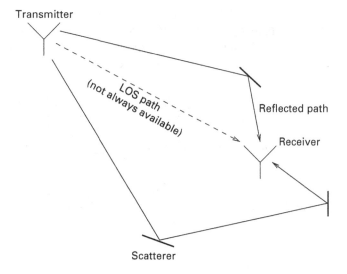

Figure 8.1 The signal from transmitter to receiver in a wireless system may undergo reflections from multiple scatterers, creating a multipath channel. The line of sight (LOS) path may or may not be available, depending on the propagation environment.

received signal y is obtained by passing the transmitted signal u through a complex baseband channel with impulse response

$$h(t) = \sum_{k=1}^{M} A_k e^{j\theta_k} \delta(t - \tau_k) \quad \textbf{Multipath channel model}, \qquad (8.1)$$

where the phases $\{\theta_k\}$ are modeled as independent and identically distributed random variables, uniformly distributed in $[0, 2\pi]$. In general, when either the transmitter, the receiver, or the scatterers are mobile, the gains A_k, θ_k, and τ_k would vary with time. For the moment, we do not consider such time variations, so that the amplitudes and delays are constants.

Narrowband Rayleigh and Rician fading models Taking the Fourier transform of (8.1), we get the channel transfer function

$$H(f) = \sum_{k=1}^{M} A_k e^{j\theta_k} e^{-j2\pi f \tau_k}. \qquad (8.2)$$

Narrowband signaling refers to a setting in which the channel transfer function is approximately constant over the signal band, which is, say, a small band around f_0. In this case, the channel can be modeled as a scalar gain h, given by

$$h \approx H(f_0) = \sum_{k=1}^{M} A_k e^{j\gamma_k}, \qquad (8.3)$$

where $\gamma_k = \theta_k - 2\pi f_0 \tau_k \mod 2\pi$. Since $\{\theta_k\}$ are i.i.d., uniform over $[0, 2\pi]$, so are $\{\gamma_k\}$. We therefore have

$$\mathrm{Re}(h) = \sum_{k=1}^{M} A_k \cos(\gamma_k), \quad \mathrm{Im}(h) = \sum_{k=1}^{M} A_k \sin(\gamma_k).$$

If the number of multipath components is large, and the contribution of any one component is small, then we can apply the central limit theorem to approximate $\mathrm{Re}(h)$, $\mathrm{Im}(h)$ as jointly Gaussian random variables. Their joint distribution is therefore specified by computing the following means and covariances, using the identities

$$\mathbb{E}[\mathrm{Re}(h)] = \sum_{k=1}^{M} A_k \mathbb{E}[\cos(\gamma_k)] = 0, \quad \mathbb{E}[\mathrm{Im}(h)] = \sum_{k=1}^{M} A_k \mathbb{E}[\sin(\gamma_k)] = 0,$$

$$\mathrm{var}\,(\mathrm{Re}(h)) = \mathbb{E}\left[(\mathrm{Re}(h))^2\right] = \sum_{k=1}^{M} \sum_{l=1}^{M} A_k A_l \mathbb{E}[\cos(\gamma_k)\cos(\gamma_l)]$$

$$= \sum_{k=1}^{M} A_k^2 \mathbb{E}[\cos^2 \gamma_k] + \sum_{k \neq l} A_k A_l \mathbb{E}[\cos(\gamma_k)] \mathbb{E}[\cos(\gamma_l)]$$

$$= \frac{1}{2} \sum_{k=1}^{M} A_k^2.$$

Similarly, we can show $\text{var}(\text{Im}(h)) = \frac{1}{2}\sum_{k=1}^{M} A_k^2$ and $\text{cov}(\text{Re}(h), \text{Im}(h)) = 0$. Thus, under our central limit theorem based approximation, $\text{Re}(h)$ and $\text{Im}(h)$ are i.i.d. $N(0, \frac{1}{2}\sum_{k=1}^{M} A_k^2)$. That is, the channel h is proper complex Gaussian, $h \sim CN(0, \sum_{k=1}^{M} A_k^2)$. This implies that the amplitude $|h|$ is a Rayleigh random variable, leading to the term *Rayleigh* fading. The power $|h|^2$ is an exponential random variable with mean $\mathbb{E}[|h|^2] = \sum_{k=1}^{M} A_k^2$. In summary, we have

$$H(f) \sim CN(0, \sum_k A_k^2) \quad \textbf{Narrowband Rayleigh fading}. \qquad (8.4)$$

Now, suppose that one of the multipath components (say component 1) is significantly stronger than the others. A classical example is when there is a "specular" line of sight (LOS) path with the smallest delay, together with a large number of smaller, "diffuse," multipath components. Then, reasoning as above, the complex gain h_{diffuse} resulting from the sum of the diffuse components can be modeled as zero mean, proper complex Gaussian, so that the net channel gain has the form

$$h = A_1 e^{j\gamma_1} + h_{\text{diffuse}}.$$

In this case, the amplitude $|h|$ is a Rician random variable, corresponding to a *Rician fading* channel model, given by

$$H(f) \sim CN(A_1 e^{j\gamma_1}, \sum_{k=2}^{M} A_k^2) \quad \textbf{Narrowband Rician fading}. \qquad (8.5)$$

The preceding narrowband models are termed *frequency nonselective*, or *flat*, fading, because the channel gain is approximately constant over the signal frequency band.

Bandwidth-dependent tap delay line model Suppose now that the transmitted signal has bandwidth large enough that the channel can no longer be modeled as approximately constant over the signal band. Such a channel is termed *frequency selective*, and we must go back to the generality of model (8.1) in order to develop a statistical model for the channel. However, one can tailor the channel model to the characteristics of the signals we anticipate using over the channel. Specifically, if the transmitted signal has bandwidth W, then it suffices to consider delays that are spaced by $1/W$. This is because the received signal y has bandwidth bounded by W, and can therefore be represented by $1/W$-spaced samples. This in turn implies that it suffices to consider taps spaced by $1/W$ in a mathematical channel model such as (8.1). That is, *unresolvable* channel multipath components that may be spaced much closer than $1/W$ can be merged into *resolvable* $1/W$-spaced taps, leading to an equivalent tap delay line (TDL) model of the form

$$h(t) = \sum_{i=1}^{\infty} \alpha_i \delta(t - \frac{i}{W}) \quad \textbf{TDL model}. \qquad (8.6)$$

If there are a large number of unresolvable multipath components that comprise each resolvable tap, then, from (8.1), the ith resolvable tap is given by

$$\alpha_i \approx \sum_{k:\tau_k \approx \frac{i}{W}} A_k e^{j\theta_k}. \tag{8.7}$$

If the number of unresolvable taps being summed in (8.7) is large enough, then, applying a central limit theorem approximation as before, we can model $\{\alpha_i\}$ as zero mean, proper complex Gaussian random variables. Furthermore, since the phases for the unresolvable components in each resolvable tap are independent, the resolvable taps $\{\alpha_i\}$ are independent. Thus, the tap amplitudes $|\alpha_i|$ are independent Rayleigh random variables. Note also that, for specular multipath components, one or more of the resolvable taps may be modeled as Rician rather than Rayleigh.

To complete the description, we must say something about the tap strengths. Channel measurements are often summarized in terms of power-delay profiles. A *power-delay profile* $P(\tau)$ ($\tau \geq 0$) can be interpreted as a density, normalizing such that $\int_0^\infty P(\tau) \, d\tau = 1$, with $P(\tau) d\tau$ denoting the fraction of power in taps with delays in the interval $[\tau, \tau + d\tau]$. The standard deviation for this density is termed the root mean squared (RMS) delay, τ_{rms}, of the channel. Let us consider an exponential PDP of the form $P(\tau) = \mu e^{-\mu \tau}$, since these are often encountered in measurement campaigns. Noting that $\tau_{\text{rms}} = 1/\mu$, we can rewrite the exponential PDP as

$$P(\tau) = \frac{1}{\tau_{\text{rms}}} e^{-\frac{\tau}{\tau_{\text{rms}}}}, \quad \tau \geq 0. \tag{8.8}$$

Knowing the PDP, we can now compute the power in the ith resolvable tap as

$$\mathbb{E}[|\alpha_i|^2] = \int_{i/W}^{(i+1)/W} P(\tau) d\tau.$$

Instead of an exact computation as above, we may also just approximate as

$$\mathbb{E}[|\alpha_i|^2] \approx cP\left(\frac{i}{W}\right), \tag{8.9}$$

where it is often convenient to choose the normalization factor c such that $\sum_{i=0}^\infty \mathbb{E}[|\alpha_i|^2] = 1$.

> **Example 8.1.1 (TDL model for exponential PDP)** For an exponential PDP with RMS delay spread τ_{rms} and a signal bandwidth W, the taps in a bandwidth-dependent TDL model (8.6) can be modeled as independent, with
>
> $$\alpha_i \sim CN(0, ab^i),$$
>
> where $b = \exp\left(-\frac{1}{W\tau_{\text{rms}}}\right)$ and $a = 1 - b$.

The TDL model is widely used in the design of wireless systems. For example, a set of nominal channel models is defined for the design of GSM cellular systems, for settings such as "typical urban" and "hilly terrain." The taps in such a model are not necessarily evenly spaced, and may follow Rayleigh, Rician, or other fading models. However, the broad rationale for such models is similar to that in the preceding discussion.

Channel time variations Time variations for typical *wireline* channels are very slow (e.g., caused by temperature fluctuations) compared with the time scale of communication, and can be safely ignored for all practical purposes, assuming that the transmitter and receiver have autocalibration mechanisms that adapt to channel variations across different settings. For example, a DSL transceiver must be able to adapt to the different channels encountered when providing connectivity to different homes. Of course, even when a channel is time-invariant, time variations may be induced by factors such as the mismatch between the frequency references used by the transmitter and receiver. However, such mismatch can be virtually eliminated for time-invariant channels by the use of synchronization circuits such as phase locked loops. For wireless mobile channels, on the other hand, channel time variations due to relative mobility between transmitter, receiver, or scatterers, cannot be eliminated, and impose fundamental limitations on performance. Fortunately, such channel time variations are, typically, slower than the symbol rate. For example, for a carrier frequency f_c of 1 GHz and a mobile terminal moving at velocity v of 100 km/hr, the maximum Doppler frequency is $f_D = f_c v/c$, which comes to less than 100 Hz. This is two orders of magnitude smaller than even a relatively slow symbol rate of 10 ksymbol/s. This means that it is possible to track the channel without excessive overhead by employing known pilot signals, or to use noncoherent techniques that exploit the fact that the channel can be approximated as roughly constant over several symbols.

For the TDL model (8.6), we can include the effect of such time variations by making the tap gains $\alpha_i(t)$ time-varying. A popular approach is to model $\alpha_i(t)$ as independent, wide sense stationary (WSS), random processes whose power spectral density depends on the Doppler spread of the channel.

Clarke's model for time-varying Rayleigh fading A now-classical model for channel time variations in a typical urban cellular wireless system is Clarke's model. The mathematical model is that of a time-varying complex gain that is a sum of a number of gains of a complex exponential form modeling Doppler shifts, as follows:

$$X(t) = \sum_k e^{j(2\pi f_k t + \theta_k)}, \qquad (8.10)$$

where f_k is the Doppler shift seen by the kth component. Figure 8.2 depicts a typical application of Clarke's model to the channel seen by a mobile receiver surrounded by a circular field of scatterers and moving at velocity v.

Figure 8.2 Clarke's model for time variations for a narrowband Rayleigh fading channel. The waves arriving from the ring of scatterers superimpose to result in a gain modeled as a complex Gaussian random process with PSD given by (8.11).

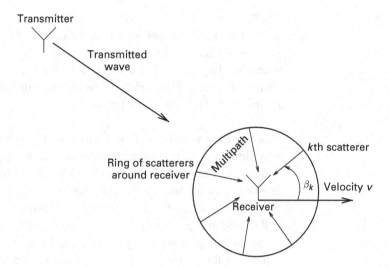

The velocity along the kth scattered component is $v\cos\beta_k$, where β_k is the angle that the path makes relative to the direction of relative motion. The resulting Doppler shift $f_k = f_D \cos\beta_k$, where $f_D = vf_c/c$ is the maximum Doppler shift, with c denoting the speed of light ($c = 3 \times 10^8$ meters per second in free space). By now familiar central limit theorem arguments, we can model $X(t)$ as a proper complex Gaussian random process with zero mean, so that it is only necessary now to specify its power spectral density. Replacing the summation over k in (8.10) by a continuum, we see that the power is a small band around frequency $f = f_D \cos\beta$ and is given by

$$S_X(f)|df| = G(\beta)|d\beta|,$$

where $G(\beta)$ is the power gain at angle of incidence β, and is influenced by both the receive antenna gain and the power-angle profile. Note that

$$\frac{df}{d\beta} = -f_D \sin\beta = -f_D\sqrt{1 - (f/f_D)^2}.$$

Now, assuming rich scattering and an omnidirectional receive antenna, we know that $G(\beta)$ is independent of β. This yields

$$S_X(f) = \frac{1}{\pi f_D \sqrt{1 - (f/f_D)^2}}, \qquad (8.11)$$

where we have normalized so that

$$\mathbb{E}[|X(t)|^2] = \int_{-f_D}^{f_D} S_X(f) df = 1.$$

The preceding development can be used to infuse a basic Rayleigh fading model with time variations in a number of different contexts. For example, Clarke's model is widely used to model time variations in the taps in a TDL model such as (8.6), since each tap can be thought of as a sum of a number of unresolvable components arriving from different directions. Clarke's model is also applied to model the time variations of a narrowband channel as in (8.2), since again the complex channel gain is a sum of a number of components, each of which could be arriving from a different direction. Simulation of Clarke's model is typically done using Jakes' simulator (or its variants), which implements the summation in (8.10) for a finite number of paths, with the Doppler shift for the kth path, $f_k = f_D \cos \beta_k$, and arrival angles β_k spaced uniformly on a circle.

In addition to the fading caused by the changes in the relative delays of various paths, the wireless channel also exhibits *shadowing* effects due to blockages of the paths between the transmitter and receiver by various obstacles.

8.2 Fading and diversity

We have shown now that Rayleigh fading is a fundamental feature of "rich scattering" channels with a large number of multipath components. We now show the potentially disastrous consequences of Rayleigh fading and discuss some remedies.

8.2.1 The problem with Rayleigh fading

Under Rayleigh fading, the noiseless received signal corresponding to a transmitted narrowband signal $s(t)$ is of the form $hs(t)$, where h is zero mean, proper complex Gaussian. It is convenient to normalize such that $h \sim CN(0, 1)$, and to define the power gain $G = |h|^2$, where G is exponential with mean one. The density of G is given by

$$p_G(g) = e^{-g} I_{\{g \geq 0\}}. \tag{8.12}$$

Furthermore, $R = \sqrt{G}$ is a Rayleigh random variable with density

$$p_R(r) = 2re^{-r^2} I_{\{r \geq 0\}}. \tag{8.13}$$

We define the average SNR parameter $\bar{S} = \bar{E}_b/N_0$, and note that the SNR $S = E_b/N_0 = G\bar{S}$. Rayleigh fading causes performance degradation because the power gain G has a high probability of taking on values significantly smaller than its mean value of one. This phenomenon is termed *fading*. The probability of a fade can be computed as follows:

$$P[G \leq \epsilon] = 1 - e^{-\epsilon} \approx \epsilon, \quad \epsilon \ll 1. \tag{8.14}$$

Thus, the probability of a 10 dB fade ($\epsilon = 0.1$) is about 10%, and the probability of a 20 dB fade ($\epsilon = 0.01$) is 1%. The bits sent during fades are very likely to be wrong, and it turns out that these events dominate the average error probability (averaged over the fading distribution) in uncoded systems. As a result, as we show below, the average error probability for uncoded communication with Rayleigh fading decreases as the *reciprocal* of the average SNR, which is much less power efficient than communication over the AWGN channel, where the error probability decays exponentially with SNR.

Averaging over fades When the fading gain fluctuates over the duration of communication, we are interested in performance averaged over fades. We would hope that such averaging would improve performance, with the good performance for large G compensating for the bad performance for small G. However, if we simply average over fades in an uncoded system, the average error probability is dominated by the performance during deep fades, which severely degrades the performance. We quantify this effect next, and then discuss more intelligent strategies for averaging over fades in Sections 8.2.2 and 8.2.3.

Performance of uncoded systems with Rayleigh fading Let us consider the error probability for noncoherent demodulation of equal-energy binary orthogonal signaling (e.g., FSK), given by $1/2\,e^{-E_b/2N_0} = 1/2\,e^{-S/2}$ for an AWGN channel. Over the Rayleigh fading channel, the error probability is a function of the normalized fading gain G as follows: $P_e(G) = 1/2\,e^{-G\bar{S}/2}$. Averaging out G using (8.12) for a Rayleigh fading channel, we get the average error probability

$$P_e = \mathbb{E}[P_e(G)] = \int P_e(g) f_G(g)\,\mathrm{d}g = \int_0^\infty \frac{1}{2} e^{-\frac{g\bar{S}}{2}} e^{-g}\,\mathrm{d}g = \frac{1}{2+\bar{S}}.$$

We therefore have

$$P_e = \left(2 + \bar{E}_b/N_0\right)^{-1} \quad \text{\textbf{Noncoherent FSK in Rayleigh fading,}} \quad (8.15)$$

which has the reciprocal decay with average SNR mentioned earlier. As shown in Problem 4.10, this result can also be deduced directly for a Rayleigh fading model, without going through the above process of averaging the error probability for an AWGN channel.

We can employ the preceding analysis to evaluate the average error probability for binary DPSK (which we know to be 3 dB better than noncoherent binary FSK on the AWGN channel), by replacing $\bar{S}$ by $2\bar{S}$ in the expression (8.15). This yields

$$P_e = \left(2 + 2\bar{E}_b/N_0\right)^{-1} \quad \text{\textbf{Binary DPSK in Rayleigh fading.}} \quad (8.16)$$

We expect similar results to hold for coherent demodulation as well. In this case, the error probability takes the form $Q(\sqrt{aS})$ (a is a constant that depends on the modulation scheme). Since the error probability is again approximately exponential in S (using $Q(x) \sim e^{-x^2/2}$), we again expect to get a reciprocal dependence on average SNR. Let us carry out the detailed computations for *coherent* binary FSK, to compare with our prior result for noncoherent FSK.

8.2 Fading and diversity

The error probability is $Q(\sqrt{E_b/N_0})$, which we rewrite as $P_e(S) = Q(\sqrt{S}) = Q(\sqrt{GS})$. We shall see in Problem 8.9 how this expectation can be evaluated using the Gamma function introduced in Problem 8.7. Here, however, we evaluate the expectation by expressing the error probability in terms of the Rayleigh random variable $R = \sqrt{G}$. We have $P_e(R) = Q(R\sqrt{S})$, and average out R using (8.13), as follows:

$$P_e = \mathbb{E}[P_e(R)] = \int P_e(r)f_R(r)\,dr = \int_0^\infty Q(r\sqrt{\bar{S}})2re^{-r^2}\,dr$$
$$= -e^{-r^2}Q(r\sqrt{\bar{S}})\Big|_{r=0}^{r=\infty} + \int_0^\infty e^{-r^2}\frac{d}{dr}Q(r\sqrt{\bar{S}})\,dr, \qquad (8.17)$$

where we have integrated by parts. We have

$$\frac{d}{dr}Q(r\sqrt{\bar{S}}) = -\sqrt{\bar{S}}\,\frac{e^{-r^2\bar{S}/2}}{\sqrt{2\pi}}.$$

Substituting into (8.17), we have

$$P_e = \frac{1}{2} - \sqrt{\bar{S}}\int_0^\infty \frac{1}{\sqrt{2\pi}}e^{-r^2(1+\frac{\bar{S}}{2})}\,dr. \qquad (8.18)$$

We can now evaluate the second term by recognizing that the integral is of the form of an $N(0, v^2)$ density integrated over positive reals. That is, setting

$$\frac{1}{2v^2} = 1 + \frac{\bar{S}}{2}$$

we can write the second term in (8.18) as

$$\sqrt{\bar{S}}v\int_0^\infty \frac{1}{\sqrt{2\pi v^2}}e^{-\frac{r^2}{2v^2}}\,dr = \frac{\sqrt{\bar{S}}v}{2}.$$

We therefore get

$$P_e = \frac{1}{2} - \frac{\sqrt{\bar{S}}v}{2} = \frac{1}{2}\left(1 - \frac{\sqrt{\bar{S}}}{\sqrt{2+\bar{S}}}\right) = \frac{1}{2}\left(1 - \left(1+\frac{2}{\bar{S}}\right)^{-\frac{1}{2}}\right).$$

At high SNR, $2/\bar{S}$ is small. Using $(1+x)^{-1/2} \approx 1 - x/2$ for x small, we find that $(1+2/\bar{S})^{-1/2} \approx 1 - 1/\bar{S}$ at high SNR. The error probability is therefore given by

$$P_e = \frac{1}{2}\left(1 - (1+2N_0/\bar{E}_b)^{-\frac{1}{2}}\right) \underset{\text{SNR}}{\overset{\text{high}}{\approx}} (2\bar{E}_b/N_0)^{-1}$$

Coherent FSK in Rayleigh fading. (8.19)

The preceding analysis for $P_e(S) = Q(\sqrt{S})$ can be employed to deduce the average error probability for any digital modulation scheme with error probability $Q(\sqrt{kE_b/N_0})$, by replacing S by kS, and hence $\bar{S}$ by $k\bar{S}$. For BPSK, which has error probability $Q(\sqrt{2E_b/N_0})$, we can replace $\bar{S}$ by $2\bar{S}$ in the above to get

$$P_e = \frac{1}{2}\left(1 - \left(1 + N_0/\bar{E}_b\right)^{-\frac{1}{2}}\right) \underset{\text{SNR}}{\overset{\text{high}}{\approx}} (4\bar{E}_b/N_0)^{-1}$$

Coherent BPSK in Rayleigh fading. (8.20)

Comparing coherent FSK performance (8.19) with noncoherent FSK performance (8.15), or comparing coherent BPSK performance (8.20) with binary DPSK performance (8.16), we note that, for large $\bar{E}_b/N_0$, coherent demodulation is 3 dB better for the Rayleigh fading channel. This is in contrast to our earlier results for AWGN channels, in which we noted that the error probabilities for coherent and noncoherent FSK have the same exponent of decay at large SNR as do those for BPSK and binary DPSK. This is because high SNR asymptotics for the AWGN channel are not an accurate predictor of performance for uncoded communication over Rayleigh fading channels, since the error probability for the latter is dominated by deep fades, where the power gain G is small. On the other hand, in a well engineered system for dealing with Rayleigh fading, we would like the performance to be dominated by the performance when the power gain G is large, which is the case most of the time: while a 10 dB fade occurs with 10% probability, by the same token, the power gain is no worse than 10 dB below its average value with 90% probability. As discussed in Section 8.2.2, coding across fades is one way to accomplish this: this is one form of *diversity*, in which the system exploits the variation of fading gains across code symbols. The code is designed to have enough redundancy to deal with the anticipated fraction of badly faded symbols. While we focus on coherent coded systems in Section 8.2.2, as shown by a simple example in Problem 8.4, the performance gap between coherent and noncoherent systems does narrow when we employ coding.

In Section 8.2.3, we discuss another form of diversity, where each transmitted symbol sees several different fading gains (e.g., if the receiver has multiple antennas). In this case, a suitable combining scheme is used to ensure that the effective gain seen by the symbol fluctuates less than a single fading gain.

8.2.2 Diversity through coding and interleaving

We have shown that multipath fading channels vary across frequency, and, when there is mobility, across time. Thus, if we employ a wide enough bandwidth, or a large enough time interval, for signaling, then we should see good channel realizations as well as bad fades. Furthermore, good channel realizations occur far more often than bad fades. Symbols that see good channel realizations are likely to be correctly demodulated. By employing error correction coding over a large enough time–frequency span, we can therefore compensate for the small fraction of symbols that see bad fades and

Figure 8.3 A simple row–column interleaver.

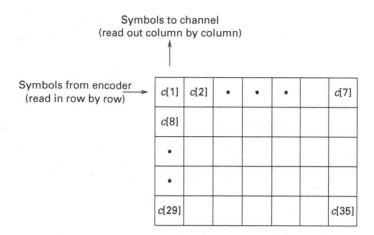

are incorrectly demodulated. Thus, coding is a way of exploiting frequency and time diversity. Since typical codes are optimized for random errors, while fades are correlated in time and frequency, the correlations are broken up by scrambling, or *interleaving*, the code symbols before modulation. A simple example of an interleaver is the row–column structure depicted in Figure 8.3, but more sophisticated interleavers may be employed in practice.

Rather than discuss specific codes, we illustrate the power of coding and interleaving using a simple information-theoretic analysis for the idealized Rayleigh faded system described later.

Ergodic capacity In a system with ideal interleaving and codewords that are long enough to see all possible channel realizations, the nth received sample is given by

$$y[n] = h[n]x[n] + w[n], \tag{8.21}$$

where $h[n] \sim CN(0, 1)$ are i.i.d. Rayleigh faded gains, $x[n]$ are transmitted symbols subject to a power constraint $\mathbb{E}[|x[n]|^2] \leq P$, and $w[n] \sim CN(0, 2\sigma^2)$ are i.i.d. WGN samples. Because we average over the Rayleigh fading distribution across the symbols of each codeword, we term the maximum achievable rate the *ergodic capacity* of the channel. Conditioned on $h[n]$, the channel is an AWGN channel at each time instant, which can be used to show, using the techniques of Chapter 6, that i.i.d. white Gaussian input, $x[n] \sim CN(0, P)$, is optimal. The average SNR is therefore SNR $= \mathbb{E}[|h|^2]P/(2\sigma^2) = P/N_0$ (with the normalization $\mathbb{E}[|h[n]|^2] = 1$). The resulting mutual information averaged over a codeword of length N is given by using the AWGN formula with SNR at time n given by $|h[n]|^2$ SNR. This yields

$$\frac{1}{N} I(x[1], \ldots, x[n]; y[1], \ldots, y[n]) = \frac{1}{N} \sum_{n=1}^{N} \log\left(1 + |h[n]|^2 \text{SNR}\right)$$
$$\to \mathbb{E}\left[\log\left(1 + |h|^2 \text{SNR}\right)\right], \quad N \to \infty,$$

where $h \sim CN(0, 1)$. Since $|h|^2 = G$ is an exponential random variable with mean one, the capacity in bits per channel use is given by

$$C_{\text{Rayleigh}} = \mathbb{E}\left[\log\left(1 + G\,\text{SNR}\right)\right] = \int_0^\infty \log\left(1 + g\,\text{SNR}\right) e^{-g} dg,$$
Ergodic capacity with Rayleigh fading. (8.22)

Using Jensen's inequality (see Appendix C and Problem 8.3), it can be shown that $C_{\text{Rayleigh}} < C_{\text{AWGN}}$ for a given SNR. However, as shown in the numerical computations of Problem 8.3, the degradation in performance is relatively modest, in contrast to the huge performance loss due to fading in uncoded systems.

Outage capacity Now, let us consider a different scenario in which each codeword sees a single fading gain. Under this *slow fading* model, the nth received sample sees the model

$$y[n] = h x[n] + w[n], \qquad (8.23)$$

where $h \sim CN(0, 1)$. The power constraint and noise model are as before. Conditioned on h, the channel is an AWGN channel, and the capacity is given by

$$C(h) = \log\left(1 + |h|^2 \text{SNR}\right) = \log\left(1 + G\,\text{SNR}\right), \qquad (8.24)$$

where $G = |h|^2$ is exponential with mean one, as before. Since G can take arbitrarily small values, so can $C(h)$, so that there is no rate at which we can always guarantee reliable communication. Instead, we might opt to communicate at rate R, where R is chosen so as to be smaller than $C(h)$ "most of the time." When $C(h) < R$, we have *outage*. The maximum rate R such that the probability of outage is ϵ is termed the *outage capacity* for that outage probability. For example, we may design a packetized system with a 10% outage probability (also called outage rate): if different packets see different enough channel realizations, then retransmissions are effective for dealing with packet failures caused by outage.

Given the monotone increasing relationship between the gain G and the capacity $C(h)$, we see that

$$C_{\text{out}} = \log\left(1 + G_{\text{out}} \text{SNR}\right),$$

where G_{out} satisfies the outage criterion

$$P[G < G_{\text{out}}] = \epsilon.$$

From (8.14), we see that $G_{\text{out}} \approx \epsilon$ for Rayleigh fading. Thus, we need to increase the SNR by a factor of $1/\epsilon$ relative to the AWGN channel. For example, the SNR penalty is 10 dB for a 10% outage rate, but it increases to 20 dB if we wish to reduce the outage rate to 1%.

8.2.3 Receive diversity

One approach to combating Rayleigh fading is to reduce the fluctuations in the power gain G seen by the transmitted signal. We illustrate this for a system with *receive diversity*, in which the receiver gets multiple versions of the transmitted signal (e.g., at different receive antenna elements that are spaced widely enough apart) which are unlikely to all fade badly at the same time. Given the sufficiency of restricting to a finite-dimensional signal space for AWGN channels, the following development applies to both vector and continuous-time observations.

For two branch receive diversity, if signal s is sent, we obtain two received signals:

$$y_1 = h_1 s + n_1, \\ y_2 = h_2 s + n_2, \quad (8.25)$$

where h_1, h_2 are complex fading gains, and n_1 and n_2 are independent complex WGN, each with variance $\sigma^2 = N_0/2$ per dimension.

To infer the optimal receiver structure, we begin, as usual, with the likelihood ratio. Denoting $y = (y_1, y_2)$ as the pair of received signals and conditioning on $\mathbf{h} = (h_1, h_2)$, we have

$$L(y|s, \mathbf{h}) = L(y_1|s, h_1) L(y_2|s, h_2), \quad (8.26)$$

where the conditional independence of y_1 and y_2 follows from the independence of the noise. Furthermore, for $i = 1, 2$,

$$L(y_i|s, h_i) = \exp\left(\frac{1}{\sigma^2}[\mathrm{Re}(\langle y, h_i s \rangle) - \frac{1}{2}||h_i s||^2]\right). \quad (8.27)$$

Coherent diversity combining If h_1 and h_2 are known, then we see from (8.26) that the LR depends on the observation only through the following decision statistic:

$$Z = \sum_{i=1}^{2} \mathrm{Re}\left(\langle y_i, h_i s \rangle\right) = \mathrm{Re}\left(\sum_{i=1}^{2} h_i^* \langle y_i, s \rangle\right). \quad (8.28)$$

We can therefore restrict attention to Z for both receiver implementation and performance analysis. (We have ignored the usual energy correction term $1/2 \sum_{i=1}^{2} |h_i|^2 ||s||^2$ in order to focus on the observation-dependent term in the LR exponent (8.27). For hypothesis testing with unequal energy signals, this term must be subtracted as usual from the decision statistics.) The combining rule that results in Z is referred to as *coherent maximal ratio combining*. Note that this can be interpreted as standard coherent matched filtering against the signal, summing over both branches, and accounting for the complex gains $\mathbf{h}$. Thus, all the results that we already know for coherent demodulation apply, with E_b computed by summing the signal energy across branches.

For example, if the signal s is employed for on–off keying, then the received energy per bit

$$E_b = \frac{1}{2}||s||^2(|h_1|^2 + |h_2|^2),$$

and the error probability is $Q(\sqrt{E_b/N_0})$. Note that the signal energy increases due to summing across branches.

Diversity and beamforming gains Let us now examine what happens when we generalize to N receive branches. Summing the signal energy across branches, we have a net energy per bit of

$$E_b = \frac{1}{2}||s||^2 \sum_{i=1}^{N} |h_i|^2 = N G^{(N)} \bar{e}_b = G^{(N)} \bar{E}_b, \qquad (8.29)$$

where $\bar{e}_b = 1/2||s||^2$ is the statistical average of the energy per bit for each branch, $\bar{E}_b$ is the average energy per bit summed over branches, and $G^{(N)}$ is the fading gain averaged over branches:

$$G^{(N)} = \frac{1}{N} \sum_{i=1}^{N} |h_i|^2. \qquad (8.30)$$

From (8.29) and (8.30), we can now observe that maximal ratio combining leads to performance improvement in two ways.

Diversity The averaging of $|h_i|^2$ in (8.30) means that fluctuations of the channel strengths in different branches will be averaged out, assuming that the gains on different branches are approximately independent. For example, for i.i.d. $\{h_i\}$, in the limit as $N \to \infty$, the empirically averaged gain tends to its statistical average:

$$G^{(N)} \to \mathbb{E}[|h_1|^2] = 1, \quad N \to \infty,$$

so that we recover the performance without fading. For finite N, the improvement of outage performance using diversity is explored in Problem 8.11.

Beamforming The summing of signal energies across branches to get an N-fold gain in the effective energy per bit leads to a beamforming gain. This happens because the signal terms add up in amplitude (in the mean of the decision statistic) across branches, while the noise terms add up in power (in the variance of the decision statistic).

Performance with coherent diversity combining As shown in Problems 8.8 and 8.9, the performance with N diversity branches decays as SNR^{-N} for large SNR, which, even for small N of two to four branches, yields significant improvement over the SNR^{-1} decay for Rayleigh fading ($N = 1$).

8.2 Fading and diversity

For example, for binary coherent FSK, we obtain the following upper bound for the error probability by applying the results of Problem 8.8:

$$P_e \leq \frac{1}{2}\left(1+\frac{1}{N}\frac{\bar{E}_b}{2N_0}\right)^{-N} = \frac{1}{2}\left(1+\frac{\bar{e}_b}{2N_0}\right)^{-N} \qquad (8.31)$$

FSK with coherent diversity combining.

Noncoherent diversity combining Now, consider the model (8.25) with the complex gains h_1 and h_2 unknown. The conditional density of the received signal given $\{h_i\}$ is given by (8.26) and (8.27), from which we conclude that it suffices to restrict attention to the complex correlator outputs

$$Z_i = \langle y_i, s \rangle \quad i = 1, 2.$$

Now, consider M-ary signaling. The received signal $y = (y_1, y_2)$ obeys one of M hypotheses specified by

$$H_k : \begin{matrix} y_1 = h_1 s_k + n_1 \\ y_2 = h_2 s_k + n_2 \end{matrix} \quad k = 1, \ldots, M. \qquad (8.32)$$

Reasoning as above, it suffices to restrict attention to the decision statistics obtained by complex correlation of the received signal in each branch with each possible transmitted signal as follows:

$$z_{ik} = \langle y_i, s_k \rangle \quad i = 1, 2, \ k = 1, \ldots, M.$$

Letting $\mathbf{z} = \{z_{ik}, i = 1, 2, \ k = 1, \ldots, M\}$ denote the collection of decision statistics, the ML rule can be written as

$$\delta_{\text{ML}}(y) = \arg \max_{1 \leq j \leq M} p(\mathbf{z}|H_j).$$

It remains to find the conditional density of $\mathbf{z}$ given each hypothesis in order to specify the ML rule in detail. Let us condition on H_j. Then $y_i = h_i s_j + n_i$, $i = 1, 2$, and

$$z_{ik} = \langle h_i s_j + n_i, s_k \rangle = h_i \langle s_j, s_k \rangle + \langle n_i, s_k \rangle. \qquad (8.33)$$

Quadratic form of ML rule We can now show that the ML rule is quadratic in $\mathbf{z}$ under rather general assumptions. Let us assume that $\mathbf{h}$ is proper complex Gaussian, and is independent of the noise. Then, conditioned on hypothesis H_k, $\mathbf{z}$ is proper complex Gaussian, say $\mathbf{z} \sim CN(\mathbf{m}_k, \mathbf{C}_k)$, where $\mathbf{m}_k$, $\mathbf{C}_k$ depend on the mean and covariance of $\mathbf{h}$, the signal correlations $\langle s_j, s_k \rangle$, and the noise PSD. Without any explicit computations, we can write

$$p(\mathbf{z}|H_k) = \frac{1}{\pi^n \det(\mathbf{C}_k)} \exp\left(-(\mathbf{z}-\mathbf{m}_k)^H \mathbf{C}_k^{-1}(\mathbf{z}-\mathbf{m}_k)\right).$$

Taking logarithms and maximizing over k, we obtain

$$\delta_{\text{ML}}(y) = \arg \min_{1 \leq k \leq M} (\mathbf{z}-\mathbf{m}_k)^H \mathbf{C}_k^{-1}(\mathbf{z}-\mathbf{m}_k) - \log \det(\mathbf{C}_k). \qquad (8.34)$$

Binary orthogonal equal-energy signaling Let us now get some explicit results by specializing to binary FSK and i.i.d. Rayleigh fading on the branches. Set $M = 2$ in (8.32), and set $\langle s_k, s_j \rangle = E\delta_{kj}$. For convenience, we set $\mathbb{E}[|h_1|^2] = \mathbb{E}[|h_2|^2] = 1$. Now, let us condition on H_1. Plugging into (8.33), we have

$$z_{11} = \langle y_1, s_1 \rangle = h_1 E + \langle n_1, s_1 \rangle, \quad z_{21} = \langle y_2, s_1 \rangle = h_2 E + \langle n_2, s_1 \rangle,$$
$$z_{12} = \langle y_1, s_2 \rangle = \langle n_1, s_2 \rangle, \quad z_{22} = \langle y_2, s_2 \rangle = \langle n_2, s_1 \rangle.$$

Under our model, it is easy to see that the decision statistics $\mathbf{Z}$ are zero mean, proper complex Gaussian, with all components uncorrelated and hence independent (this follows from the independence of h_1, h_2, n_1, n_2, and orthogonality of s_1 and s_2). We can also show that, conditioned on H_1,

$$\begin{aligned} \mathbb{E}[|z_{11}|^2] = \mathbb{E}[|z_{21}|^2] &= E^2 + 2\sigma^2 E \\ \mathbb{E}[|z_{12}|^2] = \mathbb{E}[|z_{22}|^2] &= 2\sigma^2 E \end{aligned} \quad \text{(conditioned on } H_1\text{)}.$$

Similarly, conditioned on H_2, the roles of the correlator outputs for s_1 and s_2 get interchanged, so that

$$\begin{aligned} \mathbb{E}[|z_{11}|^2] = \mathbb{E}[|z_{21}|^2] &= 2\sigma^2 E \\ \mathbb{E}[|z_{12}|^2] = \mathbb{E}[|z_{22}|^2] &= E^2 + 2\sigma^2 E \end{aligned} \quad \text{(conditioned on } H_2\text{)}.$$

Substituting into (8.34), and dropping the $\log \det(\mathbf{C}_k)$ term because it is equal for $k = 1, 2$, we have

$$\frac{|z_{11}|^2}{E^2 + 2\sigma^2 E} + \frac{|z_{21}|^2}{E^2 + 2\sigma^2 E} + \frac{|z_{12}|^2}{2\sigma^2 E} + \frac{|z_{22}|^2}{2\sigma^2 E}$$

$$\overset{H_2}{\underset{H_1}{\gtrless}} \frac{|z_{11}|^2}{2\sigma^2 E} + \frac{|z_{21}|^2}{2\sigma^2 E} + \frac{|z_{12}|^2}{E^2 + 2\sigma^2 E} + \frac{|z_{22}|^2}{E^2 + 2\sigma^2 E},$$

which simplifies to

$$|z_{12}|^2 + |z_{22}|^2 \overset{H_2}{\underset{H_1}{\gtrless}} |z_{11}|^2 + |z_{21}|^2.$$

Rewording to express the decision rule in terms of the received signal y, we have

$$\delta_{\mathrm{ML}}(y) = \arg \max_k |\langle y_1, s_k \rangle|^2 + |\langle y_2, s_k \rangle|^2. \qquad (8.35)$$

Thus, the ML decision rule leads to the intuitively plausible strategy of picking the signal for which the sum of the correlator energies across branches is the largest. While we have derived (8.35) for $M = 2$, this applies to M-ary equal energy orthogonal signaling for $M > 2$ as well, since the ML rule can be interpreted as a pairwise comparison among different hypotheses. Similarly, the result generalizes to a larger number of diversity branches simply by correlating and summing over all branches.

Summing of energies as in (8.35) is termed *noncoherent combining*, in contrast to maximal ratio combining (8.28), where knowledge of the channel gains is used to combine *coherently* across branches.

Performance with noncoherent diversity combining The noncoherent combining rule (8.35) generalizes to N branches as follows:

$$\delta_{ML}(y) = \arg\max_k \sum_{i=1}^{N} |\langle y_i, s_k \rangle|^2. \tag{8.36}$$

Considering binary FSK ($k = 1, 2$), we note that, conditioned on H_1, $U_i = |\langle y_i, s_1 \rangle|^2$ are i.i.d. exponential random variables, each with mean $1/\mu_U = E^2 + 2\sigma^2 E$, and $V_i = |\langle y_i, s_2 \rangle|^2$ are i.i.d. exponential random variables, each with mean $1/\mu_V = 2\sigma^2 E$, with these two sets of random variables also being independent of each other. The conditional error probability (also equal to the unconditional error probability P_e by symmetry), conditioned on H_1, is

$$P_e = P_{e|1} = P[V_1 + \cdots + V_N > U_1 + \cdots + U_N | H_1]. \tag{8.37}$$

It is possible to compute this error probability exactly, since $\sum_{i=1}^N U_i$ and $\sum_{i=1}^N V_i$ are independent Gamma random variables whose densities and cdfs can be characterized as in Problems 8.5 and 8.6. However, insight into the dependence of error probability on diversity level is more readily obtained using the Chernoff bound derived in Problem 8.12:

$$P_e \leq \left(\frac{1 + \frac{1}{N}\bar{E}_b/N_0}{\left(1 + \frac{1}{N}\bar{E}_b/2N_0\right)^2} \right)^N = \left(\frac{1 + \bar{e}_b/N_0}{(1 + \bar{e}_b/2N_0)^2} \right)^N \stackrel{\text{high}}{\underset{\text{SNR}}{\approx}} \left(\frac{\bar{e}_b}{4N_0} \right)^{-N}$$

FSK with noncoherent diversity combining, (8.38)

where we have used the high SNR approximation

$$\frac{1 + \bar{e}_b/N_0}{(1 + \bar{e}_b/2N_0)^2} \approx \left(\frac{\bar{e}_b}{4N_0} \right)^{-1}.$$

As with coherent systems, the error probability exhibits an SNR^{-N} decay.

8.3 Orthogonal frequency division multiplexing

The Nyquist criterion for ISI avoidance for linear modulation over a linear time-invariant channel is equivalent to requiring that the waveforms $\{x(t - kT)\}$ used to modulate the transmitted symbols $\{b[k]\}$ are orthogonal, where $x(t) = (g_T * g_C * g_R)(t)$ is the cascade of the transmit, channel, and receive filters. The system designer typically has control over the transmit and receive filters, but often must operate in scenarios in which the channel is unknown. Thus, for nontrivial channel responses $g_C(t)$, linear modulation in the time domain incurs ISI. Can we use modulating waveforms that stay orthogonal when they go through *any* LTI channel? The answer is yes: this is achieved by choosing the modulating waveforms to be eigenfunctions of the channel: a waveform $s(t)$ is an eigenfunction of a channel g_C if, when s is the input to the channel, the output $s * g_C$ is a scalar multiple of s. Thus, orthogonal eigenfunctions stay orthogonal as they go through the channel. As stated in the following theorem, which provides the conceptual basis for OFDM, there is a set of *universal* eigenfunctions that work for all LTI channels: the complex exponentials.

Theorem 8.3.1 *Consider a linear time-invariant channel with impulse response $g_C(t)$ and transfer function $G_C(f)$. Then the following statements are true:*

(a) The complex exponential waveform $e^{j2\pi ft}$ is an eigenfunction of the channel with eigenvalue $G_C(f)$. That is,

$$e^{j2\pi ft} * g_C(t) = G_C(f)e^{j2\pi ft}.$$

(b) Complex exponentials at different frequencies are orthogonal.

Proof The eigenfunction property (a) is verified as follows:

$$e^{j2\pi ft} * g_C(t) = \int_{-\infty}^{\infty} g_C(u) e^{j2\pi f(t-u)} \, du$$
$$= e^{j2\pi ft} \int_{-\infty}^{\infty} g_C(u) e^{-j2\pi fu} \, du = G_C(f) e^{j2\pi ft}.$$

We now verify statement (b) on orthogonality for different frequencies:

$$\langle e^{j2\pi f_1 t}, e^{j2\pi f_2 t} \rangle = \int_{-\infty}^{\infty} e^{j2\pi f_1 t} e^{-j2\pi f_2 t} = \delta(f_2 - f_1) = 0, \quad f_1 \neq f_2,$$

using the fact that the Fourier transform of a constant is the delta function. □

The complex exponential waveforms in Theorem 8.3.1 are defined over an infinite time horizon, and there is no restriction on the selection of the frequencies f to be used for modulation. In practice, an OFDM system employs a discrete set of subcarriers over a symbol interval of finite length T. The transmitted complex baseband waveform is given by

$$u(t) = \sum_{n=0}^{N-1} B[n] e^{j2\pi f_n t} I_{[0,T]}(t) = \sum_{n=0}^{N-1} B[n] p_n(t), \qquad (8.39)$$

where $B[n]$ is the symbol transmitted using the modulating signal $p_n(t) = e^{j2\pi f_n t} I_{[0,T]}$, using the nth subcarrier at frequency f_n. We now have to check how the results of Theorem 8.3.1 must be modified to deal with the finite signaling interval. The timelimited tone $p_n(t)$ has Fourier transform $P_n(f) = T\text{sinc}((f - f_n)T)e^{-\pi fT}$, which decays quickly as $|f - f_n|$ takes on values of

8.3 Orthogonal frequency division multiplexing

the order of k/T. If T is large compared with the channel delay spread, then $1/T$ is small compared with the channel coherence bandwidth, so that the gain seen by $P_n(f)$ is roughly constant, and the eigenfunction property is roughly preserved. That is, when $P_n(f)$ goes through a channel with transfer function $G_C(f)$, the output

$$Q_n(f) = G_C(f)P_n(f) \approx G_C(f_n)P_n(f).$$

The orthogonality of different subcarriers holds over an interval of length T if they are spaced apart by an integer multiple of $1/T$:

$$\int_0^T e^{j2\pi f_n t} e^{-j2\pi f_m t} dt = \frac{e^{j2\pi (f_n - f_m)T} - 1}{j2\pi(f_n - f_m)} = 0, \text{ for } (f_n - f_m)T = \text{ nonzero integer.}$$

We can therefore rewrite the transmitted waveform as follows:

$$u(t) = \sum_{n=0}^{N-1} B[n] p_n(t) = \sum_{n=0}^{N-1} B[n] e^{j2\pi n t/T} I_{[0,T]}. \qquad (8.40)$$

We would typically subtract $(N-1)/2T$ from all the frequencies above to center the baseband frequency content around the origin, but this is not shown, in order to keep the notation simple.

When the signal $u(t)$ goes through the channel, the nth subcarrier at frequency f_n sees a gain of $G_C(f_n)$, so that the receiver sees a noisy version of $G_C(f_n)B[n]$ when demodulating the nth subcarrier. The effect of the channel can therefore be undone separately for each subcarrier, unlike the equalization techniques (see Chapter 5) required for single-carrier modulation. While these advantages of OFDM have been well known for many decades, it is the ability to push the complexity of OFDM into the digital domain, thereby allowing cost-effective implementation by exploiting Moore's law, which has been key to its growing adoption in a multitude of applications. These include digital subscriber loop (DSL), digital video broadcast from satellites, as well as a multitude of wireless communication systems, including IEEE 802.11 based WLANs, and IEEE 802.16/20 and other fourth generation cellular systems. We discuss the digital realization of OFDM next.

Digital-signal-processing-centric implementation of OFDM For T large enough, the bandwidth of the OFDM signal u is approximately N/T. Thus, we can represent $u(t)$ accurately by sampling at rate $1/T_s = N/T$, where T_s is the sampling interval. From (8.40), the samples can be written as

$$u(kT_s) = \sum_{n=0}^{N-1} B[n] e^{j2\pi n k/N}.$$

We can recognize this simply as the inverse DFT of the symbol sequence $\{B[n]\}$. We make this explicit in the notation as follows:

$$b[k] = u(kT_s) = \sum_{n=0}^{N-1} B[n] e^{j2\pi n k/N}. \qquad (8.41)$$

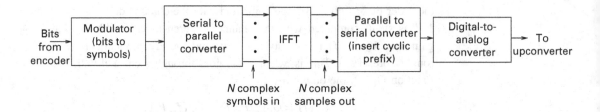

Figure 8.4 DSP-centric implementation of an OFDM transmitter.

If N is a power of two (which can be achieved by zeropadding if necessary), the samples $\{b[k]\}$ can be efficiently generated from the symbols $\{B[n]\}$ using an inverse fast Fourier transform (IFFT). The complex baseband waveform $u(t)$ can now be obtained from its samples by digital-to-analog (D/A) conversion. This implementation of an OFDM transmitter is as shown in Figure 8.4: the bits are mapped to symbols, the symbols are fed in parallel to the inverse FFT (IFFT) block, and the complex baseband signal is obtained by D/A conversion of the samples (after insertion of a cyclic prefix, to be discussed after we motivate it in the context of receiver implementation). Typically, the D/A converter is an interpolating filter, so that its effect can be subsumed within the channel impulse response.

Note that the relation (8.41) can be inverted as follows:

$$B[n] = \frac{1}{N} \sum_{k=0}^{N-1} b[k] e^{-j2\pi nk/N}. \tag{8.42}$$

This is exploited in the digital implementation of the OFDM receiver, discussed next.

We know that, once we limit the signaling duration to be finite, the ISI avoidance property of OFDM is approximate rather than exact. However, as we now show, orthogonality between subcarriers can be restored *exactly* in discrete time by using a cyclic prefix, which allows for efficient demodulation using an FFT. The noiseless received OFDM signal is modeled as

$$v(t) = \sum_{k=0}^{N-1} b[k] p(t - kT_s),$$

where the "effective" channel impulse response $p(t)$ includes the effect of the D/A converter at the transmitter, the physical channel, and the receive filter. When we sample this signal at rate $1/T_s$, we obtain the discrete-time model

$$v[m] = \sum_{k=0}^{N-1} b[k] h[m - k], \tag{8.43}$$

8.3 Orthogonal frequency division multiplexing

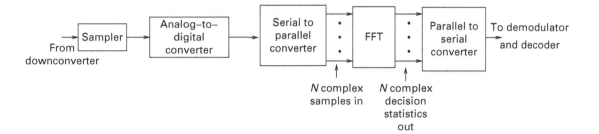

Figure 8.5 DSP-centric implementation of an OFDM receiver. Carrier and timing synchronization blocks are not shown.

where $\{h[l] = p(lT_s)\}$ is the effective discrete-time channel of length L, assumed to be less than N. We assume, without loss of generality, that $h[l] = 0$ for $l < 0$ and $l \geq L$. We can rewrite (8.43) as

$$v[m] = \sum_{l=0}^{L-1} h[l]b[m-l]. \tag{8.44}$$

Let H denote the N-point DFT of h:

$$H[n] = \sum_{l=0}^{N-1} h[l]e^{-j2\pi nl/N} = \sum_{l=0}^{L-1} h[l]e^{-j2\pi nl/N}. \tag{8.45}$$

As noted in (8.42), the DFT of $\{b[k]\}$ is the symbol sequence $B[n]$ (the normalization is chosen differently in (8.42) and (8.45) to simplify the forthcoming equations). If we could replace the linear convolution of (8.44) by the circular convolution $\tilde{v} = h \odot b$ given by

$$\tilde{v}[m] = (h \odot b)[m] = \sum_{l=0}^{N-1} h[l \bmod N]b[(m-l) \bmod N], \tag{8.46}$$

then the corresponding N-point DFTs would satisfy

$$\tilde{V}[n] = H[n]B[n], \quad n = 0, \ldots, N-1.$$

See Problem 8.13 for a review of the relationship between cyclic convolution and the DFT. Since $L < N$, we can write the circular convolution (8.46) as

$$\tilde{v}[m] = \sum_{l=0}^{\min(L-1,m)} h[l]b[m-l] + \sum_{l=m+1}^{L-1} h[l]b[m-l+N]. \tag{8.47}$$

Comparing the linear convolution (8.44) and the cyclic convolution (8.47), we see that they are identical *except* when the index $m-l$ takes negative values: in this case, $b[m-l] = 0$ in the linear convolution, while $b[(m-l) \bmod N] = b[m-l+N]$ contributes to the circular convolution. Thus, we can emulate a cyclic convolution using the physical linear convolution by sending a *cyclic prefix*; that is, by sending

$$b[k] = b[N+k], \quad k = -(L-1), -(L-2), \ldots, -1,$$

before we send the samples $b[0], \ldots, b[N-1]$. That is, we transmit the samples

$$b[N-L+1], \ldots, b[N-1], b[0], \ldots, b[N-1],$$

incurring an overhead of $(L-1)/N$ which can be made small by choosing N to be large.

At the receiver, the complex baseband signal is sampled at rate $1/T_s$ to obtain noisy versions of the samples $\{b[k]\}$. The FFT of these samples then yields the model

$$Y[n] = H[n]B[n] + N[n], \tag{8.48}$$

where the frequency domain noise samples $N[n]$ are modeled as i.i.d. $CN(0, 2\sigma^2)$, being the DFT of i.i.d. $CN(0, 2\sigma^2)$ time domain noise samples. If the receiver knows the channel, then it can implement ML reception based on the statistic $H^*[n]B[n]$. Thus, the task of channel equalization has been reduced to compensating for scalar channel gains for each subcarrier. This makes OFDM extremely attractive for highly dispersive channels, for which time domain single-carrier equalization strategies would be difficult to implement.

The PSD of OFDM Assuming that the symbols $\{B[n]\}$ transmitted on the different frequencies are uncorrelated, the PSDs corresponding to different subcarriers in the transmitted signal u in (8.39) add up, and we get, using the PSD expression in Theorem 2.5.1 from Chapter 2,

$$S_u(f) = \sum_{n=0}^{N-1} \mathbb{E}[|B[n]|^2] \frac{|P_n(f)|^2}{T}$$

$$= T \sum_{n=0}^{N-1} \sigma_B^2[n] |\text{sinc}((f - f_n)T)|^2,$$

where $\sigma_B^2[n] = \mathbb{E}[|B[n]|^2]$ is the symbol energy for subcarrier $f_n = 1/T(n - (N-1)/2)$. The PSD is plotted against the normalized frequency (normalized by the symbol rate) in Figure 8.6 for $N = 64$. Since we send N symbols over time T, the symbol rate $R_s = N/T$, and the normalized frequency is $f/R_s = fT/N$. Note the exceptionally efficient use of spectrum by OFDM, which essentially achieves Nyquist rate signaling if we ignore the overhead due to the cyclic prefix.

Peak-to-average ratio A key implementation problem with OFDM is that it has a high peak-to-average ratio (PAR); that is, the peak power of the

8.3 Orthogonal frequency division multiplexing

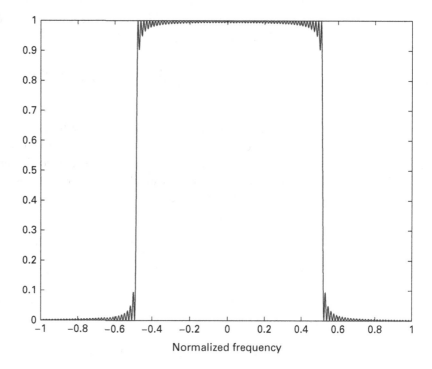

Figure 8.6 The PSD of OFDM with $N = 64$ subcarriers as a function of normalized frequency fT/N.

transmitted signal can be much larger than its average power. The basic problem is that, depending on the transmitted symbol sequence, the subcarriers can constructively interfere to yield amplitudes that scale as N, the number of subcarriers. The peak power therefore scales as N^2, while the average power only scales as N. This implies that the transmit power amplifier must exhibit a linear characteristic over a large dynamic range in order to avoid significant distortions of the OFDM signal. Since power amplifiers operate most efficiently in a nonlinear saturation regime, backing off into a linear regime leads to a loss in power efficiency. To get a quantitative idea of the PAR problem, let us develop a statistical model for the transmitted samples $\{b[k]\}$ in (8.41), obtained by taking the inverse DFT of the symbols $\{B[n]\}$. Symbols drawn from standard PSK and QAM constellations can be modeled as zero mean, proper random variables, i.e., $\mathbb{E}[B[n]] = 0$ and $\mathbb{E}[B^2[n]] = 0$. These properties are preserved under linear transformation, and are therefore inherited by $\{b[k]\}$. Thus, since $\{b[k]\}$ are the sum of N independent terms (assuming that $\{B[n]\}$ are independent), they are well approximated, by the central limit theorem, as zero mean, proper complex Gaussian. This central limit theorem approximation is accurate even for moderate values of N (e.g., $N = 8$ or more). Under the preceding assumption, $\{b[k]\}$ are uncorrelated. Furthermore, using a vector central limit theorem to model $\{b[k]\}$ as *jointly* proper complex Gaussian, we infer that they can be approximated as independent. The power of the samples $P[k] = |b[k]|^2$ can therefore be modeled as i.i.d. exponential random variables with mean $\mathbb{E}[P[k]] = 1$, normalized to unity for convenience of notation. The sample peak-to-average ratio (PAR) is defined as

$$\mathrm{PAR} = \frac{\max_{0 \leq k \leq N-1} P[k]}{\frac{1}{N} \sum_{k=0}^{N-1} P[k]} \approx \max_{0 \leq k \leq N-1} P[k], \tag{8.49}$$

replacing the empirical average of the sample power in the denominator by the statistical average. Note that the PAR for the continuous-time signal $u(t)$ obtained by D/A conversion of the samples is actually a little larger than this sample PAR, but we restrict attention to the sample PAR because it is easier to compute, and follows the same trends as we vary the symbol sequence $\{B[n]\}$. Based on our central limit theorem approximations, we can now compute the cdf of the PAR as follows:

$$P[\mathrm{PAR} \leq x] = P[\max_{0 \leq k \leq N-1} P[k] \leq x] = P[P[0] \leq x, \ldots, P[N-1] \leq x]$$

$$= (P[P[0] \leq x])^N = (1 - e^{-x})^N. \tag{8.50}$$

We can also approximate the complementary cdf of the PAR using a union bound as follows:

$$P[\mathrm{PAR} > x] = P[\max_{0 \leq k \leq N-1} P[k] \leq x] \leq \sum_{k=0}^{N-1} P[P[k] > x] = N e^{-x}. \tag{8.51}$$

This clearly shows how the PAR problem is exacerbated as we increase the number of subcarriers.

PAR reduction strategies The implementation difficulties caused by high PAR have motivated significant effort in developing PAR reduction strategies. One strategy for dealing with OFDM's high PAR is simply to clip the signal at some level. The formulas (8.50) or (8.51) can be used to infer, for example, the probability that clipping at, say, 5 dB above the average power will cause distortion of the signal. Similarly, given the input–output characteristics of the power amplifier and a desired upper bound on the probability of distortion, we can use (8.50) to infer the required backoff from the nonlinear regime. Errors caused by distortion can be handled using standard error correction and detection codes. For example, one approach is to design for a relatively small PAR, based on the probabilistic view that "most" transmitted sequences do not result in high PAR (at least for moderate N). A packet error may occur for a data sequence that leads to high PAR, and hence large distortion from a transmitter front end optimized for a smaller dynamic range. The data sequence can be scrambled pseudorandomly when such a packet is retransmitted, based on the reasoning that it is unlikely that the scrambled sequence also corresponds to a high PAR. Strategies for explicitly reducing PAR include the design of codes for PAR reduction: since the PAR depends on the data sequence $\{B[n]\}$, it is possible to restrict the set of transmitted sequences $\{B[n]\}$ so as to limit the PAR. Finally, another approach is to

8.3 Orthogonal frequency division multiplexing

reserve a set of subcarriers for PAR reduction: once the symbols on the data-carrying subcarriers have been determined, the reserved set of subcarriers is modulated so as to reduce the PAR.

Inter-carrier interference OFDM is designed to avoid ISI by signaling in the frequency domain. However, imperfect carrier synchronization can cause inter-carrier interference (ICI) between the symbols transmitted on different subcarriers. To understand why, let us go back to the noiseless samples $\tilde{v}[m]$, and note that a frequency offset Δf corresponds to multiplying, sample by sample, by the discrete-time tone

$$g[m] = e^{j(\delta m + \theta)}, \qquad (8.52)$$

where $\delta = 2\pi \Delta f \, T_s$ is the discrete-time frequency shift, and θ is a phase shift. This yields

$$\hat{v}[m] = \tilde{v}[m]g[m] \leftrightarrow \hat{V}[n] = (H[n]B[n]) \odot G[n],$$

where $\{G[n]\}$ is the DFT of $\{g[m]\}$. This convolution with G destroys the orthogonality of the subcarriers, and reintroduces the ISI that the OFDM designer has worked so hard to avoid. Thus, accurate carrier synchronization is a very high design priority. Typically, a number of known pilot (or training) symbols are allocated to aid in both carrier synchronization and channel estimation.

Wireless versus wireline applications If the transmitter knows the channel $\{H[n]\}$, then it could use waterfilling-style techniques to optimize throughput, such as using larger constellations to send more bits per symbol over subcarriers seeing stronger channel gains. Indeed, this is the strategy used for OFDM signaling over the wireline digital subscriber loop (DSL) channel; in the latter context, the term discrete multitone (DMT) is often employed instead of OFDM. Since DSL employs a physical baseband channel, the time domain samples $\{b[k]\}$ must be real-valued, which implies that the symbols $\{B[n]\}$ must obey conjugate symmetry. Problem 8.14 illustrates some aspects of DMT system design for a real baseband channel. For a passband wireless channel, the baseband samples $\{b[k]\}$ are complex-valued, so that there is no restriction on $\{B[n]\}$. Typically, it is assumed in wireless systems that the transmitter does not know the channel, which is understandable in view of the rapid time variations of a cellular mobile channel. In this case, the common strategy is to use the same constellation for each subcarrier. However, several current and emerging wireless networks consist of links with very slow time variations; examples include wireless local area networks and in-room wireless personal area networks. In this case, it may be useful to reexamine the assumption that the transmitter does not know the channel, and to explore OFDM system design when some form of channel feedback is available to the system designer.

8.4 Direct sequence spread spectrum

Spread spectrum is a broad term for modulation formats in which the bandwidth is significantly larger than the information rate. Typically, spread spectrum is employed for one or more of the following reasons:

Diversity Using a larger bandwidth provides frequency diversity for combating fading.

Multiple access Appropriately designed spread spectrum waveforms enable multiple users to access the channel simultaneously, without necessarily coordinating as required for orthogonal multiple access techniques such as TDMA or FDMA.

Low probability of intercept or detect and antijam Spreading the signal energy over a wide bandwidth makes it more difficult for an adversary to detect, intercept or jam it. These are desirable features for applications such as military communication.

We discuss here direct sequence (DS) spread spectrum, followed by a discussion of frequency hop (FH) spread spectrum in Section 8.5. Direct sequence signaling is based on linear modulation. As we know from Chapter 2, for standard linear modulation, a stream of symbols $\{b[m]\}$ is transmitted at symbol rate $1/T$ by transmitting the waveform $\sum_m b[m]p(t - mT)$, where $p(t)$ is a symbol waveform whose bandwidth is of the order of the Nyquist rate $1/T$ (typical excess bandwidths range from 25–100%). In a DS system, in order to send a symbol b, instead of using one complex dimension, we use $N > 1$ dimensions, sending instead the elements of the vector $b\mathbf{s}$, where $\mathbf{s} = (s[0], \ldots, s[N-1])^T$ is a *spreading vector*. The factor N is termed the *processing gain*. Since we send N "chips" using linear modulation to transmit a single symbol, the bandwidth of the linearly modulated system scales with the chip rate $1/T_c = N/T$ (in practice, excess bandwidth beyond this minimum Nyquist bandwidth would be employed). The modulating pulse $\psi(t)$ for this chip rate linearly modulated system is termed the *chip waveform*. Typical examples of chip waveforms include a rectangular pulse timelimited to an interval of length T_c, or a bandlimited square root Nyquist pulse at rate $1/T_c$, such as a square root raised cosine pulse.

To represent a DS signal compactly as a chip rate linearly modulated signal, it is convenient to express the symbols $\{b[n]\}$ at the chip rate. At the chip rate, each symbol can be viewed as being repeated N times. Accordingly, we define the chip rate symbol sequence $\{\tilde{b}[l]\}$ as

$$\tilde{b}[l] = b[m], \quad mN \leq l \leq (m+1)N - 1.$$

The DS transmitted complex baseband signal therefore takes the following general form:

$$u(t) = \sum_l \tilde{b}[l]s[l]\psi(t - lT_c), \quad (8.53)$$

8.4 Direct sequence spread spectrum

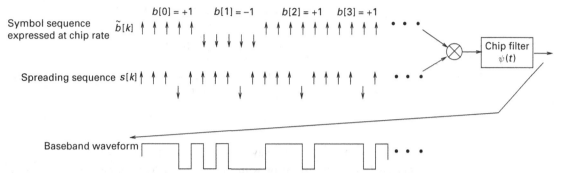

Figure 8.7 Example of DS modulation, with processing gain $N = 5$ and a short spreading sequence. In this chip rate implementation, the data sequence is converted to the chip rate and is multiplied chip-by-chip with the spreading sequence before modulating the chip waveform. These discrete-time chip rate sequences are indicated by delta functions in the figure. These are then passed through a chip filter (which may itself be implemented in DSP at a multiple of the chip rate, followed by digital-to-analog conversion) to obtain the baseband analog waveform to be upconverted.

where $\{s[l]\}$ is the *spreading sequence*. Figure 8.7 shows a typical implementation of a DS modulator, where, for convenience in drawing the figure, we have assumed the chip waveform to be a rectangular timelimited pulse rather than a bandwidth-efficient pulse. The multiplication of bits and chips taking values ± 1 shown in the figure might often be implemented using exclusive or of bits and chips taking values in $\{0, 1\}$, followed by 0 being mapped to $+1$ and 1 to -1. The N-dimensional spreading vector for the symbol $b[n]$ is given by $\mathbf{s}[n] = (s[nN], \ldots, s[(n+1)N-1])^T$. The spreading sequence is typically chosen so as to have good autocorrelation properties: in operational terms, this means that the spreading vector for a given symbol has small normalized inner products with shifts of itself, as well as with shifts of spreading vectors corresponding to adjacent symbols. We shall soon see how this property, which holds for a variety of pseudorandom spreading sequences, alleviates ISI and simplifies the receiver structure in DS systems.

While chip rate operations form the core of DS transceiver implementations, it is also important to keep in mind what is happening at the symbol rate. The transmitted signal in (8.53) can be rewritten as

$$u(t) = \sum_m b[m]s(m; t - mT), \tag{8.54}$$

where

$$s(m; t) = \sum_{l=0}^{N-1} s[mN+l]\psi(t - lT_c)$$

is the *spreading waveform* modulating the mth transmitted symbol.

Short spreading sequences If the spreading sequence $\{s[l]\}$ is periodic with period N, we call it a *short* spreading sequence. In this case, the spreading waveforms $s(m; t)$ for different symbols are all identical, with

$$s(m; t) \equiv s(t) = \sum_{l=0}^{N-1} s[l]\psi(t - lT_c).$$

The transmitted waveform therefore has exactly the form that we considered in Chapter 2 for linearly modulated signals

$$u(t) = \sum_m b[m]s(t - mT),$$

except that the modulating waveform $s(t)$ has bandwidth scaling with the chip rate $1/T_c = N/T$ instead of with the symbol rate $1/T$. For short spreading sequences, the waveform $u(t)$ is cyclostationary with period T. The consequences of this are discussed later.

Long spreading sequences If the spreading sequence $\{s[l]\}$ is aperiodic, or has a period much longer than N, we call it a *long* spreading sequence. In this case, the N chips corresponding to a given symbol are often well modeled as i.i.d. and randomly chosen. This *random spreading sequence model* is a useful tool for performance analysis that is extensively used in system design.

Code division multiple access In a DS code division multiple access (CDMA) system, multiple users access the wireless medium at the same time. If the spreading waveforms for these users were orthogonal, then multiple-access interference (MAI) could be completely eliminated. This strategy is effective when a single transmitter sends different DS waveforms to different receivers. Assuming an ideal channel between the transmitter and any particular receiver, the spreading waveforms retain their orthogonality when arriving at the receiver, thus eliminating MAI. In practice, passing the spreading waveforms through a multipath channel destroys their orthogonality. However, if a particular path is dominant, then designing the waveforms to be orthogonal can significantly reduce MAI along this path, and improve overall performance. This strategy is used on the base station to mobile "downlink" of CDMA-based digital cellular systems.

For the "uplink" of a CDMA-based cellular system, different mobiles transmitting to a base station see different channels. It is not possible to design spreading waveforms so as to maintain orthogonality at the base station receiver for all possible channel realizations. Thus, the strategy is to choose spreading waveforms that have small normalized inner products on average, regardless of the relative delays of the waveforms. This is also an appropriate design approach for peer-to-peer wireless communication, where communication for a particular transmitter–receiver pair incurs interference due to communication between other transmitter–receiver pairs.

The choice of spreading waveforms is discussed in Section 8.4.2. For the moment, suffice it to say that spreading waveforms are typically designed to have small normalized inner products with delayed versions of themselves (i.e., they have good *autocorrelation* properties) and with delayed versions of other spreading waveforms (i.e., they have good *crosscorrelation* properties). The conventional approach to receiver design for DS systems, therefore, is to ignore ISI and MAI when determining the *structure* of the receiver, but to take interference into account when evaluating the *performance* of the receiver. This leads to the rake receiver discussed in Section 8.4.1. An alternative approach is to employ more sophisticated *multiuser detection* strategies which exploit the structure of the MAI. The methods used are similar to those used for channel equalization in Chapter 5, and are discussed briefly in Section 8.4.4.

8.4.1 The rake receiver

The rake receiver is the name given to matched filter reception of a wideband waveform over a multipath channel. Such a matched filter receiver is optimal for the AWGN channel, but it ignores the effect of ISI and MAI, relying on the good autocorrelation and crosscorrelation properties of DS spreading waveforms.

Effect of multipath Suppose now that the signal u passes through a multi-path channel with impulse response

$$h(t) = \sum_{i=1}^{M} \alpha_i \delta(t - \tau_i),$$

where, for $1 \leq i \leq L$, α_i is the complex gain of the *i*th path, which has delay τ_i, and where we ignore channel time variations. The effective spreading waveform seen by symbol k is given by

$$\tilde{s}(m; t) = s(m; t) * h(t) = \sum_{i=1}^{M} \alpha_i s(m; t - \tau_i).$$

The received signal is given by

$$y(t) = (u * h)(t) + n(t) = \sum_m b[m] \tilde{s}(m; t - mT) + n(t),$$

where n is WGN with PSD $\sigma^2 = N_0/2$ per dimension. For well-designed spreading waveforms, we can typically ignore the ISI from adjacent symbols in receiver design. In this case, the model for demodulating $b[m]$ is given by

$$y(t) = b[m] \tilde{s}(m; t - mT) + n(t) \quad \text{Reduced model ignoring ISI.}$$

Based on the reduced model, the optimal statistic for deciding on $b[m]$ is to correlate y against $\tilde{s}^{(m)}$, as follows:

$$Z[k] = \int y(t) \tilde{s}^*(m; t - mT) dt = \sum_{i=1}^{L} \alpha_i^* \int y(t) s^*(m; t - mT - \tau_i) dt. \quad (8.55)$$

This receiver structure is called the *rake receiver*, because of the way in which it "rakes up" the energy in the multipath components. The L correlators used in the rake receiver are often termed "fingers" of the rake.

Chip rate implementation of the rake receiver The basic operation needed for the rake receiver is to correlate the received signal against a shifted version of a spreading waveform. Let us consider a generic operation of this form, which is called *despreading*:

$$Z(\tau) = \int y(t)s^*(t-\tau)dt, \qquad (8.56)$$

where $s(t) = \sum_{l=0}^{N-1} s[l]\psi(t-lT_c)$ is the spreading waveform, and τ is the delay. Define the chip matched filter as

$$\psi_{\mathrm{MF}}(t) = \psi^*(-t).$$

We can now rewrite the decision statistic Z as

$$Z(\tau) = \sum_{l=0}^{N-1} s^*[l] \int y(t)\psi^*(t-lT_c-\tau)dt = \sum_{l=0}^{N-1} s^*[l]Y_\tau[l], \qquad (8.57)$$

where

$$Y_\tau[l] = \int y(t)\psi^*(t-lT_c-\tau)dt = (y*\psi_{\mathrm{MF}})(lT_c+\tau) \qquad (8.58)$$

is the output of the chip matched filter at a suitable delay. Writing the delay $\tau = (D+\delta)T_c$, where $D = \lfloor \tau/T_c \rfloor$ and $\delta = \tau - DT_c \in [0,1)$, we have

$$Y_\tau[l] = (y*\psi_{\mathrm{MF}})((l+D+\delta)T_c). \qquad (8.59)$$

Thus, despreading involves a discrete correlation (8.57) of the spreading sequence $s[l]$ with suitably chosen chip rate samples $\{Y_\tau[l]\}$ at the output of the chip matched filter. The offset for the required samples is determined by the fractional delay parameter δ. We can now implement the rake receiver (8.55) as shown in Figure 8.8, by combining the outputs of L despreaders, one for each significant multipath component.

Defining $D_i = \lfloor \tau_i/T_c \rfloor$ and $\delta_i = (\tau_i/T_c) - D_i \in [0,1)$ as the delay parameters for the ith multipath component, we realize that the required sampling offsets $\delta_i T_c$ vary across multipath components. Thus, a single chip rate sampler cannot directly generate the chip rate samples $\{Y_{\tau_i}[l]\}$ required for the different multipath components. There are several approaches typically used in practice to handle this problem: (a) sample at the chip rate, or perhaps at twice the chip rate, and then interpolate the samples to approximately synthesize samples corresponding to the desired sampling times, (b) sample significantly faster than the sampling rate (e.g., at eight times the chip rate), and choose, for a given multipath component, the subset of chip rate samples whose timing

8.4 Direct sequence spread spectrum

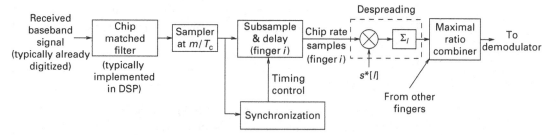

Figure 8.8 Digital-signal-processing-centric implementation of the rake receiver. The output of the chip matched filter is sampled faster than the chip rate. The synchronization circuit provides the timing control required to select a chip rate substream of samples, together with an appropriate delay, before despreading.

most closely approximates the desired delay, (c) use L different chip rate samplers: estimate the delay for each multipath component coarsely, and then lock onto it using a closed loop tracking loop. Such a loop used to be implemented using an analog delay locked loop, but in modern transceivers, DSP-based discrete-time tracking loops might be used, using samples as in (a) or (b).

DSP-centric implementations usually rely on samples obtained using a free-running clock, which is independent of the received signal parameters. These samples are then processed in a manner that depends on received signal parameters such as the delays of the multipath components. Thus, the design of algorithms for such processing requires models for how the samples depend on the received signal parameters. The following example illustrates such modeling.

Example 8.4.1 (Modeling chip rate samples) Let us consider the output of a chip matched filter in response to a delayed spreading waveform $s(t - \tau)$. Setting $s(t) = \sum_l s[l]\psi(t - lT_c)$, we see that the output $y_s = s(t - \tau) * \psi(t)$ can be written as

$$y_s(t) = (s * \psi_{MF})(t - \tau) = \sum_l s[l](\psi * \psi_{MF})(t - lT_c - \tau)$$
$$= \sum_l s[l] r_\psi(t - lT_c - \tau),$$

where

$$r_\psi(t) = (\psi * \psi_{MF})(t)$$

is the autocorrelation function of the chip waveform. Writing $\tau = (D + \delta)T_c$, D an integer and $\delta \in [0, 1)$, we obtain

$$y_s(t) = \sum_l s[l] r_\psi(t - (l+D)T_c - \delta T_c) = \sum_l s[l-D] r_\psi(t - lT_c - \delta T_c)$$

(replacing l by $l+D$ as the dummy variable for summation to get the second equality). For a rectangular chip waveform timelimited to an interval of length T_c, we have

$$r_\psi(t) = \left(1 - \frac{|t|}{T_c}\right) I_{[-T_c, T_c]}$$

(normalizing the chip waveform to unit energy, so that $r_\psi(0) = 1$). Suppose now that we sample the output of the chip matched filter at the chip rate. The signal contribution to these samples is given by

$$y_s(mT_c) = \sum_l s[l - D] r_\psi((m-l)T_c - \delta T_c).$$

For the special case of $\delta = 0$ (i.e., the sample times are chip-aligned with the delayed spreading waveform), we have

$$y_s(mT_c) = s[m - D], \qquad (8.60)$$

since $r_\psi((m-l)T_c) = \delta_{ML}$. For $0 < \delta < 1$, we have

$$r_\psi((m-l)T_c - \delta T_c) = \begin{cases} 1 - \delta, & l = m, \\ \delta, & l = m-1, \end{cases}$$

so that

$$y_s(mT_c) = (1-\delta)s[m - D] + \delta s[m - 1 - D]. \qquad (8.61)$$

That is, each chip rate sample has contributions from two adjacent elements of the spreading sequence. Suppose, for example, that we desire that $y_s(mT_c)$ to be as close to $s[m - D]$ as possible. Then we must make δ, the chip offset, small. For example, if we make $\delta = 1/16$, corresponding to the worst-case chip offset when sampling at rate $8/T_c$, we get that the contribution due to the desired chip is $1 - \delta = 15/16$. The loss in SNR due to the reduced amplitude seen by the desired chip $s[m - D]$ is therefore $20 \log_{10}(16/15) = 0.56$ dB. We ignore the inter-chip interference induced by the contribution of the chip $s[m - 1 - D]$, since this is usually small compared to the dominant sources of impairment that we design for in DS systems, such as multiple-access interference or jamming.

More generally, if ψ is square root Nyquist at rate $1/T_c$, the equality (8.60) holds for *chip-synchronous* sampling (i.e., $\delta = 0$), since $r_\psi(t)$ is Nyquist at rate $1/T_c$. For chip-asynchronous sampling, the signal contribution to the chip rate samples can be written as a discrete-time convolution between the spreading sequence and an equivalent chip rate "channel," as shown in Problem 8.20.

Estimation of the multipath delays $\{\tau_i\}$ is usually done as a two stage process, *timing acquisition*, aimed at obtaining a coarse estimate of the delay, and *tracking*, aimed at refining the delay estimate and updating it as it varies (e.g., due to relative mobility between transmitter and receiver). Timing acquisition algorithms employ hypothesis testing to obtain a coarse estimate of the delay, by quantizing the uncertainty in the delay into a discrete set of hypotheses. Timing acquisition typically occurs prior to carrier synchronization, so that noncoherent reception techniques are often required. We discuss this further in Problem 8.18.

Both rake reception and timing acquisition require the basic operation of correlating the received waveform with a spreading waveform. In practice, a receiver may have a limited number of correlators, or "rake fingers," and may assign a subset of them to demodulation, and a subset to a timing acquisition algorithm whose task is to discover new multipath components when they appear.

8.4.2 Choice of spreading sequences

We have mentioned that DS signaling is designed so that the normalized inner products between delayed versions of spreading waveforms are small. We now show that these inner products can be made small by proper design of the discrete-time correlation functions for the corresponding spreading sequences. The problem of waveform design thus reduces to one of spreading sequence design, and we briefly mention some common approaches for the latter.

Consider two spreading waveforms $u(t)$ and $v(t)$, given by

$$u(t) = \sum_{l=0}^{N-1} u[l]\psi(t - lT_c), \tag{8.62}$$

$$v(t) = \sum_{l=0}^{N-1} v[l]\psi(t - lT_c).$$

Crosscorrelation functions Let us define a continuous-time crosscorrelation function for these spreading waveforms as follows:

$$R_{u,v}(\tau) = \int u(t)v^*(t-\tau)dt. \tag{8.63}$$

In addition, let us define a discrete-time crosscorrelation function for the corresponding spreading sequences as

$$R_{u,v}[n] = \sum_l u[l]v^*[l-n]. \tag{8.64}$$

For convenience, we set $u[l] = v[l] = 0$ for $l < 0$ and $l > N-1$. This allows us to leave unspecified the range of the index l in the summation (8.64): in fact, nontrivial elements of u and v feature in the summation only over the range $n \leq l \leq N-1$.

Autocorrelation functions Setting $u = v$, we specialize to the continuous-time and discrete-time autocorrelation functions,

$$R_u(\tau) = R_{u,u}(\tau) = \int u(t)u^*(t-\tau)dt, \tag{8.65}$$

$$R_u[n] = R_{u,u}[n] = \sum_l u[l]u^*[l-n]. \tag{8.66}$$

We now express the continuous-time crosscorrelation function in terms of the discrete-time crosscorrelation function for the special case of a rectangular chip waveform timelimited to an interval of length T_c.

$$R_{u,v}(\tau) = \sum_l \sum_k u[l]v^*[k] \int \psi(t - lT_c)\psi^*(t - kT_c - \tau)dt$$

$$= \sum_l \sum_k u[l]v^*[k] r_\psi((k-l)T_c + \tau).$$

Setting $\tau = (D+\delta)T_c$ as before, where $D = \lfloor \tau/T_c \rfloor$ is an integer, and $\delta \in [0, 1)$, we get

$$R_{u,v}(\tau) = \sum_{l,k} u[l]v^*[k] r_\psi((k+D-l)T_c + \delta T_c). \tag{8.67}$$

For the rectangular chip waveform, we have

$$r_\psi((k+D-l)T_c + \delta T_c) = \begin{cases} 1-\delta, & k+D-l = 0, \\ \delta, & k+D-l = -1, \\ 0, & \text{else.} \end{cases} \tag{8.68}$$

This means that the only nonzero terms in (8.67) correspond to $k = l - D$ and $k = l - D - 1$. Substituting (8.68) into (8.67), we obtain

$$R_{u,v}(\tau) = (1-\delta)\left(\sum_l u[l]v^*[l-D]\right) + \delta\left(\sum_l u[l]v^*[l-D-1]\right)$$

$$= (1-\delta)R_{u,v}[D] + \delta R_{u,v}[D+1]. \tag{8.69}$$

The preceding expression shows that the continuous-time crosscorrelation function can be made small for an arbitrary delay τ by making the discrete-time crosscorrelation function small (on average) for all integer delays different from zero. By specializing to $u = v$, we see that this observation holds for autocorrelation functions as well. While these conclusions are based on a rectangular timelimited chip waveform, Problem 8.20 generalizes (8.69) to arbitrary chip waveforms ψ.

Pseudorandom spreading sequences A common approach is to employ spreading sequences which are either aperiodic, or periodic with period much longer than the processing gain N. In this case, the section of the spreading sequence corresponding to a single symbol is well modeled as consisting of randomly chosen, i.i.d., elements (e.g., ± 1 with equal probability for binary spreading sequences). We term this the *random spreading sequence model*. A convenient way to generate sequences of very long periods is to use maximum length shift register (MLSR) sequences: a shift register of length m, together with some combinatorial logic, can be used to generate a periodic sequence of period $2^m - 1$ (the IS-95 digital cellular standard employs $m = 42$, which generates a sequence of period $2^{42} - 1$ in excess of 4 billion). Such sequences are often termed *pseudorandom*, or pseudonoise (PN) sequences. See Problem 8.19 for a simple example of an MLSR sequence. The CDMA-based digital cellular standard IS-95 employs long spreading waveforms, with the spreading waveforms for multiple users generated from a single PN sequence of very long period simply by assigning different delayed versions of the sequence to different users.

For the random spreading sequence model, the crosscorrelation function between two pseudorandom sequences u and v of length N can be modeled as follows:
$$R_{u,v}[n] = \sum_{l} u[l]v^*[l-n] = \sum_{l=n}^{N-1} u[l]v^*[l-n].$$
When $\{u[l]\}$, $\{v[l]\}$ are modeled as i.i.d., symmetric Bernoulli, $R_{u,v}[n]$ can be modeled as a sum of $N-n$ i.i.d., symmetric Bernoulli random variables.

Note that PN sequences of shorter period (e.g., period 31 corresponding to $m = 5$), and sequences constructed from PN sequences such as Gold sequences, can be useful for CDMA systems employing short spreading waveforms.

Barker sequences For a binary sequence u, the smallest possible magnitudes for the autocorrelation function $R_u[n]$ correspond to $R_u[n] = 0$ for even n, and $R_u[n] = \pm 1$ for odd n. Barker sequences are binary sequences that achieve these lower bounds. Binary Barker sequences are only known to exist for lengths 5, 7, 11, 13. However, if the sequence elements are allowed to take on complex values, then it is possible to find Barker-like sequences with excellent autocorrelation properties for other lengths as well. A Barker sequence of length 11 is employed in the DS waveform used for the 1 and 2 Mbps rates in the IEEE 802.11b WLAN standard. This is an example of a DS system in which all users employ the same short spreading waveform. If different users arrive at the receiver with different delays, it is possible that the receiver can still demodulate the user of interest successfully by locking onto the spreading waveform at the delay corresponding to that user. The good autocorrelation properties of the Barker sequence help with this so-called *capture* effect.

8.4.3 Performance of conventional reception in CDMA systems

The conventional rake receiver is derived ignoring ISI and MAI. However, in a CDMA system, the performance of the rake receiver is significantly affected by MAI (self-interference due to ISI is small, and can typically be neglected). We now illustrate this in an idealized scenario, via the example of a *synchronous* CDMA system in which the received signal for a given symbol is given by
$$y(t) = \sum_{k=1}^{K} b_k A_k s_k(t) + n(t), \qquad (8.70)$$

where $\{b_k\}$ are BPSK symbols, $s_k(t) = \sum_{l=0}^{N-1} s_k[l]\psi(t - lT_c)$ are spreading waveforms assigned to K users, and A_k is a complex gain corresponding to user k. Assume for simplicity that all signals are real-valued, and that the elements $\{s_k[l]\}$ of the spreading sequences are i.i.d., taking values ± 1 with probability 1/2. Suppose that User 1 is the *desired* user, and that the receiver employs the matched filter statistic

$$Z_1 = \int y(t)s_1(t)dt = \sum_{k=1}^{K} b_k \int s_k(t)s_1(t)dt + \int n(t)s_1(t)dt.$$

For ψ square root Nyquist at rate $1/T_c$, normalized to $r_\psi(0) = ||\psi||^2 = 1$, we have

$$Z_1 = \sum_{k=1}^{K} b_k A_k R_{s_k, s_1}[0] + N_1 = b_1 A_1 N + \sum_{k=2}^{K} b_k A_k R_{s_k, s_1}(0) + N_1,$$

where $N_1 \sim N(0, \sigma^2 N)$. The MAI is determined by the crosscorrelation functions

$$R_{s_k, s_1}[0] = \sum_{l=0}^{N-1} s_k[l]s_1[l], \quad k = 2, \ldots, K.$$

Note that, if one of the interferers has a significantly larger power than that of the desired user, then the sign of the decision statistic can be dominated by the signal of b_k (we need $|A_k R_{s_k, s_1}[0]| \gg |A_1 N|$ for this to happen). This is the so-called *near–far problem* incurred by conventional matched filter reception. The CDMA-based digital cellular systems are particularly vulnerable to the near–far problem on the uplink, since different mobiles may incur different propagation losses to the base station. For example, the d^{-4} propagation loss often encountered in wireless systems (where d is the distance between transmitter and receiver) means that a factor of four difference in distance translates to a 24 dB difference in received power if there is no power control. Typical processing gains of the order of 10 dB certainly cannot overcome this difference. Such systems must, therefore, employ power control to ensure that the signals from different mobiles are received at the base station with roughly equal powers.

Let us now develop a Gaussian approximation to the error probability. For $\{s_k[l]\}$ i.i.d., taking values ± 1 with equal probability, the terms $s_k[l]s_1[l]$, $k = 2, \ldots, K$, are also i.i.d., taking values ± 1 with equal probability. Thus, for N large, we can approximate, using the central limit theorem, each $R_{s_k, s_1}(0)$ as a Gaussian random variable, with mean zero and variance N. Thus, we can approximately model $b_k A_k R_{s_k, s_1}(0)$ as $N(0, A_k^2 N)$. Under our model, the contribution to the MAI due to different users is independent, and the MAI is independent of the noise. The sum of the MAI and noise can therefore be modeled as a Gaussian random variable $N(0, v^2)$ whose variance is the sum of the variances of its components, given by

$$v^2 = \sum_{k=2}^{K} A_k^2 N + \sigma^2 N.$$

We therefore have the following model for the matched filter decision statistic

$$Z_1 = b_1 A_1 N + N(0, v^2).$$

The decision rule

$$\hat{b}_1 = \text{sign}(Z_1)$$

therefore incurs an error probability $P_e \approx Q(A_1 N/v)$. Noting that $E_b = A^2 N$ and that $\sigma^2 = N_0/2$, the preceding expression simplifies to

$$P_e \approx Q\left(\sqrt{\frac{1}{(2E_b/N_0)^{-1} + \frac{1}{N}\sum_{k=2}^{K} A_k^2/A_1^2}}\right). \qquad (8.71)$$

We can see the effect of the near–far problem from (8.71): if $A_k^2/A_1^2 \to \infty$ for some $k \neq 1$, then $P_e \to 1/2$.

Under perfect power control (all amplitudes equal), the expression (8.71) specializes to

$$P_e \approx Q\left(\sqrt{\frac{1}{\left(\frac{2E_b}{N_0}\right)^{-1} + \frac{K-1}{N}}}\right). \qquad (8.72)$$

Note that the system performance is interference-limited: even if E_b/N_0 is large, the probability of error cannot be better than $Q\left(\sqrt{\frac{N}{K-1}}\right)$.

In practice, the signals corresponding to different users are not aligned at the receiver at either the chip or the symbol level. Assuming that the receiver is locked to the desired User 1, the interference due to the interfering users is attenuated by this lack of chip alignment, in a manner that depends on the chip waveform ψ. Performance analysis of matched filter reception for such *asynchronous* CDMA systems is explored in Problem 8.21.

8.4.4 Multiuser detection for DS-CDMA systems

Multiuser detection refers to reception techniques that exploit the structure of the MAI in receiver design, rather than ignoring it as in conventional rake reception. An idealized setting suffices to illustrate the basic concepts, hence we consider a K-user discrete-time, real baseband, synchronous CDMA system with BPSK modulation. The N-dimensional received vector $\mathbf{r}$ over a given symbol interval is given by

$$\mathbf{r} = \sum_{k=1}^{K} A_k b_k \mathbf{s}_k + \mathbf{W}, \qquad (8.73)$$

where, for $1 \leq k \leq K$, $b_k \in \{-1, +1\}$ is the symbol for user k, A_k its amplitude, and $\mathbf{s}_k$ its spreading vector, normalized for convenience to unit energy. The vector $\mathbf{W} \sim N(0, \sigma^2 \mathbf{I})$ is WGN.

The analogy between MAI and ISI is immediate, from a comparison of the MAI model (8.73) and the ISI model (5.25) developed in Chapter 5. However, there are two major differences between the MAI and ISI models:

(i) In the MAI model, the interference vectors can be arbitrary. In the ISI model, they are restricted to being acyclic shifts of the channel impulse response.

(ii) In the MAI model, the amplitudes $\{A_k, k \neq 1\}$ for the interferers can scale independently of the amplitude A_1 of the desired user, and can in fact be much larger than A_1 (e.g., if an interfering transmitter is closer to the receiver than the desired transmitter in a system without power control). We would therefore like our multiuser detection schemes to be *near–far resistant*, i.e., to provide good performance even in the presence of such a near–far problem.

Example 8.4.2 (ML reception for a two-user system) We take as our running example a two-user system with received vector given by

$$\mathbf{r} = A_1 b_1 \mathbf{s}_1 + A_2 b_2 \mathbf{s}_2 + \mathbf{W}, \tag{8.74}$$

where b_1, b_2 are ± 1 BPSK symbols, and the signal vectors are normalized to unit energy: $\|\mathbf{s}_1\|^2 = \|\mathbf{s}_2\|^2 = 1$. We denote the signal correlation as $\rho = \langle \mathbf{s}_1, \mathbf{s}_2 \rangle$. The matched filters for the two users produce the outputs

$$\begin{aligned} z_1 &= \langle \mathbf{r}, \mathbf{s}_1 \rangle = A_1 b_1 + \rho A_2 b_2 + N_1, \\ z_2 &= \langle \mathbf{r}, \mathbf{s}_2 \rangle = \rho A_1 b_1 + A_2 b_2 + N_2, \end{aligned} \tag{8.75}$$

where $N_1 \sim N(0, \sigma^2)$, $N_2 \sim N(0, \sigma^2)$ are jointly Gaussian with $\text{cov}(N_1, N_2) = \sigma^2 \rho$. Conventional reception simply takes the sign of the matched filter outputs:

$$\hat{b}_{1,\text{MF}} = \text{sign}(z_1), \quad \hat{b}_{2,\text{MF}} = \text{sign}(z_2). \tag{8.76}$$

We henceforth term such a receiver the *matched filter (MF) receiver*. For $\rho \neq 0$, each matched filter output is corrupted by interference, so that the matched filter receiver is suboptimal. Let us now consider joint ML reception for the two users; that is, we wish to decide on $\mathbf{b} = (b_1, b_2)^T$. To this end, rewrite (8.74) as

$$\mathbf{r} = \mathbf{s_b} + \mathbf{W}, \tag{8.77}$$

where

$$\mathbf{s_b} = A_1 b_1 \mathbf{s}_1 + A_2 b_2 \mathbf{s}_2.$$

The ML rule must maximize the log likelihood ratio, which is proportional to

$$\Lambda(\mathbf{b}) = \langle \mathbf{r}, \mathbf{s_b} \rangle - \frac{1}{2} \|\mathbf{s_b}\|^2. \tag{8.78}$$

Since

$$\begin{aligned} \langle \mathbf{r}, \mathbf{s_b} \rangle &= A_1 b_1 \langle \mathbf{r}, \mathbf{s}_1 \rangle + A_2 b_2 \langle \mathbf{r}, \mathbf{s}_2 \rangle \\ &= A_1 b_1 z_1 + A_2 b_2 z_2, \end{aligned} \tag{8.79}$$

we realize that, in order to compute $\Lambda(\mathbf{b})$ for all possible $\mathbf{b}$ (four possible values in our case), it is necessary and sufficient to compute the matched filter statistics z_1 and z_2. The problem with conventional reception is that these statistics are being used separately as in (8.76), rather than jointly according to the ML rule

$$\hat{\mathbf{b}}_{\text{ML}} = \arg \max_{\mathbf{b}} \langle \mathbf{r}, \mathbf{s_b} \rangle - \frac{1}{2} \|\mathbf{s_b}\|^2. \tag{8.80}$$

Let us get the ML rule into a more explicit form. Consider the second term above:

$$\begin{aligned}
\|\mathbf{s_b}\|^2 &= \|A_1 b_1 \mathbf{s}_1 + A_2 b_2 \mathbf{s}_2\|^2 \\
&= A_1^2 b_1^2 \|\mathbf{s}_1\|^2 + A_2^2 b_2^2 \|\mathbf{s}_2\|^2 + 2A_1 A_2 b_1 b_2 \langle \mathbf{s}_1, \mathbf{s}_2 \rangle \\
&= A_1^2 + A_2^2 + 2A_1 A_2 b_1 b_2 \rho.
\end{aligned} \tag{8.81}$$

Throwing away terms independent of $\mathbf{b}$, we obtain, using (8.79) and (8.81) in (8.80), that

$$(\hat{b}_1, \hat{b}_2)_{\text{ML}} = \arg \max_{b_1, b_2} A_1 b_1 z_1 + A_2 b_2 z_2 - A_1 A_2 b_1 b_2 \rho. \tag{8.82}$$

By writing out and comparing the terms above for the four possible values of $(b_1, b_2) = (\pm 1, \pm 1)$, we get the ML decision regions as a function of z_1, z_2. For example, consider $A_1 = 1, A_2 = 2$ and $\rho = -1/2$. The ML decision is $(+1, +1)$ if the following three inequalities hold:

$$\begin{aligned}
z_1 + 2z_2 + 1 &> z_1 - 2z_2 - 1, \\
z_1 + 2z_2 + 1 &> -z_1 - 2z_2 + 1, \\
z_1 + 2z_2 + 1 &> -z_1 + 2z_2 - 1,
\end{aligned}$$

which reduces to $z_2 > -1/2$, $z_1 + 2z_2 > 0$ and $z_1 > -1$. By doing this for all possible values of (b_1, b_2), we obtain the decision regions shown in Figure 8.9. In contrast, the MF decision regions are simply the four quadrants in the figure, since they do not account for the correlation between the spreading waveforms. The relative performance of the MF and ML rules is discussed in Example 8.4.3.

One of the goals of multiuser detection is to ensure that, as long as there are enough dimensions in the signal space, the receiver performance is limited by noise rather than by interference. Performance measures that are useful for quantifying progress towards this goal are the asymptotic efficiency and near–far resistance, defined in the following for a multiuser system over an AWGN channel with noise variance σ^2 per dimension.

Asymptotic efficiency Let P_e denote the error probability attained (for a specific desired user) by a given receiver, and let $P_{e,\text{su}}$ denote the single-user

Figure 8.9 Decision regions for ML reception with two users: $A_1 = 1$, $A_2 = 2$ and $\rho = -1/2$ in Example 8.4.2.

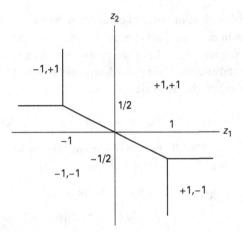

error probability for that user if there were no MAI. The asymptotic efficiency η is the ratio of the exponent of decay with MAI, relative to that in the single-user system, as the SNR gets large. That is,

$$\eta = \lim_{\sigma^2 \to 0} \frac{\log P_e}{\log P_{e,su}}. \tag{8.83}$$

For example, if $P_e \doteq e^{-a/\sigma^2}$ and $P_{e,su} \doteq e^{-b/\sigma^2}$, where $a, b \geq 0$ are exponents of decay, then $\eta = a/b$. If P_e does not decay exponentially with SNR, then we have $a = 0$ and hence $\eta = 0$.

Example 8.4.3 (Asymptotic efficiency of ML and MF reception) Consider the two-user system in Example 8.4.2, with $A_1 = 1$, $A_2 = 2$ and $\rho = -1/2$. We wish to compute the asymptotic efficiency for User 1. Without loss of generality, condition on $b_1 = +1$. In this case, we see from (8.75) that

$$z_1 = A_1 + \rho A_2 b_2 + N_1.$$

Since we compare z_1 to zero to make our decision, the asymptotic error probability is given by the minimum distance to the decision boundary. This equals A_1 if there is no MAI (i.e., if $\rho = 0$ or $A_2 = 0$), so that

$$P_{e,su} = Q\left(\frac{A_1}{\sigma}\right) \doteq e^{-A_1^2/\sigma^2}.$$

If there is MAI, then the distance from the decision boundary depends on b_2. Averaging over b_2, we get

$$P_e = \frac{1}{2} Q\left(\frac{A_1 + \rho A_2}{\sigma}\right) + \frac{1}{2} Q\left(\frac{A_1 - \rho A_2}{\sigma}\right).$$

Since $Q(x) \doteq e^{-x^2/2}$, the asymptotic performance is achieved by the worst-case argument of the Q function, which is given by $A_1 - |\rho A_2|$. If this is negative, then the error probability is bounded away from zero, and the asymptotic efficiency is zero. We can now infer that the asymptotic efficiency of matched filter reception for User 1 is given by

$$\eta_{\text{MF}} = \begin{cases} \frac{(A_1 - |\rho A_2|)^2}{A_1^2} = \left(1 - |\rho|\frac{A_2}{A_1}\right)^2, & |\rho| < \frac{A_1}{A_2}, \\ 0, & \text{else}. \end{cases} \quad (8.84)$$

For the ML receiver, the error probability of User 1 at high SNR is determined by the minimum distance $d_{\min}$ between the two signal sets $\mathbf{S}_{+1} = \{\mathbf{s_b} : b_1 = +1\}$ and $\mathbf{S}_{-1} = \{\mathbf{s_b} : b_1 = -1\}$. That is, $P_e \doteq Q\left(\frac{d_{\min}}{2\sigma}\right)$. Note that, if there is no MAI, we have $\mathbf{S}_{+1} = \{A_1 \mathbf{s}_1\}$ and $\mathbf{S}_{-1} = \{-A_1 \mathbf{s}_1\}$, so that $d_{\min,\text{su}} = 2A_1 \|\mathbf{s}_1\| = 2A_1$, which gives us $P_{e,\text{su}} = Q(A_1/\sigma)$ as before. When MAI is present, a typical element of $\mathbf{S}_{+1}$ takes the form $\mathbf{u} = A_1 \mathbf{s}_1 + b_2 A_2 \mathbf{s}_2$ and a typical element of $\mathbf{S}_{-1}$ takes the form $\mathbf{v} = -A_1 \mathbf{s}_1 + b'_2 A_2 \mathbf{s}_2$. The distance between these two signals is given by

$$d^2 = \|u - v\|^2 = \|2A_1 \mathbf{s}_1 + (b_2 - b'_2) A_2 \mathbf{s}_2\|^2.$$

For $b_2 = b'_2$, we have $d^2 = 4A_1^2 = d_{\min,\text{su}}^2$. For $b_2 = -b'_2$, we get

$$d^2 = 4(A_1^2 + A_2^2 + 2b_2 \rho A_1 A_2) = 4(A_1^2 + A_2^2 \pm 2\rho A_1 A_2),$$

whose minimum value is $4A_1^2 \left(1 + A_2^2/A_1^2 - 2|\rho|A_2/A_1\right)$. Thus, the asymptotic efficiency for ML reception is given by

$$\eta_{\text{ML}} = \min\{1, 1 + A_2^2/A_1^2 - 2|\rho|A_2/A_1\}. \quad (8.85)$$

It can be checked that, as the interference strength increases (i.e., as $A_2/A_1 \to \infty$), we have $\eta_{\text{MF}} \to 0$ and $\eta_{\text{ML}} \to 1$ (see Figure 8.11). That is, while the MF receiver performs poorly when there is a near–far problem, the ML receiver approaches single-user performance. An intuitive interpretation of the latter is that, when User 2 is much stronger, the ML receiver can first make a reliable decision on b_2 based on z_2 (since the MAI due to User 1 is weak), and then subtract out the MAI due to User 2 from z_1. Thus, the reliability of demodulation for b_1 becomes noise-limited rather than interference-limited.

The preceding example shows the power of multiuser detection: by exploiting the structure of the MAI, we can get exponential decay of error probability with SNR, rather than encountering an interference floor as with MF reception. The asymptotic efficiency of the ML receiver, however, depends on the relative amplitudes of the users, as shown in the example of Figure 8.11. Notice that the asymptotic efficiency achieves a minimum value as we vary A_2/A_1. We term this the *near–far resistance*, and formally define it as follows.

Near–far resistance The near–far resistance is the minimum value of the asymptotic efficiency as we vary the amplitudes of the interfering users.

For the two-user system in Examples 8.4.2 and 8.4.3, it can be shown that the minimum value of the asymptotic efficiency occurs at $A_2/A_1 = |\rho|$, which yields

$$\eta_{\text{ML,nf}} = 1 - |\rho|^2. \tag{8.86}$$

Thus, the near–far resistance is strictly positive for $|\rho| < 1$. For $\rho = 1$, $\mathbf{s}_1$ and $\mathbf{s}_2$ are scalar multiples of each other, and the near–far resistance is equal to zero.

The preceding results are easily generalized to more than two users, as shown in Problem 8.25. Note here that the complexity of ML reception for K users is exponential in K, so that ML reception does not scale well as the number of users increases. It therefore becomes necessary to consider suboptimal strategies, just as we did for equalization. In particular, let us consider a linear receiver for the K-user system (8.73), the decision statistic for which is of the form $Z = \langle \mathbf{c}, \mathbf{r} \rangle$, where $\mathbf{c}$ is a correlator to be designed. For BPSK signaling, hard decisions for User 1 would be of the form

$$\hat{b}_1 = \text{sign}(Z) = \text{sign}(\langle \mathbf{c}, \mathbf{r} \rangle).$$

(Since the correlator scaling is unimportant for BPSK signaling, we are not careful about it in the following development.)

For the model (8.73), the output of a linear receiver is of the form

$$Z = A_1 b_1 \langle \mathbf{c}, \mathbf{s}_1 \rangle + \sum_{k \neq 1} A_k b_k \langle \mathbf{c}, \mathbf{s}_k \rangle + \langle \mathbf{c}, \mathbf{W} \rangle. \tag{8.87}$$

The first term is the desired term, the second term is the MAI contribution, and the third term is the noise contribution. As in Chapter 5, we define the signal-to-interference ratio (SIR) as the ratio of the energy of the desired term to those of the undesired interference and noise terms, as follows:

$$\text{SIR} = \frac{A_1^2 |\langle \mathbf{c}, \mathbf{s}_1 \rangle|^2}{\sum_{k \neq 1} A_k^2 |\langle \mathbf{c}, \mathbf{s}_k \rangle|^2 + \sigma^2 \|\mathbf{c}\|^2}. \tag{8.88}$$

Let us also define the signal space $\mathbf{S}$ as the space spanned by the signal vectors $\{\mathbf{s}_1, \mathbf{s}_2, \ldots, \mathbf{s}_K\}$, and the interference subspace $\mathbf{S}_I$ as the space spanned by the interference vectors $\{\mathbf{s}_2, \ldots, \mathbf{s}_K\}$.

Zero-forcing, or decorrelating, detector The zero-forcing (ZF) detector for User 1 is a correlator $\mathbf{c} = a\mathbf{P}_I^\perp \mathbf{s}_1$ (a is an arbitrary scalar), where $\mathbf{P}_I^\perp \mathbf{s}_1$ is the projection of $\mathbf{s}_1$ orthogonal to $\mathbf{S}_I$. The latter projection exists if and only if the desired signal $\mathbf{s}_1$ is linearly independent of $\mathbf{s}_2, \ldots, \mathbf{s}_K$. The ZF detector knocks out the MAI at the expense of noise enhancement (exactly as for the ZF equalizer in Chapter 5), as shown in Figure 8.10. For $\mathbf{c} = \mathbf{P}_I^\perp \mathbf{s}_1$, we have

$$\langle \mathbf{c}, \mathbf{r} \rangle = A_1 b_1 \|\mathbf{P}_I^\perp \mathbf{s}_1\|^2 + N(0, \sigma^2 \|\mathbf{P}_I^\perp \mathbf{s}_1\|^2),$$

Figure 8.10 The ZF correlator is a scalar multiple of the projection of the desired spreading vector orthogonal to the interference subspace.

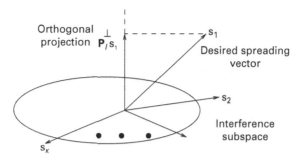

from which it can be inferred that the asymptotic efficiency is

$$\eta_{ZF} = \frac{||\mathbf{P}_I^\perp \mathbf{s}_1||^2}{||\mathbf{s}_1||^2} = ||\mathbf{P}_I^\perp \mathbf{s}_1||^2. \tag{8.89}$$

That is, the asymptotic efficiency of the ZF detector is the fraction of the desired signal energy orthogonal to the interference subspace. It is also equal to the near–far resistance, since it does not depend on the interference amplitudes.

Example 8.4.4 (ZF detector for two users) For the two-user system in Example 8.4.2, we have

$$\mathbf{P}_I^\perp \mathbf{s}_1 = \mathbf{s}_1 - \langle \mathbf{s}_1, \mathbf{s}_2 \rangle \mathbf{s}_2 = \mathbf{s}_1 - \rho \mathbf{s}_2$$

and the asymptotic efficiency or near–far resistance of the ZF detector is

$$\eta_{ZF} = ||\mathbf{P}_I^\perp \mathbf{s}_1||^2 = 1 - |\rho|^2. \tag{8.90}$$

Comparing with (8.85) and (8.86), we see that the ML and ZF detectors have the same near–far resistance (i.e., they have the same worst-case asymptotic performance). However, the ML detector has a higher asymptotic efficiency than the ZF detector for all values of A_2/A_1 except for the minimizing value of $A_2/A_1 = \rho$. See Figure 8.11. This is the price we pay for suboptimal reception.

Note that both the ML and ZF detectors require knowledge of the signal vectors for both the desired and interfering users. (In addition, the ML receiver requires the signal amplitudes and noise variance.) The matched filter receiver for a given user, on the other hand, only needs to know the signal vector corresponding to that user. The centralized knowledge of the system parameters required by multiuser detection schemes is problematic in many settings. We now discuss the linear MMSE receiver, a multiuser detection scheme which is amenable to decentralized adaptive implementation.

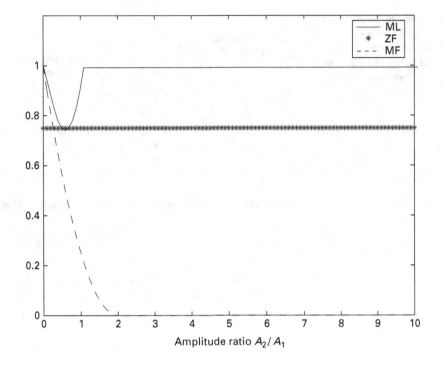

Figure 8.11 Asymptotic efficiency for User 1 as a function of the ratio of amplitudes A_2/A_1 for normalized correlation $\rho = \pm 0.5$. The ML and ZF detectors are near–far resistant, while the near–far resistance for the MF goes to zero once $A_2 \geq 2A_1$.

The MMSE receiver for User 1 is a linear correlator minimizing the MSE, given by

$$\text{MSE} = \mathbb{E}[|\langle \mathbf{c}, \mathbf{r} \rangle - b_1|^2].$$

The MMSE solution can be shown (see Chapter 5) to be of the form

$$\mathbf{c}_{\text{MMSE}} = \mathbf{R}^{-1}\mathbf{p}, \qquad (8.91)$$

where (specializing to real-valued signals and symbols)

$$\mathbf{R} = \mathbb{E}[\mathbf{rr}^T], \quad \mathbf{p} = \mathbb{E}[b_1 \mathbf{r}]. \qquad (8.92)$$

For the model (8.73), we obtain (assuming i.i.d. BPSK symbols with equal probability of ± 1)

$$\begin{aligned}\mathbf{R} &= \sum_{k=1}^{K} A_k^2 \mathbf{s}_k \mathbf{s}_k^T + \sigma^2 \mathbf{I}, \\ \mathbf{p} &= A_1 \mathbf{s}_1.\end{aligned} \qquad (8.93)$$

Two important properties of the MMSE receiver are as follows:

- The MMSE receiver tends to the ZF receiver as the noise variance gets small (we had already noted this for ISI channels in Chapter 5). Thus, its asymptotic efficiency equals that of the ZF receiver.

8.4 Direct sequence spread spectrum

- Even when the noise variance is finite, if the relative strength of an interferer relative to the desired signal gets large, then the MMSE receiver forces its contribution to the output to zero: if $A_k/A_1 \to \infty$, then $\langle \mathbf{c}_{\text{MMSE}}, A_k \mathbf{s}_k \rangle \to 0$.

These properties show that the MMSE receiver is near–far resistant, like a good multiuser detection scheme should be. They are derived for a two-user system in Problem 8.22.

The MMSE correlator (8.91) depends on the signal vectors, amplitudes, and noise variance, and hence can be implemented if we have centralized knowledge of all of these parameters. However, a more attractive decentralized implementation is to adaptively compute it based on received vectors over multiple observation intervals. The received vector for observation interval m is given by

$$\mathbf{r}[m] = \sum_{k=1}^{K} A_k b_k[m] \mathbf{s}_k + \mathbf{W}[m]. \qquad (8.94)$$

This corresponds to a system with short spreading sequences, where the signal vectors for each observation interval are the same. We can now estimate $\mathbf{R}$ and $\mathbf{p}$ in (8.92) as empirical averages for a least squares implementation (again, see Chapter 5 for more detailed discussion):

$$\hat{\mathbf{R}} = \frac{1}{M} \sum_{m=1}^{M} \mathbf{r}[m](\mathbf{r}[m])^T, \quad \hat{\mathbf{p}} = \frac{1}{M} \sum_{m=1}^{M} b_1[m] \mathbf{r}[m]. \qquad (8.95)$$

Other adaptive implementations such as LMS and RLS are also possible. The computation of $\hat{\mathbf{R}}$ is purely based on the received vectors, while the computation of $\hat{\mathbf{p}}$ requires a known training sequence $\{b_1[m]\}$ for the desired user. Since no explicit knowledge of interference parameters is required, the MMSE receiver can be implemented in completely decentralized fashion for each user.

Other suboptimal techniques include *interference cancellation* schemes combining linear reception with decision feedback. See Problem 8.23 for a simple example.

While the preceding development has been carried out for a synchronous, discrete-time system, the ideas apply more generally, as illustrated in some of the problems. However, there remain significant implementation challenges before the promise of multiuser detection can be realized for CDMA systems. In a system with many users (more than the signal space dimension), the benefits of multiuser detection over conventional matched filter reception become less pronounced. On the other hand, in systems without power control (for which multiuser detection provides the greatest performance gains over conventional reception), the dynamic range of analog-to-digital conversion must be large enough that the signal from weak users is not washed out by the quantization noise. Indeed, the first major commercial application of multiuser detection techniques may well be spatial multiplexing in single-user space–time communication systems, as discussed in Section 8.7.3, where the multiple "users" correspond to different transmit antenna elements at the same node.

8.5 Frequency hop spread spectrum

Frequency hop (FH) spread spectrum is conceptually quite simple: use a standard narrowband signaling scheme, but change the carrier frequency according to some *hopping pattern*. If multiple symbols are sent over each hop, we use the term "slow" FH, while if there is one symbol per hop, we use the term "fast" FH. Frequency hopping can provide frequency diversity to combat frequency-selective fading, as in the GSM digital cellular standard. It can also be used to provide randomized multiple access, in which different users use different hopping patterns (or different phases of the same hopping pattern), as in the Bluetooth wireless personal area networking standard. Frequency hop systems are also difficult to jam for an adversary who does not know the hopping pattern (thereby forcing the adversary to spread its jamming energy over a wide band), which makes them attractive for military communication.

Phase synchronization is difficult to acquire in FH systems because of the loss of phase coherence when hopping from one frequency to the next. Thus, noncoherent or differentially coherent modulation are attractive design choices for FH systems, since they avoid the expenditure of pilot overhead at the beginning of each hop for acquiring explicit phase estimates. Another common design choice is to apply error and erasure correction coding across hops. This is especially effective for combating frequency-selective impairments due to fading, multiple-access interference or "partial band" jamming (in which a fraction of the band is jammed). In particular, concatenated coding strategies can be very useful for FH systems with frequency-selective impairments. If we have a mechanism for detecting hops that are severely impaired, then it is efficient to erase such hops, since it is less expensive to deal with erasures than with errors. An outer code optimized primarily for erasures can be used to handle these frequency-selective impairments. Example 8.5.1 shows how Reed–Solomon (RS) codes can be used for this purpose. The random errors within hops that see "normal conditions" (e.g., background noise) can be cleaned up by an inner code. For example, an inner convolutional code, together with an outer RS code, might be an effective strategy. Of course, with the recent developments in turbo-like coded modulation, it is possible to design a single code for correcting both errors and erasures (e.g., by suitably optimizing the degree sequence for an irregular LDPC code).

We consider a simple example illustrating the efficacy of coding across hops in the following example, in which RS coding across hops is used to combat multiple-access interference. Figure 8.12 illustrates that "hits" are rare for well designed hopping patterns.

Figure 8.12 Frequency hop users employ hopping patterns that collide occasionally.

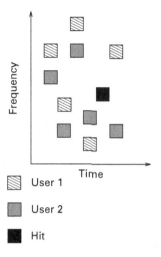

Example 8.5.1 (Coded frequency hop multiple access) Consider a K-user FH multiple-access system in which each user chooses a random hopping pattern, with the frequency for each hop being chosen equiprobably and independently from among q frequencies. Assume that all users hop in synchronized fashion, and focus on the performance of a specific user, say User 1. A "hit" occurs in a particular hop for User 1 when some other user has hopped to the same frequency. The probability of a hit is given by

$$p_{\text{hit}} = 1 - \left(\frac{q-1}{q}\right)^{K-1}. \tag{8.96}$$

Assuming that we design the modulation scheme and link budget for background noise, a hit would lead to very poor demodulation performance in that hop. Thus, it makes sense to simply erase all symbols in a hop that is hit. There are a number of mechanisms that can be used for detecting hits, which we do not discuss here. For simplicity, let us assume that hits can be detected perfectly, and that there are no demodulation errors in hops that are not hit (e.g., by the use of an error correction code across the symbols within a hop). Then a code operating across hops simply has to correct erasures due to hits. An example of such a code, that has been employed in the FH-based SINCGARS packet radio used by the US military, is a Reed–Solomon code (see Section 7.5): an (n, k) RS code has n code symbols, k information symbols, all drawn from an alphabet of size Q. The block length n is constrained by the alphabet size: $n \leq Q - 1$ (we can actually go up for $n = Q$ for an extended RS code). Thus, RS codes operate on fairly large alphabets; for example, we can use eight bits per code symbol, corresponding to $Q = 2^8 = 256$, and constraining $n \leq 255$. An (n, k) RS code can correct up to $n - k$ erasures; that is, we

> can decode correctly as long as we get any k of the n symbols. If the number of erasures is larger than $n-k$, then the RS decoder gives up. The probability P_F of such decoding failure is given by
>
> $$P_F = \sum_{h=n-k+1}^{n} \binom{n}{h} p_{\text{hit}}^h (1-p_{\text{hit}})^{n-h}.$$
>
> Problem 8.26 discusses numerical examples illustrating design choices for RS code parameters as a function of K and q.

8.6 Continuous phase modulation

Continuous phase modulation (CPM) encodes data in a signal of the form $e^{j\theta(t)}$, where $\theta(t)$ is a continuous function of t. The corresponding passband signal is of the form $u_p(t) = \cos(2\pi f_c t + \theta(t))$. The constant envelope of this signal means that we can recover it accurately even when it passes through severe nonlinearities. This is a significant advantage over power-limited channels, such as satellite and cellular wireless communication, since power amplifiers operate most efficiently in a nonlinear saturation regime. For example, let us consider the effect of an extreme nonlinearity, a hardlimiter, on the passband signal u_p (such a nonlinearity would severely distort a signal with time-varying envelope, since envelope information is destroyed by hardlimiting). The hardlimited output is given by

$$v_p(t) = \text{sign}\left(u_p(t)\right) = \text{sign}\left(\cos(2\pi f_c t + \theta(t))\right).$$

This corresponds to replacing a sinusoid by a square wave. From a local Fourier series expansion ($\theta(t)$ varies slowly compared with the carrier frequency, and hence can be approximated as constant over many cycles), the square wave is a sum of the odd harmonics:

$$v_p(t) \approx a_1 \cos(2\pi f_c t + \theta(t)) + a_3 \cos(3(2\pi f_c t + \theta(t)))$$
$$+ a_5 \cos(5(2\pi f_c t + \theta(t))) + \cdots$$

where $a_1, a_3, a_5, \ldots$ are determined as in Problem 8.27. Thus, we can recover the original signal $u_p(t)$ up to a scalar multiple simply by rejecting the harmonics at $3f_c, 5f_c, \ldots$ (a transmit antenna tuned to f_c typically suffices for this purpose).

In theory, the constant envelope property can hold even without insisting that $\theta(t)$ be continuous: for example, QPSK with an ideal rectangular time-limited pulse has a constant envelope. However, bandlimited circuits cannot implement ideal timelimited signals, and hence cannot produce instantaneous jumps in phase. Thus, if the phase jumps from θ_1 to θ_2, then the complex envelope traces a path from $e^{j\theta_1}$ to $e^{j\theta_2}$, passing through, for example,

8.6 Continuous phase modulation

Figure 8.13 The envelope of a PSK signal passes through zero during a 180° phase transition, and gets distorted over a nonlinear channel.

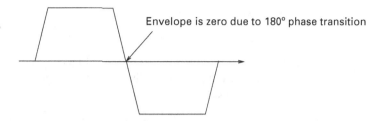

$(e^{j\theta_1}+e^{j\theta_2})/2$. Since $|(e^{j\theta_1}+e^{j\theta_2})/2| \neq 1$ for $\theta_1 \neq \theta_2$, we see that the constant envelope property no longer holds. Figure 8.13 shows the example of a PSK signal going through a 180° phase shift: even though the pulse was intended to be an ideal rectangular pulse, practical implementations have nonideal transitions as shown, so that the envelope goes through zero.

Building on our familiarity with linear modulation, we begin our discussion with minimum shift keying (MSK), a modulation format that can be interpreted both as linear modulation and as CPM. As shown in Figure 8.14 (see also Problem 2.24), MSK can be represented as offset QPSK linear modulation with bit rate $1/T_b$ with complex baseband transmitted signal

$$u(t) = \sum_k \left(b_c[k] p(t - 2kT_b) + j b_s[k] p(t - 2kT_b - T_b) \right),$$

where $b_c[k]$, $b_s[k]$ are ± 1 bits sent on the I and Q channels, and

$$p(t) = \sin \frac{\pi t}{T_s} \, I_{[0, 2T_b]}$$

is a sine pulse. The symbol duration on each of the I and Q channels is $2T_b$.

We now wish to show that u is a CPM signal (i.e., that it is a constant envelope signal with continuous phase). Since u is cyclostationary with period $2T_b$, it suffices to discuss its structure over a typical interval $[0, 2T_b]$. It is helpful to refer to the typical sample path shown in Figure 8.14.

Consider first $0 \leq t \leq T_b$. From Figure 8.14, we see that we can write

$$u(t) = b_c[0] \sin \frac{\pi t}{2T_b} + j b_s[-1] \cos \frac{\pi t}{2T_b}, \quad 0 \leq t \leq T_b,$$

Figure 8.14 MSK interpreted as offset QPSK using a sinusoidal pulse. The I and Q channel symbols are offset by T_b, as shown.

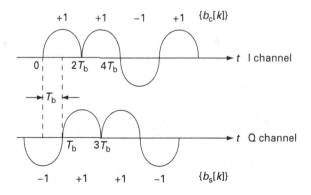

which can be rewritten as

$$u(t) = jb_s[-1]e^{ja[0]\frac{\pi t}{2T_b}}, \quad 0 \le t \le T_b,$$

where $a[0] = -\frac{b_c[0]}{b_s[-1]} = -b_c[0]b_s[-1]$ takes values in ± 1. Note that u has a constant envelope, and its phase is a linear function of t over this interval.

Now, consider $T_b \le t \le 2T_b$. From Figure 8.14, we have

$$u(t) = b_c[0]\sin\frac{\pi t}{2T_b} - jb_s[1]\cos\frac{\pi t}{2T_b}, \quad T_b \le t \le 2T_b.$$

This can be rewritten as

$$u(t) = -jb_s[1]e^{ja[1]\frac{\pi t}{2T_b}}, \quad T_b \le t \le 2T_b$$

where $a[1] = (b_c[0])/(b_s[1]) = b_c[0]b_s[1]$ takes values in ± 1. Again, note the constant envelope and linear phase.

Of the I and Q bitstreams $\{b_c[k]\}$, $\{b_s[k]\}$, exactly one bit changes sign at integer multiples of T_b. The modulating pulse for that bit vanishes at the time at which it changes sign, and $u(t)$ is determined by the bit that is *not changing sign*, which leads to continuity in phase: for example, $u(0) = jb_s[-1]$, $u(T_b) = b_c[0]$, $u(2T_b) = jb_s[0]$, and so on. Thus, the phase of $u(t)$ follows a piecewise linear trajectory over intervals of length T_b, while remaining continuous at the edges of these intervals. A trellis representation of the phase evolution is shown in Figure 8.15: the states are defined as the phase values at integer multiples of T_b, as shown. The change in phase over an interval $[iT_b, (i+1)T_b]$ is of the form $a[i]\pi/2$, where I showed by example $(i = 0, 1)$ how the bits $a[i] \in \{-1, +1\}$ are defined. We can now interpret the phase as being a result of linearly modulating the bits $\{a[i]\}$ at rate $1/T_b$, using the pulse $\phi(t)$, given by

$$\phi(t) = \begin{cases} 0, & t < 0, \\ \frac{\pi t}{2T_b}, & 0 \le t \le T_b, \\ \frac{\pi}{2}, & t \ge T_b. \end{cases} \quad (8.97)$$

Note that ϕ is a continuous function, which implies the continuity of the phase of the CPM signal, given by

$$\theta(t) = \sum_n a[n]\phi(t - nT), \quad (8.98)$$

Figure 8.15 A trellis representation for the phase evolution of MSK. The path in bold corresponds to the realization shown in Figure 8.14.

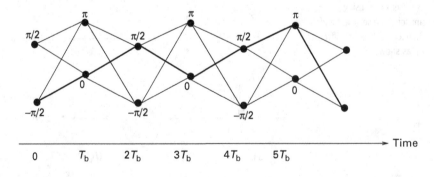

8.6 Continuous phase modulation

where we replace T_b by T to obtain the general form taken by the phase of a CPM signal modulated at a symbol rate $1/T$. In general, $\{a[n]\}$ can be chosen from a nonbinary alphabet, and ϕ may be nonlinear, and can span an interval larger than T. For MSK, $\{a[n]\}$ are binary, so that $T = T_b$, and ϕ is given by (8.97).

Defining the instantaneous frequency $f(t) = (1/2\pi)(\mathrm{d}\phi(t)/\mathrm{d}t)$, we obtain by differentiating (8.98) that the instantaneous frequency is also linearly modulated:

$$f(t) = \sum_n a[n] g(t - nT), \qquad (8.99)$$

where the frequency pulse $g(t) = (1/2\pi)(\mathrm{d}\theta(t)/\mathrm{d}t)$. The phase pulse ϕ can be written in terms of the frequency pulse g as follows:

$$\phi(t) = 2\pi \int_{-\infty}^{t} g(s)\,\mathrm{d}s. \qquad (8.100)$$

We can therefore describe a CPM system by specifying the alphabet for the $\{a[n]\}$, and specifying either the phase or the frequency pulse.

For MSK, we obtain from (8.97)

$$g(t) = \frac{1}{4T} I_{[0,T]} \quad \textbf{MSK frequency pulse} \qquad (8.101)$$

(recall that $T = T_b$). That is, the instantaneous frequency in MSK is modulated by a rectangular pulse, leading to discrete frequency shifts over intervals of length T_b. This is an example of continuous phase FSK (CPFSK), which is a special case of CPM corresponding to piecewise linear phase variation. For MSK, there are two possible frequency shifts of $\pm 1/4T$: the difference in frequencies is therefore $1/2T$, the minimum separation required to guarantee orthogonality of FSK for coherent demodulation. This is the source of the term *minimum* shift keying.

A DSP-centric realization of a CPM modulator is depicted in Figure 8.16.

In the following, we continue to focus on binary alphabets for $\{a[n]\}$, since CPM systems are usually designed for power efficiency. We now seek to improve upon MSK in terms of bandwidth efficiency. For $\{a[n]\}$ i.i.d., taking values ± 1 with equal probability, the PSD of a CPM signal with phase given by (8.98) clearly depends only on ϕ or g. An analytical derivation

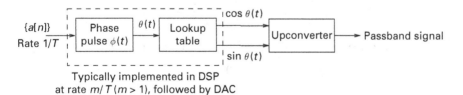

Figure 8.16 Digital signal processing-centric realization of a CPM modulator. The linear modulation of the phase $\theta(t)$, followed by look-up of the sine and cosine, is often implemented in DSP at an integer multiple of the symbol rate, followed by digital-to-analog conversion prior to upconversion.

Figure 8.17 Phase pulse for CPFSK has a linear increase. Partial response CPFSK corresponds to $L > 1$, and h is the modulation index.

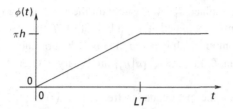

of the PSD is possible, but the resulting expressions are too messy to yield insight (numerical estimates of the PSD based on applying the FFT to a sampled version of the CPM signal are often more computationally efficient). However, it is intuitively reasonable that the bandwidth of a CPM signal can be reduced by using a smoother phase pulse ϕ (or equivalently, a smoother frequency pulse g).

To reduce the bandwidth, one approach is to extend the time duration over which a symbol $a[n]$ affects the instantaneous frequency beyond T. This is termed *partial response* signaling, since only part of the response due to a given symbol lies within an interval of length T. A simple extension of MSK, therefore, is simply to stretch out the change in phase due to a symbol over a longer period, while keeping the linear variation with time. The phase pulse for partial response CPFSK is shown in Figure 8.17. The maximum phase change due to a symbol is πh, where h is termed the *modulation index* ($0 < h < 1$ in practice). The modulation index for MSK is $h = 1/2$. The signal bandwidth can be reduced further by smoothing out the sharp transitions in the phase or frequency pulses. We now illustrate this by discussing in some detail Gaussian MSK (GMSK), which is the modulation format used in the GSM cellular system.

8.6.1 Gaussian MSK

The frequency pulse $g(t)$ for GMSK is obtained by "smoothing" the rectangular frequency pulse of MSK by passing it through a Gaussian filter. We first discuss some facts and nomenclature related to Gaussian filters.

Gaussian filter A Gaussian filter has transfer function of the form

$$H(f) = \exp\left(-\left(\frac{f}{B}\right)^2 \frac{\log 2}{2}\right), \qquad (8.102)$$

where B is the (one-sided) 3 dB bandwidth, defined as

$$\frac{|H(B)|^2}{|H(0)|^2} = \frac{1}{2}.$$

($H(f)$ is maximum at $f = 0$, and decays monotonically with $|f|$.)

It turns out that the impulse response of a Gaussian filter is also Gaussian. We can see this by massaging the inverse Fourier transform of (8.102)

8.6 Continuous phase modulation

into the form of a Gaussian density, as follows (we set $A = \ln 2/(2B^2)$ for convenience):

$$h(t) = \int_{-\infty}^{\infty} e^{-Af^2} e^{j2\pi ft} df = e^{A(\pi t/A)^2} \int_{-\infty}^{\infty} e^{-A(f+\pi t/A)^2} df,$$

completing squares. Recognizing that the integrand has the form of an $N(-\pi ft/A, a^2 = 1/2A)$ density, we see that the integral on the extreme right-hand side evaluates to $\sqrt{2\pi a^2} = \sqrt{\pi/A}$. We therefore obtain

$$h(t) = \sqrt{\frac{\pi}{A}} e^{-\pi^2 t^2/A},$$

which takes the form of an $N(0, v^2)$ density with $v^2 = A/2\pi^2 = \log 2/4\pi^2 B^2$. To summarize, the impulse response of the Gaussian filter (8.102) is given by

$$h(t) = \frac{1}{2\pi v^2} e^{-\frac{t^2}{2v^2}}, \quad v^2 = \frac{\ln 2}{4\pi^2 B^2}. \tag{8.103}$$

The normalization $H(0) = \int h(t)dt = 1$ is important, as we see shortly.

Convolving the MSK frequency pulse $g_{MSK}(t) = \frac{1}{4T} I_{[0,T]}$ with the Gaussian filter in (8.103) yields

$$g_{GMSK}(t) = \frac{1}{4T}\left[Q\left(\frac{t-T}{v}\right) - Q\left(\frac{t}{v}\right)\right]$$

$$= \frac{1}{4T}\left[Q\left(2\pi\beta \frac{t-T}{T\sqrt{\ln 2}}\right) - Q\left(2\pi\beta \frac{t}{T\sqrt{\ln 2}}\right)\right]. \tag{8.104}$$

The parameter v^2 of the Gaussian filter is related to the normalized 3 dB bandwidth $\beta = BT$ as follows: $v^2 = T^2 \ln 2/4\pi^2 \beta^2$. The GSM system specifies $\beta = 0.3$. By virtue of our normalization of the Gaussian filter such that $H(0) = \int h(t)dt = 1$, the area under the frequency pulse, and hence the net phase deviation due to a given symbol, remains the same as in MSK:

$$\int g_{GMSK}(t)dt = G_{GMSK}(0) = G_{MSK}(0)H(0) = G_{MSK}(0) = \int g_{MSK}(t)dt = \frac{1}{4}.$$

That is, each bit in GMSK causes a phase change of $\pm 2\pi \int g_{GMSK}(t)dt = \pm \pi/2$.

We plot g_{GMSK} in Figure 8.18. Note how the Gaussian filtering spreads out and smooths the rectangular MSK pulse over time. In the frequency domain, the slow sinc function decay of the MSK frequency pulse is multiplied by the rapidly decaying Gaussian transfer function $e^{-\pi^2 f^2/a^2}$ to reduce the spectral occupancy of the MSK frequency pulse significantly.

8.6.2 Receiver design and Laurent's expansion

The instantaneous frequency of a CPM signal is linearly modulated by the symbols $\{a[n]\}$. Thus, a suboptimal approach for demodulating CPM is to first extract the instantaneous frequency using an FM demodulator, and then use standard techniques for detecting linearly modulated signals. Alternatively,

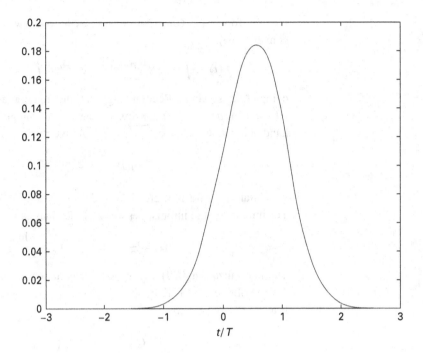

Figure 8.18 The frequency pulse $g_{GMSK}(t)$ for Gaussian MSK with $\beta = 0.3$ as in the GSM system.

one could try to use an optimal, maximum likelihood approach. Figure 8.15 shows that the phase for MSK takes one of a discrete set of values at integer multiples of T. This can be used to define a phase trellis over which to perform MLSE for the sequence $\{a[n]\}$, as shown in Problem 8.28. Of course, for MSK, it is far easier to carry out symbol-by-symbol demodulation of the sequences $\{b_c[k]\}$ and $\{b_s[k]\}$ using an OQPSK interpretation. The MLSE approach can be extended to partial response CPM for rational modulation index (in which case the phases at the end of T-length intervals take on values from a finite set, leading to a finite-state phase trellis).

The suboptimum FM demodulator does not perform well when the CPM signal passes through a dispersive channel, since the noiseless received signal can no longer be modeled using linear modulation of the instantaneous frequency. On the other hand, the MLSE approach becomes cumbersome for a dispersive channel. In addition to the explosion of the number of states required to capture the memory of the channel and the modulation, the number of correlator decision statistics per time interval T scales with the number of states. This is in contrast to linear modulation, where only one decision statistic is needed per time interval T: the matched filter output sampled at the symbol rate. Moreover, we have a suite of strategies for handling ISI for linear modulation, ranging from computationally complex MLSE to low-complexity linear and decision-feedback equalization. In contrast, it is unclear, a priori, as to how we would develop suboptimal, low-complexity equalizers for a nonlinear modulation format such as CPM.

8.6 Continuous phase modulation

An elegant solution to the preceding quandary lies in Laurent's representation of CPM signals. Laurent showed that CPM signals can be very well approximated as linear modulation in most cases of practical interest, so that we can write

$$u(t) = \exp\left(\sum_n a[n]\phi(t-nT)\right) \approx \sum_n B[n]s(t-nT) \quad \textbf{Laurent approximation,}$$
(8.105)

where $\{B[n]\}$ are "pseudosymbols" that can be computed in terms of the CPFSK symbols $\{a[n]\}$, and $s(t)$ is a function of the phase pulse $\phi(t)$. The details of the Laurent approximation are worked out for a partial response CPFSK system in Example 8.6.1. The approximation (8.105) is extremely significant for receiver design, since it allows us to exploit all of the equalization techniques that we already know for linearly modulated systems. The modulating pulse $s(t)$ is not Nyquist or square root Nyquist at rate $1/T$ for partial response CPM, so that ISI is built into this model of CPM. When we pass the CPM signal through a channel g_C, we incur further ISI from the effective modulating pulse $s * g_C$. We can now recover $\{B[n]\}$ using the equalization techniques of Chapter 5 for this effective pulse, choosing between options such as MLSE, linear MMSE or DFE, depending on our implementation constraints.

We illustrate Laurent's approach using an example, followed by a discussion of how the results generalize.

Example 8.6.1 (Laurent approximation for CPFSK with $L = 2$)
Consider a phase pulse with a linear increase over an interval of length $2T$ and modulation index $1/2$ (i.e., $L = 2$ in Figure 8.17). The modulated signal is written as

$$u(t) = \exp(j\theta(t)) = \exp\left(j\sum_{n=0}^{\infty} a[n]\phi(t-nT)\right).$$

Define the pseudosymbol sequence $\{B_0[k]\}$ as follows:

$$B_0[k] = \exp\left(j\frac{\pi}{2}\sum_{n=0}^{k} a[n]\right) = j^{\sum_{n=0}^{k} a[n]}.$$
(8.106)

We plan to approximate $u(t)$ as linear modulation with these pseudosymbols. Note that $\{B_0[k]\}$ come from a QPSK constellation. For $t \in [NT, (N+1)T)$, we can see that the phase is given by

$$\theta(t) = \phi(t-NT)a[N] + \phi(t-NT+T)a[N-1] + \frac{\pi}{2}\sum_{n=0}^{N-2} a[n].$$
(8.107)

We can use Euler's formula to write

$$e^{jab} = \cos b + j^a \sin b, \quad a = \pm 1.$$
(8.108)

Let us introduce the shorthand $\delta_N = \phi(t-NT)$ and $\gamma_N = \phi(t-NT+T)$ for $NT \leq t < (N+1)T$, suppressing the dependence on time for the moment. We can now write, for $NT \leq t < (N+1)T$,

$$u(t) = e^{j\theta(t)} = e^{ja[N]\delta_N} e^{ja[N-1]\gamma_N} B_0[N-2].$$
$$= (\cos\delta_N + j^{a[N]}\sin\delta_N)(\cos\gamma_N + j^{a[N-1]}\sin\gamma_N) B_0[N-2]. \quad (8.109)$$

Noting that $j^{a[N-1]} B_0[N-2] = B_0[N-1]$ and $j^{a[N]} j^{a[N-1]} B_0[N-2] = B_0[N]$, we can write

$$u(t) = B_0[N]\sin\delta_N \sin\gamma_N + B_0[N-1]\cos\delta_N \sin\gamma_N$$
$$+ B_0[N-2]\cos\delta_N \cos\gamma_N + B_1[N]\sin\delta_N \cos\gamma_N, \quad (8.110)$$

where $B_1[k] = j^{a[k]} B_0[k-2] = B_0[k] j^{-a[k-1]}$ is another pseudosymbol sequence. We now show that $u(t)$ can be expressed as a sum of linearly modulated signals using the pseudosymbol sequences $\{B_0[k]\}$ and $\{B_1[k]\}$, with the dominant contribution coming from the signal corresponding to $\{B[k]\}$. That is, we show that

$$u(t) = \sum_k B_0[k] s_0(t-kT) + \sum_k B_1[k] s_1(t-kT)$$

for some waveforms s_0 and s_1. The expression (8.110) holds over the length-T interval $[NT, (N+1)T]$. Summing over all such intervals, and substituting the expressions for δ_N, γ_N, we obtain

$$u(t) = \sum_N \{B_0[N]\sin\phi(t-NT)\sin\phi(t-NT+T)$$
$$+ B_0[N-1]\cos\phi(t-NT)\sin\phi(t-NT+T)$$
$$+ B_0[N-2]\cos\phi(t-NT)\cos\phi(t-NT+T)$$
$$+ B_1[N]\sin\phi(t-NT)\cos\phi(t-NT+T)\} I_{[NT,(N+1)T]}.$$

Grouping together all terms multiplying $B_0[k]$, we see that

$$u(t) = \sum_k B_0[k] s_{0,k}(t) + \sum_k B_1[k] s_{1,k}(t),$$

where

$$s_{0,k}(t) = \sin\phi(t-kT)\sin\phi(t-(k-1)T) I_{[kT,(k+1)T]}$$
$$+ \cos\phi(t-(k+1)T)\sin\phi(t-kT) I_{[(k+1)T,(k+2)T]}$$
$$+ \cos\phi(t-(k+2)T)\cos\phi(t-(k+1)T) I_{[(k+2)T,(k+3)T]}$$

and

$$s_{1,k}(t) = \sin\phi(t-kT)\cos\phi(t-(k-1)T) I_{[kT,(k+1)T]}.$$

It is now clear that we can write

$$s_{0,k}(t) = s_0(t-kT), \quad s_{1,k}(t) = s_1(t-kT)$$

8.6 Continuous phase modulation

where

$$s_0(t) = \sin\phi(t)\sin\phi(t+T)I_{[0,T)}$$
$$+ \cos\phi(t-T)\sin\phi(t)I_{[T,2T)}$$
$$+ \cos\phi(t-2T)\cos\phi(t-T)I_{[2T,3T)}$$

and

$$s_1(t) = \sin\phi(t)\cos\phi(t+T)I_{[0,T)}.$$

Figure 8.19 shows the waveforms s_0 and s_1. Clearly, s_0 is much larger, so that we can approximate the modulated waveform as QPSK linear modulation as follows:

$$u(t) \approx \sum_k B_0[k]s_0(t-kT).$$

The waveform $s_0(t)$ can be put into a more compact form by defining the extended phase function ψ, defined by piecing together the linear portion of ϕ and its reflection around $2T$:

$$\psi(t) = \begin{cases} 0, & t < 0, \\ \phi(t), & 0 \leq t \leq 2T, \\ \frac{\pi}{2} - \phi(t-2T), & 2T \leq t \leq 4T, \\ 0, & t \geq 4T. \end{cases} \quad (8.111)$$

By noting that $\cos a = \sin(\pi/2 - a)$, a little thought shows that we can write

$$s_0(t) = \sin\psi(t)\sin\psi(t+T).$$

Similarly,

$$s_1(t) = \sin\psi(t)\sin\psi(t+3T).$$

The preceding method generalizes to arbitrary phase or frequency pulses, assuming that the frequency pulse is zero outside an interval of length LT, where $L \geq 1$ is an integer. Note that, even if the frequency pulse does not have finite support, it must die out eventually in order to be integrable, so that we can approximate it well by truncation to a finite interval of length LT for some L. For example, Figure 8.20 shows the phase pulse for GMSK with $\beta = 0.3$. Clearly, truncation to $L = 4$ works well for this pulse.

For a frequency pulse of duration LT, it can be shown that the modulated signal can be decomposed into a sum of 2^{L-1} parallel linearly modulated systems, with correlated pseudosymbol sequences that are related to the CPM symbols $\{a[n]\}$. One of these parallel systems dominates, which results in the linear modulation approximation (8.105). The derivation involves simple generalizations of the arguments used in the preceding example, and we sketch some key features next.

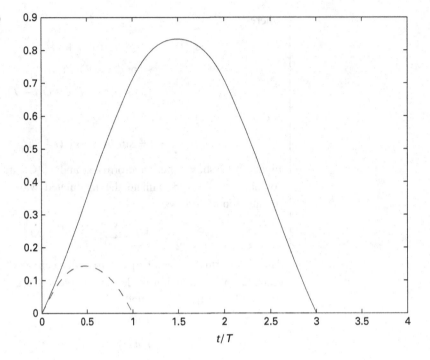

Figure 8.19 Laurent pulses s_0 and s_1 for CPFSK with $L = 2$, computed as described in Example 8.6.1. Note that the dominant pulse $s_0(t)$, shown using an unbroken line, is much larger than the pulse $s_1(t)$, shown using a dashed line.

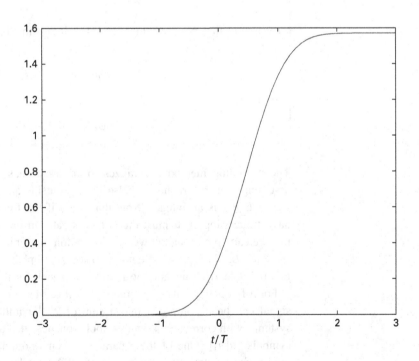

Figure 8.20 The phase pulse $\phi_{GMSK}(t)$ for Gaussian MSK with $\beta = 0.3$. Note that the variation of the phase can be captured in an interval of length $L = 4$.

The most important step in the general derivation is to account for modulation index h different from $1/2$. For general h and L, equation (8.107) becomes

$$\theta(t) = \phi(t - NT)a[N] + \phi(t - NT + T)a[N-1] +$$
$$\cdots + \phi(t - NT + (L-1)T)a[N - L + 1].$$

We define the pseudosymbols for the dominant system as

$$B_0[k] = e^{j\pi h \sum_{n=0}^{k} a[n]} = J^{\sum_{n=0}^{k} a[n]}, \quad \text{where } J = e^{j\pi h}. \tag{8.112}$$

(Note: $J = j$ for $h = 1/2$.) For rational $h = p/q$ ($p < q$, p, q integers), the pseudosymbols are drawn from a q-ary PSK alphabet.

In the example, the next critical step in the Laurent expansion was to apply the Euler-based formula (8.108) to (8.107). However, to get expressions in terms of the pseudosymbols (8.112), we must now use a more general *Euler-like* formula involving $J = e^{j\pi h}$: for $a \in \{-1, +1\}$,

$$e^{jab} = \frac{\sin(\pi h - b) + J^a \sin b}{\sin \pi h}. \tag{8.113}$$

We also need to modify the definition of the generalized phase function ψ as follows:

$$\psi(t) = \begin{cases} 0, & t < 0, \\ \phi(t), & 0 \le t \le LT, \\ \pi h - \phi(t - LT), & LT \le t \le 2LT, \\ 0, & t \ge 2LT. \end{cases} \tag{8.114}$$

We can now mimic the derivation in the example, using ψ to obtain a compact representation for the modulating signals in the 2^{L-1} parallel systems. We specify here only the modulating waveform for the dominant system, which is given by

$$s_0(t) = K(t)K(t+T)\ldots K(t+(L-1)T), \tag{8.115}$$

where

$$K(t) = \frac{\sin \psi(t)}{\sin \pi h}. \tag{8.116}$$

That is, the approximation (8.105) holds with $B = B_0$ and $s = s_0$.

8.7 Space–time communication

Space–time communication is a broad term that includes a gamut of techniques developed for communication systems with multiple transmit and receive antennas. Another term for such systems that is in widespread usage is multiple input multiple output (MIMO). The use of multiple *receive* antennas for diversity and beamforming, as described in Section 8.2.3, has been known for many decades. More recently, however, there has been increasing interest in how to use multiple *transmit* as well as receive antennas. The number of antennas at the transmitter and receiver varies across applications. For cellular systems, it is reasonable to assume that the base station is equipped with several antennas, while the mobile terminal might only have one or two antennas. In this case, the transmitter has more antennas than the receiver on the base-to-mobile *downlink*, while the receiver has more antennas on the mobile-to-base *uplink*. In contrast, for emerging IEEE 802.11n WLANs based on MIMO techniques, most nodes might have similar form factors, and a similar number of antennas (ranging from two to four).

We begin with space–time channel modeling, using a linear antenna array as our running example. We derive the array response as a function of the angle of arrival of an incoming electromagnetic wave (the same derivation holds for outgoing waves as well). In a multipath environment, the overall received signal is a sum of multiple waves arriving from different angles: for a statistical channel model, these can be statistically characterized in terms of a *power-angle profile (PAP)*. We use central limit theorem arguments to model the overall array response as complex Gaussian with covariance matrix depending on the PAP. Next, we provide an information-theoretic analysis in a "rich scattering" environment in which this spatial covariance matrix is white. This analysis motivates some of the specific techniques discussed next, which fall into one of three broad classes: spatial multiplexing, transmit diversity, and transmit beamforming.

Our discussion of space–time communication is for narrowband systems, in which the channel for a given pair of transmit and receive antenna elements can be modeled as a complex gain. Wideband channels can be converted into parallel narrowband channels using OFDM, and MIMO techniques can be used within each subcarrier; an example of such a MIMO-OFDM system is the emerging IEEE 802.11n WLAN standard.

8.7.1 Space–time channel modeling

Consider a linear array of m antenna elements with inter-element spacing of d, as shown in Figure 8.21. The reference for defining the angle of arrival of incoming waves is taken to be the broadside of the array, which is the direction perpendicular to the line of the array.

As shown in Figure 8.21, a wave arriving at angle θ incurs a path length difference between successive antenna elements of $d \sin \theta$. The corresponding difference in phase is given by

$$\phi = \phi(\theta) = \frac{2\pi d \sin \theta}{\lambda},$$

where $\lambda = c/f_c$ is the wavelength, c is the speed of light (assuming free space propagation), and f_c is the carrier frequency. Thus, taking the phase at the first antenna element as the reference, if $y(t)$ is the complex baseband signal received at the first element, then the ith element receives the signal

8.7 Space–time communication

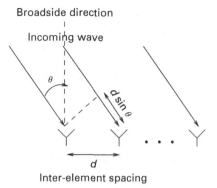

Figure 8.21 Geometry of a linear antenna array.

$y(t - i\tau)e^{j(i-1)\phi}$, where $\tau = (d\sin\theta)/c$ is the inter-element delay. For d of the order of λ, τ is of the order of $1/f_c$. Assuming that the bandwidth of y is much smaller than f_c, we can ignore the effect of the delay on $y(t)$. Under this "narrowband assumption," we can set $y(t - i\tau) \approx y(t)$, so that the received signal at the ith element is modeled as $y(t)e^{j(i-1)\phi}$. Thus, the complex baseband signal sees an m-dimensional vector gain that we term the array response, defined as

$$\mathbf{a}(\theta) = \left(1, e^{j\phi}, \ldots, e^{j(m-1)\phi}\right)^T. \tag{8.117}$$

The collection $\{\mathbf{a}(\theta)\}$ of array responses as a function of θ, as θ varies over a semicircle ($\theta \in [-\pi/2, \pi/2]$) is termed the *array manifold*. While we have derived the array manifold for a linear array, it is clear that similar calculations could be carried out to determine the array manifold for other geometries. (For two-dimensional arrays, the array manifold depends on two angular parameters.)

Now, consider a multipath channel in which the received signal is a superposition of waves with different amplitudes and phases, arriving from different directions. The composite array response is given by

$$\mathbf{h} = (h_1, \ldots, h_m)^T = \sum_{i=1}^{M} g_i \mathbf{a}(\theta_i),$$

where M is the number of multipath components, and g_i denoting the complex gain, and θ_i the angle of arrival, for the ith path. We can now argue as we did when we developed the scalar Rayleigh fading model in Section 8.1. The multipath phases $\arg(g_i)$ are well modeled as independent and uniform over $[0, 2\pi]$. As long as the magnitudes $|g_i|$ are comparable, we can apply a vector central limit theorem to model $\mathbf{h}$ as zero mean, proper complex Gaussian, with covariance matrix

$$\mathbf{C} = \sum_{i=1}^{M} |g_i|^2 \mathbf{a}(\theta_i) \left(\mathbf{a}(\theta_i)\right)^H. \tag{8.118}$$

We now introduce the PAP $P(\theta)$, with $P(\theta)d\theta$ the average power arriving from angles in the infinitesimal interval of length $d\theta$ around θ. For convenience, let us normalize $P(\theta)$ so that it plays the role of a density: $\int P(\theta)\, d\theta = 1$. Assuming that there are enough paths, we can replace the sum (8.118) by an average computed using the PAP as follows:

$$\mathbf{C} = \int \mathbf{a}(\theta)\mathbf{a}^H(\theta)\, P(\theta)\, d\theta. \tag{8.119}$$

Note that $\mathbf{v}_i$ are orthogonal for different eigenvalues, and can be chosen to be orthogonal for repeated eigenvalues. We therefore assume that they are orthogonal, and term them the channel *eigenmodes*. The random array response $\mathbf{h} \sim CN(\mathbf{0}, \mathbf{C})$ can be written as a linear combination of the eigenmodes as follows:

$$\mathbf{h} = \sum_{i=1}^{m} \tilde{h}_i \mathbf{v}_i, \tag{8.120}$$

where $\tilde{h}_i \sim CN(0, \lambda_i)$ is the complex gain along the ith eigenmode. Note that $\tilde{h}_i$ are independent under our model.

For narrow PAPs, there are a small number of dominant eigenmodes: in this case, the channel energy can be efficiently gathered by focusing communication along these eigenmodes (assuming that these are known). For a large power-angle spread, there are a larger number of eigenmodes. We dub the special case in which all eigenmodes are equally strong a "rich scattering" environment. In this case, $\mathbf{C}$ is a scalar multiple of the identity matrix, and we have i.i.d. zero mean, complex Gaussian gains at each element.

Multiple antennas at both ends Now, consider a system with N_T antenna elements at the transmitter, and N_R antenna elements at the receiver. For narrowband signaling, we can describe the channel by an $N_R \times N_T$ matrix $\mathbf{H}$, where the jth column of $\mathbf{H}$ is the receive array response to the jth transmit element. If the lth path has departure angle θ_l, arrival angle γ_l, and complex gain g_l, then the channel matrix is given by

$$\mathbf{H} = \sum_l g_l\, \mathbf{a}_R(\gamma_l)\mathbf{a}_T^T(\theta_l),$$

where $\mathbf{a}_T(\cdot)$ is the transmit array manifold and $\mathbf{a}_R(\cdot)$ is the receive array manifold. The distribution of the coefficients g_l is determined by the PAP at each end. For simplicity, we consider some special cases below.

LOS link If θ_T is the angle of departure from the transmitter, and θ_R is the angle of arrival at the receiver, then the channel matrix is rank one:

$$\mathbf{H} = \mathbf{a}_R(\theta_R)(\mathbf{a}_T(\theta_T))^T. \tag{8.121}$$

Rich scattering If there is rich scattering around both the transmitter and the receiver, then the entries of $\mathbf{H}$ can be modeled i.i.d., zero mean, proper complex Gaussian random variables.

8.7 Space–time communication

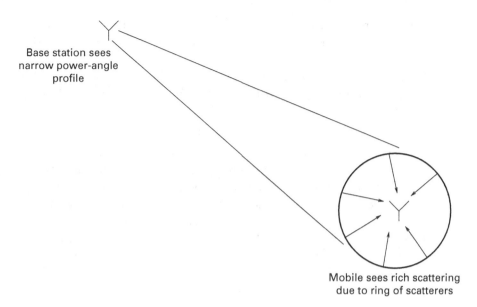

Figure 8.22 An elevated base station sees a narrow power-angle profile to a given mobile, but the mobile sees a rich scattering environment locally. The nodes may have multiple antenna elements (not shown in figure).

Narrow PAP at one end, rich scattering at other end In many cellular systems, an elevated base station would see a relatively small angular spread to a particular mobile, while the mobile may see a rich scattering environment due to buildings and other scatterers surrounding it. This scenario is depicted in Figure 8.22. Suppose that the base station sees PAP $P(\theta)$, where θ is the angle of departure. Then the ith receive element sees a $1 \times N_T$ transmit vector $\mathbf{H}(i, \cdot)$ modeled as $CN(\mathbf{0}, \mathbf{C}_P)$, where $\mathbf{C}_P$ is as in (8.119). Under the rich scattering model at the mobile receiver, we can model the transmit vectors for different receive elements as i.i.d.: thus, the rows of $\mathbf{H}$ are i.i.d. $CN(\mathbf{0}, \mathbf{C}_P)$ random vectors.

8.7.2 Information-theoretic limits

The key idea here is to decompose the channel into a number of parallel eigenmodes, with the net rate being the sum of the rates sent over the eigenmodes. For an $N_R \times N_T$ channel matrix $\mathbf{H}$, the $N_R \times 1$ channel output $\mathbf{y}$ corresponding to an $N_T \times 1$ input $\mathbf{x}$ is given by

$$\mathbf{y} = \mathbf{H}\mathbf{x} + \mathbf{w}, \tag{8.122}$$

where the noise $\mathbf{w} \sim CN(\mathbf{0}, 2\sigma^2 \mathbf{I})$ is complex WGN with variance σ^2 per dimension. As usual, the noise is i.i.d. over multiple uses of the channel. We assume that $\mathbf{H}$ is fixed.

Our information-theoretic computations rely on a singular value decomposition (SVD) to reduce the channel into a number of parallel scalar channels.

Singular value decomposition Let us first review the SVD. Assume that $N_R \geq N_T$ (otherwise replace $\mathbf{H}$ by $\mathbf{H}^H$ in the following). Define the $N_T \times N_T$ matrix

$$\mathbf{W} = \mathbf{H}^H \mathbf{H}. \tag{8.123}$$

Note that $\mathbf{W}$ is nonnegative definite, so that it has nonnegative eigenvalues: let $\mathbf{v}_i$ denote the $N_T \times 1$ eigenvector corresponding to the ith eigenvalue $\lambda_i \geq 0$, $i = 1, \ldots, N_T$. Let $\mathbf{V}$ denote a matrix whose ith column is $\mathbf{v}_i$. Note that $\{\mathbf{v}_i\}$ can be chosen to be orthonormal (eigenvectors for different eigenvalues are orthogonal, while eigenvectors for repeated eigenvalues can be chosen to be orthonormal), so that $\mathbf{V}^H \mathbf{V} = \mathbf{I}$. Suppose, now, that exactly k of these eigenvalues are strictly positive (for $i = 1, \ldots, k$, without loss of generality). Note that

$$\mathbf{H}\mathbf{v}_i = \mathbf{0}, \quad \text{if } \lambda_i = 0, \tag{8.124}$$

which occurs for $i = k+1, \ldots, N_T$. Define the $N_R \times 1$ vectors

$$\mathbf{u}_i = \lambda_i^{-\frac{1}{2}} \mathbf{H} \mathbf{v}_i, \quad i = 1, \ldots, k,$$

where $k \leq N_T \leq N_R$. Note that $\{\mathbf{u}_i\}$ are orthonormal:

$$\mathbf{u}_i^H \mathbf{u}_j = (\lambda_i \lambda_j)^{-\frac{1}{2}} \mathbf{v}_i^H \mathbf{H}^H \mathbf{H} \mathbf{v}_j = (\lambda_i \lambda_j)^{-\frac{1}{2}} \lambda_j \mathbf{v}_j = \delta_{ij}.$$

Complete the set $\{\mathbf{u}_i, i = 1, \ldots, k\}$ to form an orthonormal basis $\{\mathbf{u}_i, i = 1, \ldots, N_R\}$ in N_R dimensions, and denote by $\mathbf{U}$ the $N_R \times N_R$ matrix whose ith column is $\mathbf{u}_i$. We now claim that we can write the $N_R \times N_T$ channel matrix $\mathbf{H}$ as follows:

$$\mathbf{H} = \sum_{i=1}^{N_T} \mathbf{u}_i \sqrt{\lambda_i} \mathbf{v}_i^H = \mathbf{U} \mathbf{D} \mathbf{V}^H, \tag{8.125}$$

where $\mathbf{D} = \mathrm{diag}(\lambda_1^{\frac{1}{2}}, \ldots, \lambda_{N_T}^{\frac{1}{2}})$ is an $N_T \times N_T$ diagonal matrix; the diagonal entries $\sqrt{\lambda_i}$ are termed the *singular values* of $\mathbf{H}$. To verify (8.125), we need to check that its left and right-hand sides give the same result when operating on an arbitrary $N_T \times 1$ vector $\mathbf{x}$. To see this, note that any such vector can be expressed in terms of the orthonormal basis $\{\mathbf{v}_i\}$ as follows:

$$\mathbf{x} = \sum_{i=1}^{N_T} \mathbf{v}_i \mathbf{v}_i^H \mathbf{x},$$

so that

$$\mathbf{H}\mathbf{x} = \sum_{i=1}^{N_T} \mathbf{H} \mathbf{v}_i \mathbf{v}_i^H \mathbf{x} = \sum_{i=1}^{k} \mathbf{H} \mathbf{v}_i \mathbf{v}_i^H \mathbf{x}$$

$$= \sum_{i=1}^{k} \sqrt{\lambda_i} \mathbf{u}_i \mathbf{v}_i^H \mathbf{x} = \sum_{i=1}^{N_T} \sqrt{\lambda_i} \mathbf{u}_i \mathbf{v}_i^H \mathbf{x},$$

8.7 Space–time communication

where we have used (8.124) in the second equality, and the fact that $\lambda_i = 0$ for $i > k$ in the fourth equality. This completes the proof of (8.125).

Application of SVD If we send the vector $\mathbf{v}_i$ through the channel matrix $\mathbf{H}$, we get a nonzero output corresponding to the k nonzero singular values. For these, we get the noiseless outputs $\mathbf{H}\mathbf{v}_i = \sqrt{\lambda_i}\mathbf{u}_i$, $i = 1, \ldots, k$. Since the vectors $\{\mathbf{v}_i\}$ form an orthonormal basis for the input space, and the vectors $\{\mathbf{u}_i\}$ form an orthonormal basis for the output space, we have effectively decomposed the channel into k parallel channels, corresponding to k nontrivial "eigenmodes." More formally, let us express the $N_T \times 1$ channel input $\mathbf{x}$ in terms of the orthonormal basis $\{\mathbf{v}_i\}$, and the $N_R \times 1$ channel output $\mathbf{y}$ in terms of the orthonormal basis $\{\mathbf{u}_i\}$ as follows:

$$\hat{\mathbf{x}} = \mathbf{V}^H \mathbf{x}, \quad \hat{\mathbf{y}} = \mathbf{U}^H \mathbf{y}.$$

Using the SVD (8.125), the channel model (8.122) can now be rephrased as

$$\hat{\mathbf{y}} = \mathbf{D}\hat{\mathbf{x}} + \hat{\mathbf{w}},$$

where $\hat{\mathbf{w}}$ is complex WGN as before. That is, we now have k parallel scalar channels of the form:

$$\hat{y}_i = \sqrt{\lambda_i}\hat{x}_i + \hat{w}_i, \quad i = 1, \ldots, k,$$

corresponding to the k nonzero eigenvalues of $\mathbf{W}$.

What we do now depends on whether or not the transmitter knows the channel. If it does, then we can simply apply the waterfilling solution developed in Chapter 6 for parallel Gaussian channels. Let us therefore assume that the transmitter does not know the channel. We wish to communicate at the largest rate possible, under the input power constraint $\mathbb{E}[\|\mathbf{x}\|^2] \leq P$. A codeword spanning n time intervals is a sequence of input vectors $\mathbf{X} = \{\mathbf{x}[j], j = 1, \ldots, n\}$. The corresponding output vectors are $\mathbf{Y} = \{\mathbf{y}[j], j = 1, \ldots, n\}$, where $\mathbf{y}[j] = \mathbf{H}\mathbf{x}[j] + \mathbf{w}[j]$, where $\mathbf{W} = \{\mathbf{w}[j], j = 1, \ldots, n\}$ is complex WGN. We wish to maximize the average mutual information $(1/n)I(\mathbf{X}, \mathbf{Y})$. We first show that this can be achieved by i.i.d. zero mean, proper complex Gaussian inputs $\mathbf{x}[j]$. To show the optimality of Gaussian inputs, we can reason as follows:

- If the channel is known at the receiver, $H(\mathbf{Y}|\mathbf{X}) = H(\mathbf{W})$, independent of the input distribution. Thus, maximizing the mutual information is equivalent to maximizing the output entropy $H(\mathbf{Y})$.
- We have $H(\mathbf{Y}) \leq \sum_{j=1}^{n} H(\mathbf{y}[j])$, with equality if $\{\mathbf{y}[j]\}$ are independent. The latter can be achieved by choosing $\{\mathbf{x}[j]\}$ independent, and by maximizing the marginals $H(\mathbf{y}[j])$. Thus, the problem reduces to choosing $\mathbf{x}[j]$ i.i.d. with a distribution that maximizes $H(\mathbf{y}[j])$ for each j.
- Given an input $\mathbf{x}$ with covariance $\mathbf{C}_x$, the covariance of $\mathbf{y} = \mathbf{H}\mathbf{x} + \mathbf{w}$ is given by $\mathbf{C}_y = \mathbf{H}\mathbf{C}_x\mathbf{H}^H + 2\sigma^2 \mathbf{I}$. For a given covariance $\mathbf{C}_y$, the (differential) entropy is maximized if $\mathbf{y}$ is proper complex Gaussian (the proof is similar

to that for a scalar Gaussian channel in Chapter 6). This can be achieved by choosing the input **x** to be proper complex Gaussian.
- Since the entropy of the output does not depend on the mean, choosing a nonzero input mean represents a power expenditure which does not increase the output entropy. We can therefore set the input mean to zero.

Based on the preceding reasoning, the only remaining issue is the choice of the input spatial covariance $\mathbf{C_x}$ subject to the power constraint trace $(\mathbf{C_x}) \leq P$. Before doing this, We provide a formula (proved in Problem 8.31) for the differential entropy for an $n \times 1$ complex Gaussian vector $\mathbf{Z} \sim CN(\mathbf{m}, \mathbf{C})$:

$$h(\mathbf{Z}) = \log \det (\pi e \mathbf{C}) = \sum_{i=1}^{n} (\log \lambda_i + \log \pi e), \quad (8.126)$$

where det denotes the determinant of a matrix, and $\{\lambda_i\}$ are the eigenvalues of $\mathbf{C}$.

Capacity with proper complex Gaussian input For the channel (8.122), for complex Gaussian input (which leads to complex Gaussian output), we obtain, upon some manipulation,

$$I(\mathbf{x}; \mathbf{y}) = H(\mathbf{y}) - H(\mathbf{w}) = \log \det \left(2\sigma^2 \mathbf{I} + \mathbf{H} \mathbf{C_x} \mathbf{H}^H\right) - \log \det(2\sigma^2 \mathbf{I})$$
$$= \log \det \left(\mathbf{I} + \tfrac{1}{2\sigma^2} \mathbf{H} \mathbf{C_x} \mathbf{H}^H\right). \quad (8.127)$$

Robustness of white Gaussian input The specific choice of spatially white Gaussian input $\mathbf{C_x} = P\mathbf{I}/N_T$ has the property that it sends equal energy along all channel eigenmodes $\{\mathbf{v}_i\}$, regardless of the channel matrix $\mathbf{H}$. Furthermore, the projections $\hat{\mathbf{x}}_i$ along the eigenmodes are independent, so that the entropies of the parallel channels corresponding to the eigenmodes simply add up. The resulting mutual information is

$$C_{\text{white}} = \log \det \left(\mathbf{I} + \frac{\text{SNR}}{N_T} \mathbf{H} \mathbf{H}^H\right) = \sum_i \log \left(1 + \frac{\text{SNR}}{N_T} \lambda_i\right), \quad (8.128)$$

where SNR $= P/2\sigma^2$.

We have just presented a heuristic argument that using spatially white inputs is a good strategy when the transmitter does not know the spatial channel. Moreover, spatially white inputs can be shown to be *optimal* in certain settings. Specifically, for rich scattering environments in which the entries of **H** are i.i.d., zero mean complex Gaussian random variables, spatially white inputs maximize ergodic capacity. They are also conjectured to maximize outage capacity, assuming that the desired outage probability is small enough.

Ergodic capacity for rich scattering The ergodic capacity for a rich scattering environment in which the entries of **H** are i.i.d., zero mean, proper complex Gaussian is achieved by spatially white input with $\mathbf{C_x} = P\mathbf{I}/N_T$.

Substituting into (8.127) and taking expectation over $\mathbf{H}$, we get

$$C_{\text{rich scattering}} = \mathbb{E}\left[\log\det\left(\mathbf{I} + \frac{\text{SNR}}{N_T}\mathbf{HH}^H\right)\right] = \sum_i \mathbb{E}\left[\log\left(1 + \frac{\text{SNR}}{N_T}\lambda_i\right)\right], \tag{8.129}$$

where λ_i are the eigenvalues of the *Wishart* matrix $\mathbf{W}$, defined as

$$\mathbf{W} = \begin{cases} \mathbf{H}^H\mathbf{H}, & N_T \leq N_R, \\ \mathbf{HH}^H, & N_R \leq N_T. \end{cases}$$

That is, $\mathbf{W}$ is an $m \times m$ matrix, where $m = \min(N_T, N_R)$. We see, therefore, that

$$C = \min(N_T, N_R)\mathbb{E}\left[\log\left(1 + \frac{\text{SNR}}{N_T}\lambda\right)\right] \text{ MIMO capacity with rich scattering,}$$

where λ is a typical eigenvalue of $\mathbf{W}$. Thus, even without specifying the distribution of λ, we see that the capacity scales linearly with the minimum of the number of transmit and receive antenna elements, as long as the distribution of λ has some probability mass away from zero. This linear scaling, which is termed the *spatial multiplexing gain*, is analogous to that provided by additional bandwidth: adding antennas so as to increase $\min(N_T, N_R)$ increases the dimension of the available signal space, just as increasing the time–bandwidth product does. In Section 8.7.3, we discuss constructive techniques for exploiting these spatial degrees of freedom.

8.7.3 Spatial multiplexing

We know now that MIMO capacity in a rich scattering environment scales with $\min(N_T, N_R)$. Let us now consider how we would attain this linear scaling of capacity in practice. For simplicity, suppose that $N_R \geq N_T$. The received signal can be written as

$$\mathbf{y} = b_1\mathbf{h}_1 + \cdots + b_{N_T}\mathbf{h}_{N_T} + \mathbf{W}. \tag{8.130}$$

Comparing with the synchronous CDMA (8.73) in Section 8.4.4, we note that the receive array responses $\{\mathbf{h}_i\}$ play the same role as the scaled spreading codes $\{A_k\mathbf{s}_k\}$. That is, we can think of this as a CDMA system with N_T "users," with the ith user, or transmit element, being assigned a spreading code $\mathbf{h}_i$ by nature. We can now apply multiuser detection techniques as in Section 8.4.4 to demodulate these N_T streams. Assuming that $N_R \geq N_T$, $\{\mathbf{h}_i\}$ are likely to be linearly independent in a rich scattering environment, and linear interference suppression based on the ZF or MMSE criterion should work. Indeed, the first prototype BLAST (Bell Labs layered space–time architecture) system developed by Bell Labs employed linear MMSE reception. There are many different combinations of coded modulation at the transmitter, and multiuser detection techniques at the receiver, that can be employed. However, they are all based on the basic understanding that (a) spatial multiplexed systems are analogous to CDMA systems, with signaling waveforms provided by nature, and (b) since the N_T "users" are colocated at the transmitter, it is possible to employ a single channel code across users, thus getting the benefit of transmit diversity as well (different transmit elements can see space–time channels of different quality, depending on the strengths of the channel gains and the geometric relationship between the channel vectors $\{\mathbf{h}_i\}$).

8.7.4 Space–time coding

We have seen the benefits of time diversity for coded transmission over fading channels. Time diversity, however, is often not available for systems with quasistationary nodes, in which channel time variations might be very slow relative to the duration of the codeword; an example is the indoor WLAN channel, in which nodes such as laptops might be tetherless, but relatively immobile. In such scenarios, spatial and frequency diversity are particularly important. In this section, we focus on space–time codes for exploiting spatial transmit diversity, assuming a narrowband, time-invariant system without frequency or time diversity.

Consider a system with N_T transmit antennas and one receive antenna (any space–time code designed for one receive antenna can be used with multiple receive antennas, simply by employing maximal ratio combining at the receiver). The channel matrix is now a vector $\mathbf{h} = (h_1, \ldots, h_{N_T})$, where h_i is the gain from transmit element i to the receive element. The input to transmit element i at time m is denoted by $x_i[m]$, with $\mathbf{x}[m] = (x_1[m], \ldots, x_{N_T}[m])^T$. The received signal at time m is given by

$$y[m] = h_1 x_1[m] + \cdots h_{N_T} x_{N_T}[m] + w[m] = \mathbf{h}\mathbf{x}[m] + w[m], \quad (8.131)$$

where $w[m]$ are i.i.d. $CN(0, 2\sigma^2)$ noise samples. A space–time codeword spanning N symbol intervals is a sequence $\mathbf{X} = \{\mathbf{x}[m], m = 1, \ldots, N\}$, chosen from a set of 2^{NR} possible codewords, where R bits per channel use is the code rate. The channel $\mathbf{h}$ is fixed over the codeword duration. We consider *coherent* space–time codes here, designed under the assumption that the transmitter does not know the channel $\mathbf{h}$, but that the receiver does know the channel. We constrain the transmit power to P per unit time, splitting the power evenly among all of the transmit elements. Thus, we set $\mathbb{E}[||\mathbf{x}[n]||^2] = \sum_{l=1}^{N_T} \mathbb{E}[|x_l[n]|^2] = P$, with $\mathbb{E}[|x_l[n]|^2] = P/N_T$, $1 \le l \le N_T$.

Ideal performance From the capacity formula (8.127), we see that i.i.d. Gaussian input sent from each transmit element achieves the capacity

$$C(\mathbf{h}) = \log\left(1 + \frac{||\mathbf{h}||^2}{N_T} \text{SNR}\right). \quad (8.132)$$

This is equal to the capacity of an AWGN channel in which each symbol sees an effective channel gain of $G = ||\mathbf{h}||^2/N_T = \frac{1}{N_T}\sum_{l=1}^{N_T} |h_l|^2$. For i.i.d. $\{h_l\}$, the averaging used to produce this effective gain reduces fluctuations in $C(\mathbf{h})$, and improves the outage capacity.

While the capacity *formula* (8.132) is the same as that for an AWGN channel with gain $G = ||\mathbf{h}||^2/N_\mathrm{T}$, the *physical* space–time channel (8.131) is much messier: the symbols sent from different transmit elements at a given time superimpose at the receiver, which then has to somehow disentangle them: doing so for i.i.d. Gaussian input requires ML decoding with complexity exponential in N_T. Practical space–time code constructions, therefore, are designed to allow structured decoding at reasonable complexity. Consider, for example, space–time coding for a cellular downlink: we might want to improve performance by increasing the number of base station antennas, but we do not want the decoding complexity at the mobile to explode.

In the following, we provide two examples of space–time codes. The first, which we term antenna hopping, is suboptimal, but scales well with N_T. The second is the Alamouti code, which is optimal, but only applies to $N_\mathrm{T} = 2$. In both cases, it is straightforward to use a standard error-correcting coded modulation in conjunction with the space–time code: the map between symbols coming out of the encoder and the output of the transmit elements is a straightforward one. Thus, we discuss only the impact of the space–time code on information-theoretic limits, with the understanding that such limits are within reach using, say, an appropriate turbo-like coded modulation strategy prior to the space–time code.

Antenna hopping A trivial, but effective, space–time code that scales easily to an arbitrary number of transmit antennas is simply to hop across transmit elements, using exactly one transmit element at a time in an alternating fashion. An example for three antennas is shown in Figure 8.23. The maximum achievable rate for a given channel $\mathbf{h}$ with this strategy therefore becomes the average of the rates corresponding to each transmit element:

$$C_\text{alternating}(\mathbf{h}) = \frac{1}{N_\mathrm{T}} \sum_{l=1}^{N_\mathrm{T}} \log\left(1 + |h_l|^2 \, \text{SNR}\right). \qquad (8.133)$$

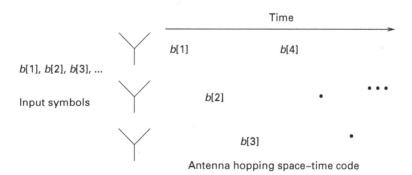

Figure 8.23 Antenna hopping space–time code.

It can be shown using Jensen's inequality (see Appendix C and Problem 8.32) that this is smaller than the capacity in (8.132) for any fixed $\mathbf{h}$, which also implies that the outage capacity is smaller, regardless of the statistics of $\mathbf{h}$. Note that, if the gains $\{h_l\}$ are i.i.d., then the terms in the average (8.133) are i.i.d. Thus, for a moderately large number of transmit elements (e.g., $N_T \geq 6$), $C(\mathbf{h})$ is well approximated as a Gaussian random variable using the central limit theorem. This allows us to compute an analytical approximation to the outage capacity, as shown in Problem 8.32. Moreover, if we let $N_T \to \infty$, then the law of large numbers implies that $C(\mathbf{h})$ tends to the ergodic capacity of the Rayleigh fading channel, given by (8.22).

Alamouti code The Alamouti code is optimal for $N_T = 2$, in the sense that it achieves the capacity (8.132). Its simplicity makes it attractive for implementation, and it has been standardized for third generation cellular systems. Suppose that a stream of symbols $\{b[n]\}$ is coming from an outer encoder. These are sent in blocks of two over the channel. For example, as shown in Figure 8.24, $b[1]$ and $b[2]$ are mapped to $\mathbf{x}[1]$ and $\mathbf{x}[2]$ as follows:

$$\begin{aligned} x_1[1] = b[1], \quad x_1[2] = -b^*[2], \\ x_2[1] = b[2], \quad x_2[2] = b^*[1]. \end{aligned} \quad (8.134)$$

The corresponding received signals are

$$\begin{aligned} y[1] &= h_1 x_1[1] + h_2 x_2[1] + w[1] = h_1 b[1] + h_2 b[2] + w[1], \\ y[2] &= h_1 x_1[2] + h_2 x_2[2] + w[2] = -h_1 b^*[2] + h_2 b^*[1] + w[2]. \end{aligned}$$

We would now like to think of these two received samples as a single vector, and see the vector response to the two symbols. Let us therefore consider the vector $\mathbf{y} = (y[1], y^*[2])^T$, which obeys the following model:

$$\begin{aligned} \mathbf{y} = \begin{pmatrix} y[1] \\ y^*[2] \end{pmatrix} &= b[1]\begin{pmatrix} h_1 \\ h_2^* \end{pmatrix} + b[2]\begin{pmatrix} h_2 \\ -h_1^* \end{pmatrix} + \begin{pmatrix} w[1] \\ w^*[2] \end{pmatrix} \quad (8.135) \\ &= b[1]\mathbf{v}_0 + b[2]\mathbf{v}_1 + \mathbf{w}. \end{aligned}$$

Note that the vectors $\mathbf{v}_0$ and $\mathbf{v}_1$ modulating $b[1]$ and $b[2]$, respectively, are orthogonal, regardless of $\mathbf{h}$. Also, the noise vector $\mathbf{w} = (w[1], w^*[2])^T$ remains complex WGN. Thus, by projecting $\mathbf{y}$ along $\mathbf{v}_0$ and $\mathbf{v}_1$, we can create two entirely separate AWGN channels for $b[1]$ and $b[2]$ with effective SNR

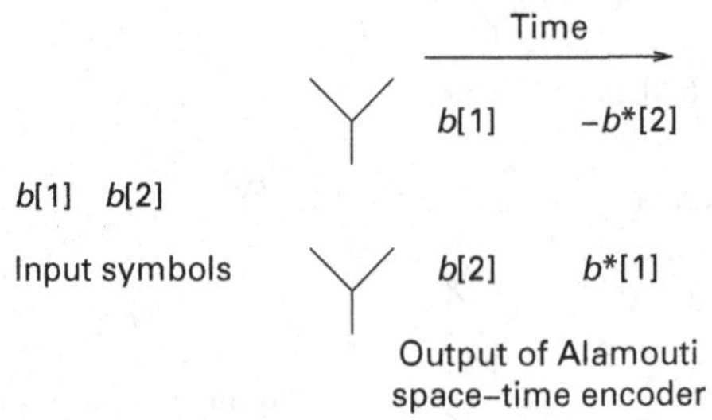

Figure 8.24 Alamouti space–time code.

$||\mathbf{v}_0||^2/2\sigma^2 = ||\mathbf{v}_1||^2/2\sigma^2 = ||\mathbf{h}||^2/2\sigma^2$. The capacity of such a channel is precisely (8.132), with the power normalization $\mathbb{E}[|b[n]|^2] = P/N_T = P/2$ for the symbols sent to the antenna elements. This shows the optimality of the Alamouti code.

Significant research effort has gone into developing space–time block codes generalizing the Alamouti code to a larger number of antennas, as well as into trellis-based "convolutional" constructions. Discussion of such strategies is beyond the scope of this book.

8.7.5 Transmit beamforming

Consider a system with N_T transmit elements and one receive element. If the space–time channel $\mathbf{h} = (h_1, \ldots, h_{N_T})$ from the transmitter to the receiver is known at the transmitter, then a spatial matched filter $\mathbf{h}^H$ can be implemented at the transmitter. This is equivalent to sending along the sole channel eigenmode, and is information-theoretically optimal from the results of Section 8.7.2. Thus, to send the symbol sequence $\{b[n]\}$, we send from transmit antenna k the sequence $x_k[n] = h_k^* b[n]$. The received signal at time n is given by

$$y[n] = \sum_{k=1}^{N_T} h_k x_k[n] + w[n] = b[n] \sum_{k=1}^{N_T} |h_k|^2 + w[n],$$

where $w[n]$ is WGN. Thus, we see an effective channel gain of $||\mathbf{h}||^2$.

Transmit beamforming is an intuitively obvious strategy if the transmitter knows the channel. It is well matched to cellular downlinks, such as the one shown in Figure 8.22, in which the base station transmitter may have a large number of antennas, but the number of eigenmodes may be small because (a) the power-angle profile from the base station's point of view is narrow, and (b) the mobile receiver may have only one or two antennas. Of course, there are many issues regarding the implementability and optimality of transmit beamforming strategies, but these are beyond the present scope. Some references for further reading on this topic are provided in Section 8.8.

8.8 Further reading

We mention a few among the many recent books devoted to wireless communication: Goldsmith [87], Rappaport [88] and Stuber [89] provide broad descriptions of the field, but with somewhat different emphases. Tse and Viswanath [90] focus on the fundamentals of fading, interference, and MIMO channels. Other useful texts include Jakes [91] and Parsons [92].

From a Shannon theory point of view, there has been a significant effort in recent years at understanding the effect of channel time variations on capacity. For a time-varying channel, it is unrealistic to assume that the channel is known a priori to the receiver, hence channel estimation, explicit or implicit, must be part of the model for which capacity should be computed. Examples of Shannon theoretic characterization of noncoherent communication over time-varying channels include Marzetta and Hochwald [93], Lapidoth and Moser [94], and Etkin and Tse [95]. Turbo constructions and comparison with Shannon theoretic limits with constellation constraints are provided by Chen *et al.* [82] and Jacobsen and Madhow [96].

There is a very large literature on OFDM, but a good starting point for getting more detail than in this chapter are the relevant chapters in [87] and [90]. For DS-CDMA with conventional reception, we recommend the book by Viterbi [97], which provides a description of the concepts underlying the CDMA-based digital cellular standards. The Rappaport text [88] also provides detail on systems aspects of these standards. Properties of spreading sequences are discussed by Pursley and Sarwate [98]. Detailed error probability analyses for direct sequence CDMA with conventional reception are contained in [99, 100, 101]. An excellent treatment of multiuser detection is found in the text by Verdu [102]. Early papers on multiuser detection include Verdu [103, 104], Lupas and Verdu [105, 106], and Varanasi and Aazhang [107, 108]. Early papers on *adaptive* multiuser detection, or adaptive interference suppression, include Abdulrahman *et al.* [109], Rapajic and Vucetic [110], Madhow and Honig [31], and Honig *et al.* [111]. The application of adaptive interference suppression techniques to timing acquisition is found in Madhow [112]. Finally, standard adaptive algorithms have difficulty keeping up with the time variations of the wireless mobile channel; see Madhow *et al.* [113], and the references therein, for variants of the linear MMSE criterion for handling such time variations.

For the Laurent decomposition of CPM, the best reference is the original paper by Laurent [114]. An early example of the use of the Laurent approximation for MLSE reception for GMSK is given in [115]. Generalization of the Laurent decomposition to M-ary CPM is provided in a later paper by Mengali and Morelli [116]. An alternative decomposition of CPM is given in the paper by Rimoldi [117]. The book by Anderson *et al.* [118] provides a detailed treatment of several aspects of CPM.

For space–time communication, the original technical report by Telatar [119] (also published in journal form in [120]) is highly recommended reading, as is the original paper by Foschini [121]. There are a number of recent books focusing specifically on space–time communication, including Paulraj *et al.* [122] and Jafarkhani [123]. A recent compilation edited by Bolcskei *et al.* [124] is a useful resource, covering a broad range of topics. The book by Tse and Viswanath [90] contains a detailed treatment of fundamental tradeoffs in space–time communication. Space–time communication plays a key role in the emerging IEEE 802.11n WLAN standard. Finally, while the channel for indoor space–time communication is often well modeled as frequency nonselective (small delay spreads imply large coherence bandwidths relative to typical bandwidths used), emerging broadband systems for outdoor space–time communication, such as IEEE 802.16 and 802.20, are frequency selective. Since these emerging standards are based on OFDM, we often term the resulting systems MIMO-OFDM. Examples of recent theoretical research on MIMO-OFDM include Bolcskei et al. [125], and Barriac and Madhow [126, 127].

8.9 Problems

Problem 8.1 (Rician fading) Consider a narrowband frequency nonselective channel with channel gain h modeled as $h \sim CN(m, 2v^2)$. Then $|h|$ is a Rician random variable with Rice factor $K = |m|^2/2v^2$ measuring the relative strengths of the LOS and diffuse components. Suppose that we are designing a system for 95% outage. If a nominal link budget is computed based on the average channel power gain $E[|h|^2]$, we define the link margin needed to combat fading as the additional power (expressed in dB) needed to attain a 95% outage rate.

(a) What is the required link margin (dB) for Rayleigh fading ($K = 0$)?
(b) What is the required link margin (dB) for the AWGN channel ($K = \infty$)?
(c) Plot the required link margin (dB) as a function of K, as K varies from 0 to ∞. What is the value of K at which the required link margin equals 3 dB?

Problem 8.2 (Power-delay profile and coherence bandwidth) In this problem, we relate the channel coherence bandwidth to the channel PDP $P(\tau)$. It is convenient to consider a limiting case of the TDL model (8.6) for large signaling bandwidth W. We therefore get

$$h(t) = \int_0^\infty h(\tau)\delta(t-\tau),$$

where we replace i/W by τ, and $\alpha(i/W)$ by $h(\tau)$, but maintain the assumption that $h(\tau)$ are independent zero mean complex Gaussian random variables, with $h(\tau) \sim CN(0, P(\tau))$, where we normalize $P(\tau)$ to unit area. Let $H(f) = \int h(t)e^{-j2\pi ft}dt$ denote the (random) channel gain at frequency f.

(a) Show that $\{H(f)\}$ are jointly complex Gaussian random variables, with $H(f) \sim CN(0, 1)$.
(b) Show that $\text{cov}(H(f_1), H(f_2)) = \hat{P}(f_1 - f_2)$, where $\hat{P}(f)$ is the Fourier transform of the PDP $P(\tau)$.
(c) Define the coherence bandwidth (somewhat arbitrarily) as the minimum separation $|f_1 - f_2|$ where the normalized correlation between $H(f_1)$ and

$H(f_2)$ is 0.1. Find the coherence bandwidth for an exponential PDP with rms delay spread of 10 microseconds.

(d) Suppose you are designing an OFDM system with subcarrier spacing much smaller than the coherence bandwidth, in order to ensure that the channel gain $H(f)$ over a subcarrier is approximately constant. If your design criterion is that the normalized correlation between adjacent subcarriers is at least 0.9, what is the minimum subcarrier spacing?

Problem 8.3 (Shannon capacity with Rayleigh fading) Consider the information-theoretic analysis of the coded Rayleigh fading channel described in Section 8.2.2.

(a) Using (8.22) and Jensen's inequality (see Appendix C), show that the ergodic capacity with Rayleigh fading is strictly smaller than that of an AWGN channel with the same average SNR. That is, fading always reduces the capacity relative to a constant channel gain.

(b) For a Rayleigh fading channel, we have $h[n] \sim CN(0, 1)$. Compute $C_{\text{fading}}(S/N)$ for this case and plot it as a function of SNR in dB. Also plot the AWGN capacity for comparison.

(c) Find (analytically) the asymptotic penalty in dB for Rayleigh fading as SNR $\to \infty$.

(d) Repeat (c) for SNR $\to 0$.

(e) If the transmitter also knew the fading coefficients $\{h[n]\}$ (e.g., this could be achieved by explicit channel feedback or, for slow fading, by the use of reciprocity), what is your intuition as to what the best strategy should be?

Problem 8.4 Suppose that you employ BPSK with hard decisions and ideal interleaving over a Rayleigh fading channel.

(a) Show that the Shannon capacity as a function of $\bar{E}_b/N_0$ is given by

$$C = \mathbb{E}[1 - H_B(p(G))],$$

where the expectation above is with respect to an exponential random variable G with mean one, H_B is the binary entropy function, and $p(G) = Q\left(\sqrt{2\bar{E}_b G/N_0}\right)$.

(b) Show that (a) also holds for binary DPSK with hard decisions, except that $p(G) = \frac{1}{2}e^{-G\bar{E}_b/N_0}$.

(c) Plot the capacities in (a) and (b) as a function of $\bar{E}_b/N_0$ (dB). Also plot for comparison the corresponding capacity for both coherent BPSK and DPSK with hard decisions over the AWGN channel. Comment on the difference in performance between the coherent and noncoherent systems, and on the difference between the AWGN and Rayleigh fading channels.

8.9 Problems

Problem 8.5 (Sum of i.i.d. exponential random variables) For a random variable X, set

$$K_X(s) = \mathbb{E}[e^{sX}] = \int_{-\infty}^{\infty} e^{sx} p(x) dx, \qquad (8.136)$$

where $p(x)$ denotes the density of X. Note that $K_X(s)$ is the Laplace transform of the density $p(x)$ (although s is replaced by $-s$ in standard signals and systems texts). The present notation is more common when taking Laplace transforms of densities. The expression (8.136) is defined wherever the integral converges, and the range of s where the integral converges is called the region of convergence.

(a) For X exponential with mean one, show that $K_X(s) = 1/(1-s)$, with region of convergence $\mathrm{Re}(s) < 1$.

(b) If $Y = X_1 + X_2$, where X_1 and X_2 are independent, show that

$$K_Y(s) = K_{X_1}(s) K_{X_2}(s).$$

(c) Differentiate (8.136) with respect to s to show that the Laplace transform of $xp(x)$ is $dK_X(s)/ds$.

(d) Consider $Y = X_1 + \cdots + X_N$, where $\{X_i\}$ are i.i.d., exponential with mean one. We term the random variable Y a standard Gamma random variable with dimension N. Show that

$$K_Y(s) = \frac{1}{(1-s)^N}, \quad \mathrm{Re}(s) < 1. \qquad (8.137)$$

(e) Use (d), and repeated applications of (c), to show that the density of Y in (d) is given by

$$p(y) = \frac{y^{N-1}}{(N-1)!} e^{-y}, \quad y \geq 0. \qquad (8.138)$$

(f) Now, suppose that the exponential random variables in (d) each have parameter μ, where $1/\mu$ is the mean. Show that the density of the sum Y is given by

$$p(y) = \mu \frac{(\mu y)^{N-1}}{(N-1)!} e^{-\mu y}, \quad y \geq 0. \qquad (8.139)$$

Hint Scale the random variables in (d) and (e) by $1/\mu$.

Problem 8.6 (Relating the Gamma random variable to a Poisson process) Consider a Poisson process $\{N(t), t \geq 0\}$, which is defined as follows: we start with $N(0) = 0$, and $N(t)$ jumps by one at times $t_1, t_2, \ldots$, so that $N(t_n) = n$ and $N(t_n^-) = n-1$. Setting $t_0 = 0$, define the inter-arrival times

$X_i = t_i - t_{i-1}$, $i = 1, 2, \ldots$. These are i.i.d. exponential random variables with mean $1/\mu$. The Poisson process is also characterized by the following properties:

- *Increments are Poisson random variables* For $u_2 \geq u_1 \geq 0$, $N(u_2) - N(u_1)$ is a Poisson random variable with mean $m = \mu(u_2 - u_1)$. That is, its pmf is given by

$$P[N(u_2) - N(u_1) = k] = \frac{m^k}{k!} e^{-m}, \quad k = 0, 1, 2, \ldots$$

- *Independent increments* Increments over nonoverlapping time intervals are independent. That is, for $u_4 \geq u_3 \geq u_2 \geq u_1 \geq 0$, the random variables $N(u_4) - N(u_3)$ and $N(u_2) - N(u_1)$ are independent.

Using these properties, we show how to infer the properties of Gamma random variables. Consider a Gamma random variable Y of dimension N: $Y = X_1 + \cdots + X_N$, where $\{X_i\}$ are i.i.d. exponential random variables with mean $1/\mu$. Interpreting the X_i as inter-arrival times of a rate μ Poisson process $N(t)$ as discussed above, we can now infer properties of Gamma random variables from those of the Poisson random process as follows.

(a) Show that the event $Y \leq y$ is equivalent to the event $N(y) \geq N$.
(b) Use (a) to infer that the cdf of Y is given by

$$P[Y \leq y] = 1 - e^{-\mu y} \sum_{k=0}^{N-1} \frac{(\mu y)^k}{k!}.$$

(c) Show that the event $Y \in [y, y+\delta)$ is equivalent to the events $\{N(y) \leq N\}$ **and** $\{N(y+\delta) > N\}$.
(d) For small δ, argue that the event in (c) is dominated by the events $\{N(y) = N\}$ **and** $\{N(y+\delta) = N+1\}$.

Hint Show that the ratio of the probability that the increment over an interval of length δ equals two or more to the probability that it equals one tends to zero as $\delta \to 0$.

(e) Using (c) and (d), find the density $p(y)$ of Y using the following relations for small δ:

$$P[Y \in [y, y+\delta)] \approx p(y)\delta$$

and

$$P[Y \in [y, y+\delta)] \approx P[N(y) = N, N(y+\delta) = N+1]$$
$$= P[N[y] = N, N(y+\delta) - N(y) = 1].$$

Use the independent increments property and simplify to obtain (8.139).

8.9 Problems

Problem 8.7 (The Gamma function) For $x > 0$, the Gamma function is defined as

$$\Gamma(x) = \int_0^\infty t^{x-1} e^{-t} dt. \tag{8.140}$$

(a) For $x > 1$, use integration by parts to show that

$$\Gamma(x) = (x-1)\Gamma(x-1). \tag{8.141}$$

(b) Show that $\Gamma(1) = 1$.

(c) Use (a) and (b) to show that, for any positive integer N,

$$\Gamma(N) = (N-1)!.$$

(d) Show that $\Gamma(1/2) = \sqrt{\pi}$.

Hint You can relate the corresponding integral to a standard Gaussian density by substituting $t = z^2/2$.

(e) Suppose that Y is a standard Gamma random variable of dimension n, as in Problem 8.5(d)–(e). Find $\mathbb{E}[\sqrt{Y}]$.

Problem 8.8 (Error probability bound for receive diversity with maximal ratio combining) Consider a coherent binary communication system over a Rayleigh fading channel with N-fold receive diversity and maximal ratio combining, where the fading gains are i.i.d. across elements, and the received signal at each element is corrupted by i.i.d. AWGN processes. Let e_b denote the average bit energy per diversity branch, and let E_b denote the average bit energy summed over diversity branches.

(a) For a BPSK system, show that the error probability with maximal ratio combining can be written as

$$P_e = \mathbb{E}\left[Q\left(\sqrt{aY}\right)\right], \tag{8.142}$$

where a is a constant, and Y is a standard Gamma random variable of dimension N, as in Problem 8.5(d)–(e). Express a in terms of both e_b/N_0 and E_b/N_0.

(b) Using the bound $Q(x) \leq \frac{1}{2} e^{-x^2/2}$ in (8.142), show that (see Problem 8.5(d))

$$P_e \leq \frac{1}{2} \frac{1}{\left(\frac{a}{2}+1\right)^N}. \tag{8.143}$$

(c) For $N = 1, 2, 4, 8$, plot the error probability bound in (b) on a log scale versus E_b/N_0 (dB) for BPSK. Also plot for reference the error probability for BPSK transmission over the AWGN channel. Comment on how the error probability behaves with N.

Problem 8.9 (Exact error probability for receive diversity with maximal ratio combining) Consider the setting of Problem 8.8. We illustrate the exact computation of the error probability (8.142) for $N = 2$ in this problem, using the Gamma function introduced in Problem 8.7.

(a) It is helpful first to redo the computation for Rayleigh fading that was done in the text (i.e., leading to expressions such as (8.20)) using the Gamma function. For $N = 1$, we have

$$P_e = \int_0^\infty Q(\sqrt{ay})\, e^{-y}\, dy = -e^{-y} Q(\sqrt{ay})\Big|_0^\infty + \int_0^\infty e^{-y} \frac{d}{dy} Q(\sqrt{ay})\, dy$$

$$= \frac{1}{2} - \frac{\sqrt{a}}{2\sqrt{2\pi}} \int_0^\infty y^{-\frac{1}{2}} e^{-(1+\frac{a}{2})y}\, dy = \frac{1}{2} - \frac{\sqrt{a}}{2\sqrt{2\pi}} \gamma\left(\frac{1}{2}\right)\left(1+\frac{a}{2}\right)^{-\frac{1}{2}}$$

$$= \frac{1}{2}\left(1 - \left(1+\frac{2}{a}\right)^{-\frac{1}{2}}\right). \tag{8.144}$$

(b) Now, use (8.138) for $N = 2$ and integrate by parts to show that

$$P_e = \int_0^\infty Q(\sqrt{ay})\, y e^{-y}\, dy = -e^{-y} y Q(\sqrt{ay})\Big|_0^\infty + \int_0^\infty e^{-y} \frac{d}{dy}(y Q(\sqrt{ay}))\, dy.$$

(c) Obtain the expression

$$P_e = \int_0^\infty e^{-y} Q(\sqrt{ay})\, dy - \frac{\sqrt{a}}{2\sqrt{2\pi}} \int_0^\infty y^{\frac{1}{2}} e^{-(1+\frac{a}{2})y}\, dy.$$

(d) The first term is exactly the same as in (a) for $N = 1$, and evaluates to (8.144). The second term can be evaluated using Gamma functions. Simplify to obtain the expression

$$P_e = \frac{1}{2}\left(1 - \left(1+\frac{2}{a}\right)^{-\frac{1}{2}} - \frac{1}{a}\left(1+\frac{2}{a}\right)^{-\frac{3}{2}}\right). \tag{8.145}$$

(e) Plot the preceding expression (log scale) as a function of E_b/N_0 (dB) for BPSK. Plot for comparison the corresponding bound in Problem 8.8.

(f) Show that the high SNR asymptotics for the error probability expressions (8.144) and (8.145) are given by

$$\begin{aligned} P_e &\approx \frac{1}{2a},\quad N = 1, \\ P_e &\approx \frac{3}{2a^2},\quad N = 2. \end{aligned} \tag{8.146}$$

Hint For x small, $(1+x)^b \approx 1 + bx$. Apply this for $x = 2/a$.

(g) Mimic the development in (c) and (d) for a general value of $N \geq 2$. That is, letting $P_e(N)$ denote the error probability with N-fold diversity, show that

$$P_e(N) = P_e(N-1) - f(N),$$

where you are to determine an expression for $f(N) > 0$.

8.9 Problems

Problem 8.10 (Error probability for selection diversity) Consider again the setting of Problem 8.8, but now assume that the receiver uses *selection diversity*. That is, it demodulates the signal coming in on the receive diversity branch with the biggest gain. The error probability again takes the form (8.142), but now

$$Y = \max(X_1, \ldots, X_N),$$

where $\{X_i\}$ are i.i.d. exponential random variables, each with mean one.

(a) Find the cumulative distribution function of Y.

(b) Find the density of Y. Specialize to $N = 2$ to show that

$$p(y) = 2\left(e^{-y} - e^{-2y}\right), \quad y \geq 0.$$

(c) For $N = 2$, evaluate the average error probability (8.142) to obtain

$$P_e = \frac{1}{2} - \left(1 + \frac{2}{a}\right)^{-\frac{1}{2}} + \frac{1}{2}\left(1 + \frac{4}{a}\right)^{-\frac{1}{2}}. \tag{8.147}$$

Hint Use (8.144).

(d) Show that the high SNR asymptotics of (8.147) give

$$P_e \approx \frac{3}{2a^2},$$

which is the same result as obtained in (8.146) for maximal ratio combining.

Hint Use the approximation $(1+x)^b \approx 1 + bx + \frac{b(b-1)}{2}x^2$ for small x.

Note The arguments used for $N = 2$ extend to general N-fold selection diversity, noting that the density of Y can be written as a linear combination of e^{-ky}, $k = 1, 2, \ldots, n$.

Problem 8.11 (Outage rates with receive diversity) Consider a system with N-fold receive diversity with maximal ratio combining, as in Problems 8.8–8.9.

(a) Find a Chernoff bound on the probability of a 10 dB fade after diversity combining. That is, consider the random variable Y in Problems 8.8–8.9, and find an upper bound on the probability $P[Y < 0.1\mathbb{E}[Y]]$ of the form $e^{-\beta N}$.

Hint For i.i.d. $\{X_i\}$, find the Chernoff bound $P[X_1 + \cdots + X_N < tN] \leq e^{-\beta N}$, where $\beta = \min_{s<0}\{M(s) - st\}$, with $M(s)$ denoting the moment generating function of X_1. See Appendix B.

(b) Using the upper bound in (a), estimate the number of diversity branches needed to reduce the probability of a 10 dB fade to 10^{-3}.

(c) For the number of diversity branches found in (b), calculate the exact probability of a 10 dB fade after diversity combining using the Gamma cdf derived in Problem 8.6(b).

Problem 8.12 (Error probability bound for noncoherent diversity combining) Recall from (8.37) that the error probability for FSK with noncoherent diversity combining is given by

$$P_e = P[V_1 + \cdots + V_N > U_1 + \cdots + U_N],$$

where $\{V_i\}$ are i.i.d. exponential random variables with mean $1/\mu_V = 2\sigma^2 E$, and $\{U_i\}$ are i.i.d. exponential random variables with mean $1/\mu_U = 2\sigma^2 E + E^2$ ($\{V_i\}$ and $\{U_i\}$ are independent). Set $X_i = V_i - U_i$, so that the error probability is given by

$$P_e = P[X_1 + \cdots + X_N > 0].$$

Note that $E[X_1] = \mathbb{E}[V_1] - \mathbb{E}[U_1] = -E^2 < 0$, so that we are computing a tail probability for which a nontrivial Chernoff bound exists (see Appendix B). Apply (B.6) from Appendix B (with $a = 0$) to obtain the bound

$$P_e \leq e^{-NM^*(0)},$$

where $M^*(0) = \min_{s>0} M(s)$, and $M(s) = \log \mathbb{E}[e^{sX_1}]$ is the moment generating function for X_1.

(a) Show that

$$\mathbb{E}[e^{sX_1}] = \frac{\mu_V}{\mu_V - s} \frac{\mu_U}{\mu_U + s}.$$

What is the range of s for which the preceding expectation exists?

(b) Show that the minimizing value of s for the Chernoff bound is

$$s_0 = \frac{\mu_V + \mu_U}{2}.$$

(c) Show that the Chernoff bound is given by

$$P_e \leq e^{-\alpha N},$$

where

$$\alpha = 2 \log \left(\frac{\frac{\mu_V + \mu_U}{2}}{\sqrt{\mu_V \mu_U}} \right).$$

(d) Substitute the values of μ_U and μ_V (note that $E = E_b$ for binary FSK) to obtain (8.38).

Problem 8.13 (Cyclic convolution and DFT) This problem shows that cyclic convolution in the time domain corresponds to multiplication in the DFT domain. That is, if h and g are vectors of length N, with DFT H and G, respectively, and if

$$y[k] = (h \odot g)[k] = \sum_{m=0}^{N-1} h[m] b[(k-m) \bmod N], \quad k = 0, 1, \ldots, N-1$$

denotes their cyclic convolution, then the DFT of y is given by

$$Y[n] = H[n]G[n]. \tag{8.148}$$

Show this result using the following steps:

(a) Define the complex exponential functions $\{g_n, n = 0, 1, \ldots, N-1\}$ as

$$g_n[k] = e^{j2\pi nk/N}, \quad k = 0, 1, \ldots, N-1.$$

Show that

$$h \odot g_n = H[n]g_n.$$

That is, the complex exponentials $\{g_n\}$ are eigenfunctions for the operation of cyclic convolution with h, with eigenvalues equal to the DFT coefficients $\{H[n]\}$.

(b) Recognize that g can be expressed as a linear combination of the eigenfunctions $\{g_n\}$ as follows:

$$g = \frac{1}{N} \sum_{n=0}^{N-1} G[n]g_n.$$

(c) Use the linearity of cyclic convolution to infer that

$$y = h \odot g = \frac{1}{N} \sum_{n=0}^{N-1} G[n](h \odot g_n).$$

Plug in the result of (a) to obtain that

$$y = h \odot g = \frac{1}{N} \sum_{n=0}^{N-1} G[n]H[n]g_n.$$

Recognize that the right-hand side is an inverse DFT to infer (8.148).

Problem 8.14 (OFDM for a real baseband channel) Suppose that you were implementing an OFDM system over a real baseband channel. Modifying (8.40) to center the transmitted signal around the origin, we have

$$u(t) = e^{-j\pi(N-1)t/T} \sum_{n=0}^{N-1} B[n]e^{j2\pi nt/T} I_{[0,T]}.$$

(a) What is the constraint on the frequency domain symbols $B[n]$ such that the time domain signal $u(t)$ is real-valued? Can you choose $\{B[n]\}$ from a complex-valued constellation? What is the bandwidth of the signal $u(t)$, as a function of N and T?

(b) For $N = 512$, a cyclic prefix of 50 samples, and a sampling rate of 2 MHz, what is the bandwidth and inter-carrier spacing of the baseband channel used? What is the maximum channel delay spread that the system is designed for?

Problem 8.15 (OFDM for a complex baseband channel) A complex baseband channel has impulse response

$$h(t) = \delta(t) - 0.7\mathrm{j}\delta(t-1) - 0.4\delta(t-2),$$

where the unit of time is microseconds.

(a) Plot $|H(f)|$ versus f over a frequency range spanning 10 times the coherence bandwidth.
(b) What is the Shannon capacity for signaling over this channel (assume AWGN) using a bandwidth of 10 MHz, assuming a white input and SNR of 10 dB? Compare it with the Shannon capacity for a nondispersive AWGN channel.
(c) Now assume that the channel is known to the transmitter, and that the transmitter employs waterfilling. Find the Shannon capacity for SNR of 10 dB. Compare this with the results from (b).
(d) How would you design an OFDM system to try to approach the information-theoretic limits of (b) and (c)? How many subcarriers? What constellations?

Problem 8.16 (Overhead in OFDM systems) Consider an OFDM system to be used over an indoor wireless channel modeled as having an exponential PDP with an rms delay spread of 100 milliseconds. Suppose that the cyclic prefix is chosen to span a length such that 90% of the channel energy falls within the length.

(a) What is the overhead for a 64 subcarrier system with a subcarrier spacing of 100 kHz?
(b) How does the answer to (a) change if we use a 128 subcarrier system with the same subcarrier spacing?
(c) How does the answer to (a) change if we use a 128 subcarrier system with half the subcarrier spacing?

Problem 8.17 (Peak-to-average ratio for OFDM systems) Consider an OFDM system with 64 subcarriers.

(a) Simulate the time domain samples (8.41) for i.i.d. frequency domain symbols $\{B[k]\}$ drawn from a QPSK constellation. Collect samples for multiple OFDM symbols and plot the histograms (separately for the I and Q components). Comment on whether the histograms look Gaussian.
(b) Compute the sample PAR defined in (8.49). Plot a histogram of the sample PAR obtained over multiple OFDM symbols. Estimate the probability that the sample PAR exceeds 10 dB from the empirical distribution thus obtained. Compare this with the analytical estimate obtained from (8.51). Plot a histogram of the PAR for multiple simulation runs, and comment on whether the histogram looks Gaussian.

(c) Repeat (a)–(b) for 16-QAM, normalizing to the same symbol energy $E[|B[k]|^2]$. Comment on how, if at all, the results depend on the constellation.

(d) Denoting the average sample power by $P = \mathbb{E}[|b[k]|^2]$, suppose that the signal is clipped by the transmitter whenever its magnitude exceeds $a\sqrt{P}$. That is, the transmitted samples are given by

$$\hat{b}[k] = \min\left\{1, \frac{a\sqrt{P}}{|b[k]|}\right\} b[k].$$

If there is no source of error other than the self-interference across subcarriers introduced by the clipping, the recovered frequency domain symbols $\{\hat{B}[k]\}$ are simply the DFT of $\{\hat{b}[k]\}$. Find, using simulations, the symbol error probability due to clipping for the 16-QAM system, for clip levels that are 0 dB, 3 dB, and 10 dB higher than the average sample power.

(e) Find a value for the clip level such that the symbol error rate in (d) is 1%.

(f) Repeat (d) and (e) for QPSK modulation, and comment on whether, and how, the sensitivity to clipping varies with constellation size.

Problem 8.18 (Timing acquisition in DS systems) Consider a DS signal $s(t) = \sum_{l=0}^{N-1} s[l]\psi(t - lT_c)$, where $\{s[l]\}$ is the spreading sequence, and ψ is the chip waveform (square root Nyquist at rate $1/T_c$). The received complex baseband signal is given by

$$y(t) = As(t - D) + n(t),$$

where D is an unknown delay taking values in $[0, MT_c)$, A is an unknown complex gain, and n is AWGN. The unknown delay can be expressed as a multiple of the chip interval as follows: $D = (K + \delta)T_c$, where K is an integer between 0 and $M - 1$, and $\delta \in [0, 1)$.

(a) Assuming a joint ML estimate of A and D, show that the delay estimate is given by

$$\hat{D}_{\text{ML}} = \max_{u \in [0, MT_c)} |Z(u)|^2,$$

where $Z(u)$ is the output of the matched filter $s_{\text{MF}}(t) = s^*(-t)$ at time u:

$$Z(u) = (y * s_{\text{MF}})(u) = \int y(t)s^*(t - u)dt.$$

(b) Express the signal contribution to $Z(u)$ in terms of the continuous-time autocorrelation function of the spreading waveform $s(t)$ defined in (8.65).

(c) Assume that the spreading sequence has an ideal discrete-time autocorrelation function (defined in (8.66)). That is, assume that $R_s[n] = N\delta_{n0}$. Further, suppose that $\delta = 0$ (i.e., the unknown delay is an integer $D = KT_c$). The integer part

$$\hat{K} = \max_{0 \le k \le M-1} |Z(kT_c)|^2.$$

What is the probability of making an error in the delay estimate (i.e., $P[\hat{K} \neq K]$) as a function of E_s/N_0, where $E_s = |A|^2 ||s||^2$ is the energy of the spreading waveform?

Hint Can you relate to a problem in noncoherent communication (see Chapter 4)?

(d) Now, suppose that $\delta \in [0, 1)$. Assume further that ψ is a rectangular waveform of duration T_c. Assuming that the output of the matched filter is still sampled at the chip rate, show that, for $u = kT_c$ (k integer), the signal contribution to $Z(u)$ is nontrivial only for $k = K$ and $k = K + 1$. Formulate the problem of joint ML estimation of A, K, and δ (A is a nuisance parameter), simplifying as much as possible. Also, discuss simple suboptimal estimators that estimate δ after estimating K, say, using the rule in (c).

Problem 8.19 (A linear feedback shift register sequence) Consider a 3 bit vector $(s[0], s[1], s[2])$ that evolves in time according to the following relationships:

$$s_n[1] = s_{n-1}[0]$$
$$s_n[2] = s_{n-1}[1]$$
$$s_n[0] = s_{n-1}[1] + s_{n-1}[2],$$

where the addition is in binary arithmetic (i.e., it is an exclusive or).

(a) Describe how you should implement the preceding time evolution using a clocked, 3 bit shift register.
(b) Starting from the initial condition $(1, 0, 1)$, describe the time evolution of the 3 bit vector. How many distinct values does the 3 bit vector take?
(c) If you take $\{s_n[2]\}$ as an "output," what binary sequence do you get, starting from the initial condition in (b)? Is the sequence periodic?
(d) Find the autocorrelation function of the sequence in (c). Is this a good choice for a DS spreading sequence?

Problem 8.20 (Asynchronous CDMA with bandlimited chip waveforms) Consider a DS system employing a square root raised cosine chip waveform ψ with 50% excess bandwidth. In this problem, we develop a model for the chip matched filter output y_s similar to that developed in Example 8.4.1 for a rectangular chip waveform.

(a) Find a time domain expression for $r_\psi(t)$.
(b) Show that $y_s(mT_c) = (s * g)[m]$, where s is the spreading sequence, and g is an "equivalent chip rate channel" which depends on ψ and δ. Specify g for $K = 2$ and $\delta = 1/4$, truncating it as appropriate when the entries become insignificant.
(c) For DS waveforms u and v as in (8.62), generalize the relation (8.69) between the continuous-time and discrete-time crosscorrelation functions to arbitrary chip waveforms, starting from (8.67). In particular, show that

8.9 Problems

the continuous-time autocorrelation function is a discrete-time convolution between the discrete-time autocorrelation function and an equivalent chip rate channel.

Problem 8.21 (Direct sequence CDMA with conventional reception) Consider a DS signal $u(t)$ of the form (8.53), where the spreading sequence $\{s[l]\}$ is modeled as having i.i.d., zero mean elements satisfying $E[|s[l]|^2] = 1$. The symbol sequence $\tilde{b}[l]$ (expressed at chip rate in (8.53)) is also normalized to unit energy.

(a) Show that the PSD of u is given by

$$S_u(f) = \frac{|\Psi(f)|^2}{T_c}.$$

(b) Now, suppose that u is passed through a chip matched filter $\psi^*(-t)$. Show that the output z has PSD

$$S_z(f) = \frac{|\Psi(f)|^4}{T_c}.$$

The random process $z(t)$ models the contribution of an interfering signal at the output of a chip matched filter for a desired signal. The delay of the interfering signal relative to the receiver's time reference is modeled as uniform over $[0, T_c]$ (this is implicit in the PSD computation; see Chapter 2). The random processes, and relative delays, corresponding to different interfering signals are independent, so that the PSDs add up.

(c) Now, consider a DS-CDMA system with complex baseband received signal

$$y(t) = A_1 u_1(t) + \sum_{k=2}^{K} A_k u_k(t) + n(t),$$

where $u_k(t)$, $k = 2, \ldots, K$ are interfering signals of the form considered in (a) and (b), $u_1(t)$ is the desired signal, and n is AWGN. Suppose that y is passed through a chip matched filter with samples synchronized to the desired signal. Let $E_c = ||\psi||^2$ denote the energy of the chip waveform. Show that the signal contribution to the lth sample at the output of the chip matched filter is given by

$$S[l] = \tilde{b}_1[l]s_1[l]E_c.$$

Evaluate the result for the rectangular chip waveform $\psi(t) = I_{[0,T_c]}(t)$.

(d) Show that the PSD of the noise plus interference at the output of the chip matched filter is given by

$$S_I(f) = N_0|\Psi(f)|^2 + \sum_{k=2}^{K} |A_k|^2 \frac{|\Psi(f)|^4}{T_c}.$$

(e) Let $I[l]$ denote the contribution of the interference plus noise to the lth sample at the output of the chip matched filter. Show that

$$\mathbb{E}[|I[l]|^2] = N_0 E_c + \frac{\sum_{k=2}^{K} |A_k|^2}{T_c} \int_{-\infty}^{\infty} |\Psi(f)|^4 df.$$

Evaluate the result for the rectangular chip waveform $\psi(t) = I_{[0,T_c]}(t)$.

Hint Use Parseval's identity.

(f) Let $Z = \sum_{l=0}^{N-1} s_1^*[l](S[l] + I[l])$ denote the output obtained by correlating over N chip samples. Setting $\tilde{b}[l] \equiv b$, where b is the desired symbol, show that

$$Z = bNE_c + \sum_{l=0}^{N-1} s_1^*[l]I[l],$$

where the first term corresponds to the desired signal, and the second to the interference plus noise. Normalizing $\mathbb{E}[|b|^2] = 1$, we see that the symbol energy $E_s = NE_c$.

(g) Show that the SIR at the output of the correlator satisfies

$$\frac{1}{\text{SIR}} = \frac{N_0}{E_s} + \frac{\sum_{k=2}^{K} |A_k|^2}{|A_1|^2} \frac{\int_{-\infty}^{\infty} |\Psi(f)|^4 df}{T_c E_c^2}. \tag{8.149}$$

(h) Specialize (8.149) for a rectangular chip waveform and QPSK. Further, assume that all users arrive at the receiver at equal power. Show that the SIR satisfies

$$\frac{1}{\text{SIR}} = \frac{N_0}{2E_b} + \frac{2(K-1)}{3N}.$$

(i) Comment on the relative performance of asynchronous and synchronous DS-CDMA with conventional reception by comparing the result with (8.72).

Problem 8.22 (Illustrating the properties of MMSE reception) Consider the two-user system in Example 8.4.2.

(a) Find the MMSE correlator for User 1, and express it in the form

$$\mathbf{c}_{\text{MMSE}} = \alpha_1 \mathbf{s}_1 + \alpha_2 \mathbf{s}_2,$$

specifying $\{\alpha_i\}$ explicitly in terms of the system parameters.

(b) Find an explicit expression for the MMSE.
(c) Show that the limit of (a) as $\sigma^2 \to \infty$ is the ZF correlator.
(d) Show that the limit of (a) as $A_2/A_1 \to \infty$ (with σ^2 fixed at some positive value) is the ZF correlator.

Problem 8.23 (Successive interference cancellation) For the two-user system in Example 8.4.2, consider the following successive interference cancellation (SIC) scheme:

$$\hat{b}_1 = \text{sign}(z_1),$$
$$\hat{b}_2 = \text{sign}\left(z_2 - A_1\rho\hat{b}_1\right).$$

That is, we decide on the first user's bit ignoring the MAI due to User 2, and then use this decision to cancel the interference due to User 1 in User 2's matched filter output.

(a) Do you think this scheme works better when User 1 has larger or smaller amplitude than User 2?
(b) Find the asymptotic efficiency for User 1. When is it nonzero?
(c) What can you say about the asymptotic efficiency for User 2 as $A_1/A_2 \to \infty$?

Problem 8.24 Consider a three-user DS-CDMA system with short spreading sequences given by

$$\mathbf{s}_1 = (1, 1, 1, -1, -1)^T,$$
$$\mathbf{s}_2 = (1, -1, 1, 1, -1)^T,$$
$$\mathbf{s}_3 = (1, 1, -1, -1, 1)^T,$$

where $\mathbf{s}_k$ denotes the spreading sequence for user k.

(a) Assuming a synchronous system, compute the near–far resistance for each user, assuming zero-forcing reception. Express the noise enhancement in each case in dB.
(b) Assuming that all users are received at equal power, find the error probability for User 1 with matched filter reception assuming BPSK modulation with E_b/N_0 of 15 dB. Compare with the error probability, assuming zero-forcing reception.
(c) Repeat (b), this time assuming that Users 2 and 3 are 10 dB stronger than User 1.
(d) Now, consider a chip-synchronous, but symbol-asynchronous, system, in which User 2 is delayed with respect to User 1 by 2 chips, while User 3 remains synchronized with User 1. Find the near–far resistance for User 1, assuming zero-forcing reception with an observation interval spanning a symbol interval for User 1. Compare with the corresponding result in (a).

Hint User 2 generates two interference vectors modulated by independent symbols over User 1's symbol interval. We must therefore consider an *equivalent synchronous system* with three interference vectors.

Problem 8.25 Consider the following real baseband vector model for a K-user synchronous CDMA system with processing gain N and BPSK modulation.

$$\mathbf{r} = \mathbf{SAb} + \mathbf{N}, \tag{8.150}$$

where $\mathbf{b}$ is the vector of bits sent by the K users; the $N \times K$ matrix $\mathbf{S}$ has the spreading waveforms as its columns: its kth column is $\mathbf{s}_k$, $1 \leq k \leq K$; $\mathbf{A}$ is a diagonal matrix whose k diagonal element is A_k, $1 \leq k \leq K$: $\mathbf{A} = \mathrm{diag}(A_1, \ldots, A_K)$; and $\mathbf{N} \sim N(0, \sigma^2 \mathbf{I})$ is AWGN.

(a) Show that the vector of matched filter outputs, $\mathbf{y} = \mathbf{S}^T \mathbf{r}$, is a sufficient statistic for deciding on $\mathbf{b}$.

Hint Show that the log likelihood ratio of $\mathbf{r}$, conditioned on $\mathbf{b}$, depends only on $\mathbf{y}$.

(b) Show that the ML decision rule for $\mathbf{b}$ is of the form

$$\hat{\mathbf{b}}_{\mathrm{ML}} = \arg\max_{\mathbf{b}} \mathbf{u}^T \mathbf{y} - \frac{1}{2} \mathbf{b}^T \mathbf{R} \mathbf{b}. \tag{8.151}$$

Specify the vector $\mathbf{u}$ and the matrix $\mathbf{R}$.

(c) For the setting of Problem 8.24(a), assuming that $A_k \equiv 1$, find the ML decision for $\mathbf{b}$ assuming that the received signal is $\mathbf{r} = (1.5, 4, -0.5, -4, 2)^T$.

Problem 8.26 (Coded frequency hop multiple access) Consider frequency hop multiple access with Reed–Solomon coding, as described in Example 8.5.1.

(a) Suppose that there are $K = 10$ simultaneous users and $q = 64$ hopping frequencies. Choose a $(31, k)$ RS code, operating over an alphabet of size $Q = 32$, which maximizes the rate subject to the constraint that the decoding failure probability is at least as good as 10^{-3}.

(b) Fixing $K = 10$, minimize the number of hopping frequencies needed so as to attain a decoding failure probability of 10^{-3} and an information rate of at least 0.8.

Problem 8.27 (Fourier series for a square wave) A periodic signal $u(t)$ of period P can be written as a Fourier series as follows:

$$u(t) = \sum_{k=-\infty}^{\infty} a[k] e^{j2\pi k f_0 t}, \tag{8.152}$$

where $f_0 = 1/P$ is the fundamental frequency, and $k f_0$ is the kth harmonic with Fourier coefficient

$$a[k] = \frac{1}{P} \int_P u(t) e^{-j2\pi k t/P}. \tag{8.153}$$

Here $\int_P$ denotes integration over any conveniently chosen integral of length P. As usual, we denote the transform relationship as $u(t) \leftrightarrow \{a[k]\}$.

(a) For a periodic square wave of period P, specified over $(0, P)$ as

$$u(t) = \begin{cases} +1, & 0 < t < \frac{P}{2}, \\ -1, & \frac{P}{2} < t < P, \end{cases}$$

find the Fourier series coefficients using (8.153).
(b) If $u(t) \leftrightarrow \{a[k]\}$, find the Fourier series for $u(t - \tau)$ and $du(t)/dt$.
(c) Find the Fourier series for the periodic impulse train $\sum_{k=-\infty}^{\infty} \delta(t - kP)$.
(d) Use (b) and (c) to redo (a) (i.e., find the Fourier series for the derivative of the square wave, and then infer that of the square wave).

Problem 8.28 (CPFSK basics) Consider CPFSK with $L = 3$ and a modulation index $h = 1/4$.

(a) Sketch the frequency pulse $g(t)$ and the phase pulse $\phi(t)$.
(b) Writing $\theta(t) = \sum_{n=0}^{\infty} a[n]\phi(t - nT)$, where $\{a[n]\}$ take values ± 1, sketch the phase trajectory for the symbol sequence $\{+1, -1, -1, -1, +1, +1\}$, assuming that $\theta(0) = 0$.
(c) What is the maximum number of possible values that the phase takes at integer multiples of T?

We now develop the MLSE receiver over a phase trellis whose states are the values of the phase at integer multiples of T, based on the received signal

$$y(t) = Ae^{j\theta(t)} + n(t), \quad t \geq 0,$$

where A is a known gain, and n is AWGN.
(d) Show that the sufficient statistics for MLSE reception are

$$Z_n[\alpha] = \int_{nT}^{(n+1)T} y(t)e^{-j\alpha t}dt,$$

where α takes a discrete set of values. Specify the values for α.
(e) What are the minimal number of states required for MLSE reception using the Viterbi algorithm based on the sufficient statistics in (d)?

Hint What are the parameters you need to predict the phase evolution between nT and $(n+1)T$, in addition to the initial condition $\theta(nT)$? How many values can $\theta(nT)$ take?

Problem 8.29 (PSD simulation for CPM waveforms) We discuss how to compute the PSD of a communication waveform by simulation, and then apply it to compute the PSD for GMSK. Suppose that $u(t), 0 \leq t < T_o$ is a communication waveform observed over an interval T_o (spanning a large number of symbol intervals). We know from Chapter 2 that we can estimate the PSD as $|U(f)|^2/T_o$. We can now average over several simulation runs to smooth out the estimate. In this problem, we discuss how to implement this using a discrete-time simulation.

(a) Sample u at rate T_s to obtain $\{u[k] = u(kT_s), k = 0, \ldots, N-1\}$, where $N = T_o/T_s$. Letting $U[k]$ denote the DFT of $\{u[k]\}$, show that it is related to $U(f)$ as follows:
$$U\left(\frac{k}{NT_s}\right) \approx U[k]T_s.$$

(b) For Gaussian MSK at rate $1/T$, suppose that the sampling rate is $8/T$, and that you want a frequency resolution at least as good as $1/(10T)$. To estimate the PSD by simulation using an N-point FFT (i.e., N is a power of 2), what is the smallest required value of N, and the corresponding observation interval T_o (as a multiple of T)?

(c) By averaging over multiple simulation runs, estimate and plot the PSD of Gaussian MSK with $\beta = 0.3$. Plot for comparison the PSD of MSK, computed either analytically or by simulation. Express $S_u(f)$ in dB (with $S_u(0)$ normalized to 0 dB) and a linear scale on the f-axis. Comment on the relative spectral occupancy of the two schemes.

Problem 8.30 (Laurent approximation) Use Example 8.6.1, and the comments following it, as a guide for developing a Laurent approximation for CPFSK with $L = 3$ and $h = 1/2$. Plot the waveforms $K(t)$ and $s_0(t)$. (Set $T = 1$ for convenience.) How many symbols does the ISI span in the corresponding linearly modulated system? Specify the set of values that the pseudosymbols $B_0[n]$ can take.

Problem 8.31 (Differential entropy for complex Gaussian random vector) We wish to derive the expression (8.126) for the differential entropy for an $n \times 1$ complex Gaussian vector $\mathbf{Z} \sim CN(\mathbf{m}, \mathbf{C})$. Let $\{\lambda_i\}$ denote the eigenvalues of $\mathbf{C}$ and $\{\mathbf{v}_i\}$ the corresponding orthonormal eigenvectors.

(a) Argue that we can set the mean vector $\mathbf{m}$ to zero without changing the differential entropy.

(b) Directly compute $h(\mathbf{Z}) = \mathbb{E}[-\log p(z)]$ using the density of $\mathbf{Z}$ to infer that
$$h(\mathbf{Z}) = \log \det(\pi e \mathbf{C}).$$
Now relate the determinant of $\mathbf{C}$ to the eigenvalues $\{\lambda_i\}$ to infer that
$$h(\mathbf{Z}) = \sum_{i=1}^{n}(\log(\lambda_i) + \log(\pi e)). \qquad (8.154)$$

(c) As an alternative proof of (8.154), define the complex random variables $Y_i = \mathbf{v}_i^H \mathbf{Z}$. Show that $\{Y_i\}$ are independent $CN(0, \lambda_i)$. Show that $h(\mathbf{Z}) = \sum_i h(Y_i)$. Now use the results of Chapter 6 on scalar proper complex Gaussian random variables to infer (8.154).

8.9 Problems

Problem 8.32 (Capacity of the antenna-hopping space–time code) Consider the antenna-hopping space–time code with capacity C given by (8.133).

(a) Use Jensen's inequality (Appendix C) to show that the capacity of the antenna-hopping code is smaller than the capacity (8.127) with optimal space–time coding.

(b) Assuming that $h_i \sim CN(0, 1)$ are i.i.d., apply the CLT to approximate the capacity given by (8.133) as a Gaussian random variable. Compute the mean and variance by simulation or numerical integration for SNR of 10 dB and six transmit antennas.

(c) Using the Gaussian approximation for the distribution of C, and the results from (b), predict the 1% outage capacity for the setting of (b). Compare with the corresponding capacity for the AWGN channel, and with the ergodic capacity of the Rayleigh fading channel.

We now develop an analytical approximation for the variance of C.

(d) For a real random variable X with $\mathbb{E}[X] = 1$, show that

$$\log(1+aX) = \log(1+a) + \log\left(1 + \frac{a}{1+a}(X-1)\right).$$

(e) Using the approximation $\log(1+x) \approx x$ in the preceding equation, derive the following estimate for the variance of $\log(1+aX)$:

$$\operatorname{var}(\log(1+aX)) = \frac{a^2}{(1+a)^2}\operatorname{var}(X).$$

(f) Apply the result of (e) to estimate the variance of the capacity as

$$\operatorname{var}(C) = \frac{1}{N_T}\left(\frac{\text{SNR}}{\text{SNR}+1}\right)^2,$$

for $h_i \sim CN(0, 1)$ i.i.d.

(g) Compare the 1% outage capacity for SNR of 10 dB using this variance estimate with that obtained in (c).

Problem 8.33 (Capacity with ideal space–time coding) Consider the capacity C given by (8.127) for $h_i \sim CN(0, 1)$ i.i.d. Assume an SNR of 10 dB and $N_T = 6$, unless specified otherwise.

(a) Specify the density of the averaged channel gain $G = \frac{1}{N_T}\sum_{l=1}^{N_T}|h_l|^2$.

(b) Use numerical integration using (a) to estimate the 1% outage capacity. Compare with the results from Problem 8.32.

(c) Approximate G as a Gaussian random variable, and use analytical estimates of its mean and variance to estimate the 1% outage capacity. Compare with the results from (b).

Problem 8.34 (Deterministic MIMO channel) Consider a MIMO system with $t = 2$ transmit elements and $r = 3$ receive elements. The 3×2 channel matrix is given by

$$\mathbf{H} = \begin{pmatrix} 1+j & j \\ \sqrt{2}j & \frac{1+j}{\sqrt{2}} \\ -1+j & 1 \end{pmatrix}.$$

Assume that the net transmitted power per channel use is P, and that the noise per receive element is i.i.d. $CN(0, 1)$.

(a) Specify the eigenmodes and eigenvalues associated with the matrix $\mathbf{W} = \mathbf{H}^H \mathbf{H}$.
(b) Assuming that the transmitter knows the channel, specify, as a function of P, the transmit strategy that maximizes mutual information as a function of P. Plot the mutual information thus attained as a function of P.
(c) Repeat (b) for two different cases:

   (i) the second transmit element is not used, but all three receive elements are used.
   (ii) the third receive element is not used, but both transmit elements are used.

Plot the mutual information as a function of P for both these cases on the same plot as (b), and discuss the results.

(d) Now, suppose the transmitter does not know the channel. Plot, as a function of P, the mutual information attained by transmitting at equal power from both antennas. Compare with the result of (b), and with (c), case (i). Discuss the results.

Problem 8.35 (Deterministic MIMO channel) Consider a 2×2 MIMO system with channel matrix

$$\mathbf{H} = \begin{pmatrix} 1+j & 1 \\ 1-j & -1 \end{pmatrix}.$$

If $\mathbf{x}$ is the 2×1 transmitted vector, then $\mathbf{y} = \mathbf{H}\mathbf{x} + \mathbf{n}$ is the 2×1 received vector, where $\mathbf{n}$ is a noise vector with i.i.d. $CN(0, 1)$ entries. The net transmitted power $E[||\mathbf{x}||^2] = P$. If both the transmitter and receiver know $\mathbf{H}$, describe in detail the transmit strategy that maximizes the ergodic capacity for $P = 1$. What is the corresponding capacity in bits per channel use?

Problem 8.36 (MIMO capacity with rich scattering) Consider an $n \times n$ MIMO system with i.i.d. $CN(0, 1)$ channel entries. Let SNR denote the average SNR in a 1×1 system with the same transmit power. Assume that the transmitter does not know the channel, and splits its power evenly among all transmit elements.

(a) Find the ergodic capacity and the outage capacity corresponding to 10% outage rate for a 2×2 system and a 3×3 system. Plot these as a function of SNR (dB).

(b) Use either analysis or simulation (e.g., for a 6×6 system) to estimate $C(n)/n$, where $C(n)$ is the ergodic capacity for an $n\times n$ system.

Problem 8.37 (MIMO with covariance feedback) A MIMO system has six transmit elements and one receive element. The transmit elements are arranged in a linear array with inter-element spacing d. The carrier wavelength is denoted by λ. The spatial channel from transmitter to receiver has a power-angle profile that is uniform between $[-\pi/12, \pi/12]$. Letting **h** denote the 6×1 channel, define the spatial covariance matrix as $\mathbf{C} = E[\mathbf{hh}^H]$, where the expectation can be viewed as an average across subcarriers.

(a) Compute **C** and find its eigenvalues $\gamma_1, \ldots, \gamma_6$, ordered as $\gamma_1 \geq \ldots \geq \gamma_6$, for $d/\lambda = 0.5$. Do you have any physical interpretation for the corresponding eigenvectors?

(b) Suppose that the transmitter knows **C**, and sends with power P_i along the ith eigenmode, $i=1,\ldots,6$, where $\sum_i P_i = P$. Show that the ergodic capacity as a function of the power allocation vector $\mathbf{P} = (P_1, \ldots, P_6)$ is given by $C(\mathbf{P}) = E[\log(1+\sum_{i=1}^{6} P_i \lambda_i X_i)]$ where $\{X_i\}$ are i.i.d. exponential random variables with mean one.

Appendix A Probability, random variables, and random processes

In this appendix, we provide a quick review of some basics of probability, random variables, and random processes that are assumed in the text. The purpose is to provide a reminder of concepts and terminology that the reader is expected to be already familiar with.

A.1 Basic probability

We assume that the concept of sample space and events is known to the reader. We simply state for the record some important relationships regarding the probabilities of events. For events A and B, the union is denoted by $A \cup B$ and the intersection by $A \cap B$.

Range of probability For any event A, we have $0 \leq P[A] \leq 1$.

Complement The complement of an event A is denoted by A^c, and

$$P[A] + P[A^c] = 1. \tag{A.1}$$

Independence Events A and B are independent if

$$P[A \cap B] = P[A]P[B]. \tag{A.2}$$

Mutual exclusion Events A and B are *mutually exclusive* if $P[A \cap B] = 0$.

Probability of unions and intersections The following relation holds:

$$P[A \cup B] = P[A] + P[B] - P[A \cap B]. \tag{A.3}$$

Conditional probability The conditional probability of A given B is defined as (assuming that $P[B] > 0$)

$$P[A|B] = \frac{P[A \cap B]}{P[B]}. \tag{A.4}$$

Law of total probability For events A and B, we have

$$P[A] = P[A \cap B] + P[A \cap B^c] = P[A|B]P[B] + P[A|B^c]P[B^c]. \quad (A.5)$$

This generalizes to any partition of the entire probability space: if $B_1, B_2, \ldots$ are mutually exclusive events such that their union covers the entire probability space (actually, it is enough if the union contains A), then

$$P[A] = \sum_i P[A \cap B_i] = \sum_i P[A|B_i]P[B_i]. \quad (A.6)$$

Bayes' rule Given $P[A|B]$, we can compute $P[B|A]$ as follows:

$$P[B|A] = \frac{P[A|B]P[B]}{P[A]} = \frac{P[A|B]P[B]}{P[A|B]P[B] + P[A|B^c]P[B^c]}, \quad (A.7)$$

where we have used (A.5). Similarly, in the setting of (A.6), we can compute $P[B_j|A]$ as follows:

$$P[B_j|A] = \frac{P[A|B_j]P[B_j]}{P[A]} = \frac{P[A|B_j]P[B_j]}{\sum_i P[A|B_i]P[B_i]}. \quad (A.8)$$

A.2 Random variables

We summarize important definitions regarding random variables, and also mention some important random variables other than the Gaussian, which is discussed in detail in Chapter 3.

Cumulative distribution function (cdf) The cdf of a random variable X is defined as

$$F(x) = P[X \leq x].$$

Any cdf $F(x)$ is nondecreasing in x, with $F(-\infty) = 0$ and $F(\infty) = 1$. Furthermore, the cdf is right-continuous.

Complementary cumulative distribution function (ccdf) The ccdf of a random variable X is defined as

$$F^c(x) = P[X > x] = 1 - F(x).$$

Continuous random variables X is a continuous random variable if its cdf $F(x)$ is differentiable. Examples are the Gaussian and exponential random variables. The probability density function (pdf) of a continuous random variable is given by

$$p(x) = F'(x). \quad (A.9)$$

For continuous random variables, the probability $P[X = x] = F(x) - F(x^-) = 0$ for all x, since $F(x)$ is continuous. Thus, the probabilistic interpretation of pdf

is that it is used to evaluate the probability of infinitesimally small intervals as follows:
$$P[X \in [x, x+\Delta x]] \approx p(x)\, \Delta x,$$
for Δx small.

Discrete random variables X is a discrete random variable if its cdf $F_X(x)$ is piecewise constant. The jumps occur at $x = x_i$ such that $P[X = x_i] > 0$. Examples are the Bernoulli, binomial, and Poisson random variables. That is, the probability mass function is given by
$$p(x) = P[X = x] = \lim_{\delta \to 0^+} F(x) - F(x - \delta).$$

Density We use the generic term "density" to refer to both pdf and pmf, relying on the context to clarify our meaning.

Expectation The expectation of a function of a random variable X is defined as
$$\mathbb{E}[g(X)] = \int g(x) p(x) \mathrm{d}x \quad \textbf{X continuous},$$
$$\mathbb{E}[g(X)] = \sum g(x) p(x) \quad \textbf{X discrete}.$$

Mean and variance The mean of a random variable X is $\mathbb{E}[X]$ and its variance is $\mathbb{E}\left[(X - \mathbb{E}[X])^2\right]$.

Gaussian random variable This is the most important random variable for our purpose, and is discussed in detail in Chapter 3.

Exponential random variable The random variable X has an exponential distribution with parameter μ, if its pdf is given by
$$p(x) = \mu e^{-\mu x} I_{[0,\infty)}(x).$$
The cdf is given by
$$F(x) = (1 - e^{-\mu x}) I_{[0,\infty)}(x).$$

For $x \geq 0$, the ccdf is given by $F^c(x) = P[X > x] = e^{-\mu x}$. Note that $\mathbb{E}[X] = \mathrm{var}(X) = 1/\mu$.

Random variables related to the Gaussian and exponential random variables are the Rayleigh, Rician, and Gamma random variables. These are discussed as they arise in the text and problems.

Bernoulli random variable X is Bernoulli if it takes values 0 or 1. The Bernoulli distribution is characterized by a parameter $p \in [0, 1]$, where $p = P[X = 1] = 1 - P[X = 0]$.

Binomial random variable X is a binomial random variable with parameters n and p if it is a sum of independent Bernoulli random variables, each with parameter p. It takes integer values from 0 to n, and its pmf is given by
$$P[X = k] = \binom{n}{k} p^k (1-p)^{n-k}, \quad k = 0, 1, \ldots, n.$$

A.2 Random variables

Poisson random variable X is a Poisson random variable with parameter $\lambda > 0$ if it takes values from the nonnegative integers, with pmf given by

$$P[X = k] = \frac{\lambda^k}{k!} e^{-\lambda}, \quad k = 0, 1, 2, \ldots$$

Note that $\mathbb{E}[X] = \mathrm{var}(X) = \lambda$.

Joint distributions For multiple random variables $X_1, \ldots, X_n$ defined on a common probability space, which can also be represented as an n-dimensional *random vector* $\mathbf{X} = (X_1, \ldots, X_n)^T$, the joint cdf is defined as

$$F(\mathbf{x}) = F(x_1, \ldots, x_n) = P[X_1 \leq x_1, \ldots, X_n \leq x_n].$$

For jointly continuous random variables, the joint pdf $p(\mathbf{x}) = p(x_1, \ldots, x_n)$ is obtained by taking partial derivatives above with respect to each variable x_i, and has the interpretation that

$$P[X_1 \in [x_1, x_1 + \Delta x_1], \ldots, X_n \in [x_n, x_n + \Delta x_n]] \approx p(x_1, \ldots, x_n) \Delta x_1 \ldots \Delta x_n,$$

for $\Delta x_1, \ldots, \Delta x_n$ small.

For discrete random variables, the pmf is defined as expected:

$$p(\mathbf{x}) = p(x_1, \ldots, x_n) = P[X_1 = x_1, \ldots, X_n = x_n].$$

Marginal densities from joint densities This is essentially an application of the law of total probability. For continuous random variables, we integrate out all arguments in the joint pdf, except for the argument corresponding to the random variable of interest (say x_1):

$$p(x_1) = \int \ldots \int p(x_1, x_2, \ldots, x_n) \, dx_2 \ldots dx_n.$$

For discrete random variables, we sum over all arguments in the joint pmf except for the argument corresponding to the random variable of interest (say x_1):

$$p(x_1) = \sum_{x_2} \ldots \sum_{x_n} p(x_1, x_2, \ldots, x_n).$$

Conditional density The conditional density of Y given X is defined as

$$p(y|x) = \frac{p(x, y)}{p(x)}, \tag{A.10}$$

where the definition applies for both pdfs and pmfs, and where we are interested in values of x such that $p(x) > 0$. For jointly continuous X and Y, the conditional density $p(y|x)$ has the interpretation

$$p(y|x) \approx \Delta y P\left[Y \in [y, y + \Delta y] \big| X \in [x, x + \Delta x]\right],$$

for $\Delta x, \Delta y$ small. For discrete random variables, the conditional pmf is simply the following conditional probability:

$$p(x|y) = P[X = x | Y = y].$$

Bayes' rule for conditional densities Given the conditional density of Y given X, the conditional density for X given Y is given by

$$p(x|y) = \frac{p(y|x)p(x)}{p(y)} = \frac{p(y|x)p(x)}{\int p(y|x)p(x)\mathrm{d}x} \quad \text{Continuous random variables,}$$

$$p(x|y) = \frac{p(y|x)p(x)}{p(y)} = \frac{p(y|x)p(x)}{\sum_x p(y|x)p(x)} \quad \text{Discrete random variables.}$$

A.3 Random processes

A random process X is a collection of random variables $\{X(t), t \in \mathcal{T}\}$ defined on a common probability space, where the index set $\mathcal{T}$ often, but not always, has the interpretation of time (for convenience, we often refer to the index as time in the remainder of this section). Since the random variables are defined on a common probability space, we can talk meaningfully about the joint distributions of a finite subset of these random variables, say $X(t_1), \ldots, X(t_n)$, where the sampling times $t_1, \ldots, t_n \in \mathcal{T}$. Such joint distributions are said to be the *finite-dimensional distributions* for X, and we say that we know the statistics of the random process X if we know all possible joint distributions for any number and choice of the sampling times $t_1, \ldots, t_n$.

In practice, we do not have a complete statistical characterization of a random process, and settle for partial descriptions of it. In Chapters 2 and 3, we mainly discuss second order statistics such as the mean function and the autocorrelation function, which are typically easy to compute analytically, or to measure experimentally or by simulation. Furthermore, if the random process is Gaussian, then second order statistics provide a complete statistical characterization. In addition to the focus on Gaussian random processes in the text, other processes such as Poisson random processes are introduced on a "need-to-know" basis in the text and problems.

For the purpose of this appendix, we supplement the material in Chapters 2 and 3 by summarizing what happens to random processes through linear systems. We restrict our attention to wide sense stationary (WSS) random processes, and allow complex values: X is WSS if its mean function $m_X(t) = \mathbb{E}[X(t)]$ does not depend on t, and its autocorrelation function $\mathbb{E}[X(t+\tau)X^*(t)]$ depends only on the time difference τ, in which case we denote it by $R_X(\tau)$. The Fourier transform of $R_X(\tau)$ is the power spectral density $S_X(f)$.

A.3.1 Wide sense stationary random processes through LTI systems

Suppose, now, that a WSS random process X is passed through an LTI system with impulse response $h(t)$ (which we allow to be complex-valued) to obtain an output $Y(t) = (X * h)(t)$. We wish to characterize the joint second order statistics of X and Y.

Defining the crosscorrelation function of Y and X as

$$R_{YX}(t+\tau,t) = \mathbb{E}[Y(t+\tau)X^*(t)],$$

we have

$$R_{YX}(t+\tau,t) = \mathbb{E}\left[\left(\int X(t+\tau-u)h(u)\mathrm{d}u\right)X^*(t)\right] = \int R_X(\tau-u)h(u)\mathrm{d}u, \quad (A.11)$$

interchanging expectation and integration. Thus, $R_{YX}(t+\tau,t)$ depends only on the time difference τ. We therefore denote it by $R_{YX}(\tau)$. From (A.11), we see that

$$R_{YX}(\tau) = (R_X * h)(\tau).$$

The autocorrelation function of Y is given by

$$\begin{aligned}R_Y(t+\tau,t) &= \mathbb{E}[Y(t+\tau)Y^*(t)] \\ &= \mathbb{E}\left[Y(t+\tau)\left(\int X(t-u)h^*(u)\mathrm{d}u\right)\right] \\ &= \int \mathbb{E}[Y(t+\tau)X^*(t-u)]h^*(u)\mathrm{d}u \\ &= \int R_{YX}(\tau+u)h^*(u)\mathrm{d}u.\end{aligned} \quad (A.12)$$

Thus, $R_Y(t+\tau,t)$ depends only on the time difference τ, and we denote it by $R_Y(\tau)$. Recalling that the matched filter $h_{\mathrm{MF}}(u) = h^*(-u)$, we can see, replacing u by $-u$ in the integral at the end of (A.12), that

$$R_Y(\tau) = (R_{YX} * h_{\mathrm{MF}})(\tau) = (R_X * h * h_{\mathrm{MF}})(\tau).$$

Finally, we note that the mean function of Y is a constant given by

$$m_Y = m_X * h = m_X \int h(u)\mathrm{d}u.$$

Thus, X and Y are jointly WSS: X is WSS, Y is WSS, and their crosscorrelation function depends on the time difference. The formulas for the second order statistics, including the corresponding power spectral densities obtained by taking Fourier transforms, are collected below:

$$\begin{aligned}R_{YX}(\tau) &= (R_X * h)(\tau), \\ R_Y(\tau) &= (R_{YX} * h_{\mathrm{MF}})(\tau) = (R_X * h * h_{\mathrm{MF}})(\tau), \\ S_{YX}(f) &= S_X(f)H(f), \\ S_Y(f) &= S_{YX}(f)H^*(f) = S_X(f)|H(f)|^2.\end{aligned} \quad (A.13)$$

A.3.2 Discrete-time random processes

We have emphasized continuous-time random processes in this appendix, and in Chapters 2 and 3. Most of the concepts, such as Gaussianity and (wide sense) stationarity, apply essentially unchanged to discrete-time random processes, with the understanding that the index set is now finite or countable. However, we do need some additional notation to talk about discrete-time random processes through discrete-time linear systems. Discrete-time random processes are important because these are what we deal with when using DSP in communication transmitters and receivers. Moreover, while a communication system may involve continuous-time signals, computer simulation of the system must inevitably be in discrete time.

z-transform The z-transform of a discrete-time signal $s = \{s[n]\}$ is given by

$$S(z) = \sum_{n=-\infty}^{\infty} s[n]z^{-n}.$$

The operator z^{-1} corresponds to a unit delay. Given the z-transform of $S(z)$ expressed as a power series in z, you can read off $s[n]$ as the coefficient multiplying z^{-n}. We allow the variable z to take complex values. We are often most interested in $z = e^{j2\pi f}$ (on the unit circle), at which point the z-transform reduces to a discrete-time Fourier transform (see below).

Discrete-time Fourier transform (DTFT) The DTFT of a discrete-time signal s is its z-transform evaluated at $z = e^{j2\pi f}$; i.e., it is given by

$$S(e^{j2\pi f}) = S(z)|_{z=e^{j2\pi f}} = \sum_{n=-\infty}^{\infty} s[n]e^{-j2\pi fn}.$$

It suffices to consider $f \in [0, 1]$, since $S(e^{j2\pi f})$ is periodic with period 1.

Autocorrelation function For a WSS discrete-time random process X, the autocorrelation function is defined as

$$R_X[k] = \mathbb{E}[X[n+k]X^*[n]].$$

The crosscorrelation between jointly WSS processes X and Y is similarly defined:

$$R_{XY}[k] = \mathbb{E}[X[n+k]Y^*[n]].$$

Power spectral density For a WSS discrete-time random process X, the PSD is defined as the DTFT of the autocorrelation function. However, it is often also convenient to consider the z-transform of the autocorrelation function. As before, we use a unified notation for the z-transform and DTFT, and define the PSD as follows:

$$\begin{aligned}S_X(z) &= \sum_{n=-\infty}^{\infty} R_X[n]z^{-n}, \\ S_X(e^{j2\pi f}) &= \sum_{n=-\infty}^{\infty} R_X[n]e^{-j2\pi fn}.\end{aligned} \quad (A.14)$$

Similarly, for X, Y, jointly WSS, the cross-spectral density S_{XY} is defined as the z-transform or DTFT of the crosscorrelation function R_{XY}.

Convolution If $s_3 = s_1 * s_2$ is the convolution of two discrete-time signals, then $S_3(z) = S_1(z)S_2(z)$.

Matched filter Let $h_{\text{MF}}[n] = h^*[-n]$ denote the impulse response for the matched filter for h. It is left as an exercise to show that

$$H_{\text{MF}}(z) = H^*((z^*)^{-1}). \tag{A.15}$$

This implies that $H_{\text{MF}}(e^{j2\pi f}) = H^*(e^{j2\pi f})$. Note that, if h is a real-valued impulse response, (A.15) reduces to $H_{\text{MF}}(z) = H(z^{-1})$.

Discrete-time random processes through discrete-time linear systems
Let $X = \{X[n]\}$, a discrete-time random process, be the input to a discrete-time linear time-invariant system with impulse response $h = \{h[n]\}$, and let $Y = \{Y[n]\}$ denote the system output. If X is WSS, then X and Y are jointly WSS with

$$\begin{aligned} R_{YX}[k] &= (R_X * h)[k], \\ R_Y[k] &= (R_X * h * h_{\text{MF}})[k]. \end{aligned} \tag{A.16}$$

The corresponding relationships in the spectral domain are as follows:

$$\begin{aligned} S_{YX}(z) &= H(z) S_X(z), \\ S_Y(z) &= H(z) H^*((z^*)^{-1}) S_X(z), \\ S_{YX}(e^{j2\pi f}) &= H(e^{j2\pi f}) S_X(e^{j2\pi f}), \\ S_Y(e^{j2\pi f}) &= |H(e^{j2\pi f})|^2 S_X(e^{j2\pi f}). \end{aligned} \tag{A.17}$$

A.4 Further reading

Expositions of probability and random processes sufficient for our purposes are provided by a number of textbooks on "probability for engineers," such as Yates and Goodman [128], Woods and Stark [129], and Leon-Garcia [130]. A slightly more detailed treatment of these same topics, still with an engineering focus, is provided by Papoulis and Pillai [131].

Those interested in delving deeper into probability than our present requirements may wish to examine the many excellent texts on this subject written by applied mathematicians. These include the classic texts by Feller [132], Breiman [133], and Billingsley [134]. Worth reading is the excellent text by Williams [135], which provides an accessible yet rigorous treatment to many concepts in advanced probability. Mathematically rigorous treatments of stochastic processes include Doob [136], and Wong and Hajek [137].

Appendix B   The Chernoff bound

We are interested in finding bounds for probabilities of the form $P[X > a]$ or $P[X < a]$ that arise when evaluating the performance of communication systems.

Our starting point is a weak bound known as the Markov inequality.

Markov inequality If X is a random variable which is nonnegative with probability one, then, for any $a > 0$,

$$P[X > a] \leq \frac{\mathbb{E}[X]}{a}. \tag{B.1}$$

Proof of Markov inequality For $X \geq 0$ (with probability one), we have

$$\mathbb{E}[X] = \int_0^\infty x p_X(x)\,dx \geq \int_a^\infty x p_X(x)\,dx \geq a \int_a^\infty p_X(x)\,dx = a P[X > a],$$

which gives the desired result (B.1). $\square$

Of course, the condition $X \geq 0$ is not satisfied by most of the random variables we encounter, so the Markov inequality has limited utility in its original form. However, for any arbitrary random variable X, the random variable e^{sX} (s a real number) is nonnegative. Note also that the function e^{sx} is strictly increasing in x if $s > 0$, so that $X > a$ if and only if $e^{sX} > e^{sa}$. We can therefore bound the tail probability as follows:

$$P[X > a] = P[e^{sX} > e^{sa}] \leq e^{-sa}\mathbb{E}[e^{sX}] = e^{M(s)-sa}, \quad s > 0, \tag{B.2}$$

where

$$M(s) = \log \mathbb{E}[e^{sX}] \tag{B.3}$$

is the moment generating function (MGF) of the random variable X. Equation (B.2) gives a family of bounds indexed by $s > 0$, and the Chernoff bound is obtained by finding the best bound in the family by minimizing the exponent on the right-hand side of (B.2) over $s > 0$. Specifically, define

$$M^*(a) = \min_{s>0} M(s) - sa. \tag{B.4}$$

Then the Chernoff bound for the tail probability is

$$P[X > a] \leq e^{M^*(a)}. \tag{B.5}$$

Note that similar techniques can also be used to bound probabilities of the form $P[X < a]$, except that we would now consider $s < 0$ in obtaining a Chernoff bound:

$$P[X < a] = P[e^{sX} > e^{sa}] \leq e^{-sa}\mathbb{E}[e^{sX}] = e^{M(s)-sa}, \quad s < 0.$$

We do not pursue this separately, since we can always write $P[X < a] = P[Y = -X > -a]$ and apply the techniques that we have already developed.

Theoretical exercise Show that $M^*(a) < 0$ for $a > \mathbb{E}[X]$, and $M^*(a) = 0$ for $a < \mathbb{E}[X]$. That is, the Chernoff bound is nontrivial only when we are finding the probability of intervals that do not include the mean.

Hint Show that $F(s) = M(s) - sa$ is concave, and that $F'(0) = \mathbb{E}[X] - a$. Use this to figure out the shape of $F(s)$ for the two cases under consideration.

Chernoff bound for a Gaussian random variable Let $X \sim N(0, 1)$. Find the Chernoff bound for $Q(x)$, $x > 0$. The first step is to find $M(s)$. We have

$$\mathbb{E}[e^{sX}] = \int_{-\infty}^{\infty} e^{sx} \frac{e^{-x^2/2}}{\sqrt{2\pi}} dx = e^{s^2/2} \int_{-\infty}^{\infty} \frac{e^{-(x-s)^2/2}}{\sqrt{2\pi}} dx = e^{s^2/2},$$

where we have completed squares in the exponent to get an $N(s, 1)$ Gaussian density that integrates out to one. Thus, $M(s) = s^2/2$ and $F(s) = M(s) - sx = s^2/2 - sx$ is minimized at $s = x$ to get a minimum value $M^*(x) = -x^2/2$. Thus, the Chernoff bound on the Q function is given by

$$Q(x) \leq e^{-x^2/2}, \quad x > 0.$$

Note that the Chernoff bound correctly predicts the exponent of decay of the Q function for large $x > 0$. However, as we have shown using a different technique, we can improve the bound by a factor of 1/2. That is,

$$Q(x) \leq \frac{1}{2} e^{-x^2/2}, \quad x > 0.$$

An important application of Chernoff bounds is to find the tail probabilities of empirical averages of random variables. By the law of large numbers, the empirical average of n i.i.d. random variables tends to their statistical mean as n gets large. The probability that the empirical average is larger than the mean can be estimated using Chernoff bounds as follows (a Chernoff bound can be similarly derived for the probability that the empirical average is smaller than the mean).

Chernoff bound for a sum of i.i.d. random variables Let $X_1, \ldots, X_n$ denote i.i.d. random variables with MGF $M(s) = \log \mathbb{E}[e^{sX_1}]$. Then the tail probability for their empirical average can be bounded as

$$P[\frac{X_1 + \cdots + X_n}{n} > a] \leq e^{nM^*(a)}, \tag{B.6}$$

where $M^*(a) < 0$ for $a > \mathbb{E}[X_1]$. Thus, the probability that the empirical average of n i.i.d. random variables is larger than its statistical average decays exponentially with n.

Proof We have, for $s > 0$,

$$P[\frac{X_1 + \cdots + X_n}{n} > a] = P[X_1 + \cdots + X_n > na] \leq e^{-na}\mathbb{E}[e^{s(X_1 + \cdots + X_n)}]$$
$$= e^{n(M(s) - sa)},$$

using the independence of the $\{X_i\}$. The bound is minimized by minimizing $M(s) - sa$ as for a single random variable, to get the value $M^*(a)$. The result that $M^*(a) < 0$ for $a > \mathbb{E}[X_1]$ follows from the theoretical exercise. $\square$

The event whose probability we estimate in (B.6) is a *large deviation*, in that the sum $X_1 + \cdots + X_n$ is deviating from its mean $n\mathbb{E}[X_1]$ by $n(a - \mathbb{E}[X_1])$, which increases linearly in n.

Comparison with central limit theorem The preceding "large" deviation is in contrast to the $\sqrt{n}$-scaled deviations from the mean that the central limit theorem (CLT) can be used to estimate. The CLT says that

$$\frac{X_1 + \cdots + X_n - n\mathbb{E}[X_1]}{\sqrt{n}\sigma_X} \to N(0, 1) \text{ in distribution,}$$

where $\sigma_X^2 = \text{var}(X_1)$. Thus, we can estimate tail probabilities as

$$P[X_1 + \cdots + X_n > n\mathbb{E}[X_1] + a\sqrt{n}\sigma_X] \approx Q(a).$$

That is,

$$P[X_1 + \cdots + X_n > b] \approx Q\left(\frac{b - n\mathbb{E}[X_1]}{\sqrt{n}\sigma_X}\right). \tag{B.7}$$

It is not always clear cut when to use the Chernoff bound (B.6), and when to use the CLT approximation (B.7), when estimating tail probabilities for a sum of i.i.d. (or, more generally, independent) random variables, but it is useful to have both these techniques in our arsenal when trying to get design insights.

Appendix C Jensen's inequality

We derive Jensen's inequality in this appendix.

Convex and concave functions Recall the definition of a convex function from Section 6.4.1: specializing to scalar arguments, f is a *convex, or convex up, function* if it satisfies

$$f(\lambda x_1 + (1-\lambda)x_2) \leq \lambda f(x_1) + (1-\lambda)f(x_2), \tag{C.1}$$

for all x_1, x_2, and for all $\lambda \in [0,1]$. The function is *strictly convex* if the preceding inequality is strict for all $x_1 \neq x_2$, as long as $0 < \lambda < 1$.

For a *concave, or convex down, function*, the inequality (C.1) is reversed. A function f is convex if and only if $-f$ is concave.

Tangents to a convex function lie below it For a differentiable function $f(x)$, a tangent at x_0 is a line with equation: $y = f(x_0) + f'(x_0)(x - x_0)$. For a convex function, any tangent always lies "below" the function. That is, regardless of the choice of x_0, we have

$$f(x) \geq f(x_0) + f'(x_0)(x - x_0), \tag{C.2}$$

as illustrated in Figure C.1. If the function is not differentiable, then it has multiple tangents, all of which lie below the function. Just like the definition (C.1), the property (C.2) also generalizes to higher dimensions. When x is a vector, the tangents become "hyperplanes," and the vector analog of (C.2) is called the *supporting hyperplane* property. That is, convex functions have supporting hyperplanes (the hyperplanes lie below the function, and can be thought of as holding it up, hence the term "supporting").

To prove (C.2), consider a convex function satisfying (C.1), so that

$$f(\lambda x + (1-\lambda)x_0) \leq \lambda f(x) + (1-\lambda)f(x_0), \quad \text{for convex } f.$$

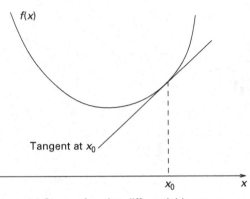
(a) Convex function differentiable at x_0

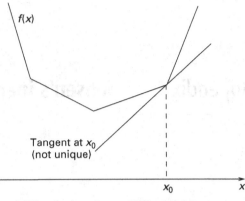
(b) Convex function not differentiable at x_0

Figure C.1 Tangents for convex functions lie below it.

This can be rewritten as

$$f(x) \geq \frac{f(\lambda x + (1-\lambda)x_0) - (1-\lambda)f(x_0)}{\lambda}$$
$$= f(x_0) + (x - x_0)\frac{f(x_0 + \lambda(x - x_0)) - f(x_0)}{\lambda(x - x_0)}.$$

Taking the limit as $\lambda \to 0$ of the extreme right-hand side, we obtain (C.2).

It can also be shown that, if (C.2) holds, then (C.1) is satisfied. Thus, the supporting hyperplane property is an alternative definition of convexity (which holds in full generality if we allow nonunique tangents corresponding to nondifferentiable functions). We are now ready to state and prove Jensen's inequality.

Theorem (Jensen's inequality) *Let X denote a random variable. Then*

$$\mathbb{E}[f(X)] \geq f(\mathbb{E}[X]), \quad \text{for convex } f, \tag{C.3}$$
$$\mathbb{E}[f(X)] \leq f(\mathbb{E}[X]), \quad \text{for concave } f. \tag{C.4}$$

If f is strictly convex or concave, then equality occurs if and only if X is constant with probability one.

Proof We provide the proof for convex f: for concave f, the proof can be applied to $-f$, which is convex. For convex f, apply the supporting hyperplane property (C.2) with $x_0 = \mathbb{E}[X]$, setting $x = X$ to obtain

$$f(X) \geq f(\mathbb{E}[X]) + f'(\mathbb{E}[X])(X - \mathbb{E}[X]). \tag{C.5}$$

Taking expectations on both sides, the second term drops out, yielding (C.3). If f is strictly convex, the inequality (C.5) is strict for $X \neq \mathbb{E}[X]$, which leads to a strict inequality in (C.3) upon taking expectations, unless $X = \mathbb{E}[X]$ with probability one. □

Example applications Since $f(x) = x^2$ is a strictly convex function, we have
$$\mathbb{E}[X^2] \geq (\mathbb{E}[X])^2,$$
with equality if and only if X is a constant almost surely. Similarly,
$$\mathbb{E}[|X|] \geq |\mathbb{E}[X]|.$$
On the other hand, $f(x) = \log x$ is concave, hence
$$\mathbb{E}[\log X] \leq \log \mathbb{E}[X].$$

References

1. J. D. Gibson, ed., *The Communications Handbook*. CRC Press, 1997.
2. J. D. Gibson, ed., *The Mobile Communications Handbook*. CRC Press, 1995.
3. J. G. Proakis, *Digital Communications*. McGraw-Hill, 2001.
4. S. Benedetto and E. Biglieri, *Principles of Digital Transmission: With Wireless Applications*. Springer, 1999.
5. J. R. Barry, E. A. Lee, and D. G. Messerschmitt, *Digital Communication*. Kluwer Academic Publishers, 2004.
6. S. Haykin, *Communications Systems*. Wiley, 2000.
7. J. G. Proakis and M. Salehi, *Fundamentals of Communication Systems*. Prentice-Hall, 2004.
8. M. B. Pursley, *Introduction to Digital Communications*. Prentice-Hall, 2003.
9. R. E. Ziemer and W. H. Tranter, *Principles of Communication: Systems, Modulation and Noise*. Wiley, 2001.
10. J. M. Wozencraft and I. M. Jacobs, *Principles of Communication Engineering*. Wiley, 1965. Reissued by Waveland Press in 1990.
11. A. J. Viterbi, *Principles of Coherent Communication*. McGraw-Hill, 1966.
12. A. J. Viterbi and J. K. Omura, *Principles of Digital Communication and Coding*. McGraw-Hill, 1979.
13. R. E. Blahut, *Digital Transmission of Information*. Addison-Wesley, 1990.
14. T. M. Cover and J. A. Thomas, *Elements of Information Theory*. Wiley, 2006.
15. K. Sayood, *Introduction to Data Compression*. Morgan Kaufmann, 2005.
16. D. P. Bertsekas and R. G. Gallager, *Data Networks*. Prentice-Hall, 1991.
17. J. Walrand and P. Varaiya, *High Performance Communication Networks*. Morgan Kaufmann, 2000.
18. T. L. Friedman, *The World is Flat: A Brief History of the Twenty-first Century*. Farrar, Straus, and Giroux, 2006.
19. H. V. Poor, *An Introduction to Signal Detection and Estimation*. Springer, 2005.
20. U. Mengali and A. N. D'Andrea, *Synchronization Techniques for Digital Receivers*. Plenum Press, 1997.
21. H. Meyr and G. Ascheid, *Synchronization in Digital Communications*, vol. 1. Wiley, 1990.
22. H. Meyr, M. Moenclaey, and S. A. Fechtel, *Digital Communication Receivers*. Wiley, 1998.
23. D. Warrier and U. Madhow, "Spectrally efficient noncoherent communication," *IEEE Transactions on Information Theory*, vol. 48, pp. 651–668, Mar. 2002.
24. D. Divsalar and M. Simon, "Multiple-symbol differential detection of MPSK," *IEEE Transactions on Communications*, vol. 38, pp. 300–308, Mar. 1990.

25. M. Abramowitz and I. A. Stegun, eds., *Handbook of Mathematical Functions: With Formulas, Graphs and Mathematical Tables*. Dover, 1965.
26. I. S. Gradshteyn, I. M. Ryzhik, A. Jeffrey, and D. Zwillinger, eds., *Table of Integrals, Series and Products*. Academic Press, 2000.
27. G. Ungerboeck, "Adaptive maximum-likelihood receiver for carrier-modulated data-transmission systems," *IEEE Transactions on Communications*, vol. 22, pp. 624–636, 1974.
28. G. D. Forney, "Maximum-likelihood sequence estimation of digital sequences in the presence of intersymbol interference," *IEEE Transactions on Information Theory*, vol. 18, pp. 363–378, 1972.
29. G. D. Forney, "The Viterbi algorithm," *Proceedings of the IEEE*, vol. 61, pp. 268–278, 1973.
30. S. Verdu, "Maximum-likelihood sequence detection for intersymbol interference channels: a new upper bound on error probability," *IEEE Transactions on Information Theory*, vol. 33, pp. 62–68, 1987.
31. U. Madhow and M. L. Honig, "MMSE interference suppression for direct-sequence spread spectrum CDMA," *IEEE Transactions on Communications*, vol. 42, pp. 3178–3188, Dec. 1994.
32. U. Madhow, "Blind adaptive interference suppression for direct-sequence CDMA," *Proceedings of the IEEE*, vol. 86, pp. 2049–2069, Oct. 1998.
33. D. G. Messerschmitt, "A geometric theory of intersymbol interference," *Bell System Technical Journal*, vol. 52, no. 9, pp. 1483–1539, 1973.
34. W. W. Choy and N. C. Beaulieu, "Improved bounds for error recovery times of decision feedback equalization," *IEEE Transactions on Information Theory*, vol. 43, pp. 890–902, May 1997.
35. G. Ungerboeck, "Fractional tap-spacing equalizer and consequences for clock recovery in data modems," *IEEE Transactions on Communications*, vol. 24, pp. 856–864, 1976.
36. S. Haykin, *Adaptive Filter Theory*. Prentice-Hall, 2001.
37. M. L. Honig and D. G. Messerschmitt, *Adaptive Filters: Structures, Algorithms and Applications*. Springer, 1984.
38. A. Duel-Hallen and C. Heegard, "Delayed decision-feedback sequence estimation," *IEEE Transactions on Communications*, vol. 37, pp. 428–436, May 1989.
39. J. K. Nelson, A. C. Singer, U. Madhow, and C. S. McGahey, "BAD: bidirectional arbitrated decision-feedback equalization," *IEEE Transactions on Communications*, vol. 53, pp. 214–218, Feb. 2005.
40. S. Ariyavisitakul, N. R. Sollenberger, and L. J. Greenstein, "Tap-selectable decision-feedback equalization," *IEEE Transactions on Communications*, vol. 45, pp. 1497–1500, Dec. 1997.
41. D. Yellin, A. Vardy, and O. Amrani, "Joint equalization and coding for intersymbol interference channels," *IEEE Transactions on Information Theory*, vol. 43, pp. 409–425, Mar. 1997.
42. R. G. Gallager, *Information Theory and Reliable Communication*. Wiley, 1968.
43. I. Csiszar and J. Korner, *Information Theory: Coding Theorems for Discrete Memoryless Systems*. Academic Press, 1981.
44. R. E. Blahut, *Principles and Practice of Information Theory*. Addison-Wesley, 1987.
45. R. J. McEliece, *The Theory of Information and Coding*. Cambridge University Press, 2002.
46. J. Wolfowitz, *Coding Theorems of Information Theory*. Springer-Verlag, 1978.
47. C. E. Shannon, "A mathematical theory of communication, part I," *Bell System Technical Journal*, vol. 27, pp. 379–423, 1948.

References

48. C. E. Shannon, "A mathematical theory of communication, part II," *Bell System Technical Journal*, vol. 27, pp. 623–656, 1948.
49. S. Arimoto, "An algorithm for calculating the capacity of an arbitrary discrete memoryless channel," *IEEE Transactions on Information Theory*, vol. 18, pp. 14–20, 1972.
50. R. E. Blahut, "Computation of channel capacity and rate distortion functions," *IEEE Transactions on Information Theory*, vol. 18, pp. 460–473, 1972.
51. S. Boyd and L. Vandenberghe, *Convex Optimization*. Cambridge University Press, 2004.
52. M. Chiang and S. Boyd, "Geometric programming duals of channel capacity and rate distortion," *IEEE Transactions on Information Theory*, vol. 50, pp. 245–258, Feb. 2004.
53. J. Huang and S. P. Meyn, "Characterization and computation of optimal distributions for channel coding," *IEEE Transactions on Information Theory*, vol. 51, pp. 2336–2351, Jul. 2005.
54. R. E. Blahut, *Algebraic Codes for Data Transmission*. Cambridge University Press, 2003.
55. S. Lin and D. J. Costello, *Error Control Coding*. Prentice-Hall, 2004.
56. E. Biglieri, *Coding for Wireless Channels*. Springer, 2005.
57. C. Heegard and S. B. Wicker, *Turbo Coding*. Springer, 1998.
58. B. Vucetic and J. Yuan, *Turbo Codes: Principles and Applications*. Kluwer Academic Publishers, 2000.
59. C. Berrou, A. Glavieux, and P. Thitimajshima, "Near Shannon limit error-correcting coding and decoding," in *Proc. 1993 IEEE International Conference on Communications (ICC'93)*, vol. 2 (Geneva, Switzerland), pp. 1064–1070, May 1993.
60. C. Berrou and A. Glavieux, "Near optimum error correcting coding and decoding: turbo codes," *IEEE Transactions on Communications*, vol. 44, pp. 1261–1271, Oct. 1996.
61. J. Pearl, *Probabilistic Reasoning in Intelligent Systems: Networks of Plausible Inference*. Morgan Kaufmann Publishers, 1988.
62. R. J. McEliece, D. J. C. Mackay, and J. F. Cheng, "Turbo decoding as an instance of Pearl's 'belief propagation' algorithm," *IEEE Journal on Selected Areas in Communications*, vol. 16, pp. 140–152, Feb. 1998.
63. S. ten Brink, "Convergence behavior of iteratively decoded parallel concatenated codes," *IEEE Transactions on Communications*, vol. 49, pp. 1727–1737, Oct. 2001.
64. A. Ashikhmin, G. Kramer, and S. ten Brink, "Extrinsic information transfer functions: model and erasure channel properties," *IEEE Transactions on Information Theory*, vol. 50, pp. 2657–2673, Nov. 2004.
65. H. E. Gamal and A. R. Hammons, "Analyzing the turbo decoder using the Gaussian approximation," *IEEE Transactions on Information Theory*, vol. 47, pp. 671–686, Feb. 2001.
66. S. Benedetto and G. Montorsi, "Unveiling turbo codes: some results on parallel concatenated coding schemes," *IEEE Transactions on Information Theory*, vol. 42, pp. 409–428, Mar. 1996.
67. S. Benedetto and G. Montorsi, "Design of parallel concatenated convolutional codes," *IEEE Transactions on Communications*, vol. 44, pp. 591–600, May 1996.
68. S. Benedetto, D. Divsalar, G. Montorsi, and F. Pollara, "Serial concatenation of interleaved codes: performance analysis, design, and iterative decoding," *IEEE Transactions on Information Theory*, vol. 44, pp. 909–926, May 1998.

69. H. R. Sadjadpour, N. J. A. Sloane, M. Salehi, and G. Nebe, "Interleaver design for turbo codes," *IEEE Journal on Selected Areas in Communications*, vol. 19, pp. 831–837, May 2001.
70. R. Gallager, "Low density parity check codes," *IRE Transactions on Information Theory*, vol. 8, pp. 21–28, Jan. 1962.
71. D. J. C. MacKay, "Good error correcting codes based on very sparse matrices," *IEEE Transactions on Information Theory*, vol. 45, pp. 399–431, Mar. 1999.
72. T. J. Richardson and R. L. Urbanke, "The capacity of low-density parity-check codes under message-passing decoding," *IEEE Transactions on Information Theory*, vol. 47, pp. 599–618, Feb. 2001.
73. S.-Y. Chung, T. J. Richardson, and R. L. Urbanke, "Analysis of sum-product decoding of low-density parity-check codes using a Gaussian approximation," *IEEE Transactions on Information Theory*, vol. 47, pp. 657–670, Feb. 2001.
74. T. J. Richardson, M. A. Shokrollahi, and R. L. Urbanke, "Design of capacity-approaching irregular low-density parity-check codes," *IEEE Transactions on Information Theory*, vol. 47, pp. 619–637, Feb. 2001.
75. T. J. Richardson and R. L. Urbanke, "Efficient encoding of low-density parity-check codes," *IEEE Transactions on Information Theory*, vol. 47, pp. 638–656, Feb. 2001.
76. M. Luby, M. Mitzenmacher, A. Shokrollahi, and D. Spielman, "Efficient erasure correcting codes," *IEEE Transactions on Information Theory*, vol. 47, pp. 569–584, Feb. 2001.
77. M. Luby, "LT codes," in *Proc. 43rd Annual IEEE Symposium on Foundations of Computer Science (FOCS 2002)*, pp. 271–282, 2002.
78. A. Shokrollahi, "Raptor codes," *IEEE Transactions on Information Theory*, vol. 52, pp. 2551–2567, Jun. 2006.
79. C. Douillard, M. Jézéquel, C. Berrou, et al., "Iterative correction of intersymbol interference: turbo equalization," *European Transactions on Telecommunications*, vol. 6, pp. 507–511, Sep.–Oct. 1995.
80. R. Koetter, A. C. Singer, and M. Tuchler, "Turbo equalization," *IEEE Signal Processing Magazine*, vol. 21, pp. 67–80, Jan. 2004.
81. X. Wang and H. V. Poor, "Iterative (turbo) soft interference cancellation and decoding for coded CDMA," *IEEE Transactions on Communications*, vol. 47, pp. 1046–1061, Jul. 1999.
82. R.-R. Chen, R. Koetter, U. Madhow, and D. Agrawal, "Joint noncoherent demodulation and decoding for the block fading channel: a practical framework for approaching Shannon capacity," *IEEE Transactions on Communications*, vol. 51, pp. 1676–1689, Oct. 2003.
83. G. Caire, G. Taricco, and E. Biglieri, "Bit-interleaved coded modulation," *IEEE Transactions on Information Theory*, vol. 44, pp. 927–946, 1998.
84. G. D. Forney and G. Ungerboeck, "Modulation and coding for linear Gaussian channels," *IEEE Transactions on Information Theory*, vol. 44, pp. 2384–2415, Oct. 1998.
85. M. V. Eyuboglu, G. D. Forney Jr, P. Dong, and G. Long, "Advanced modulation techniques for V.Fast," *European Transactions on Telecomm.*, vol. 4, pp. 243–256, May 1993.
86. G. D. Forney Jr, L. Brown, M. V. Eyuboglu, and J. L. Moran III, "The V.34 high-speed modem standard," *IEEE Communications Magazine*, vol. 34, pp. 28–33, Dec. 1996.
87. A. Goldsmith, *Wireless Communications*. Cambridge University Press, 2005.
88. T. S. Rappaport, *Wireless Communications: Principles and Practice*. Prentice-Hall PTR, 2001.

89. G. L. Stuber, *Principles of Mobile Communication*. Springer, 2006.
90. D. Tse and P. Viswanath, *Fundamentals of Wireless Communication*. Cambridge University Press, 2005.
91. W. C. Jakes, *Microwave Mobile Communications*. Wiley–IEEE Press, 1994.
92. J. D. Parsons, *The Mobile Radio Propagation Channel*. Wiley, 2000.
93. T. Marzetta and B. Hochwald, "Capacity of a mobile multiple-antenna communication link in Rayleigh flat fading," *IEEE Transactions on Information Theory*, vol. 45, pp. 139–157, Jan. 1999.
94. A. Lapidoth and S. Moser, "Capacity bounds via duality with applications to multi-antenna systems on flat fading channels," *IEEE Transactions on Information Theory*, vol. 49, pp. 2426–2467, Oct. 2003.
95. R. H. Etkin and D. N. C. Tse, "Degrees of freedom in some underspread MIMO fading channels," *IEEE Transactions on Information Theory*, vol. 52, pp. 1576–1608, Apr. 2006.
96. N. Jacobsen and U. Madhow, "Code and constellation optimization for efficient noncoherent communication," in *Proc. 38th Asilomar Conference on Signals, Systems and Computers* (Pacific Grove, CA), Nov. 2004.
97. A. J. Viterbi, *CDMA: Principles of Spread Spectrum Communication*. Pearson Education, 1995.
98. M. B. Pursley and D. V. Sarwate, "Crosscorrelation properties of pseudorandom and related sequences," *Proceedings of the IEEE*, vol. 68, pp. 593–619, May 1980.
99. M. B. Pursley, "Performance evaluation for phase-coded spread-spectrum multiple-access communication–part I: system analysis," *IEEE Transactions on Communications*, vol. 25, pp. 795–799, Aug. 1977.
100. J. S. Lehnert and M. B. Pursley, "Error probabilities for binary direct-sequence spread-spectrum communications with random signature sequences," *IEEE Transactions on Communications*, vol. 35, pp. 85–96, Jan. 1987.
101. J. M. Holtzman, "A simple, accurate method to calculate spread-spectrum multiple-access error probabilities," *IEEE Transactions on Communications*, vol. 40, pp. 461–464, Mar. 1992.
102. S. Verdu, *Multiuser Detection*. Cambridge University Press, 1998.
103. S. Verdu, "Minimum probability of error for asynchronous Gaussian multiple-access channels," *IEEE Transactions on Information Theory*, vol. 32, pp. 85–96, Jan. 1986.
104. S. Verdu, "Optimum multiuser asymptotic efficiency," *IEEE Transactions on Communications*, vol. 34, pp. 890–897, Sep. 1986.
105. R. Lupas and S. Verdu, "Linear multiuser detectors for synchronous code-division multiple-access channels," *IEEE Transactions on Information Theory*, vol. 35, pp. 123–136, Jan. 1989.
106. R. Lupas and S. Verdu, "Near-far resistance of multiuser detectors in asynchronous channels," *IEEE Transactions on Communications*, vol. 38, pp. 496–508, Apr. 1990.
107. M. K. Varanasi and B. Aazhang, "Multistage detection in asynchronous code-division multiple-access communications," *IEEE Transactions on Communications*, vol. 38, pp. 509–519, Apr. 1990.
108. M. K. Varanasi and B. Aazhang, "Near-optimum detection in synchronous code-division multiple-access systems," *IEEE Transactions on Communications*, vol. 39, pp. 725–736, May 1991.
109. M. Abdulrahman, A. U. H. Sheikh, and D. D. Falconer, "Decision feedback equalization for CDMA in indoor wireless communications," *IEEE Journal on Selected Areas in Communications*, vol. 12, pp. 698–706, May 1994.

110. P. B. Rapajic and B. S. Vucetic, "Adaptive receiver structures for asynchronous CDMA systems," *IEEE Journal on Selected Areas in Communications*, vol. 12, pp. 685–697, May 1994.
111. M. Honig, U. Madhow, and S. Verdu, "Blind adaptive multiuser detection," *IEEE Transactions on Information Theory*, vol. 41, pp. 944–960, Jul. 1995.
112. U. Madhow, "MMSE interference suppression for timing acquisition and demodulation in direct-sequence CDMA systems," *IEEE Transactions on Communications*, vol. 46, pp. 1065–1075, Aug. 1998.
113. U. Madhow, K. Bruvold, and L. J. Zhu, "Differential MMSE: a framework for robust adaptive interference suppression for DS-CDMA over fading channels," *IEEE Transactions on Communications*, vol. 53, pp. 1377–1390, Aug. 2005.
114. P. A. Laurent, "Exact and approximate construction of digital phase modulations by superposition of amplitude modulated pulses," *IEEE Transactions on Communications*, vol. 34, pp. 150–160, Feb. 1986.
115. D. E. Borth and P. D. Rasky, "Signal processing aspects of Motorola's pan-European digital cellular validation mobile," in *Proc. 10th Annual International Phoenix Conf. on Computers and Communications*, pp. 416–423, 1991.
116. U. Mengali and M. Morelli, "Decomposition of M-ary CPM signals into PAM waveforms," *IEEE Transactions on Information Theory*, vol. 41, pp. 1265–1275, Sep. 1995.
117. B. E. Rimoldi, "A decomposition approach to CPM," *IEEE Transactions on Information Theory*, vol. 34, pp. 260–270, 1988.
118. J. B. Anderson, T. Aulin, and C.-E. Sundberg, *Digital Phase Modulation*. Plenum Press, 1986.
119. E. Telatar, "Capacity of multi-antenna Gaussian channels," *AT&T Bell Labs Internal Technical Memo # BL0112170-950615-07TM*, Jun. 1995.
120. E. Telatar, "Capacity of multi-antenna Gaussian channels," *European Transactions on Telecommunications*, vol. 10, pp. 585–595, Dec. 1999.
121. G. Foschini, "Layered space–time architecture for wireless communication in a fading environment when using multi-element antennas," *Bell-Labs Technical Journal*, vol. 1, no. 2, pp. 41–59, 1996.
122. A. Paulraj, R. Nabar, and D. Gore, *Introduction to Space–Time Wireless Communications*. Cambridge University Press, 2003.
123. H. Jafarkhani, *Space–Time Coding: Theory and Practice*. Cambridge University Press, 2003.
124. H. Bolcskei, D. Gesbert, C. B. Papadias, and A. J. van der Veen, eds., *Space–Time Wireless Systems: From Array Processing to MIMO Communications*. Cambridge University Press, 2006.
125. H. Bolcskei, D. Gesbert, and A. Paulraj, "On the capacity of OFDM-based spatial multiplexing systems," *IEEE Transactions on Communications*, vol. 50, pp. 225–234, Feb. 2002.
126. G. Barriac and U. Madhow, "Characterizing outage rates for space–time communication over wideband channels," *IEEE Transactions on Communications*, vol. 52, pp. 2198–2208, Dec. 2004.
127. G. Barriac and U. Madhow, "Space–time communication for OFDM with implicit channel feedback," *IEEE Transactions on Information Theory*, vol. 50, pp. 3111–3129, Dec. 2004.
128. R. D. Yates and D. J. Goodman, *Probability and Stochastic Processes: A Friendly Introduction for Electrical and Computer Engineers*. Wiley, 2004.
129. J. W. Woods and H. Stark, *Probability and Random Processes with Applications to Signal Processing*. Prentice-Hall, 2001.
130. A. Leon-Garcia, *Probability and Random Processes for Electrical Engineering*. Prentice-Hall, 1993.

131. A. Papoulis and S. U. Pillai, *Probability, Random Variables and Stochastic Processes*. McGraw-Hill, 2002.
132. W. Feller, *An Introduction to Probability Theory and its Applications*, vols. 1 and 2. Wiley, 1968.
133. L. Breiman, *Probability*. SIAM, 1992 (Reprint edition).
134. P. Billingsley, *Probability and Measure*. Wiley-Interscience, 1995.
135. D. Williams, *Probability with Martingales*. Cambridge University Press, 1991.
136. J. L. Doob, *Stochastic Processes*. Wiley Classics, 2005 (Reprint edition).
137. E. Wong and B. Hajek, *Stochastic Processes in Engineering Systems*. Springer-Verlag, 1985.

Index

adaptive equalization, 223
  least mean squares (LMS), 225
  least squares, 223
  recursive least squares (RLS), 224
antipodal signaling, 113
asymptotic efficiency
  MLSE, 236
  of multiuser detection, 419
asymptotic equipartition property (AEP)
  continuous random variables, 268
  discrete random variables, 266
autocorrelation function
  random process, 32, 36
  signal, 15
  spreading waveform, 413
AWGN channel
  M-ary signaling over, 94
  optimal reception, 101

bandwidth, 30
  fractional energy containment, 17
  fractional power containment, 48
  normalized, 48
bandwidth efficiency, 42
  linear modulation, 53
  orthogonal modulation, 56
Barker sequence, 415
baseband channel, 15
baseband signal, 15
BCH codes, 366
BCJR algorithm, 312
  backward recursion, 317
  forward recursion, 316
  log BCJR algorithm, 320
  summary, 319
  summary of log BCJR algorithm, 324
Bhattacharya bound, 310, 371
binary symmetric channel (BSC), 264
  capacity, 272
biorthogonal modulation, 57

bit interleaved coded modulation (BICM), 357
  capacity, 359
Blahut–Arimoto algorithm, 284
block noncoherent demodulation
  DPSK, 188
bounded distance decoding, 365

capacity
  bandlimited AWGN channel, 253
  binary symmetric channel, 272
  BPSK over AWGN channel, 274
  discrete time AWGN channel, 259
  optimal input distributions, 282
  plots for AWGN channel, 276
  power–bandwidth tradeoffs, 276
  power-limited regime, 255
  PSK over AWGN channel, 275
Cauchy–Schwartz inequality, 10
  proof, 60
channel coding theorem, 270
coherent receiver, 29
complex baseband representation, 18
  energy, 22
  filtering, 26
  for passband random processes, 40
  frequency domain relationship, 22
  inner product, 22
  modeling phase and frequency offsets, 28
  role in transceiver implementation, 27
  time domain relationship, 19
complex envelope, 19
complex numbers, 8
composite hypothesis testing, 171
  Bayesian, 171
  GLRT, 171
concave function, 281
conditional error probabilities, 90
continuous phase modulation (CPM), 428
  Laurent approximation, 434

convex function, 281
convolution, 10
convolutional codes, 294
  generator polynomials, 296
  nonrecursive nonsystematic encoder, 295
  performance of ML decoding, 303
  performance with hard decisions, 310
  performance with quantized observations, 309
  recursive systematic encoder, 296
  transfer function, 307
  trellis representation, 296
  trellis termination, 317
correlation coefficient, 80
correlator
  for optimal reception, 104
Costas loop, 190
covariance
  matrix, 80
  properties, 81
crosscorrelation function
  random process, 34, 36
  spreading waveform, 413

dBm, 88
decision feedback equalizer (DFE), 228
decorrelating detector, 422
delta function, 10
differential demodulation, 173
differential entropy, 266
  Gaussian random variable, 267
differential modulation, 57
differential PSK, *see* DPSK
direct sequence, 405
  CDMA, 408
  long spreading sequence, 408
  rake receiver, 409
  short spreading sequence, 407
discrete memoryless channel (DMC), 263
divergence, 269
diversity combining
  maximal ratio, 393
  noncoherent, 396
downconversion, 24
DPSK, 57
  binary, 59
  demodulation, 173
  performance for binary DPSK, 187

energy, 9
energy per bit (E_b)
  binary signaling, 111
energy spectral density, 14
entropy, 265
  binary, 265
  concavity of, 282
  conditional, 268
  joint, 268
equalization, 199
  fractionally spaced, 220
  model for suboptimal equalization, 215
error event, 235
error sequence, 232
Euler's identity, 9
excess bandwidth, 51
EXIT charts, 329
  area property, 335
  Gaussian approximation, 333

FDMA, 379
finite fields, 365
Fourier transform, 13
  important transform pairs, 13
  properties, 14
  time–frequency duality, 13
frequency hop, 426
frequency shift keying (FSK), 55
Friis formula, 133

Gaussian filter, 432
Gaussian random vector, 81
generalized likelihood ratio test (GLRT), 171
Gramm–Schmidt orthogonalization, 98
Gray coding, 127, 129
  BER with, 130

Hamming code, 344
hypothesis testing, 88
  irrelevant statistic, 93
  sufficient statistic, 94

I and Q channels
  orthogonality of, 21
I component, 19
in-phase component, *see* I component
indicator function, 13
inner product, 9
intersymbol interference, *see* ISI
ISI, 199
  eye diagrams, 203

Kuhn–Tucker conditions, 282
Kullback–Leibler (KL) distance, 269

law of large numbers (LLN), 253
  interpretation of differential entropy, 267
  interpretation of entropy, 265
  large deviations, 253
LDPC codes, 342
  belief propagation, 352
  bit flipping, 349
  degree distributions, 347
  Gaussian approximation, 354

Index

message passing, 349
  rate for irregular codes, 348
  Tanner graph, 345
likelihood function, 162
likelihood ratio, 92
  signal in AWGN, 162
line codes, 44
linear code, 343
  dual code, 343
  generator matrix, 343
  parity check matrix, 344
linear equalization, 216
  performance, 226
linear modulation, 43
  example, 25
  power spectral density, 34, 47, 69
link budget analysis, 133
  example, 135
link margin, 134
low Density Parity Check codes, *see* LDPC codes
lowpass equivalent representation, *see* complex baseband representation

MAP
  decision rule, 91
  estimate, 159
matched filter, 12
  delay estimation, 61
  for optimal reception, 104
  optimality for dispersive channel, 202
matrix inversion lemma, 224
maximum a posteriori, *see* MAP
maximum likelihood (ML)
  application to multiuser detection, 148
  decision rule, 90
  decoding of convolutional codes, 298
  estimate, 159
  geometry of decision rule, 106
  multiuser detection, 418
  sequence estimation, 204
maximum likelihood sequence estimation, *see* MLSE
MIMO, *see* Space–time communication, 439
minimum mean squared error, *see* MMSE
minimum probability of error rule, 91
minimum Shift Keying (MSK), 429
  Gaussian MSK, 432
  preview, 71
MLSE, 204
  performance analysis, 231
  whitened matched filter, 212
MMSE
  adaptive implementation, 223, 425
  linear MMSE equalizer, 220
  linear multiuser detection, 424
  properties, 424
modulation
  degrees of freedom, 41
MPE rule, *see* minimum probability of error rule
multipath channel, 381
multiuser detection, 417
  asymptotic efficiency, 419
  linear MMSE, 424
  ML reception, 418
  near–far resistance, 421
mutual information, 268
  as a divergence, 269
  concavity of, 282

near–far problem, 416
nearest neighbors approximation, 121, 130
noise figure, 87, 133
noncoherent communication, 153
  block demodulation, 187
  high SNR asymptotics, 182
  optimal reception, 171
  performance for binary orthogonal signaling, 182
  performance with M-ary orthogonal signaling, 185
  receiver operations, 29
norm, 9
Nyquist
  criterion for ISI avoidance, 49, 66
  sampling theorem, 41
Nyquist pulse, 51

OFDM, 397
  cyclic prefix, 401
  peak-to-average ratio, 402
  power spectral density, 402
offset QPSK, 71
on–off keying, 112
orthogonal modulation
  bandwidth efficiency, 56
  BER, 130
  binary, 113
  coherent, 55
  noncoherent, 55

parallel Gaussian channels, 277
  waterfilling, 279
parameter estimation, 159
  amplitude, 160
  delay, 167
  phase, 166
Parseval's identity, 14
passband channel, 15
passband filtering, 26
passband signal, 15
  time domain representation, 19

performance analysis
  16-QAM, 123
  M-ary orthogonal signaling, 124
  ML reception, 109
  QPSK, 117
  rotational invariance, 115
  scale-invariance, 112
  scaling arguments, 116
  union bound, 118
phase locked loop (PLL), 155
  ML interpretation, 169
power efficiency, 112, 122
power spectral density, 32
  analytic computation, 60
  linear modulation, 34, 47, 69
  WSS random process, 37
power-delay profile, 384
principle of optimality, 209, 300
proper complex Gaussian
  density, 178
  random process, 179
  random vector, 177
  WGN, 179
proper complex random vector, 177

Q component, 19
Q function, 77
  asymptotic behavior, 79
  bounds, 78, 137, 138
quadrature component, *see* Q component

raised cosine pulse, 51, 67
random coding, 270
random processes
  autocorrelation function, 36
  autocovariance function, 36
  baseband and passband, 33
  crosscorrelation function, 36
  crosscovariance function, 36
  cyclostationary, 39, 65
  ergodicity, 38
  Gaussian, 85
  jointly WSS, 37
  mean function, 36
  power spectral density, 32
  spectral description, 31
  stationary, 36
  wide sense stationary (WSS), 37
random variables
  Gaussian, 76
  joint Gaussianity, 81
  Rayleigh, 137
  Rician, 137
  standard Gaussian, 76
  uncorrelated, 83
Rayleigh fading, 382
  Clarke's model, 385

ergodic capacity, 391
interleaving, 391
Jakes' simulator, 387
performance with diversity, 394, 397
preview, 148
receive diversity, 392
uncoded performance, 388
receiver sensitivity, 133
Reed–Solomon codes, 366
Rician fading, 383

sampling theorem, 41
Shannon, 252
signal space, 42, 94
  basis for, 98
signal-to-Interference Ratio (SIR), 222
sinc function, 13
Singleton bound, 365
singular value decomposition (SVD), 444
soft decisions
  bit level, 131
  symbol level, 106
space–time communication, 439
  Alamouti code, 450
  BLAST, 447
  capacity, 446
  channel model, 440
  space–time codes, 448
  spatial multiplexing gain, 447
  transmit beamforming, 451
spatial reuse, 379
spread spectrum
  direct sequence, 405
  frequency hop, 426
square root Nyquist pulse, 52
square root raised cosine (SRRC) pulse, 52
synchronization, 153
  transceiver blocks, 155

Tanner graph, 345
tap delay line, 383
TDMA, 379
transfer function bound
  ML decoding of convolutional
    codes, 308
  MLSE for dispersive channels, 237
trellis coded modulation, 360
  4-state code, 362
  Ungerboeck set partitioning for 8-PSK,
    360
turbo codes
  BER, 328
  design rules, 341
  EXIT charts, 329
  parallel concatenated, 325
  serial concatenated, 327
  weight enumeration, 336
two-dimensional modulation, 45

Index

typicality, 266
    joint, 270
    joint typicality decoder, 271

union bound, 118
    intelligent union bound, 120
upconversion, 24

Viterbi algorithm, 210, 301

Walsh–Hadamard codes, 56
WGN, *see* white Gaussian noise
white Gaussian noise, 86
    geometric interpretation, 96
    through correlator, 180
    through correlators, 95

zero-forcing detector, 422
zero-forcing equalizer, 216
    geometric interpretation, 217

Printed in the United States
By Bookmasters